FACT, FAITH AND FICTION IN THE DEVELOPMENT OF SCIENCE

BOSTON STUDIES IN THE PHILOSOPHY OF SCIENCE

VOLUME 205

R. HOOYKAAS
Emeritus Professor of the History of Science
in Utrecht

Fact, Faith and Fiction in the Development of Science

The Gifford Lectures Given in the University of St Andrews 1976

KLUWER ACADEMIC PUBLISHERS
DORDRECHT / BOSTON / LONDON

A C.I.P. Catalogue record for this book is available from the Library of Congress.

ISBN 0-7923-5774-4

Published by Kluwer Academic Publishers,
P.O. Box 17, 3300 AA Dordrecht, The Netherlands.

Sold and distributed in North, Central and South America
by Kluwer Academic Publishers,
101 Philip Drive, Norwell, MA 02061, U.S.A.

In all other countries, sold and distributed
by Kluwer Academic Publishers,
P.O. Box 322, 3300 AH Dordrecht, The Netherlands.

Printed on acid-free paper

Printed in the Netherlands.

Frontispiece: Preparation of Experiments, from: Thomas Norton, *Ordinall of Alkimy* (British Library, Norton additional ms 10302, fol.37v). By permission of the British Library.

TABLE OF CONTENTS

Editors' Foreword ix

Selected Bibliography of R. Hooykaas xv

AUTHOR'S PREFACE 1

INTRODUCTION 5

I. ON NATURAL THEOLOGY 17

II. HARMONY IN NATURE 27

III. THE PHILOSOPHER'S STONE 79

IV. THE UNDYING FIRE 99

V. A TUNNEL THROUGH THE EARTH 117

VI. 'AND THE SUN STOOD STILL' 147

VII. THINKING WITH THE HANDS 183

VIII. PHYSICAL AND MATHEMATICAL THEORIES 229

IX. WORKS OF NATURE, WORKS OF ART 265

X. CLEOPATRA'S NOSE 319

XI. THE 'THINKING REED' 343

NOTES 373

LIST OF ILLUSTRATIONS 425

The Text and Editorial Actions with Regard to It 429

Acknowledgements 433

Index 435

EDITORS' FOREWORD
THE BOOK AND ITS AUTHOR

This is a book in which the past of science comes alive in a way both personally coloured and generally accessible. Like all publications by the late Professor R. Hooykaas, it is at the same time a scholarly work, based from start to finish upon original, thorough research. *Fact, Faith and Fiction in the Development of Science* is the title he himself gave to the book. Its chapters started life as the Gifford Lectures he delivered at the University of St Andrews in 1976. From then on until his death in 1994 he worked on and off to expand these lectures into even more cogently argued, well-annotated and illustrated chapters. Out of that effort came what now lies before us — a book that, more comprehensively so than any other work he wrote, exemplifies the unique Hooykaas approach to the history of science.

The title is meant to express what Hooykaas regarded as the broad, threefold constitution of the scientific endeavour. 'Faith' is not to be equated here with religion, the relations of which to science were often the subject of Hooykaas's investigations. Rather, as he explains in the Introduction, in the context of the present book he meant by that term such broad, *a priori* held conceptions as may lead the scientist on his or her adventurous path towards the unknown, like the idea of unity, of order, of simplicity, or of harmony. Science in its ongoing advance, so we believe we may sum up Hooykaas's final convictions on the matter, never ceases to display quite variously mingled contributions from *faith* (in the sense just indicated), from *facts* (given by nature yet entirely subject to our mode of interpreting them), and from *fictions* in the sense of those daring intellectual tools, such as theories and hypotheses and models, which reflect the scientist's creative imagination. However much Hooykaas loved to point at recurrent patterns in scientific thought through the ages, he was also convinced that no generalized rule can tell the scientist how the three basic elements here distinguished ought to be combined in any given case — the historian of science must learn to cope with the endless and, as such, quite captivating *variety* of ways in which this triad (facts; broad, leading ideas, and specific theorizing) has gone into the making of what at any given point in time could with justice be called 'science'.

Such, then, is the leading idea of the present book.

Between 1976 and 1998 a good deal of work on the book has taken place. During Hooykaas's remaining lifetime, VMcK, who together with her late husband had been a friend of the Hooykaas family for decades, did her utmost to encourage

Hooykaas, to put chapters on disk file as they emerged from his study, to suggest and (with his consent) carry out improvements in both the content and the language of the book, and to seek for the final touch wherever possible. Out of that joint effort came eleven, either fully or almost completed chapters — a loss of five as compared to the total number of lectures, which is a great pity but not irreparable in view of the set-up of the book, which is intended not so much as a fully integrated argument, but rather as a range of case studies, each delving deeply but essentially on its own, and meant to illustrate the broad, underlying thesis outlined above.

After Hooykaas died, HFC, who had first been his student and later became his colleague, was entrusted with his intellectual legacy. A good portion of Hooykaas's well-stocked library is now in the Museum Boerhaave at Leyden; his notes and correspondence make up the Hooykaas collection at the North Holland National Archives in Haarlem, and the book project on which he was working is now in the reader's hands. In order to get it there, final touches still had to be applied. While sticking meticulously to Hooykaas's own written text, we had to make sure about hosts of leftover details like note numbers or captions or stray inconsistencies or the status of some corrections. The efforts involved were carried out in good part by HFC's one-time graduate student, JCB, who was financially enabled to do so thanks to a 3-months-grant, here gratefully acknowledged, from the Netherlands Organisation for Scientific Research NWO. Further acknowledgements and expressions of gratitude, as well as a more detailed history of the text and an account of editorial actions with regard to it, are to be found at the back of the book, starting on page 429. The remainder of this Foreword is devoted to the author and his scholarly endeavour. (Most of what follows appeared previously in an eloge written by HFC in *Isis* 89 (1) of March 1998, pages 181-184).

In the field of history of science, Hooykaas was among the pioneers. Along with a range of other gifted loners in the 1920s and 1930s, he taught himself how to leave behind customary roots seekers' and philosophers' history of science, in a fresh search for ways and means to treat the past in its own right, for the various frames in which scientists have set themselves problems and struggled for solutions. The effort pervades the programmatic introduction to the book Hooykaas published in 1933 as his doctoral dissertation, *Het begrip element in zijn historisch-wijsgeerige ontwikkeling* ('The concept of element in its historical-philosophical development'). This introduction takes its place among several other pieces of the interbellum period in laying down rules — or, rather, sketching a mindset — required for turning the history of science into real history. The sense of personal discovery exuded by the piece, and indeed by the entire book that followed, was never to leave Hooykaas. It contributed in no small measure to the captivating liveliness of his teaching — the message and the maxims he tirelessly sought to convey to his students were the fruit of his own, personal quest.

Reijer Hooykaas was born in 1906 in Schoonhoven, into a Calvinist family of local silversmiths. He studied chemistry at the University of Utrecht. The choice of a topic for his dissertation was hardly an obvious one — the more so since, in the 1930s, it was not at all clear whether he would ever have a chance to pursue the history of science other than as a spare time occupation. Hooykaas became a chemistry teacher at two grammar schools in succession, using his spare time first for completing his PhD thesis and then for writing a range of articles published mainly in two Dutch periodicals: a chemists' weekly, for which he elaborated findings of his dissertation, and a journal for Protestant scientists and physicians, which became the somewhat unlikely recipient of profoundly original studies on Kepler's concept of hypothesis and on Pascal's science and religion. The latter publications, in particular, aroused sufficient interest in the history of science at the Free University of Amsterdam (the intellectual centre of Calvinism in the Netherlands) that a chair was created immediately after World War II — the first anywhere in the country — which Hooykaas was called to occupy. From 1948 to 1960 he added mineralogy and crystallography to his university teaching schedule. In 1967 he moved to the University of Utrecht in order — after an interregnum involving Struik and Ravetz — to succeed E.J. Dijksterhuis. Here he stayed until he retired in 1976, using the time not only to pursue his own studies and to create a certain amount of *Nachwuchs* but to turn what had been little more than a room in a large building into a sparingly housed yet full-fledged and well-equipped Institute for the History of the Exact Sciences. In no way did his retirement mark a break in his research — he undauntedly went on to make contributions to the field almost to the end.

As the reader of the present book will readily appreciate, the most striking aspect of Hooykaas's historiographical legacy is its versatility. Atomism both ancient and modern, alchemy, iatrochemistry, the Chemical Revolution — all broached in the dissertation — were the subjects of many articles, some written in English for *Janus* or the *Archives Internationales*, some in German for *Sudhoffs Archiv*. The history of science in the Netherlands naturally received some in-depth study — for example, essays on Isaac Beeckman, on Christiaan Huygens, on Dutch scientific societies, on Copernicanism in the Netherlands, and on what science meant for government policy in the Golden Age of the Dutch Republic. Historical problems of science and religion provided food for many articles and some books, including an authoritatively edited and annotated edition of Rheticus' lost theological defence of heliocentrism rediscovered by Hooykaas. In the early 1960s he taught himself Portuguese, going on to do pioneering work on the significance for science of the Voyages of Discovery and the intellectual upheaval these created in a humanist culture — here *Science in Manueline Style*, which appeared both separately and as a part of Volume 4 of the *Obras completas de D. João de Castro* was the principal fruit. The history of crystallography caught his attention, leading to lengthy studies of, among others, Romé de Lisle and Haüy. Uniformitarianism and other basic issues in the history of geology likewise made for several path-breaking books and articles, most of them written in English. Petrus Ramus was the subject of a book, this one in French. Methodological issues came in for scrutiny in yet other contexts. Nor is this quite an exhaustive list. Still, it should go far to explain how in 1971 Hooykaas could publish (in Dutch, to be sure) a 'History of the sciences from Babel to Bohr,' which provides a continuous story by blending summaries of his own publications and little else. It seems safe to say that no other serious historian of science could have pulled off such a feat.

Where, amidst all this variety, was the heart of the history of science located in Hooykaas's view? Not in one single feature; not in some all too beautifully fitting system. He detested monolithic thinking and, for all his love of order and orderliness, abhorred grand systems. 'The Bible has no system', he wrote repeatedly of that primary source of inspiration to him. Rather, basic to Hooykaas was his concrete, almost tangible conception of nature. He was certainly not a simple-minded apologist for Protestantism in his many studies on the historical relations between science and religion, as a glance at his masterful 1939 article on Pascal or his 1943 book on Boyle or many a passage in the present book will reveal, yet it is clear that a biblical conception of nature as freely created by a sovereign God in accordance with His own ways (as distinct from the ways of human beings) stood at the centre of Hooykaas's historical thought. Hooykaas found the historical drama of humanity's ongoing search to gain a grasp on nature to reside in this: how the human mind, with all its subjectivity and its capacity for self-deception, but also its piercing gifts of discovery, has through the ages come to terms with what he once felicitously called 'the whimsical tricks of nature'. Time and again our intellect seems to have caught nature in its net; time and again nature, induced by our very search to do so, breaks out of the net by displaying novel features that time and again may fail to comply with what we had been led to expect. And it is precisely this ever-recurring interplay between nature and the scientist, between 'facts' on the one hand and 'faith' and 'fiction' on the other, which Hooykaas never sought to bring out more cogently than in the present work.

Its chapters, accordingly, achieve this aim in quite different ways. Some are mostly centred on one individual (e.g., Copernicus and contemporary reasons for accepting or rejecting heliocentrism). Others pursue one topic through the centuries (e.g., how one side-issue treated by Newton in the *Principia* both stimulated and blocked research up to Dalton and beyond). Again others focus on a particular period (e.g., medieval efforts to turn Aristotelian natural philosophy into mechanical science). The joint effect these varied chapters are likely to have on the reader is to get across a feeling for the boundless richness of humankind's efforts to come to grips with nature. This is accomplished without drowning the reader in an equally endless sea of historical facts, but rather by confronting him or her with a range of very learned yet quite accessible variations upon one common theme. The inspiration Hooykaas's oral presentations so abundantly exuded has been preserved here in written form. What the reader may certainly *not* expect is an up-to-date inclusion and/or discussion of the secondary literature on topics here treated. What he or she *may* expect is a coherent, erudite vision of science through the centuries, vividly illustrated by means of instructive examples chosen from all over its many branches, spiced by countless flashes of typical Hooykaas insight, and invariably taken straight from original, published sources. Whether a historian or not, every reader with an interest in the realities of the scientific pursuit may benefit from Hooykaas's wisdom here collected in concentrated form.

Enschede, Netherlands J. Christiaan Boudri
Enschede, Netherlands H. Floris Cohen
Keele, Staffordshire, Great Britain Valerie MacKay

SELECTED BIBLIOGRAPHY OF R. HOOYKAAS

[The complete publication list of R. Hooykaas contains over hundred and thirty books and papers.]

- 'Pascal: his Science and his Religion' (originally in Dutch, 1939; translation: *Tractrix* 1 (1989), pp.115-139).
- *Robert Boyle. A Study in Science and Christian Belief* (originally in Dutch, 1943; translation: Lanham (MD): University Press of America 1997).
- 'The Discrimination between 'Natural' and 'Artificial' Substances and the Development of Corpuscular Theory'. *Archives Internationales d'Histoire des Sciences* 1 (1948), pp.640-651.
- 'The Species Concept in 18th Century Mineralogy'. *Archives Internationales d'Histoire des Sciences* 5 (1952), pp.45-55.
- *Humanisme, Science et Réforme. Pierre de la Ramée (1515 - 1572)*. Leiden: Brill 1958.
- *Natural Law and Divine Miracle. A Historical-critical Study of the Principle of Uniformity in Geology, Biology and Theology*. Leiden: Brill 1959.
- *L'Histoire des Sciences, ses Problèmes, sa Méthode, son But*. Coimbra 1963. 35 pp. Also in: *Revista da Faculdade de Ciências da Universidade de Coimbra* 32 (1963), pp.5-35, and in —, *Selected Studies*. Coimbra 1983, pp.9-41.
- *Continuité et discontinuité en géologie et biologie*. Paris 1970.
- [entries for the *Dictionary of Scientific Biography* on Isaac Beeckman, René-Just Haüy, Jean-Baptiste-Louis Romé de Lisle].
- *Religion and the Rise of Modern Science*. Edinburgh: Scottish Academic Press 1972.
- 'The Reception of Copernicanism in England and the Netherlands', in: *The Anglo-Dutch Contribution to the civilization of early modern society*. London 1976, pp. 33-44.
- 'Humanism and the Voyages of Discovery in 16th Century Portuguese Science and Letters', in: *Mededelingen Koninklijke Nederlandse Academie van Wetenschappen, afdeling Letterkunde* 42, 4. Amsterdam 1979; 67 p.
- *Science in Manueline Style. The historical context of D. João de Castro's Works*. Coimbra: Academia Internacional da Cultura Portuguesa 1980 (separate edition of pp.231-426 in vol.4 of: A. Cortesão and L. de Albuquerque ed., *Obras Completas de D. João de Castro*, Coimbra 1968 - 1980; 4 vols.).

- 'Pitfalls in the Historiography of Geological Science', in: *Histoire et Nature* 19-20 (1981-2), pp.21-34.
- 'Wissenschaftsgeschichte, eine Brücke zwischen Natur- und Geisteswissenschaften', in: *Berichte zur Wissenschaftsgeschichte* 5 (1982), pp.153-172.
- *Selected Studies in History of Science.* Coimbra 1983. 667 p.
- *G.J. Rheticus' Treatise on Holy Scripture and the Motion of the Earth.* Amsterdam: North Holland 1984.
- 'The Rise of Modern Science: When and Why?', in: *British Journal for the History of Science* 20 (1987), pp.453-473.
- 'Humanities, Mechanics and Painting (Petrus Ramus; Francisco de Holanda); in: *Revista da Universidade de Coimbra* 36 (1991), pp.1-31.
- 'The Historical and Philosophical Background of Haüy's Theory of Crystal Structure', in: *Academiae Analecta, Mededelingen van de Koninlijke Academie voor Wetenschappen, Letteren en Schone Kunsten van België, Klasse der Wetenschappen* 56, 1 (1994), pp.1-108.

[NB Aspects of Hooykaas's life and work are examined in a range of obituaries, e.g., one in *Perspectives on Science and Christian Faith* 48, 2 (June 1996) by Arie Leegwater, or the one in *Isis* already mentioned; also in several sections of HFC's book *The Scientific Revolution. A Historiographical Inquiry*. Chicago: University of Chicago Press 1994, and in J.H. Brooke's and M. Hunter's 1997 Foreword to the translated version of Hooykaas's book on Boyle listed above].

Professor R. Hooykaas
(1 August 1906 – 4 January 1994)

AUTHOR'S PREFACE

When I received the invitation from this ancient and renowned university of St Andrews to deliver the Gifford lectures on Natural Theology, I felt not only honoured but also somewhat embarrassed. Lord Gifford enjoins (1885) his lecturers to 'promote ... and diffuse the study of Natural Theology, in the widest sense of that term, in other words, "The Knowledge of God, the Infinite, the All, ... the Sole Reality, and the Sole Existence, the Knowledge of His Nature and Attributes, the Knowledge of the Relations which men and the whole universe bear to Him, the Knowledge of the Nature and Foundation of Ethics or Morals, and of all Obligations and Duties thence arising."'[1]

Natural Theology does not, in general, play any considerable role in the thoughts of a modern scientist, and though I would not wish wholly to repudiate it, I am hardly the man to invite to Scotland on a mission to defend it.

Reading further in Lord Gifford's Trust Disposition did nothing to stimulate my enthusiasm for Natural Theology. The lecturers are enjoined 'to treat their subject as a strictly natural science, the greatest of all possible sciences, indeed, in one sense, the only science, that of Infinite Being, without any reference to or reliance upon any supposed special exceptional or so-called miraculous revelation', and to consider Natural Theology as a science, 'just as astronomy or chemistry is.' Most scientists, non-Christian as well as Christian, would share my doubts about the possibility of complying with these injunctions.

Then again, I wondered, how can one say anything new on this subject? Already in 1803 the Edinburgh Review wrote: 'On the subject of Natural Theology no one looks for originality and no one pretends to discovery.'[2] This is hardly an encouragement, and the repetitiousness of the arguments for natural theology down the years shows that the Edinburgh Review's assessment was not far from the truth.

Even the words of Lord Gifford that surprised me most, *viz.* that Natural Theology is a science like astronomy and chemistry, find a parallel in earlier authors. In the early 1830's another Scotsman, Lord Henry Brougham, wrote that 'the fundamental branch of Natural Theology' and Physics and Psychology 'are not only closely allied one to the other, but are to a very considerable extent identical.'[3] Brougham's choice of physics and psychology was not made without good reason. Natural theology has two main aspects. In the first place it claims to be the knowledge of God innate in the souls of all people of all times, as a result of each individual mind's being a part of the universal Mind (Stoicism), or as made in the image of God (Judeo-Christian tradition): it claims to be a rational or intuitive knowledge of God, quite apart from any 'special' revelation through prophecy, visions, etc. Secondly, it is the knowledge of God acquired by observing and by profoundly investigating Nature, and that not only human nature but that of the whole universe around us. The former aspect relates to humanistic disciplines like logic, ethics and metaphysics, whereas the latter would be a subject more fit for contemplation by the natural scientist.

On both aspects the arguments bolstering up Natural Theology have not become stronger in the course of time, so that one wonders that lectures on the topic could have continued for almost a century at the four ancient Scottish universities.

The answer is that Lord Gifford took the subject in a wide sense. He even allowed agnostics and freethinkers to be elected to deliver the lectures, 'provided they be able and reverent men', a strange provision as they would have to deal with a 'science' whose existence they doubted or denied. It allowed the Committees supervising the lectures so wide an interpretation that it was possible to invite lecturers like Werner Heisenberg (St Andrews 1955/56) — who hardly touched on religion at all — and Karl Barth (1937/38), who reminded the Senate of Aberdeen University that he was 'an avowed opponent of all natural theology' and, while speaking about nothing but religion, flatly declared his antipathy to natural theology. Barth's misgivings about a science of natural theology were of a kind I share to a large extent, but he received a reassurance similar to that which the present St Andrews committee kindly gave to me. Nevertheless, he rightly felt that we owe it to the founder not to ignore natural theology completely. He pointed out that, though more than one of those who had given these lectures 'must have had to rack their brain over these requirements of Lord Gifford … to none they must have given so much trouble as to me.' It was a 'radical error', and it could not be 'the business of a Reformed theologian to raise so much as his little finger to support Lord Gifford's undertaking.' So he decided to do 'justice at least indirectly to his intention.' Natural theology is thrown into relief by the dark background of a totally different theology to which it exists as an antithesis. 'And when its conflict with this adversary no longer attracts attention, ... interest in "Natural Theology" too, soon tends to flag. Why then should the service not be rendered it of presenting to it once more this its indispensible opponent, since the requirement is, that "Natural Theology" shall here be served?' [4]

The 'background' mentioned is, of course, God's revelation in Jesus Christ as it is attested in Holy Scripture, which, says Barth, is 'the clear antithesis to that form of teaching which declares that man himself possesses the capacity and the power to inform himself about God, the world and man.'[5] And so he could with a clear conscience leave Natural Theology aside after his introductory lecture.

Such a holy trick is not, however, at the disposal of a mere scientist. There is no revealed science in contrast to natural science in the way that there is revealed theology over against natural theology. Moreover, the fact that some of the greatest scholars of our age took the Gifford injunctions not too literally did nothing to alleviate my doubts as to whether I could do so, too: 'Quod licet Jovi, non licet bovi.'

So, Lord Gifford's ghost did not leave me at peace, until I read the lectures delivered more than fifty years ago by a former St Andrews professor, the classicist A.E. Taylor.[6] They reminded me that many problems which the present generation of theologians and scientists deem quite the latest thing, are just the revival of old errors about science and religion, which were fashionable in my own student days and which now, in the disguise of a different jargon, parade as brand new. As a professor of 'moral philosophy' and an expert on Plato he found occasion to point out that many tenets of modern theologians are closer to the natural theology of ancient pagan philosophers than to Christianity.

The other branch of natural theology, which is based on the contemplation of the world around us, has (since the rise of modern science) extended, but not strengthened its arsenal of arguments by the results of the scientific investigation of nature, which shows the wonderful order of the universe and the no less wonderful adaptations of the animal kingdom.

So, although these lectures deal with the patterns of thought and the methods of research in different scientific disciplines, we shall inevitably come across cases of mutual influence between science and natural theology, for instance because they share a belief in the order of nature. Moreover, by inserting (as a kind of work of supererogation) before we tackle the subject of this course, a short chapter on what natural theology was about, we hope to have complied with the wishes of the founder of this lectureship.

INTRODUCTION

THE THEME OF THIS WORK: FACTS, FAITH AND FICTION

An open mind is generally recommended for the scientist. But an open mind is not the same as an empty mind: what will be brought out from facts by the mind that is confronted by them depends to a great extent upon what that mind already contained by way of experience, convention, education. The following chapters trace a number of influential ideas and methods that have been part of the mental furniture of scientists over the centuries.

We shall be looking, not so much at the achievements of science as upon its legitimate foundations: facts upon which science is built, beliefs by which it has been inspired and fictions or hypotheses on which its theories were constructed.

FACTS

What do we mean by 'facts'? At first sight the answer seems simple enough. Objectively facts are the real things and events. But then we forget that all we say about facts are human statements which have been filtered through coloured glasses: the senses and the mind of man.

Small wonder then that the definitions of 'facts' vary, as is evident, for example, from the statements made by two eminent scholars working at the University of St Andrews in the 19th century. The physicist Sir David Brewster writing in 1845 about 'the real meaning and sterling value of *a fact in science'*, said a fact is 'that eternal and immutable truth which every man must believe, and which all men may possess, that indestructible element of knowledge, which time cannot alter, nor power crush, nor fire subdue — that self-luminous atom, which shines brightest in the dark, and whose vestal fire an intellectual priesthood will ever struggle to maintain.'[1]

This rhetoric contrasts in style with the sober words of the geologist and physicist James David Forbes (1848). He reminded his readers of Newton's confession that to him truth, 'like the first dawning opens slowly, by little and little into a clear light'. 'But', says Forbes, 'the superficial thinker has no such dawn. Facts [to him] are things which admit of no degrees. He knows them or he does not. He is in a blaze of what he calls light, or else he is in total darkness. He evidently never thought in the way that Newton thought, but knowledge, if knowledge it be, comes to him through some different avenue.' In Forbes's opinion 'facts are not knowledge, any more than boots possess understanding.'[2]

These statements from Brewster and Forbes depict the two-fold character of facts. Brewster speaks of objective things and events, which are as they are, independently of the ideas we have about them. He is perfectly right in claiming that they are the basis of science, though he seems rather naïvely to confound them with our conceptions of facts. Forbes, on the other hand, has a keen eye for this latter aspect: it is not the objective 'fact' of nature but the human conception of it that finds its place in 'science'. Their different approaches show in a nutshell the tension existing between the objective fact of nature and its translation — as adequately as possible — into the 'fact' of the scientific system. It is the never-ending dialogue between man and nature, in which the signs yielded by nature have to be translated and interpreted into the language of man. Consequently Forbes rightly stressed that this translation shows degrees of correctness, i.e. that there are degrees of conformity of the 'facts' of science to the facts of nature. It is a fact that a dog is a mammal, but that water molecules consist of hydrogen and oxygen atoms is a fact in a different sense, and that light consists of electromagnetic vibrations is a fact in still another sense.

It must be said that Forbes was writing not against Brewster but against the historian Macaulay, who had, he thought, been over-ready to identify knowledge with being-acquainted-with-facts (see section X.9). Forbes saw clearly that we should not confuse nature with the science of nature. In his opinion we ought not to forget that facts have nothing great or little in themselves, but that the greatness lies in the ideas that they disclose, and that it is in the individual mind which originates and grasps the knowledge that the greatness truly dwells.[3] With keen historical insight he pointed out that it would be false and arrogant to disparage great thinkers of the past because they did not know many facts which we now know; 'this is a temptation that is greatest in sciences that, like natural sciences are of an essentially progressive character.'[4]

Forbes's attack was directed against a superficial historical evaluation — against the identification of the quality of scientific insight or knowledge of our predecessors with the factual basis on which it had been founded. Brewster, on the other hand, attacked superficial and insufficient factual knowledge, which proclaims — as in his opinion the anonymous author of 'The Vestiges of Creation' did — phantasies as facts; knowledge, that is, of fewer facts than are available at the time, or of more pretended facts than are really known, and therefore leading (in the context of the period) to false statements. Consequently the one warned against identification of merely factual knowledge with genuinely scientific knowledge, the other against a scientific knowledge that did not sufficiently keep to the facts.

Correct knowledge of facts is not in itself science, whereas the adjective 'scientific' could hardly be denied to some theories that were wrong as to the facts but right in their structure (see chapters III and IV on alchemy and the phlogiston theory). Of course, both scholars would have recognized that the basis of science consists of facts and the final test of its truth lies in facts. But true conclusions can be reached only when both reliable facts and correct scientific arguments go together.

It would require long digressions of a philosophical (ontological and epistemological) character to deal with the definition of a 'fact'. It is a fact that we see the sun rise; and even if it remains invisible because of the clouds, we recognize as a fact that it is rising, with an appeal to the analogy of nature. But it is also a fact that at the same moment, on the other side of the earth, it is not rising. Science tries to objectify these facts, so that all people may agree about them. The medieval astronomer objectified direct solar observations into the scientific fact that the sun goes around the earth, whereas Copernicans reduce the observations to the scientific fact that the earth is revolving on its axis while travelling around the sun. What is a fact to one generation may become a fiction — and that in a deprecatory sense — to the next.

As we cannot know all facts by our own experience, we have to accept many of them on the authority of other people. On the authority of those who travelled in foreign countries, or pretended to have done so, our ancestors believed that what the travellers told was the truth: fact and not phantasy. The 16th century zoologist William Turner — himself a keen and critical observer — when dealing with the strange origin of the barnacle goose, says that it grows on drift wood.[5] Lest anyone might think this a fable, he adds that it is the common testimony of all the coastal folk of Britain and that the famous historian Giraldus Cambrensis confirms it. Yet Turner was careful, and did not consider it safe to trust common report; he says that 'for the rarity of the thing' he did not wholly believe Giraldus.

He then tells how quite recently he 'asked a man whose obvious integrity won my confidence, a man by profession a theologian, by race an Irishman, by name Octavian, whether he thought Giraldus reliable in this respect. He took oath by the very gospel that he professed, and replied that it was the perfect truth ... that he had seen with his own eyes and handled with his own hands birds still half-formed.' Canon Raven, who quoted this story, considered Turner 'could hardly be blamed', as spontaneous generation was universally believed in at that time,[6] and — we might add — the testimonies were overwhelming. Indeed it would not be difficult to refer to scientific hoaxes generally believed on fewer testimonies, just because they fitted in with cherished expectations (e.g. the Piltdown case*). In particular on the authority of great scholars — Aristotle, Descartes (and even Newton) — some of their fictions were believed to be facts.

In addition we have to take into account that statements about facts of nature are sometimes delivered to us by people who at all costs wanted their theories confirmed: scholars are not immune to the temptation to steer the 'facts' for their own interests. I read that in 1697 the Masters of the University of St Andrews recommended its transference to Perth, because the inhabitants of St Andrews were brutish and quarrelsome, with a particular aversion to learning and learned men, and because the air there was 'thin and piercing' so that 'old men coming to the place were instantly cut off.'[7]

But in the early 1860's Robert Chambers, the Edinburgh printer, publisher and geologist, took up residence at St Andrews, a place to which, twenty years previously, he had become attached not only 'on account of its agreeable society', but also for its 'bracing atmosphere.'[8] We must either assume that not only the character of the inhabitants, but also the climate had changed since the late 17th century, or else — which is more probable — that the university people of 1697 adapted the truth to their personal interests.

Pictures in books have an especial power to turn phantasies into 'facts'. If, in the 16th century, one had heard voyagers' tales about rhinoceroses, dragons and wild men with tails in Africa and had seen in books by respectable authors pictures of these alongside those of more familiar animals, one might readily have assumed that all these pictures had the same value. Only more evidence could reveal the first

* [*Editors' note:* The Piltdown hoax was a deliberate fraud in the study of human evolution. It was perpetrated in 1908 - 1915. 'Finds' were made, among gravel workings at Piltdown in Sussex, of fragments of a thick skull and half of a simian lower jawbone containing two molars (apparently) worn flat in human fashion. Flints and bones (some worked by modern implements), together with genuine mastodon fossils were discovered nearby.

From this evidence the Dawn Man, *Eoanthropus dawsonii,* was deduced, a creature with a large brain, yet possessed of a jaw similar to that of a modern orang utan. At the time this 'Piltdown Man' was readily accepted by most specialists in the field, because he fitted their prior hopes and expectations of a 'missing link'. The hoax was not exposed until 1953 by fluorine testing (invented in 1949), although clear evidence of fakery was present all the time and would have been visible to a suspicious eye. See: C. Dawson and A.S. Woodward, 'On the Discovery of a Paleolithic Human Skull and Mandible in a Flint Bearing Gravel at Piltdown (Fletching) Sussex', in: *Quart. J. Geol. Soc.* 69 (London 1913), pp.117-151; Ronald Millar, *The Piltdown Men*, London: Gollancz 1972. Also: Stephen Jay Gould, *The Panda's Thumb*, London: Penguin 1980, pp.92-104.]

to be a fact and no phantasy, the second a phantasy and no fact and the third a wrongly interpreted fact.

Something similar happened with atoms, or quasi-atomic particles in general. The fictions of the Cartesians became facts to themselves and phantasies to their critics (Pascal, Huygens and Newton). The atoms of Dalton have turned out to represent not only some facts, but also a large number of unwarranted hypotheses, as has become particularly clear since the discovery of radio-activity and the development of modern atomic physics. Moreover these Daltonian atoms are even now not 'facts' in the same way as substances like sugar, brimstone, diamond, which are said to consist of them.

Let us say then that facts in science are for us phenomena, things or events, which are held to exist and whose existence or occurence all people sufficiently equipped with senses, reason and instruments recognize. Most people will believe them to exist independently of the human mind. Our problem, however, is that many so-called facts turn out to be spurious; and also that our inner thoughts and feelings, however subjective they may seem to others, are to ourselves objective facts of experience.

Facts do not speak for themselves. If there were no human mind to make them speak, they would remain dumb. And in letting them speak we are in danger of confounding them with our own fictions.

Faith

Faith, according to the writer of the Epistle to the Hebrews, is 'the evidence of things not seen.'[9] The impressive list of the faithful in that chapter is introduced by the statement that through faith we understand that the world is framed by the Word of God and that things that are seen are made from things that are invisible.

There is some analogy between religious faith and scientific faith, or — if one thinks this word too heavily loaded with religious connotations — with scientific 'belief'. In the first place the scientist has faith or belief in things that are seen in that he trusts that his senses and his reason — if rightly used — do not deceive him and that the world does exist. Subjectively the certainty of such a faith is the basis of the recognition of facts as such. It is really a matter of trust (fides) that what we observe and talk about in science has an objective value. Even those scientists who, for philosophical reasons, have some doubt about this, at least behave as if they had this belief.

Secondly, there is in science a faith regarding things not seen: the preconceptions behind and precedent to all scientific endeavour, the firmly believed axioms and

Fig. 1. Grotesque (Peter Floetner 1546).

postulates and an often quasi-religious belief in the truth of generally accepted theories. Axioms in particular are not tested directly, but only by finding that — with the help of observation and reasoning — they form the basis of a coherent system that is conformable to experience, or at any rate not contradicted by it.

Thirdly, there is a faith *a priori* — metaphysical in its origin — in the order of natural phenomena, in the simplicity (proportionality; analogy) of nature. It is one of the great incentives to science. And scientists inspired by this belief claimed to find it confirmed *a posteriori* by the results of their research.

Science would be impossible without a faith in order in nature. And yet, at first sight, this order is not so very evident. Many natural events — tempests, floods, pestilence — happen unexpectedly and haphazardly. Their very unexpectedness, however, suggests that as a rule our expectations are fulfilled and that there is rule and order in nature. This is most evident in astronomical phenomena. The alternation of day and night accompanies the repetition of the motion of the starry sky; the alternation of seasons is closely related to the annual repetition of the sun's course among the stars. In the organic kingdom there is the endless repetition of the generations; wolves bringing forth wolves, sheep bringing forth sheep. This is not to deny, of course, that there are a multitude of phenomena which seem rather unorderly — earthquakes, comets and volcanic eruptions. These defy, or for a long time defied, human efforts to bring them under a rule. So it was not experience alone but also a belief in an order as yet undiscovered — that is, in a certain uniformity of nature — which played, and still plays an important role in science. Even the rather unsophisticated Portuguese navigator Dom João de Castro (1500 - 1548), who was neither a philosopher nor a professional 'scientist', when all his efforts to discover a law connecting the variation (declination) of the magnetic needle and geographical longitude had been frustrated, continued to believe, by following a certain scientific instinct, that some law or other 'must lie hidden in the secret chambers of nature.'[10]

More or less implied in the belief in 'order' in nature is a belief also in simplicity. In science it is of no use to 'explain' with the ancient atomists an infinite diversity of phenomena by an infinite diversity of atoms. There is a sheer necessity for simplification if scientific explanation wants to avoid licence. Epicurus was satisfied by his ability to show that our world might be, and that it might have arisen in the way he depicted; and if this were not the case for this world, his story would doubtless fit the case for some other world among the infinite number of possible worlds. This infinite variety of sizes, forms, arrangements and motions of atoms, may have relieved some people's anxieties (about a future life), but it did not solve scientific problems; it could not predict future events from knowledge of present events. Atomism became a really scientific doctrine only when this 'infinity' was reduced to a relative simplicity, and chaos to order.

The more complicated 'order' is, the less orderly it seems to be. There is always a tendency in science to reduce a complicated 'order' to a higher, a more fundamental one, i.e. to a greater simplicity. 'Simplicity is the mark of truth' (*simplex sigillum veri*) was a principle guiding Aristotle, Ptolemy, Copernicus,

Galileo, Newton, Lavoisier, Dalton, Haüy, Fresnel and a host of other outstanding scientists. Newton's adage that 'the analogy of nature should not be deviated from' (*ab analogia naturae non est recedendum*), is of a similar character. Ptolemy held that the simplest of the competing answers to a scientific problem is probably closest to the reality of nature.[11]

Similarly, geologists in reconstructing the past, try to maintain as much as possible, the uniformity of nature throughout the ages. Even when such a metaphysical or ontological belief is lacking, there are at least methodological, practical reasons for choosing the simplest possible theory (Ockham). Those who do not claim to arrive at realistic theories, will at least prefer the simplest or most 'elegant' description and explanation.

The idea of simplicity is also connected with those of Beauty, Symmetry and Harmony. The order of Kosmos (ornament, world order) stands over against the disorder of Chaos. 'Beauty' implies a cooperation of parts within the Whole, and also a mathematical proportion of the parts, welding them together into a harmonious totality.

The faith in order, law, simplicity, harmony, beauty has often been connected with the faith that there is logos, reason, mind at work in the universe. This was fertile ground for natural theology — that is, for belief in a Mind to which the human mind has, however remotely, some resemblance, so that it is able to recognize these attributes in a creation which is the work of that Mind. Alternatively it may be held that it is the human mind which imposes this order and harmony upon nature, as this is the only way it can create an orderly system. The ideas of law and harmony (proportionality) are borrowed respectively from human law and human aesthetics and architecture, so it might be said that we project our ideas into nature — and onto the Divine Mind — and thus unwittingly pull out of nature what we had ourselves put into it. But whether we put our ideas into nature or conversely extract from nature, our ideas about nature are inevitably part of our mental framework. The belief that order and rule can be found in nature is in practice indispensible for the cultivation of science. Even those who deny that there is any Idea or plan realized in the universe, or who hold that, if there is one, it far surpasses human understanding, even these scientists in practice behave as if there is order in the universe.

FICTION

By fiction in science we do not mean arbitrary inventions, such as those met with in science fiction. Fiction is here rather a response by human imagination to experience about nature. These fictions we do not feel compelled to accept by intuition, by what Pascal called the 'heart', or by some methodological necessity, as was the case with *a priori* beliefs. Nor are we sure of them by critical observation as is the case with facts.

To the 'fictions' in science belong its working hypotheses — those of a mathematical as well as those of a non-mathematical type. Among scientific fictions are the thought models, the supposed analogies between what happens in the world of phenomena and what goes on behind the scenes in the hypothetical, rather abstract world of atoms, electric particles, fields and forces. Sometimes these are accorded the status of 'facts'; in other cases they are only posited as a means of finding and relating facts. 'If A be true, B must follow', and this conclusion or prediction must then be tested by observation and experiment.

Whether a scientific fiction proves to be a boon or a bane depends upon the reliability of our observations and interpretations of phenomena which give rise to it. Scientific imagination, whatever the value attributed to its products, plays an essential role in reconnoitring nature but its effect is negative when it hampers unprejudiced observation of nature.

Structure of the Book

The choice of 'Fact, Faith and Fiction' as the title of this book means that we shall be concerned more with scientific procedures than with the special achievements of scientists. We shall consider these from the point of view of the history rather than the philosophy of science, though of course the latter cannot be left out. In considering scientific procedures the intention is not in the first place to express opinions as to how things should be done, but just to relate how they in fact have been done.

It may be said beforehand that unorthodox methods have sometimes led to great results, whereas a more orthodox approach remained fruitless. Small wonder: in many cases today's scientific orthodoxy was yesterday's heresy. What the scientist does and what he thinks, depends on the facts available, on the method practised, on the beliefs held about nature and natural science, on the indoctrination received through education and the climate of opinion of the place and time in which he lived. Nevertheless, however different these may be for different places and times, we will recognize that certain patterns of thought, certain approaches to nature, were common to all men in all times: they are in the literal sense 'catholic'.

What makes a theory 'scientific' is not so much the 'truth' contained in it as its inner structure and the method used in building it. Methods and results which seem acceptable as 'scientific' in one age, may seem 'unscientific' in the next: the scientists' ways of thought, however, show a much greater continuity. It should not, however, be forgotten that science does not have a monopoly of making true statements. Many conceptions are scientific without being true; and many beliefs are true without being scientific.

I would have liked to make a systematic division in which the emphasis could be laid successively on fact, faith and fiction one after the other. History, however, is no systematic discipline like philosophy or science; its subject is the intricate texture into which throughout the ages the phenomena of human life and the forces, ideas and motives behind them, are woven together. When we look at science we find that fact, faith and fiction are always present together, though in proportions which vary greatly.

The procedure followed in these chapters will be a mixed one. In some chapters a certain theme is chosen, and we shall see how it comes back again and again in the course of history; in others we have concentrated on the interaction of fact, faith and fiction in the work of some particular scientist.

Our first topic will be the part played by a faith in mathematical harmony in nature. We shall look at the numerical speculations of the Pythagoreans; at Kepler's geometrical speculations in astronomy; at some interesting cases provided by less well-known 18th and 19th century scientists and finally at the ideas of one of the great pioneers of modern atomic theory, Niels Bohr (chapter II).

Next we shall deal with the attempt to build scientific theory upon direct observation by the senses: Aristotle's naïve realism and the qualitative theories of the alchemists, which lived on even among the founders of modern chemistry. Instead of seeking a world *behind* the visible world, this approach attempts to build theory upon direct sensory experience; taking an organic view of nature, it moves by subtle reasoning from sensed qualities to substances presumed to be responsible for them (chapters III and IV).

In chapter V we shall deal with a topic bearing on the relation between mechanics and physics in the Middle Ages. Starting from Aristotle's theory of motion (which was based on naïve observation), we shall look at the efforts at mathematization by some medieval philosophers, the impetus theory, the method of latitudes and the 'calculationes'. One of the most delightful examples, the problem of a body falling down an imaginary tunnel through the earth, will give us occasion to discuss thought-experiments and the conflict between organistic and mechanistic images of nature.

The case of Copernicus will lead us to further discussion of these two world views and also to an investigation of the influences hampering or furthering the introduction of the new world system (chapter VI).

The conflict between the organistic and the mechanistic approach will raise the methodological question, whether experimentation is a legitimate method of research into nature (chapter VII). This has been one of the most troublesome issues for the founders of modern science: is it legitimate to experiment with nature? Can any worthwhile knowledge be gained by means of manual labour? If results do not fit with theory, which should give way?

In chapter VIII we will meet another issue, which is still debated up to the present time. Is the purpose of scientific theory to represent reality, or is it only to offer a useful way of bringing phenomena together, without making any pretence at a realistic interpretation?

The classical opposition between 'Nature' and 'Art', a problem raised in the humanities by Peter Ramus, and in the sciences by Francis Bacon, is discussed in chapter IX. Was it perhaps impossible in principle to surpass or even to imitate nature successfully, as for example by chemical synthesis of minerals, or by making artificial models of natural processes?

How much a fiction may blind us and lead to a refusal to accept facts, will be illustrated in chapter X under the title 'Cleopatra's nose'. Here we shall see how one of Newton's very few 'hypotheses' from his *Principia* hindered the development of chemistry for the first half of the 19th century.

Finally in chapter XI we shall deal with scientific method and with natural theology from the standpoint of Pascal.

In the book it will become evident that all scholars and scientists, when divested of their 'robes de pédant' (to use Pascal's expression), are very 'human'. The humanities and the sciences are not separate islands but peninsulae of the same mainland: when we go back to their origins it turns out that they are branches of the same trunk. Natural science, like literature, historiography, philosophy, etc. is a very human affair. The large gap is not between the sciences of nature and the humanities but between the objects of their studies: nature and humankind. Just as natural science, the study of nature, should not be confused with nature, so the humanities, the self-contemplation and self-study of man, are not the same as mankind. There is no antagonism between the sciences and the humanities for they are all humanities.

Historiography of science, which is one of the links between the sciences and the humanities, should not be confused with the historiography of the relations between science (and technology) and the political, economic, social, geographical circumstances. They differ more or less as the history of economics differs from economic history (respectively the study of the history of the theories of economics and general history in its economic aspects). But the borderline is vague, and those who occupy themselves with the history of science are unavoidably confronted with the question how the personal character and the political, social, theological environment of the scientist influenced the contents of his science.

To study all these aspects gives an immense pleasure, as the study of nature herself is ever again a source of delight to those who devote their life to it and regard it as more than a mere matter of bread-winning. Kepler, one of the most 'humanistic' and 'humane' of scientists, once wrote: 'It seems to me that the ways in which people come to the knowledge of natural things are almost as much to be admired as the nature of these things itself.'

I. ON NATURAL THEOLOGY

1. THE BOOK OF NATURE 17
2. CORPUSCULAR THEORY 20
3. ZOOLOGICAL 'PROOFS' 21
4. REVELATION IN NATURE AND SCRIPTURE 22
5. SABUNDUS 23

It has been claimed that Natural Theology has been an incentive to the scientific investigation of nature and that, conversely, this investigation yields arguments in favour of it. Certainly some of the greatest scientists (Kepler, Boyle, Newton, Hutton) used their discipline to point to the power, wisdom and goodness of the Creator of all things.

It is not the task of the historian to decide whether their arguments were valid, but rather to provide data about the character and development of science and its relations to other aspects of culture. Accordingly we shall first say something about Natural Theology, mainly that of two of its outstanding advocates, — both non-scientists; the one, Cicero, a pagan philosopher and statesman, the other, Ramundus Sabundus, a Christian apologist of the late Middle Ages.[1] Having done that we shall concentrate these lectures upon widely divergent topics from the history of science, and of the scientists, concerning the structure of science, the discoveries made, the methods employed, the goals desired.

1. THE BOOK OF NATURE

Already in Antiquity we find natural theology taking a stand in opposition to the doctrines that there is no rationality in nature, and that there are no objective moral principles distinguishing between Good and Evil (as might, for example, be asserted from the observation that what is right in one country is wrong in a neighbouring one).

Thus all idealist philosophers agreed in criticizing the materialist position of the ancient Greek atomists. Democritus (circa 760 – circa 370 BC), Epicurus (341 – 270 BC) and Lucretius (circa 99 – 55 BC) held that the only reality is to be found in atoms, — indivisible, indestructible and unchanging, moving in empty space and making, by chance, collisions which lead to aggregates which might ultimately coalesce into legs, arms, heads and so on, and that these then unite into bodies.

Some of these bodies were without eyes, others without mouths, but at last some fortuitous combinations turned out to be viable, that is they could feed, defend and propagate themselves.

Although the Greek atomists made Chance into Necessity (*anankè*), it was a blind necessity, not representing a rational plan. They were not looking for a fixed order (though they did have to admit some fixed principles in nature such as the indivisibility of atoms and the intrinsic heaviness of matter). Their system did not purport to further scientific creativity, but rather to take away the fear of life hereafter and of the power of the gods, it being held that at death our atoms are dispersed throughout the world, and our individuality annihilated and that the gods themselves are but aggregates of atoms.

Plato, Varro (116 - 27 BC) and Cicero expressed their opposition to these views. It was Varro who, according to St Augustine, coined the term 'natural theology', distinguishing it from mythical theology (the stories of the poets about the Olympic gods) and from civic theology (the official cult instituted by the State to keep society together). He recognized, like Aristotle and others before him, that the gods of the Egyptians, Greeks and Romans were personified powers of nature. These *numina*, being de-mythologized by the philosophers, were brought back, on a higher level, to their original status of de-personalized forces, bound together by the World Soul or World Intellect.[2]

The second book of Cicero *On the Nature of the Gods* (45 BC) has, down the ages, exerted a great influence on natural theologies. In a downright attack on the Epicureans Cicero declares that the regularity of the stars is full of Reason and not of Chance. It is impossible that this beautiful world should result from the fortuitous collision of indivisible particles of matter,[3] devoid of any property except shape, size and motion: copies of the twenty one letters of the alphabet could not produce by themselves the *Annals* of Ennius,[4] nor does the random concourse of atoms ever produce a house or a city.[5]

In order to prove the existence of a planning deity, in contrast to the powerless 'dieux fainéants' of the Epicureans, Cicero refers first to astronomy. 'When we gaze upward to the sky and contemplate the heavenly bodies, what can be so obvious and so manifest as that there must exist some Power (*numen*), possessing the highest intelligence, by whom these things are ruled?';[6] a deity omnipotent and omnipresent.[7] The sympathetic agreement, interconnexion and affinity of things — so evident in natural processes — could not exist, and all the parts of the world would not be singing together in harmony, were they not held together and pervaded by a single divine spirit,[8] which possesses sensation and reason.[9] The whole world is better than any of its constituent parts:[10] therefore Man, who is a part of the world, participates in that Intelligence, though on a lower level.[11] In particular the stars are divine and possess intelligence,[12] as is evident by their order and regularity (*ordo et constantia*) for nothing can move by reason and rule, without design (*consilium*). The world, then, is a vast organism, governed by an intelligent Nature. It is eternal: there are cyclic transmutations of the elements one into another,[13] cyclic motions of the stars and cyclic relations of the stars to the earth.[14] The world is governed in a way most wise, rational, beautiful[15] and providential.[16]

Plants and animals are so wonderfully adapted to the way of life of their kind and their parts cooperate so miraculously to preserve their life, that they, too, display this rational design.[17]

Finally, the whole of nature is adapted to the preservation and convenience of Man,[18] so that we are led to believe that in the last resort 'the world and all things contained in it, were made for the sake of gods and men.'[19]

This sketch of Stoic natural theology demonstrates that, however much particular illustrations may have changed, Cicero's arguments for plan and providence in nature have remained a stock in trade of natural theology throughout the ages. They purport to demonstrate the existence of a rational Mind in the universe as well as the main attributes of that Mind — its power, wisdom and goodness as displayed in the works of nature. Its wisdom is shown in the structure of the universe and of the animals. Its power is displayed in the overwhelming greatness of the world and the forces working in it. Its goodness is usually connected with the teleological argument: the beneficence that is evident from the design by which plants, animals and in particular Men and their environment fit together, and also from that inner adaptation and cooperation of the parts of the living body for the maintenance of the whole.

The early Christian apologists shared with the ancient idealist philosophers the belief in a divine moral and physical order, but they avoided any identification of the divine Mind with the universe in which it manifested itself. Rejecting a pantheistic immanent deity (God - Nature), they believed in a transcendent Creator, who could not lack what his highest creatures possessed, namely personality.

But there remained some ambiguity when, as frequently happened, the expression 'God *or* Nature' (*Deus sive Natura*) found itself replaced by 'God *and* Nature', as if the natural order, once created, had an existence and power of itself apart from God. Theological voluntarism therefore stressed that the creator God did not have to obey anything, not even the eternal Ideas, but created all things according to His free will and not in obedience to an eternal Reason. In their wake the late-medieval nominalist philosophers contributed to the de-deification of nature and so to the liberation of scientific research.[20]

2. Corpuscular Theory

When, in the 17th century modern science developed, many of its protagonists were ardent advocates of corpuscular or even of atomistic theories. Consequently the Christian adherents of the idealistic philosophies (Platonism, Aristotelianism, Stoicism) accused these 'new' philosophers of lapsing into materialism and of depriving nature of its spiritual principle, thereby reducing it to a lifeless machine. Even nowadays it is often maintained (in a tone of compliment rather than of criticism) that the rise of modern science was essentially a vindication of ancient materialism against philosophical idealism.[21]

The adherents of the 'new', or as it was often called, 'mechanical' philosophy, however, were actually as far away from the one as from the other. They maintained that the 'old' school was tainted by pagan beliefs, for it attributed to universal Nature and also to special 'natures' such as Forms, Ideas or species, a self-sufficient existence apart from God. The self-perpetuating Ideas, the substantial Forms and seminal principles of the idealist philosophers were, said Malebranche (1638 - 1715), 'little gods' (*numina*), whereas, according to corpuscular theory they were just *nomina*, names for combinations that had no power in themselves but worked only according to the will of God. Again, to Robert Boyle the current adage of the scholastic teleologists that 'God and Nature do nothing in vain' sounded almost blasphemous, as if they shared power like the two Roman consuls.[22]

Boyle considered materialistic atomism less dangerous to Christian religion than the seemingly pious idealistic systems. He was one of the great advocates not only of the atomism of the mechanical philosophy but also of natural theology and the teleological view of nature. For this mechanicism of the 17th century, in spite of borrowing its *physical* notions from ancient atomism, was *metaphysically* its opposite. This is seen even in its name, for a mechanism requires a mechanic who made it according to a rational plan (μηχανη: clever device) and who did so with a definite purpose. By contrast with the eternal atoms of the Ancients which gave rise to the world by chance collisions, the atoms of the mechanical philosophers have been endowed by their Creator with movements according to laws that lead to the realization of His plan for them. If the world is not an organism but a mechanism, its Creator is, so to say, a Mechanician, and this is indeed how God is referred to (1599) by a commentator on Aristotle's *Mechanica*.[23] In the great mechanical artefact, the Universe, the natural theologian recognized the hand of the great Artificer, who makes things according to his free planning.

On the other hand, Boyle's views were far from an idealistic standpoint, as can be seen from his comparison of the world to the ingenious machinery of the great clock of Strasbourg Cathedral. It is true that Cicero, too, used this image when comparing the starry heavens to clockwork because of its regularity. And he, too, spoke of the *machina mundi* but to him this meant only that there was ingenuity displayed in it and not that it was a mechanical instrument. For Cicero it was animated by the Divine Mind. For the 'Christian virtuosi' of the new philosophy, however, no trace of divinity was left within nature. Its rational character was the result not of an inner logical and vital Necessity but of the free will of a transcendent God.

This view had an interesting consequence. Though the new philosophy recognized rationality in Nature, man's reason could not decide beforehand which kind of rationality was laid down in Nature. So one had to go and look rather than reason it out. Most pioneers of modern science were thus 'rational empiricists' rather than rationalists.[24] Their natural theology, too, did not bear an aprioristic character. God's existence was not demonstrated deductively. Newton claimed that natural philosophy (i.e. science) found God *a posteriori* by ascending by means of observation and experimentation through the hierarchy of causes until it was inevitably led to the Cause of Causes, the Creator of all things.[25]

Summing up, we conclude that in their metaphysics Greek idealistic philosophies as well as 17th century mechanistic philosophy recognized plan in nature, but that to the latter this was the plan of a transcendent Creator whose reason far surpassed human reason. And in their physics both ancient materialism and mechanistic philosophy explained phenomena by material principles — matter, motion, structure — though the mechanicists held that physical reality was not the only reality.

3. ZOOLOGICAL 'PROOFS'

In the 17th century there was a shift in the focus of natural theology from astronomy to zoology. And in the 18th century, when natural history became extremely popular, this tendency became even stronger.

In Antiquity biological teleology had already been much used by the great physician Galen (1st century). In the late 17th century, Boyle, though himself no biologist, held that living beings provide us with more convincing proofs of the wisdom and providence of God than non-living beings. Newton, also a non-biologist, declared himself most impressed by the design demonstrated by the eye, which could not but have been devised by a Being with a supreme knowledge of optics. In the 19th century Lord Brougham, in his treatise on the beehive,[26] paid most attention to the adaptation and constitution of animals and in particular to their way of life. And the Duke of Argyll let his discourse on 'Law in Nature' culminate in an exposition on the adaptedness of a bat's wing to the purpose which it serves.[27]

4. Revelation in Nature and Scripture

Christian theologians generally considered natural theology as a preliminary to revealed theology. 'By nature' (i.e. naturally) man could come to a theistic belief in the existence of an almighty, wise and benevolent Creator. To this 'special' revelation then added the Trinitarian and Christological tenets of religion. In the 18th century there was a growing tendency to deism, which increased the importance of Natural Theology as compared with special revelation, though downright rejection of the latter was exceptional. In the early 19th century the teleological arguments of William Paley's *Natural Theology* were very popular.

Many Darwinists were deists (for example Lyell), agnostics (Darwin himself) or atheists (Büchner). In their talk about the organic world, Natural Selection took the place of the Creator. By denying design in nature and proclaiming random variations, natural selection and the struggle for life (survival of the fittest) as the causes of evolution, they in effect assumed a metaphysical position similar to that of the ancient Greek atomists. That the same mechanism of evolution could, however, be fitted in just as well with a theistic and Christian belief (that included design in nature) is evident from the writings of some of Darwin's earliest and staunchest adherents, the Anglican clergyman Charles Kingsley and the famous American botanist Asa Gray.[28]

In general Natural Theology claimed only to demonstrate from His works in nature the existence and the power, wisdom and goodness of God. There has been, however, another kind of natural theology, which held that in the works of creation — in man (the microcosm) as well as in the universe (the macrocosm) — was also expressed the Trinitarian essence of the Creator. This conception was a combination of the Christian belief that man is made in the image of God with the Platonic belief that man is the image of the macrocosm. Such a natural theology may be found in Paracelsus and occasionally even in Kepler, who held that the structure of the universe shows the triune character of its Creator: the sun corresponding to the Father, the sphere of the fixed stars to the Son and the intermediate space to the Holy Spirit.[29]

5. SABUNDUS

The late medieval *Theologia Naturalis* (before 1432) of the Catalan Raymond de Sébonde, seems at first sight to go much further in this direction.[30] It exerted a wide influence during the next two centuries; in the 16th century Michel de Montaigne translated it into French, and Pascal in the late 17th century was quite familiar with it.

It shows an uneasy relation between natural and revealed theology. Indeed in 1559 it was put on the Roman Index but later (from 1564 onwards) only its prologue was prohibited, so that later printed editions appeared in a curtailed version.

In the 'Prologue' Sabundus says that Man knows himself, and God, and also what is necessary for salvation through the study and experience of all creatures and of human nature.[31] This study confirms Holy Scripture for us (for man will believe thanks to this study) and for us it precedes Scripture: 'Therefore God has given us two Books, namely the Book of all creatures or the Book of Nature, and the Book of Holy Scripture. The former was given to Man from the beginning ... for each creature is nothing but a letter written by God's finger ... Man himself ... is the principal letter in this book.'[32] Ramundus then points out that the second book, the Book of Scripture, is given to man because he could not read the first one, and that whereas the first book is common to all people, clergy and laity, the second can be read only by the clergy. Moreover the first cannot be falsified or wrongly interpreted, whereas the second can be falsely interpreted and thus lead to heresy.[33] But they never really contradict each other as they have been written by the same God.[34]

There appears to be some inner contradiction in this prologue for — having praised the excellence and general accessibility of the book of nature and its rationality as compared with the Bible — Sabundus declares that none can read this always open book unless he is illuminated by God and cleansed from original sin. For this reason none of the ancient philosophers could read this 'science' (*scientia*), being blinded as to their own salvation. They could not read in it the true wisdom that leads to eternal life, even though it is written there.[35]

Most Christian believers followed St Paul in attributing a more modest scope to the knowledge of God which can be obtained from nature. In his letter to the Romans, the apostle says that 'the invisible things of God ... are clearly seen, being understood by the things that are made, even His eternal power and Godhead.'[36] At first sight this seems to be pure Natural Theology: but then we note that despite the objective presence of signs of God's being and power in the creatures, Paul speaks of a subjective 'clearly seeing' of them only in the light of special revelation; for the corruption of the nature of fallen man mars and blurs the picture, so that it is totally disfigured. The true knowledge of God in nature thus does not precede revealed theology but rather follows as a fruit of it. It is not a preparation for revealed religion but a consequence of it. As the letter to the Hebrews puts it: '*through faith* we understand that the worlds were founded by the Word of God, so that things which are seen were made not of things which do appear.'[37] Any 'natural theology', in the sense of an independent 'science', then, would be of little value, and it would be built on the weak ground of human investigation.

It is remarkable that Sabundus' main text remains within the bounds of this biblical orthodoxy. It emphasizes that God creates the world not by necessity but voluntarily, and that He does so by art, not letting it emanate from His own nature, but out of nothingness.[38] The relation of God to the world is that of builder to house: the latter is not of the same nature as its artificer.[39]

This is an important statement, for it repudiates pantheism in the sense of the Stoa by implying that the world is *fabricated* and not *generated* by God.[40] Moreover the author is careful to point out that the analogy with the builder and building should not be pushed too far. A house can exist without the artificer who built it, whereas the world needs God's unceasing presence. If God were to cease to sustain and to will it, it would vanish back into nothing.[41] Accordingly Sabundus terms this divine upholding a 'continual creation.'[42]

The world that God created to the glory of his own power, wisdom and goodness is in the highest degree 'proportionate, orderly, congruent and useful.'[43] This usefulness is not a mere utilitarian one: it is stressed that among the good things mankind receives from the world are his enjoyment (*gaudium*) of its beauty and the instruction he gets from its contemplation.[44] In the hierarchy of nature, things serve one another, and consequently they are all made in the last resort for the highest creature, 'for the sake of man' (*propter hominem*).[45] In accordance with this position Sabundus is of the opinion that man can know God better by studying his own nature than by contemplation of external nature: in the hierarchy of matter-plants-animals-human beings he is the highest and the only one made in God's image.[46]

The treatise devotes much attention to the differences in the manner of teaching of the two Books. Whereas the Book of Nature and the philosophers speak 'rationally', the Book of Holy Scripture speaks apodictically and 'authoritatively', not attempting to 'prove' anything. But he who has no notion of God cannot know whether the Bible is from God, and therefore man has to read first in the Book of the creatures (which teaches him to know God) and then come to the Book of Holy Scripture. 'The book of creatures (*liber creaturarum*) is the gate, the way, the door giving entrance ... to the book of Holy Scripture'.[47]

But then it seems that Sébonde realizes that all the time he was busy 'demonstrating', he knew beforehand what would be the result, as from the start he possessed a strong Christian belief. In his opinion man is able to show that it is reasonable that there are three persons in the Godhead, 'but that it is really so that the one God is in three distinct persons ... that could not come up first in the heart of man' but has to be spoken by God.[48] In the same way nobody can excogitate that God became man, and even if he could speculate about the possibility, he could not say that it must happen as a fact, or *when*, or *how*, or *where*. 'No creature could know by himself;'[49] God had to reveal it to us in the New Testament.[50]

Similarly, though man can prove that the world has been created and why it has been created, he cannot prove *when* or *how* it has been created; this he can only know from the Old Testament. Nor could he know that man was created on the sixth day,[51] if it had not been told him in the Bible.

It is evident that there is a tension between on the one hand Sébonde's rational and 'natural' theology with its timeless verdicts and on the other hand his insight into the strongly historical character of the Christian religion, whose core consists in events that occurred once, at a certain place and time, especially the life, death and resurrection of Jesus of Nazareth.

The rationalism of Ramundus' natural theology is totally reversed in the fideism of his Christian theology: he concludes that the words God speaks in the Bible are 'beyond all reason' and that 'the more they surpass human nature, the more they should be believed, and the more difficult it is to believe them, the more they should be believed, for this is a sign that they are spoken by God and not by man.' From this it follows that the articles of the Christian faith transcend human understanding and that they are therefore the more credible and certain.[52]

In the light of this conclusion we may now read afresh Sébonde's statements in his Prologue: that God gave to man His own book, not falsifiable, written by His fingers, in which every creature is a letter demonstrating God's wisdom and the doctrine needed for man's salvation. We noted already that he qualifies this by saying that the heathen philosophers could not read that true wisdom in it, since for this one needs to be enlightened by God and cleansed from original sin. Having heard him to the end, it seems that to him natural theology was logically an introduction to revealed theology, but that psychologically it came after it.

In other words, Christian apologists can never efface their Christian education. Even the later deists who discarded revealed theology, produced a natural theology that bore the stamp of their Christian environment.

II. Harmony in Nature

1. The Pythagoreans 28
1.1. Numbers and the Structure of the World 28
1.2. Consonance and Numerical Ratios 29
2. Plato 31
2.1. Plato on Music 31
2.2. Plato on Astronomy 33
3. Jabir-ibn Hayyan 35
4. The Structure and Harmony of the Universe 36
4.1. Kepler 36
4.2. Kepler's Natural Theology 41
4.3. The Law of Titius-Bode 43
5. Crystal Theory 46
5.1. Abbé Haüy 46
5.2. C.S. Weiss 51
6. Series of Chemical Compounds 53
6.1. Richter's Laws of Acids and Bases 53
6.2. Homologous Series 55
7. Series of Chemical Elements 57
7.1. The Law of Triads 57
7.2. Newlands's Law of Octaves 58
7.3. The Periodic System 59
8. The Harmony of Light: Balmer to Bohr 68
9. Conclusion 74

One of the most quoted texts in the literature of ancient astronomy and chemistry is taken from the apocryphal *Book of Wisdom*. It runs: God 'has ordered all things according to measure, number and weight',[1] or, as the 15th century alchemist Thomas Norton phrased it:

> by Ponders right, With Number and Measure wisely sought,
> In which three resteth all that God has wrought.[2]

That mathematics plays a fundamental role in the universe had been a general belief since antiquity. Not only was it held that the volumes, weights and lengths in nature are arranged according to measures which are *fixed*, but also that these measures are in proportions which are *meaningful* (though hidden from the vulgar). A faith in a mathematical harmony has inspired numerical speculations — some fruitful, others wildly fanciful — from the time of Pythagoras up to the present day. In various degrees the influence of the philosophies of Pythagoras (572 - 492 BC) and Plato (429 - 348 BC) contributed to the mathematization of science.

6. The Pythagoreans

6.1. Numbers and the Structure of the World

To the Pythagoreans 'number' was more than just one aspect of things; they held that things *consist* of number. Thus the number one is conceived to be a point; a line consists of at least *two* points, a plane of *three*, and a solid body of at least *four* points. These four numbers are contained in the most perfect of all numbers, ten, which is their sum, as can be satisfyingly displayed:

```
      •
    •   •
  •   •   •
•   •   •   •
```

Fig. 2. Tetraktys

This quaternary (*tetraktys*) played an important role in their number speculations. Thus they held that — since the universe is most perfect, most symmetrical and most beautiful — it must also contain the most perfect number of moving bodies, namely ten. According to the Pythagoreans there is, in the centre of the spherical universe,[3] a fire — the 'hearth of the universe', the 'watch tower of Zeus' — around which the earth, the sun, the moon and the five planets revolve. Together with the boundary of the universe, on which the fixed stars are placed, this makes nine bodies. The sacred number ten is reached by assuming that there is a counter-earth, which remains invisible to us because the hemisphere on which we live is always turned away from it (as also from the hearth of the universe). Aristotle, disapprovingly characterized their way of proceeding as follows:

> They held that the whole heaven is harmony and number ...: facts ... they adapted to the properties and parts of the heaven and its whole arrangement. And if there was anything wanting anywhere, they left no stone unturned to make their whole system coherent. For example, regarding as they do the number ten as perfect and as embracing the whole nature of numbers, they say that the bodies moving in the heaven are also ten in number and as those which we see are only nine they make the counter-earth a tenth.'[4]

He accused the Pythagoreans of 'not seeking accounts and explanations in conformity with the phenomena, but trying by violence to bring the phenomena into line with reason (*logos*) and opinion (*doxa*) of their own.'[5]

The Pythagoreans, then, belonged to those who — in Pascal's phrase — 'make blind windows in a house for the sake of symmetry': a weakness which Aristotle detected more easily in other systems than in his own.

There is a precarious balance between our preconceptions as to what we would like the world to be and what experience tells us it is in fact. This balance may be disturbed by giving in too much to *a priori* rational prejudice at the cost of *a posteriori* interpretation of facts. Nevertheless, it occasionally happens that fictions or hypotheses are confirmed afterwards by facts — facts which perhaps would not have been sought had it not been for those fictions. We shall come later in this chapter to the striking recent example of an expectation giving rise to an important discovery — the case of the 'omega-minus particle', predicted in 1961 and discovered in 1964 (see also page 75).[6] There is, of course, no question of putting Pythagorean speculation on the same level with the work of modern nuclear physicists, yet it is interesting to recognize at least some affinity between them. They share the enthusiasm generated by the discovery of a law thought to be one of the fundamentals of nature. In both cases the existing theoretical framework gives great weight to aesthetic notions of symmetry, simplicity and completeness. In each case, by odd coincidence, nine out of the ten entities expected by the theory had already been found.[7] In both cases the belief in order was more specific than a mere proclamation that there must be some mathematical law or other. The difference is that in the modern case the hypothetical entity was later found, whereas the counter-earth remained only an article of faith.[8] The possibility of new discoveries makes the modern physicist reluctant to claim any final truth for his theories however aesthetically satisfying. The Pythagorean metaphysical faith demanded that there *must* be the tenth heavenly body and for them it needed no confirmation by facts, the modern physicist deemed it no more than highly probable that a tenth particle would exist and only after experimental confirmation did it become accepted.

1.2. Consonance and Numerical Ratios

Number played an essential role, not only in Pythagorean astronomy, but also in the Pythagorean theory of music. The Pythagoreans held that the harmony of tones consists in simple ratios, 1 : 2 (octave), 2 : 3 (fifth), and 3 : 4 (fourth). These ratios they generalized back to the cosmos, where the heavenly bodies were said to produce musical tones depending upon their velocities, which in turn depended upon their distances from the centre of the world. The distances could be likened to

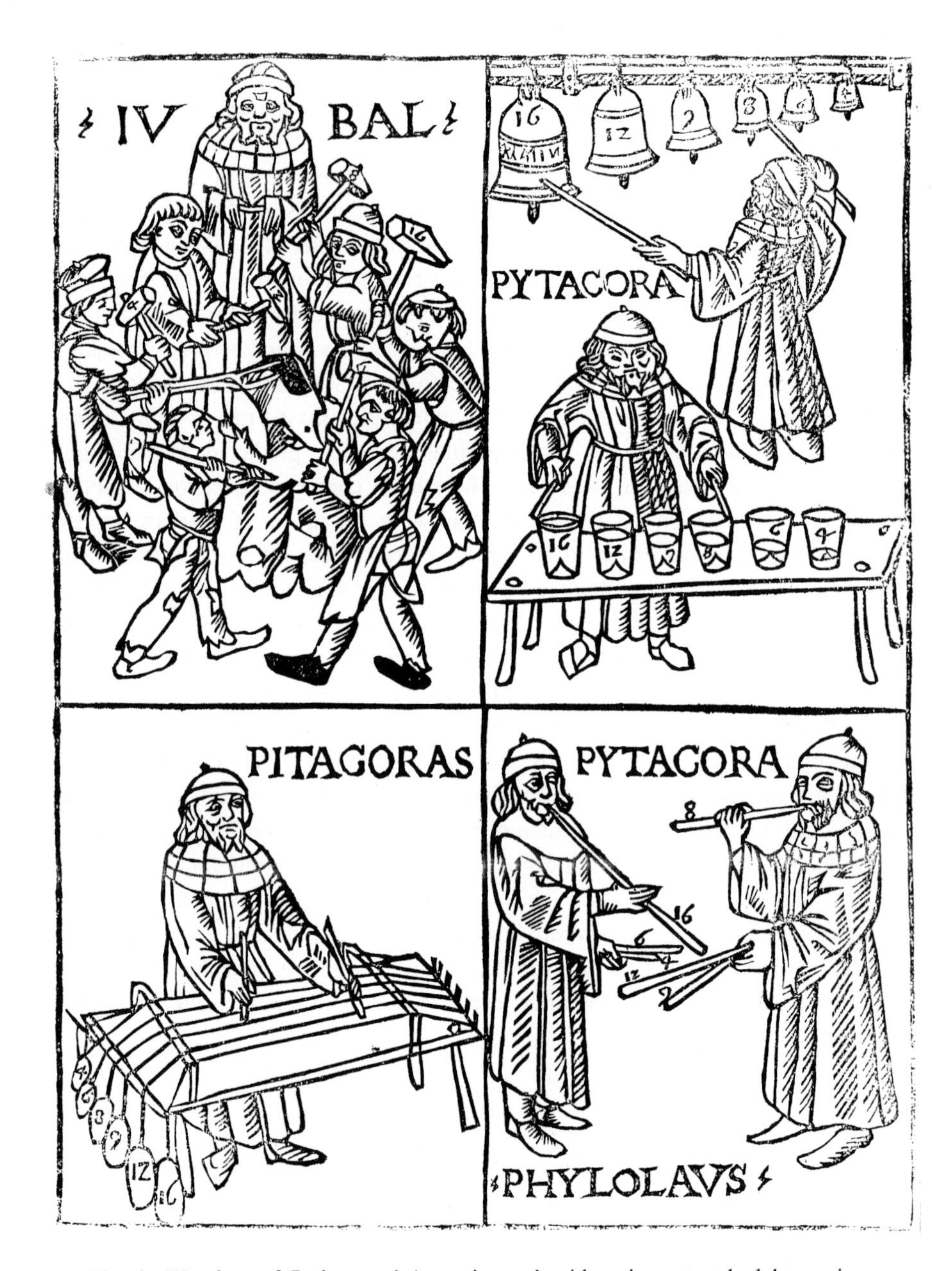

Fig. 3. Woodcut of Pythagoras' 'experiments' with strings stretched by various weights, flutes of various lengths, bells filled with differing amounts of water, etc. (Franchino Gaffurio, *Theoria Musica* 1492, p.127.)

the lengths of the strings of an instrument:[9] the slower motions give the deeper tones; the whole choir together produces the 'harmony of the spheres', inaudible to mortals in whose ears it has sounded from their birth.

Pythagoras is said to have discovered the numerical relations of harmonious sounds by means of experiments. The story goes that as he passed by a smithy and heard the harmonious clang of the hammers of differing weights striking the anvil, he noticed that the smallest produced the highest pitch. He then ascertained that their weights were in corresponding ratios: a hammer with double the weight of another was said to form an octave with it. He was reported to have discovered the same for the notes sounded by equal strings stretched by a set of weights: if one weight was twice another the notes formed an octave.[10] At first sight this seems to be a nice example of a working hypothesis confirmed by experimental verification. Unfortunately, however, the frequency emitted is actually not inversely proportional to the weight of the hammer, and in the case of the strings, it is for the square roots of the weights stretching the strings and not the weights themselves that the proportionality holds. One suspects that these were 'thought experiments' invented to fit the theory but not rigorously tested by experiment.

More credible is the claim that Pythagoras discovered that there is a proportionality between the lengths of strings (under equal tensions), and the pitches emitted, as this is indeed confirmed by experience. We conclude that faith *a priori* in some aesthetically appealing quantitative relation may lead to false experiments: sometimes 'results' get adapted to the preconception and sometimes, perhaps, the experiment is never actually performed because the outcome seems so obvious. But on the other hand an aesthetic conviction may lead to correct experiments, as in the case of the law of the inverse proportionality of length and pitch.

2. Plato

2.1. Plato on Music

Plato (429 - 348 BC), was strongly influenced by the mathematical emphasis of the Pythagoreans, to the extent that he thought poorly of those theoreticians of music who believed that audible sounds could teach them the true ratios of harmonious sounds. Indeed he mocked people who measure tones and harmonies by their ears,

> those good people who inflict a thousand tortures on the strings and put them on the rack and wring them with their pegs. Their work will not reach their goal; they give precedence to their ears before their intellect.[11]

He is presumably referring here to those whose methods were akin to those of Aristoxenus (fl. circa 350 BC), a theoretician of music who said:

> We hold that the voice follows the natural law in its motion ... therefore we try to give proofs in agreement with the phenomena. In this we are unlike our predecessors. Some of these ... reject sensual observations as being inaccurate and posit rational principles saying that height and depth consist in certain numerical ratios and relative rates of velocity, thus enouncing theories ... wholly at variance with the phenomena ... Our method rests, in the last resort, on an appeal to the two faculties of hearing and intellect ....[12]

To Aristoxenus sensual observation comes first; measurement may come afterwards: it is by hearing that we should decide which tones form consonances, and in practical music-making no other method produces acceptable results. In contrast to the Pythagoreans, he did not wish to restrict the ratios of consonance to the three fixed *a priori*.

For Plato, however, even the Pythagoreans relied too much on the sensible world. Their conception might be superior to that of the experimental musicians: like himself, they fixed the harmonic proportions *a priori* but he criticized the fact that they then thought it worthwhile to verify them by experiment. Their work, 'like that of the astronomers, must remain without result.'[13] What Plato wanted was something more radical. The whole problem of consonances was not a question of those consonances 'which are heard'; he wanted to ask which numbers are consonant and which are not, and why not.[14]

We have now met with three different ideas as to the proper use of mathematics:

a) mathematics has an ontological value, and empirical methods are always inferior, because the visible world is inferior and not wholly conformable to the Ideal one (Plato),
b) mathematics has an ontological value for the world observed by the senses, and therefore the mathematical picture stands the test of experience (Pythagoreans),
c) mathematics is useful to provide us with exact language to describe the facts obtained through sensual observation (Aristoxenus).

In the ontological conceptions (a and b), Number itself is considered as a cause (*causa formalis*), an archetype, an eternal model after which the physical things have been made.

Modern science is closer to the third conception: it recognizes numbers and geometrical forms as indispensible means for exact description of nature, as they are associated with things and events. The modern scientist may enjoy the aesthetic satisfaction given by the regular forms of a crystal or a flower, or the regular sequences of compounds met with in chemistry (as manifested in homologous series of organic substances), or the law of multiple proportions in general or in the periodic system of the elements, but he does not consider the 'beautiful' numbers he discovers as *causes* but rather as *effects*.

These differing attitudes to number are clearly illustrated by discussions over the theory of music in the 16th century. The antagonism between adherents of the numerological conception of harmony and the empirical one[15] flared up again in the dispute between Vincenzo Galilei (circa 1520 - 1591) and Zarlino. Vincenzo Galilei had believed the theory of the Pythagoreans until he 'ascertained the truth by means of experiment, the teacher of all things'. Two weights in the ratio 2 : 1 attached to two equal strings, do not yield an octave. For the octave the ratio of the weights must be 4 : 1.[16] Vincenzo did not believe that the beautiful, simple ratios are more natural than less simple ones; in his opinion 81 : 64 is as 'natural' as 5 : 4.[17] Moreover the ear can hardly distinguish between them.[18] His son, Galileo Galilei, was later to point out that the ratios 2 : 3 and 1 : 2 are *in themselves* not decisive for the 'essence' (the 'natural Form') of the fifth and the octave, as the ratios for obtaining these consonances by variation of the thickness, or the tension or the length follow different laws.[19] 'Truth' then, at least in the field of musical harmony, could not be found by numerological speculations; it must be found by the ear and by instruments.

It gradually dawned upon the theoreticians of music that the ear may base its judgments upon something other than the hard and fast mathematical relations found by measurement of lengths of strings or weights stretching strings, or by establishing the wavelengths and amplitudes obtained by more refined experimentation. Acoustics and musical theory became two ever more divergent empirical approaches.

2.2. Plato on Astronomy

Since Plato considered musical theory and astronomy to be cognate disciplines, his approach to astronomy is similar to that towards music: true astronomy, he claimed, cannot be found by looking at a 'painted ceiling' (the sky!), any more than true harmony can be found by listening to the sounds of strings.[20] So while Plato agreed with the Pythagoreans that astronomy and harmonics are sister disciplines,[21] he denied (at least on this occasion) that the senses — eye in the one case and ear in the other — have anything to do with them: 'ideal' astronomy and 'ideal' harmony are not to be found in the world of phenomena.[22]

In a similar way he was more interested in the everlasting heavenly bodies (their observed motions duly reduced to ideal motions) than in changeable earthly bodies. In studying the heavenly motions, however, he was confronted with a grave and obvious problem. Whereas the fixed stars share in the daily rotation of the heaven without changing their relative positions, the planets have, besides the daily motion, an irregular and much slower motion across the sky. After a certain period they return to the same place among the fixed stars, but in the course of their journeying they move forward, stand still, and turn backward before resuming their first direction. It seemed incredible to Plato that these most divine heavenly beings should move in such a disorderly fashion and at such irregular rates. It is scandalous, he wrote in the *Laws*, that 'these great deities' including the Sun and the Moon should be called 'planets' — that is, 'wanderers' — as if they never follow a fixed course but wander hither and thither. 'The Athenians call it an impiety to investigate the greatest god (Zeus) and to rack the mind to find the cause of this motion.'[23] Plato is here alluding to the reputation for impiety which astronomers had incurred by their irreverent penetration into the mystery of the gods. He intends now to restore the dignity of the planets themselves, and also of their study, which had fallen into disrepute. If they can be shown to have only regular and circular movements they are not vagabonds, and the study of these divinities becomes worthwhile.[24]

Plutarch (46 - 126 AD) describes how, the philosophers having brought astronomy into discredit by explaining away the divine, replacing it by irrational causes, blind forces and Necessity, Plato restored confidence because he made natural laws subservient to the authority of divine principles.[25] The philosophers might have dethroned the Olympic gods but Plato brought them back in a more sophisticated form by his rational 'natural' theology.

Plato therefore set the astronomers the task of finding out what are the *true* motions of the planets, i.e. the motions fitting to their divine character. They were to find a set of uniform and circular movements which, in combination, would account for their seemingly irregular motions. They were to 'save the appearances' (*salvare apparentias*: σωζειν τα φαινόμενα.)[26] Henceforth the great problem of astronomy was 'how can the phenomena be saved?' rather than — as we moderns would expect — 'how can the hypotheses be saved?' The visible *fact* of the irregular courses of the divine bodies could not be a *true* fact; yet it would be going too far to deny these observed phenomena altogether (as the Eleatics did when they said that 'change' is incomprehensible and thus cannot exist). So the phenomena have to be 'saved'; that is, they have to be reduced to motions that *must* be true since they have the divine properties of being without beginning or end (circular paths) and of being immutable (uniform velocities). Here then we see how a quasi-religious *faith* leads to a lofty *fiction* which is supposed to reveal the fundamental and 'objective' *facts* hidden behind the veil of the so-called 'facts' observed by the senses.

Plato's challenge was met by his friend and pupil Eudoxus (circa 408 - 355 BC) in his theory of concentric spheres and later by Hipparchus and Ptolemy in their systems of eccentrics and epicycles.[27]

3. JABIR-IBN-HAYYAN

In the Islamic world the influence of Greek number speculation can be recognized in the works attributed to Jabir-ibn-Hayyan. The older Jabirian writings (from the second half of the 9th century) deal with practical alchemy; the later ones (first half of the 10th century) occupy themselves in the main with numerology — the art of extracting secrets from numbers in order to establish *a priori* the constitution of things.[28]

According to Jabir the constitution of each body can be ascertained by the 'method of Balance' (*mizan*).[29] For the constitution is subject to laws of quantity — a matter of having the right proportions of the weights of the four elements or of the intensities of the four qualities (hot, cold, moist and dry).[30] This balance of the four 'natures' in the compound consists in numerical relations, by analogy with the musical harmonies of the universe.[31] As sounds were reduced to numbers by the Pythagoreans, so Jabir reduced qualitative differences of bodies to quantities.[32]

But how are these ratios to be found? Jabir's answer is: by the balance (*mizan*) of the letters of their names. The letters and the 'natures' (hot, cold, moist and dry) are coordinated; the reason is that the words are formed after the image of the things they represent; hence the elements of words (the letters) correspond with the elements of things (the four 'natures'). The consonants of each word are divided into four groups; each consonant belongs to one of the four natures to a certain degree, a 'weight' to be found in the tables. Thus, by a rather complicated method, the name reveals the quantities or intensities of the four natures which enter into the compound.[33] As Jabir put it: 'Look how the letters are traced in the natures, and how the natures are traced in the letters; how the natures are transformed into letters and the letters into natures.'[34]

The Jabirian theory of mizan was one of several vain efforts made under the influence of Pythagorizing and Platonizing philosophies, to set natural science on a mathematical basis of quantity and measure. Like other similar efforts, however, it was not at all founded on *measuring*.[35] It was a wholly aprioristic 'science', a very 'cobweb of learning'. The doctrine of mizan had little or nothing to do with weighing by means of a balance. Small wonder then that Jabir's efforts had no practical and lasting results. Tying up numbers with substances, on the basis of a numerical value arbitrarily attributed to the letters of their name, seems as meaningless as classifying people on the basis of their telephone numbers. We cannot say that Pythagoras' speculations were devoid of good effect, for they were a source of inspiration to Kepler; but their Jabirian products were sterile.

It is interesting, nevertheless, to see the analogy between the Jabirian and Pythagorean methods: according to the Pythagoreans the world is written in mathematical language; Jabir too held that God wrote the world in numbers — but at the same time in letters. The essence of a thing is expressed in its name.[36] But this name is in Arabic; what if another language had been used? This question is not to be asked: God dictated the Koran in Arabic; it is God's language. When creating different substances God spoke his 'fiat', thus giving them their names in Arabic as he 'wrote' them at the same time in the mathematical proportions of their elements.[37] There is no difference in the degree of ontological objectivity between the two modes. Starting from the faith that every thing is a concretisation of the word of God, there is nothing inherently irrational in seeking a deeper meaning behind its name. All ways are open to trial. There is nothing ridiculous in walking into a blind alley. It becomes foolish only if one does so after the notice 'no thoroughfare' has been erected, that is, after previous investigations have convincingly shown that that direction goes nowhere.

It is worthwhile trying to identify ourselves for a moment with those Pythagoreans, alchemists, neoplatonists, etc. and to look through their eyes at the principles of modern science. If we can do that, however inadequately, sharing in imagination their lack of factual information, and imagining also the indoctrinating influence of the then prevailing climate of opinion, we might then fleetingly experience a feeling of astonishment at the foolishness of modern science (e.g. the solution of Schrödinger's equation for a series of integers and use of the Pauli exclusion principle as a basis for understanding the elements and their chemistry) not dissimilar to that which we usually have when looking at the oddities of ancient scientific (or pseudo-scientific, if you prefer) conceptions. The important difference is that the wisdom built on our foolishness has an incomparably broader basis of facts and is acquired through a great variety of methods that have proved themselves successful. They are successful in the sense that our science gives power over nature. It was this power that the Jabirians yearned after and boasted to have acquired though, as we know, they did not possess it. Wishing to command nature, they failed to realize that they must first obey her.

4. The Structure and Harmony of the Universe

4.1. Kepler

The issue raised in connection with the Pythagoreans is a perpetual one in scientific method. The Pythagoreans used (some of) their experiments (correct or false, real or fictitious) to 'confirm' theories in which they firmly believed beforehand. They believed 'consonance' had to be found in the ratios which they held to be the most beautiful and most rational. On the other hand, such mathematical theories might also have been put forward after the testing of several hypotheses, the final selection among them, though, guided by what amount to aesthetic considerations.

The great example of such Pythagorizing working in combination with experience is that of the young German theologian, Johannes Kepler (1571 - 1630). Proceeding from his 'insight' that the profoundest reason for the existence of the three stationary components of the universe (*viz.* the Sun, the outer heaven and the intermediate space) is that the world is an image of its Triune Creator,[38] he next wished to find out why the number, distances and motions of the planets are as they are and not otherwise.[39] He spent much time 'almost playfully' (*quasi lusu*) trying out various solutions of the problem after the manner of the Pythagoreans, that is, by establishing arithmetical relations. First he tried out the idea that one planetary orbit might have two, three or four times the radius of another, but neither these ratios nor any other simple ratios met the case. Being unsuccessful he tried placing imaginary planets (supposed invisible because of their small size) between Jupiter and Mars and between Venus and Mercury.[40] His hope was that some mathematical rule might be found to fit the succession of ratios of the radii. At long last, what he deemed to be the correct answer came upon him like a flash of lightning: it bore a geometrical rather than an arithmetical character, being based on the properties of the five regular solids, which had already served Plato in geometrizing his theory of the elements. Circumscribe a regular dodecahedron around a sphere containing the orbit of the Earth, and then circumscribe a sphere around this dodecahedron; this sphere contains the orbit of Venus.[41] Circumscribe a tetrahedron around the orbit of Mars; the sphere that contains this tetrahedron will have the radius of the orbit of Jupiter. In the same way inscribe an icosahedron within the orbit of the Earth and the sphere that fits within it will have the radius of the orbit of Venus, etc. A single constructional rule thus gives at the same time the relative distances of the planets, and a reason *a priori* for their number being only six, all five regular solids having been used (figure 4).[42]

Kepler was overjoyed that he was the first mortal to whom God had revealed the mathematical plan according to which He had created the world, especially as he did the work not as an astronomer but just for intellectual diversion.[43] He believed that this was a special divine revelation, so suddenly had it come upon him after all his earlier vain labour. As he put it twenty five years later, it was as if it were dictated by an oracle descended from heaven.[44] 'With a shedding of tears like that of him who cried *Eureka,*' he realized that God had bestowed on him a particular grace in showing him His most profound wisdom in creating the world.[45]

Some deliberate plan there must have been — so much he had known all the time — for God does nothing without plan and reason.[46] He had also known in

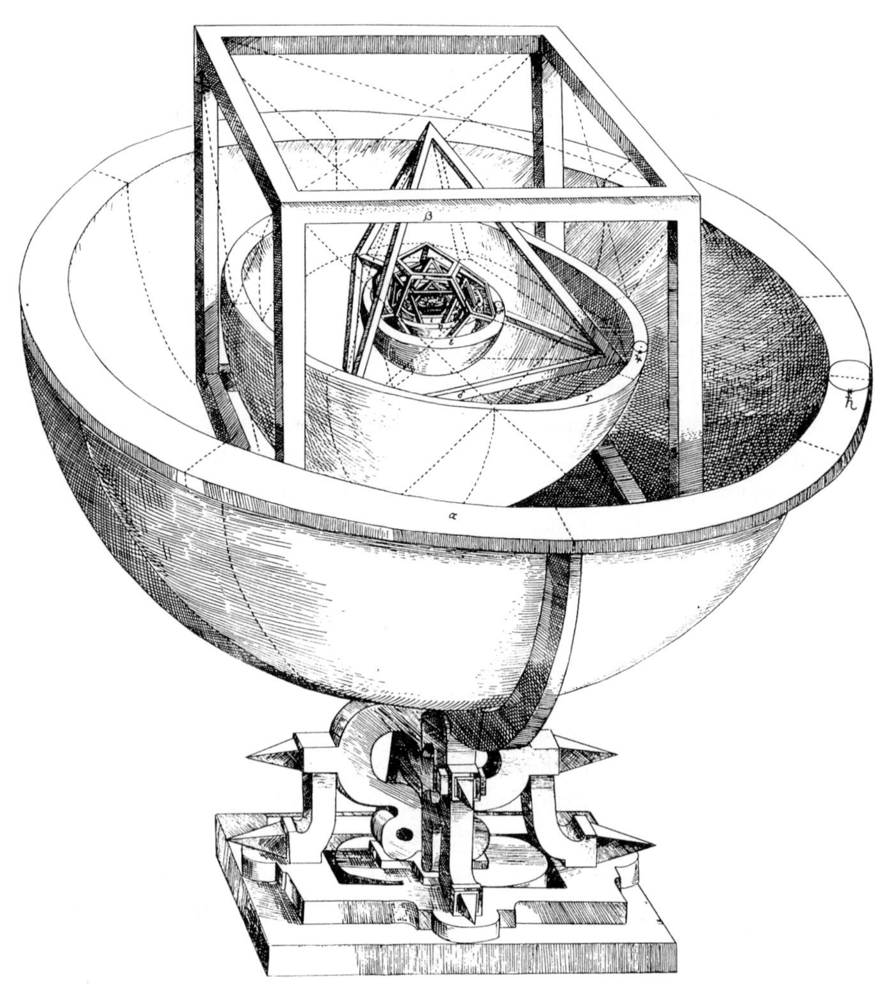

Fig. 4. Kepler's five regular geometrical bodies, representing the dimensions and the distances of the planetary orbs (Kepler, *Mysterium Cosmographicum* 1597 (*Werke* I)). α = sphere of Saturn; β = cube; γ = sphere of Jupiter; δ = tetrahedron; ε = sphere of Mars; ζ = dodecahedron; η = orb of the Earth; θ = icosahedron; ι = sphere of Venus; κ = octahedron; λ = sphere of Mercure; μ = Sun.

advance that this plan must be *mathematical*, for he believed with Plato that the Godhead always geometrizes. God created all things according to quantities and gave to us a mind capable of understanding these:

> As the eye has been created for colours, and the ear for sounds, so the mind of man is not for something, but for the understanding of quantities (*ad quanta*

> *intelligenda*); and it understands a thing more correctly the closer it is to pure quantities, or to its origin, and the more it is distant from them, the more darkness and errors occur.[47]

God chose mathematical things as causes of natural things because they served as exemplars and archetypes (*Ideas*) in Him in a divine abstraction from material quantities.[48] We know, so Kepler says, because the Ideas of quantities are already present in souls made after the image of God. 'Herein the heathen Philosophers and the Doctors of the Church agree.'[49]

But though Kepler might be a thorough-going Platonist, his Christianity was even stronger. In an afterthought he adds that Plato 'only in this respect offends piety, that he does not regard the laws of the Good and of Necessity as being in the *will* of God, but outside it, in the geometrical Idea.'[50] The Creator made what He willed[51] and what he willed was the use of mathematical causes for natural things.[52]

Finally, then, theological voluntarism is in Kepler stronger than 'intellectualism' and his empiricism is correspondingly stronger than his aesthetic apriorism. Despite his enthusiasm he did not forget to check immediately by laborious computations and measurements whether his joy was justified, before fulfilling his vow to publish this great revelation.[53]

Nevertheless, there were contemporaries who considered that Kepler's conclusions were too hastily arrived at. The greatest astronomical observer of the 16th century, Tycho Brahe (1546-1601), though he agreed with Kepler (and Copernicus!) that God made the world according to a certain harmony and proportion, considered Kepler's effort as premature, and cautiously wanted more measurements to be made before any investigation of the supposed harmonies.[54]

Kepler himself took the reality of the visible world and the necessity of empirical measurement of it very seriously, as became evident in his next great and most important work, *Astronomia Nova, seu Physica Coelestis Tradita Commentariis de Motibus Stellae Martis* (1609). After an immense labour of calculation and having tried, according to Plato's injunctions, various combinations of circular movements, he concluded that it was necessary to revise fundamentally the preconception of all previous astronomers and philosophers — not only Plato, Aristotle and Ptolemy, but also his particular heroes Aristarchus and Copernicus — *viz.* the dogma of the perfect circularity and uniformity of the heavenly motions. This had been 'the more a mischievous thief of time as it had been supported by the authority of all philosophers and apparently in agreement with metaphysics.'[55] Kepler now concluded that the paths of the planets are not circles but ellipses, of which the Sun occupies one focus (Kepler's first law).

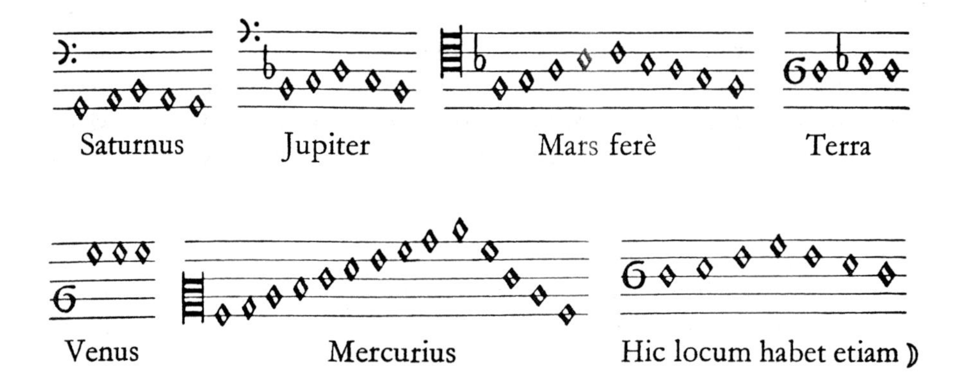

Fig. 5. The music of the planetary spheres according to Kepler (*Harmonice Mundi* 1619). From: —, *Gesammelte Werke* VI.

Thus the most 'catholic' dogma, believed at all times, at all places, by all people, was abandoned; and it was a discrepancy of 8 minutes of arc which, after long hesitation, led Kepler to this radical step. His own prejudices were firmly against it, and his respect for authorities both of the past (Copernicus) and the present (Galileo) made it all the harder. He must have felt the temptation to attribute the discrepancy to some small error of measurement. Kepler knew, however, that the measurements of Tycho (on which his work was based) were the most exact ever made and could not contain an error as great as 8 minutes. So he discarded the circle, and in so doing made the remaining epicycles, to which Copernicus had had willy nilly to resort, superfluous. He could rightly claim that 'these eight minutes paved the way for the total reformation of astronomy.' Rightfully, too, he could call his book *New Astronomy*. Whereas Copernicus had emphasized that his own new astronomy was fundamentally a very old one, Kepler's was undisguisedly presented as new and unheard of (see chapter VI).

The ellipticity of the planetary orbits was to Kepler both a gain and a loss: a loss because the velocity of the planets was no longer uniform and because an ellipse is less symmetrical and therefore less 'simple' than a circle. But the loss was more than compensated, for, the epicycles being now abolished, each planet performed only one movement. Moreover, the loss in uniformity of velocity was compensated by the discovery of another uniformity in planetary motion: the areas swept out in equal times by the line joining the planet to the focus occupied by the Sun are equal (Kepler's second law).

Kepler was not yet at the end of his task. His Pythagorean faith in the harmony of the spheres urged him on to further research. He believed he had found the musical harmonies in the numerical ratios of the velocities of the planets in particular situations in their orbits. In 1619 his *World Harmony* appeared in which he showed the melodies the planets bring forth as they go round their orbits and change from their lowest to highest velocities. Venus, whose path is almost circular and whose velocity is therefore almost constant, emits only one tone, whereas the capricious Mercury traverses the whole scale. The Earth has a flat and rather sad tone which suitably expresses, says the whimsical Kepler, that it is a 'vale of tears'.

One of the speculations in this book led to a result which would later become known as an extremely important theoretical fact: the squares of the times that the several planets take to complete one revolution are proportional to the cubes of their mean distances from the Sun (Kepler's third law).

4.2. Kepler's Natural Theology

To Kepler natural theology was the beginning and end of his work. Religious belief inspired his scientific research and this belief was strengthened by the results of that research. When in 1595 he expounded his 'discovery' of the *Cosmographical Mystery*, he admonished his readers that they should not forget the purpose of all this, namely the knowledge, admiration and veneration of the Great Maker.

> For it is nothing to proceed from the eyes to the mind, from sight to contemplation, from the visible course to the profoundest counsel of the Creator, when you want to rest with them, and not with one leap and the whole devotion of the soul soar towards the knowledge, love and admiration of the Creator.[56]

The discovery of what, he was convinced, were the laws of harmony of the universe filled him again with overwhelming joy and he invited his readers to join him in a hymn of praise:

> Great is our Lord, and great is his power, and his wisdom is without limits: praise Him, ye heavens, praise him ye Sun, Moon and Planets ... Praise Him, celestial Harmonies; praise Him, ye witnesses of the discovered harmonies ... (*And here Kepler writes in the margin: 'you above all others, Maestlin, in your happy old age, for you have again and again given me hope, and animated my efforts' — a touching testimony to his filial love for the master who had compensated for his lack of parental care.)* You, my soul, praise Him also, the Lord your Creator as long as I shall live. For of Him and through Him and to Him are all things: that which is comprehended by the senses is understood by the mind, all that which is as yet wholly unknown as well as that which we know and which is but a small part of them: for there is still more beyond it. To Him be praise, honour and glory, world without end. Amen.[57]

Like most Christians Kepler recognized two sources of revelation: the work of God's fingers and the work of God's tongue.[58] The Book of Nature shows us a material image of God's essence, and just as man's soul is an immaterial image of God, those two fit together. Thus a certain knowledge of God's being can be obtained from both: the human mind and nature are made for each other. In both of them a shadow of God may be found. Thus both kinds of natural theology play a role in Kepler's epistemology.

The divine archetypes expressed in nature are also reflected in the human mind: in a Platonic sense Kepler holds that the phenomena serve to arouse what is set already in the mind, but slumbering. The difference from Plato, however, is that though Kepler too wants to lift the soul from the phenomena up to a higher world, this does not imply that he followed Plato in leaving the phenomena behind. To him the divine Ideas are not considered as distorted images seen in the imperfect mirror of the material world; they are perfect images, for 'where matter is, there is geometry' (*ubi materia, ibi geometria*). There is then, so he says, some knowledge (*scientia*) which is innate and independent of sensual experience, namely in the mathematical science. We know it 'by instinct.' Consequently the mind rules over the senses; the eye perceives quantities because the mind does so, and not the reverse. 'Mathematics was in the divine Mind before creation, in all eternity ... and with the image of God it went into man.'[59]

This still leaves us, of course, with the task of finding out *which* geometrical archetypes have been made concrete in the creation. It is precisely because Kepler has such a firm *belief* that what he will finally find bears a divine character that he is so meticulous in his obedience to 'facts'. To him this was a high religious duty.[60] His original destination to be a minister of the divine Word had been changed, but not his dedicated zeal. He devoted himself to the ministry of the divine Work, to the glory of God who wants to be known through the Book of Nature.[61] God may be glorified by astronomy and there too one is in a Holy Presence, so that one might well exclaim with Peter: 'Depart from me, for I am a sinful man'.[62] Kepler considered himself a priest to the Book of Nature,[63] to the universe, 'that most beautiful temple of God.'[64] This made it unthinkable that he should deal with it in any superficial way, treating the Book of Nature lightly, or as he put it: changing hypotheses lightly. To him his speculations were therefore not less important than (what we would consider) his solid astronomical work.

Kepler saw himself as a singer and player before the face of the Lord. To the question: of what use is astronomy? Kepler answers: Why do we have painters and musicians who please the eye and ear? Astronomy is a pleasure to the mind. God wanted man, His image, to be master of nature and to enjoy the beauty of His world.[65] For the benefit of its future inhabitants He decorated the world: 'The final object of the world and of all creation is man.'[66] He even gave the earth its position in the middle of the planets so that there would be as many of them inside as outside its orbit. 'All things for man' (*omnia propter hominem*), then, gave a twofold effect to this priesthood of the Book of Nature: it made man rejoice in God's work and it impelled him to praise the Creator.

We are perhaps tempted to consider Kepler's speculations as rather loose, but this is hardly fair. Much of what he believed may have turned out to be wrong, but he had racked his brain and strained his eyes testing the speculations by means of complicated computations and precise observations. In the *Astronomia Nova* there are thirty nine chapters detailing his fruitless efforts to find a solution for the planetary orbits. We owe it to his *faith* in the harmony of the world that among his many *fictions* there are three, the famous Keplerian laws, which became great *facts* among the foundations of modern astronomy. Kepler's strenuous efforts to comprehend the world in a platonic-pythagorean spirit, together with his religious submission to givens (*data*) and facts (*facta*) in nature, resulted in one of the major discoveries on which the vast cosmic physical synthesis of Newton rests.

4.3. The Law of Titius-Bode

In 1766 Johann Daniel Tietz (1729 - 1796), better known under the Latinized name of Titius, announced a new law connecting the distances between the planets, though probably without having made any new observations. It was inserted in his German translation of Bonnet's *Contemplation de la Nature*. In the dedication Titius praises Bonnet's efforts to demonstrate the coherence of the world, the chain of nature and the great uniformity in nature's works, all of which show the power and wisdom of its Creator.[67]

It seems that it was Bonnet's idea of a 'ladder of nature' (*échelle de la nature*) in which all inanimate and animate beings stand in an ascending series, which inspired him to seek a regular sequence for the distances of the planets from the sun. For in the text of Bonnet's book Titius inserts without warning or explanation a passage by himself, in which he states a law according to which the distances between the planets are in a nearly regular geometric progression: taking Saturn's distance from the sun as 100, the distances of the planets from the sun are given by the series $D = 4 + 2^n \times 3$, where $n = -\infty, 0, 1, \ldots, 5$ (see table 1).

Between Mars and Jupiter 'there is a deviation from this so exact progression', for there are $4 + 24 = 28$ parts in which no planet has been seen. 'And would the Architect have left this space empty? Never! Let us confidently posit that this space without any doubt belongs to a hitherto undiscovered satellite of Mars ... What an admirable proportion!'

Table 1 *Distances of the planets from the sun, after the Law of Titius-Bode*

Planet	Distance from the sun (Saturn = 100)	number n in $D = 4 + 2^n x3$
Mercury	4 (= 4 + 0)	$-\infty$
Venus	7 (= 4 + 3)	0
Earth	10 (= 4 + 6)	1
Mars	16 (= 4 + 12)	2
–	32 (= 4 + 28)	3
Jupiter	52 (= 4 + 48)	4
Saturn	100 (= 4 + 96)	5

In 1772 the astronomer J.E. Bode (1747 - 1826) made this law widely known and in 1781 the discovery of a new planet by Herschel provided the opportunity to test it. The new planet fitted with the Titius-Bode formula, if *n* was taken as 6. Bode gave it the name Uranus. The minor planet Ceres, the first of the asteroids discovered by Piazzi (in 1801) fitted nicely into the vacant place between Mars and Jupiter, but alas, in later years the law failed completely for the outer planet Neptune.

The law of Titius-Bode is not entirely aprioristic in character. It was posited *a priori* that a certain regularity would exist, but *which* regularity it was had to be found out by study of the experimental data. Some interpolation and extrapolation had to be applied, and these were initially confirmed by new discoveries. The snag was that the law did not turn out to be universally valid.

Kepler's geometrical speculation about the regular solids, the number and distances of the planets had also seemed at first to be rather satisfactory. Although it limited the possible number of planets *a priori,* it did not rule out in advance every possibility of testing it by the discovery of *new* planets. In one of his earliest attempts to find a law connecting the distances of the planets, Kepler in fact had to assume the existence of two as yet undiscovered planets. Although this proved a blind alley, from the methodological standpoint it was a quite legitimate attempt.

Titius and Bode were doing a similar thing when they introduced their hypothetical planet between Mars and Jupiter. Bode deliberately initiated the search for it and Piazzi actually found it. Here nature was more compliant, even generous to the scientist when it yielded the new planet (Uranus) outside the range of the known planets, which fitted with the man-made law. In spite of all, however, the law which had had such a stimulating effect, and had led to such remarkable and solid discoveries eventually broke down for Neptune.*

A more recent, but somewhat similar example which we shall consider below is afforded by Mendeléev. He found that he could complete his periodic table of the elements only by leaving some vacant places in the series. In course of time all the vacant spaces were filled by the discovery of new elements. In this case, it has since long been concluded that the periodicity really does exist in nature. Will the same be said of Gell-Mann's triangle of particles (see page 75)? It certainly stimulated a hunt for the missing elementary particle, and the hunt proved successful.

We have seen how a belief in mathematical regularity, harmony and proportionality may lead to assumptions which are:

a) simply false (Kepler's regular solids and his theory involving two extra planets),
b) possibly false, yet conformable with several phenomena and productive of important results (Titius) or
c) correct (Mendeléev).

Evidently there can be no hard and fast prescriptions in such matters. The scientist can only tentatively put forward his hypotheses and then wait to see how far nature answers his assumptions. Sometimes nature graciously consents to fit in with his aesthetic or logical ideas; at other times it may seem for a while to support him, but then withdraws and leaves him empty-handed and on others it totally refuses to answer his expectations. This uncertainty is both the frustration and charm of man's dialogue with nature.

All this shows how careful we have to be in our historical evaluation of scientific ideas. Kepler's postulation of two unknown planets may now be easily dismissed as fantastic speculation, but if they had been found to exist, he would probably have been praised by some people for his profound intuition. On the other hand, if the gaps in Mendeléev's table of the elements had not filled up, his story would no doubt have served as an illustration that even an exact experimenter sometimes

* [*Editor's note* (*VMcK*): In modern times the Titius-Bode law still stirs interest. Two mathematicians in Barcelona, Llibre and Pinol, have been investigating the four-body problem with the bodies representing two planets, the sun and the centre of mass of the galaxy. They find that the effect of the latter is to render 'most stable' those solutions in which the outer planet has a period three times that of the inner planet. This corresponds to a radius ratio of 2.08 and hence they predict a law of the form $D = C \times 2.08^n$ for the distances of the planets from the sun, with C an arbitrary constant. J. Llibre and C. Pinol, 'A Gravitational Approach to the Titius-Bode Law', in: *The Astronomical Journal* 93 no.5 (1987), pp.1272-9.]

founds strange speculations on half-truths. In science success may justify theory; but it cannot be the only criterion in historical evaluation.

5. Crystal Theory

5.1. Abbé Haüy

When solid substances assume polyhedral forms the individuals of each species can occur in a multitude of forms: different combinations of plane faces, bordered by edges of different relative lengths. Yet, there is order behind this chaos. In all individuals of the same species the angles between similar faces are the same (law of constancy of angles; Romé de Lisle and Carangeot, 1783). For example, in the six-sided columns of quartz the faces, however irregular they may seem, always form dihedral angles of 120°.

Some 18th century crystallographers set themselves the task of finding a mathematical law connecting the directions of the several faces that occur (or can occur) in crystals of the same chemical species. Romé de Lisle (1736 - 1790) did so by selecting one of the simplest forms of a species (e.g. the cube of sea salt; the Iceland rhombohedron of calc spar) as the 'primitive form', and then deriving from it the more complicated 'secondary forms' by truncating the edges or the solid angles. He did not manage, however, to establish a quantitative relation between the primary and secondary forms.

The mineralogist and crystallographer abbé R.J. Haüy (1743 - 1822) was more successful. He believed that he had found the primitive form of crystals of a given species by cleavage of the various secondary forms: sea salt crystals easily split into cubes; calc spar prisms and scalenohedrons (figure 6) easily yield a rhombohedral 'nucleus' with the same angles as Iceland spar crystals. The cubes of sea salt and the rhombohedrons of calc spar may be split into smaller fragments of the same form and it may be concluded 'by analogy' that the smallest particles that still have the properties of the species — the so-called 'constituent' or 'integrant' molecules — have this same form.[68]

Haüy's first systematic attempt to form a theory of crystal structure in 1784 was based upon the notions of constancy of angles, and of primitive forms and integrant molecules. He declared that 'a Supreme Wisdom' had subjected even the smallest molecules to 'always underlying laws' which gave birth to 'harmony and regularity.'[69] The existence of such laws is proven by 'the agreement between calculation and observation.'[70] In the very simple case of a cubical primitive form he derives the secondary forms by stacking layers of similarly shaped contiguous molecules on the faces of this cubical 'nucleus' (*noyau*). If each layer recedes from the edges of the preceding one by one row of molecules and the layers have a thickness of one molecule, the resulting secondary form will be a dodecahedron

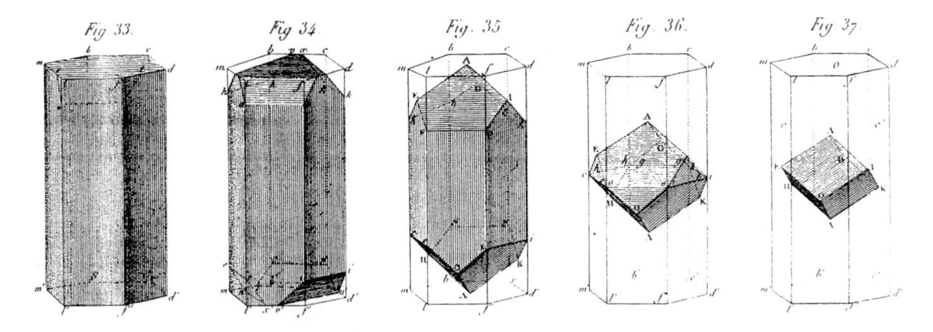

Fig. 6. Calc spar prisms and scalenohedrons; stages of formation of primitive rhombohedron. From: Haüy, *Traité de Cristallographie* I (1822), pl.3 (from these fig. 33-37).

with rhombic faces (see figure 7). The angles of the various secondary forms can easily be calculated, and Haüy found that observations confirmed his calculations. The decrement and the thickness of subsequent layers is, so he says, confined to a small number of molecules; usually one or two, occaisionally three to six. As the molecules are infinitely small, the surface of the crystal will not in fact seem to be stepped but quite smooth.

In order to test the law of simple decrements Haüy had to start with regular forms like the cube of sea salt of which the dihedral angles (90°) and the relative dimensions (1 : 1 : 1) of the nuclear form and the integrant molecules are known *a priori*. Calc spar, too, was a good start: the dimensions of the edges of the cleavage form (and of the molecules) are as 1 : 1 : 1, and the angles of the primitive form could be calculated after Haüy had (allegedly) found that the rhombic faces of the cleavage nucleus make angles of 45° with the vertical as well as with the horizontal faces of the hexagonal prism from which it could be obtained (see figure 6).

Having thus proved the law of simple decrements for crystals of a high degree of regularity, Haüy applied it 'by analogy' to crystals whose relative molecular dimensions are not predictable *a priori*. He then used the 'inverse operation', tentatively assuming a certain law of decrement. If this has been appropriately chosen for deriving certain secondary forms, it will lead back to the relative dimensions of the molecules.[71]

The agreement of calculations based on the theory of simple decrements with the angles measured using his contact goniometer (see figure 8) was to Haüy a confirmation of his belief in the 'simplicity' and 'analogy' of nature:

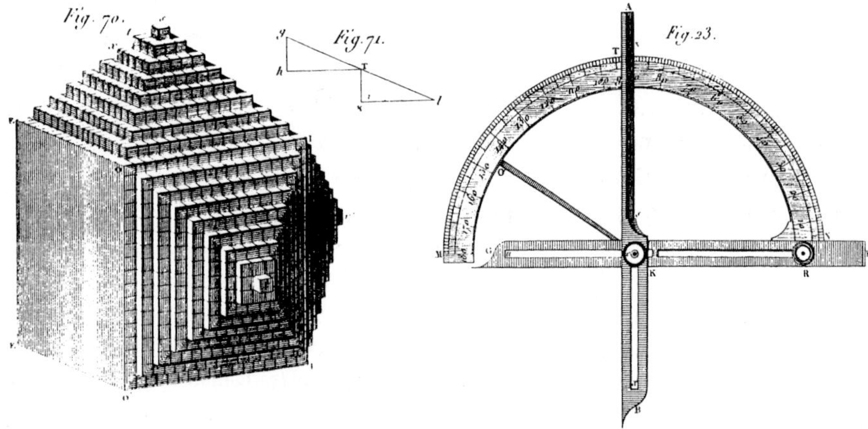

Fig. 7. Law of decrescence: development of a dodecahedron with rhombic faces, resulting from a primitive form (Haüy, *Traité de Cristallographie I* (1822), pl. 5 (from these fig. 70-71).

Fig. 8. Haüy's contact goniometer (*Traité de Cristallographie I* (1822), pl. 2, fig. 23).

> We recognize here what characterizes in general the laws emanated from the power and wisdom of a God who has created her and leads her; economy and simplicity in its means, inexhaustible richness and variety in its results.[72]

5.1.1. Limits

Since our means of observation and measurement are not absolutely perfect, the scientist is often confronted with the problem: should the apparent deviation of our measurements from the 'simple law' be attributed to an inevitable imperfection of the measurement, or is the law not wholly conformable to nature? When such doubts arise strict empiricists will be inclined to take the latter side, whereas those who, like Haüy, *a priori* foster a strong belief in the simplicity of nature will tend to attribute deviations from the 'law' to faults in the measuring procedure. Here the concept of a 'limit', in the sense of a simple value or an equality, may be an influential factor.

In 1784 Haüy pointed out that the data yielding a perfect figure of molecules often depend upon the observation of the equality of the respective inclinations of certain faces and on the equality of certain plane angles, and that this equality, though not absolute according to the numerical data obtained by measurements, should be supposed to be absolute per se, so that e.g. angles that are practically 90° may be supposed to be so in the strictest sense. Nature, so he says, seems to halt at certain *fixed points* or *limits*, like the perpendicular direction and the equality of certain quantities of the same order.[73] The same is the case with limit forms (*formes limites*): the cube, the regular octahedron and the regular tetrahedron. They have the function of fixed points, as all their dimensions and angles are equal and do not admit 'more' or 'less'; all lines in them can be calculated *a priori*.[74] In Haüy's opinion they are 'types' for all less regular forms, for they lend themselves to calculations that are more certain than any measurement.[75]

The success that this principle had enjoyed in the theory of simple decrements emboldened Haüy to apply analogical reasoning also to the relations between the diagonals of faces and in general to other lines either upon or traversing the crystals. Here again the *formes limites* served as 'types' for all other forms. In the regular octahedron the ratio of 'the perpendicular from the middle of a horizontal edge to the axis' is to 'the part of the axis it intercepts' as $1 : \sqrt{2}$; in the regular tetrahedron as $1 : \sqrt{8}$, etc. This may be summarized (in Haüy's opinion) by the rule that in the regular bodies the ratios of several dimensions are as simple integers or square roots of integers. 'By analogy' a similar mode of expression may be assumed for the ratios of dimensions in forms that do not wholly possess the character of limits: they, too, will be as the ratios of the roots of rather simple integers.[76] In the same way the ratios of the dimensions of the integrant molecules and those of their diagonals may be indicated.[77]

Here 'analogy', which had helped Haüy to make his greatest discovery, led him astray, as became most evident in the case of calc spar. This mineral had provided him with the 'mother idea' of his theory when he found (1784) that its rhombohedral nucleus (which has the form of the crystals of Iceland spar) is obtained by cleavage of a hexagonal prism of that mineral; each of its faces forms equal angles (45°) with the vertical and horizontal faces of the prism (figure 6). To Haüy this equality was a most important example of a limit set by nature; he put it on the level of 'simple limits' like the right angle. On the assumption of these simple limits (in this case the 45° angles), he calculated that the ratio of diagonals for the rhombohedral units of Iceland spar would be the 'simple ratio' of $\sqrt{(3/2)}$.

The existence of simple limits was to Haüy an unshakeable dogma: neither the fact that a century earlier Christiaan Huygens had found 45° 20' (and not 45° so that the 'simple' ratio between the diagonals cannot be found by calculation) nor the fact that in his own time, W.H. Wollaston and others had confirmed Huygens's value using the highly precise reflection goniometer, could change Haüy's opinion. He considered the calculated value of the ratio of the diagonals of the faces of the rhombohedron ($\sqrt{2} : \sqrt{3}$) as more reliable than the value obtained by direct measurement, for the fundamental 'limit value' of 45°, upon which the calculations were based, is 'absolute', whereas all measurement is imperfect.[78]

Haüy maintained that his contact goniometer (precision of half a degree (30'); exceptionally a third of a degree) yielded more reliable results than the reflection goniometer (precision 1'). Whereas Wollaston and others found the larger dihedral angles of the Iceland spar crystal to be 105° 5', Haüy stuck to 104° 28', though the difference of 37' could have been easily detected even by his less precise instrument.[79]

Haüy pointed out that if Wollaston and others were right, the ratio of the diagonals would be $\sqrt{111}/\sqrt{73}$, instead of the simple value of $\sqrt{3}/\sqrt{2}$ ($=\sqrt{111}/\sqrt{74}$) which he had measured. The contact goniometer was in his opinion of 'sufficient precision' for the purpose 'one has in mind when using it' and, moreover, it could be more easily handled, so that 'one would not be tempted to substitute the ratio $\sqrt{111}/\sqrt{73}$ for that of $\sqrt{3}/\sqrt{2}$, which on the one hand differs so little from it and on the other hand has such a great advantage over it by the simplicity of the limit (45°) from which it is derived.'[80]

Haüy was convinced that the downfall of the theory of 'simple radicals of dimensions' would imply the breakdown of the theory of simple decrements and that the use of the reflection goniometer led to results that have no coherence and undermine the simple laws of structure, transforming them into anomalies.[81] His younger contemporaries, like H.J. Brooke (1819), however, deplored the fact that the 'imaginary simplicity' which Haüy supposed 'to exist naturally in the ratios of certain lines either upon or traversing the crystal' affected the worth of his theory, and regretted that Haüy regarded the disagreement of an observed measurement with this 'simplicity' as an error of observation rather than a correction of his theoretic determination.[82]

In his pythagorizing conception of the simplicity of numerical relations Haüy adapted data of observation. He declared that the crystallographer who has calculated angles that are closely connected with a 'fundamental result' (like that of the equality of the angles between the faces of Iceland spar crystals with respect to a horizontal and a vertical plane) measures them only for satisfying himself, for 'he was certain in advance that observation, if it were exact, would speak like the theory.'[83] He turned matters topsy-turvy when speaking of the 'prejudice' of his critics who were faithful to the contingent data obtained by the use of the reflection goniometer, which made it impossible to dismiss the discrepancies between theory and observation. When Wollaston measured 107° for the larger dihedral angles of iron spar [$FeCO_3$], Haüy maintained his own value for it (104° 28'): 'the diversity between the indications of the mechanical measurement is but apparent, and it must vanish before the results of the laws of structure.'[84]

H.J. Brooke hit the nail on the head when he said that the cause of Haüy's errors was not the inaccuracy of his measurements, but his false tenet about 'limits'.

The belief in simplicity and harmony of numerical relations led Haüy to important results of lasting value in one case, whereas it misled him in another. *A priori* both of his efforts were equally legitimate, but in each case one had to find out by experience whether or not the simple relation was 'scientifically true'. The fact that no hard and fast criterion for simplicity exists adds to the difficulty of deciding whether the 'law' or the 'measurement' is wrong, though in this case the almost unanimous verdict of the next generation was that the master had gone too far in his effort to save his own erroneous theory with the help of an imprecise instrument.

5.2. *Weiss*

In the 19th century Haüy's theory of crystal structure receded into the background and the descriptive method of C.S. Weiss (1780 - 1856) took its place. Being strongly influenced by German *Naturphilosophie* Weiss opposed Haüy's 'mechanistic' theory. So he substituted for the law of simple decrements the more empirical 'law of simple intercepts'. Having imagined in each crystal a system of three (in some cases four) coordinate axes, he considered one of the crystal faces as the fundamental plane, whose intersections with these axes determines the unit lengths or parameters (which are as a : b : c). The law then states that all other faces of the crystal intercept the axes at distances from the centre of the coordinate system that are as ma : nb : pc, in which m, n and p are simple integers.[85]

Thus in spite of his critique of Haüy's theory he maintained its quantitative aspect and he even went so far as to accept the thesis that the ratios of certain dimensions in the crystal might be expressed as the square roots of simple integers. In his opinion the crystallographer 'will be grateful to Haüy that he has paved the way for this kind of supposition.'[86] Similarly Weiss held that certain parameters are to each other as the radicals of simple integers: thus the parameters intersected by the fundamental plane on the three coordinate axes might be, in the case of feldspar, for example, as $\sqrt{13} : \sqrt{(3\times13)} : \sqrt{3}$.[87]

One of the best guarantees of the real conformity to nature of this law is, in Weiss's opinion, the fact that when starting from the simplest case (i.e. the equality of the three fundamental lengths a, b and c in the regular system) the ratios of the several dimensions in the crystal to the fundamental dimensions can be expressed *a priori* by radicals.[88] Analogical reasoning similar to that of Haüy, led to the assumption that in *all* crystals the ratios of the dimensions of certain lines either upon the surface of or traversing the body of the crystals are as simple integers or as radicals of simple integers.[89]

Being a *Naturphilosoph*, Weiss went much further than Haüy in numerical speculation, in particular in his search for the analogy between musical harmony and the proportions found in crystals. He compared the ratios of certain dimensions in various forms of the 'sphaeroedrical' (i.e. the regular or tesseral) system, such as the cube {100}, the octahedron {111}, the rhombic dodecahedron {110}, and the icosahedron {210} with the harmonic relations of the tones.

> If we remember the development of the harmonic proportions in a vibrating body and in music, we will perceive many surprising similarities between the development of the harmonic proportions in these crystal forms and the musical ones.[90]

For each of the above solid bodies he compares the dimensions of various lines from the centre of the body to certain points on the periphery, *viz.* the perpendicular to the nearest face, the line to the remotest angle of the body, and the line perpendicular to the edges, which he calls respectively the smallest, the largest and the middle dimensions. (Where a body has different kinds of edges, angles and faces there are several 'middle' dimensions).

In the case of the cube the proportions of the 'smallest', 'middle' and 'largest dimensions' are as $1 : \sqrt{2} : \sqrt{3}$. For the octahedron they are as $1 : \sqrt{(3/2)} : \sqrt{3} = \sqrt{(1/3)} : \sqrt{(1/2)} : 1$, i.e. $S : M = \sqrt{2}:\sqrt{3}$; $M : L = 1:\sqrt{2}$; $S : L = 1:\sqrt{3}$.

The result shows, according to Weiss, a 'conspicuously large analogy' between the spatial proportions in these crystal forms and the 'harmonic ratios of the tones of music'. It is, however, not a direct proportionality, for one finds 'the same proportions between the spatial dimensions, when expressed as square roots, as are assumed in the simplest harmonic ratios of the tones for the *integer* numbers of the relative vibrations.'[91] The ratios of the 'dimensions' in the cube resemble the ratio of the octave plus the fifth, and those in the octahedron that of the fifth plus the octave, etc., etc. In the icosahedron ratios emerged 'which', Weiss admitted, 'do not seem to have further relation to musical proportions.'

There is no point in following Weiss further in his elaboration of the analogy, for plainly these considerations did not yield anything new: the relative dimensions in the forms of the regular system are geometrically fixed, quite apart from any measurement. In this case aesthetic preconceptions did not need any confirmation by experience, nor did they help to discover new facts in crystallography. They just served to satisfy their author's own aesthetic needs.

More examples could be given of efforts to discover analogies between music and laws of nature,[92] but they had little effect on the development of scientific knowledge. Why then should we pay any attention to such hypotheses? Because they complete the historical picture. The history of science is not a series only of success stories, but also one of failures. The scientists' search for order and harmony gave rise to some speculations that led into blind alleys; to others which, though they were mistaken, stimulated great discoveries (Kepler), and still others that largely fitted in with the reality of nature. To these latter belonged Haüy's theory of crystal structure.

The history of crystallography shows how what originally seemed to be mere speculation could develop into a successful theory. Already in the 17th and 18th centuries the apparent 'regularity' of crystal forms had given rise to theories of the form and structure of crystals. But the author of one of the most ingenious efforts to discover mathematical rules in the crystal world, Torbern Bergman (1735 - 1784), finally gave up in despair, concluding that although form is an important external character of a mineral, 'when this most important of external characters is so uncertain, ... what will then be the value of the others? ... they give support to the expert eye, but they do not convince.'[93]

The discovery of the law of constancy of angles (Carangeot and Romé de Lisle, 1783) and Haüy's discovery of symmetry relations and the law of simple rational decrements, however, gave rise to a system of idealized crystal forms, which much later found its explanation in an aprioristic system of 230 possible arrangements of crystal building units (from Bravais and Sohncke to Fedorov, Schönflies and Barlow). This eventually found experimental confirmation by means of X-rays.[94]

Plato and Kepler were convinced that the regular forms (cube, octahedron, tetrahedron, dodecahedron, icosahedron) played a key role in the construction of the universe. The real world, however, is different, and much more complicated, but it is an equally beautiful illustration of their adage that 'the Godhead always geometrizes.'

11. Series of Chemical Compounds

11.1. Richter's Laws of Acids and Bases

An early and fervent advocate of the mathematization of chemistry was J.B. Richter (1762 - 1807). He determined the weights of various alkaline substances (potash, soda, magnesia, lime) which were neutralized by a given quantity of acid (hydrochloric acid, sulphuric acid). These and other experiments made an important contribution to quantitative chemistry and thus to the development of the notion of chemical equivalents (that is of the relative masses of substances entering into a chemical combination).*

* [*Editors' note:* The chemical equivalent (or equivalent weight) of an element is the number of units of weight of it which will combine with or replace one unit of weight of hydrogen. If oxygen instead of hydrogen is taken as standard, the equivalent of an element is the number of units of weight of it which will combine with or replace 8 units of oxygen.

The equivalent weight is thus the atomic weight divided by a small integer which will frequently be equal to the valency. But it is not necessarily so, as consideration of the stable compounds of nitrogen, ammonia (NH_3), nitrogen oxide (NO), and dinitrogen oxide (N_2O), shows.

He put forward the law that the weights of various alkaline substances which neutralize a given quantity of an acid are 'in arithmetical progression': a, a + b, a + 3b, a + 5b, a + 7b ... In the series for hydrochloric acid, for example, 1000 parts by weight of the acid are neutralized by 734 parts of alumina, 858 parts of magnesia, 1107 parts of lime, 3099 parts of baryta. This yields the series:

1) alumina	a	= 734	= 734
2) magnesia	a + b	= 734 + 124½	= 858½
3) lime	a + 3b	= 734 + 3 x 124½	= 1107½
4) ?	a + 5b	= 734 + 5 x 124½	= 1356½
...	...	...	...
10) ?	a + 17b	= 734 + 17 x 124½	= 2850½
11) baryta	a + 19b	= 734 + 19 x 124½	= 3099½

When divided by b, the series from the second member onwards becomes a/b + 1; a/b + 3; a/b + 5; ... ; a/b + 17; a/b + 19. Richter therefore assumed, on the basis of this law, that the alkaline bodies corresponding to the numbers 5 through to 17 exist though as yet undiscovered: 'it is highly probable that they are lacking in the series,' which consists partly of earths that are really at hand, partly of earths that are possible.[95]

In another work Richter compared the weights of various acids which neutralize 1000 parts of a given earth. He found that the same quantity of the alkaline substance (*viz.* magnesia) is neutralized by quantities of various organic acids that are in a geometrical progression, *viz.* as a; ab^3; ab^4; ab^8; ab^{11}; ab^{14}; ab^{15}; ab^{16}. The vacant spaces ab^2, ab^5, etc. would probably be filled by yet unknown acids containing carbon.[96]

Although Richter claimed to have confirmed his 'hypotheses' by careful quantitative experiments,[97] his experiments were in fact rather crude. Moreover he quite freely 'corrected' the values obtained by experiment to make them conformable to his 'laws' of geometrical and arithmetical progression. For example, though the neutralization of 1000 parts of magnesia required 1283 parts of oxalic acid, he 'corrected' this to 1244 parts in order to make it satisfy the formula ab^4.[98] Even so, in the case of vitriolic acid (sulphuric acid) neutralizing various alkaline earths, he found it necessary to make the law of the series more complicated by a set of 'corrections' (which themselves form an arithmetical series). Such corrections, either of the data or of the law, did not shake his certainty

---

The notion originated with J.B. Richter and was obviously of practical use as long as there was uncertainty as to the valency exhibited by an element in any given compound, but has fallen into disuse since about the 1950s].

that 'the masses of the alkaline earths as yet known which combine with vitriolic and muriatic acid (hydrochloric acid) are members of arithmetical progressions.'[99]

These series provide, potentially, a stimulus to seek as yet undiscovered substances and to check numerical values already found by experiment. Richter pointed out that the law justifies an expectation that the missing alkaline earths exist in nature. For, he points out, it is only in the last half-century that the second and the eleventh members of one particular series (magnesia and barytes) have been discovered. Richter also remarked that it is doubtful whether without such recently discovered earths, it would have been possible to discover the law.[100] This is remarkably reminiscent of the periodic system (1869) where, without the prior discovery of sufficient elements and their atomic weights, the idea of periods could not have been put forward as a respectable hypothesis. It is interesting to note that laws implying the existence of some regular series have intrinsic consequences independently of their being fictitious or real, false or correct.

Richter did not go so far as to conclude the 'necessity of the existence of elements which are up till now lacking in our experience'; this would be, in his opinion, as mistaken as to claim that there 'must be another planet between Mars and Jupiter, because it fits in with the law of the distance of the planets from the sun.'[101] This statement shows the difference between this man, who seeks to be both a mathematical and an empirical chemist and those who on purely speculative grounds fill up the gaps in nature as if their supposed occupants really existed. Richter's 'intermediate acids' and 'intermediate bases' have never been found and the simple arithmetical and geometrical relations between the equivalent weights of acids and bases have turned out to be false. Yet from the logical and methodological standpoint, his idea of arithmetical series has proved of great value, especially in the notion of homologous series in organic chemistry.

6.2. Homologous Series

The homologous series of carbon compounds are a more successful example of an arithmetical series in which there is indeed a constant increment in weight between subsequent members. In them, moreover, there is a constant increment in the *composition* of the subsequent members. 'Two bodies are homologous when they differ by n times CH_2.'[102]

The hydrocarbons and the alcohols can be arranged in series in which the difference between subsequent members is one atom of carbon and two of hydrogen.

CH_4; C_2H_6; C_3H_8; C_4H_{10} ... C_nH_{2n+2}

The first homologous series, proposed in 1842 by J. Schiel, was one of 'radicals': methyl $(CH_2)H$; aethyl $(CH_2)_2H$; glyceryl $(CH_2)_3H$; amyl $(CH_2)_5H$ ..., in which the fourth member $[(CH_2)_4H]$ was supposed missing. Shortly afterwards J.B. Dumas showed that organic acids may be arranged in a similar series. Between the highest, margaric acid $C_{17}H_{34}O_2$ (according to Dumas $C_{34}H_{68}O_4$) and the lowest, formic acid CH_2O_2 (in Dumas's notation $C_2H_4O_4$) there should be 17 members of which no fewer than 9 were still lacking. Each next member was found by subtracting CH_2 (in Dumas's notation C_2H_4) from the preceding higher one. The general introduction of the idea of homologous series is due to Charles Gerhardt, who coined the term in 1843.[103]

Not surprisingly it was suggested more than once that other groups of elements such as the halogens might be homologous series (J.P. Cooke 1854, J.B. Dumas 1857). The analogy between elements and organic radicals (methyl, aethyl, etc) goes back to Lavoisier, who considered the latter as 'compound radicals', which behave like the simple radicals produced by those simple substances or elements which have so far withstood all efforts to decompose them.[104]

In 1857 - 1858 Dumas, who had already tried to develop the theory of triads of elements, compared the numerical regularities of atomic weights with those in homologous series. The weights of the halogens do not form a simple progression: the relation between their equivalent weights is as:

$$a : a + d : a + 2d + d : 2a + 2d + 2d.$$

In numbers this means that:

$$\begin{array}{lcrcrcr} F & = & 19 & & & & \\ Cl & = & 19 & + & 16.5 & & \\ Br & = & 19 & + & 2 \times 16.5 & + & 28 \\ I & = & 2 \times 19 & + & 2 \times 16.5 & + & 2 \times 28 \end{array}$$

Just like Richter before him, Dumas evidently sometimes had to make the 'arithmetical' series more complicated in order to adapt it to the experimental data. (In the case of the oxygen group the series was simpler:

$O = 8$; $S = 8 + 8$; $Se = 8 + 4x8$; $Te = 8 + 7x8$. That is, a, 2a, 5a, 8a.)

In general such theories were short-lived; but they serve well to show how scientists are always groping for a way out of a labyrinth, a situation which is obscured by stories in which the development of science is depicted as a smooth journey from one glory to the next.

The comparison of composite radicals and their homology with simple radicals ('elements') strengthened the belief that the latter, too, might be composite, and thus furthered the idea that they might possess the same constituents.

There are similarities and differences between Richter's arithmetical series of alkaline substances and the homologous series of carbon compounds. Richter's missing links cannot be found; those of the hydrocarbons, etc. could be synthesized. There is no simple relation either between the molecular weights of the acids, nor between those of the alkaline substances, whereas there is a regular increment in weight of the subsequent members of the homologous series.

But the *idea* of an arithmetical series of masses — which implies the prediction of missing members and the possibility of correcting those masses in accordance with the 'law' — unites the two series. Logically and methodologically they are of the same kind. Neither of them fixes *a priori* the number of members of the series; in both of them the gaps indicate what *possibly*, not certainly, lies undiscovered in nature. This means that both have a strongly empirical aspect.

The difference, then, is mainly that the one is found in nature, the other not: whereas their similarity consists in their common theoretical structure.

It is interesting that the chemist J.R. Partington, while recognizing Richter's great contribution to the mathematization of chemistry (in particular his research on the fixed ratios of constituents of chemical compounds and his introduction of equivalent weights), blamed him nevertheless for his 'obstinate pursuit of mathematical relations which are really non-existent' and for his never being able 'to free himself from his early mistakes of the fictitious arithmetical and geometrical series of combining ratios.'[105] On the other hand, the philosopher E. Cassirer praises Richter for putting forward an idea which may not have been empirically right, yet which is meaningful in its general tendency: 'it is ... the general Pythagorean fundamental idea of harmony of the universe, which stands here at the cradle of the new chemistry ...' Cassirer compares the general tendency of Richter's mind with that of Kepler, 'with whom he shares the basic thought of the radical numerical construction of the universe.'[106]

Both, the chemist and the philosopher, were right. But the different character of their verdicts shows that it is worth taking notice not only of the general background of the persons described by historians of science, but also of that of the historians themselves.

7. Series of Chemical Elements

7.1. The Law of Triads

Ever since Lavoisier's decision that those substances which could not be decomposed should be considered as elementary — at least until experience should show them to be decomposable — there had remained much dissatisfaction about the rather chaotic multitude of such 'simple bodies'.

Many efforts were made to bring some order into the situation, for example by grouping elements into sets of three of which the middle one had the mean atomic weight of the other two and also showed an intermediate character in its chemical properties. In 1815-16 J.W. Döbereiner (1780-1849) pointed out that in the series calcium, strontium, barium the 'equivalent weight' of strontium is about the arithmetic mean of the other two. Afterwards he found similar rules for lithium, sodium, potassium and other 'triads'. In 1829 he elaborated this 'law of triads' in a more comprehensive effort to 'group the elementary substances according to their analogy'[107] — that is, according to the similarity of their chemical properties. Shortly after the discovery of bromine (1826) he predicted that its atomic weight might be the mean of those of the atomic weights of the other two salt-formers (halogens), chlorine and iodine. The awkward fact that fluorine, too, belongs to the halogens, threatened to break the magic spell of the number three, but this danger was side-stepped by saying that fluorine had the same relation to the halogens as the alkali metals (sodium, etc) have to the alkaline earth metals (calcium, etc).

The law of triads clearly implied risks — in some cases of grouping some element with two others on weak grounds and in others of rejecting a strongly similar element because otherwise the number three would be exceeded. Although it might thus lead to artificial separation between related elements it also inspired the hope of discovering new elements which would complete a triad.[108]

It was particularly risky to sort out *natural* groups of three elements on the basis of contemporary limits to the power of *artificial* (chemical) analysis. There could be no certainty that all of them were of the same level of 'simplicity', Lavoisier's criterion of the simple body being purely empirical. The mathematical criterion of the mean atomic weight helped to reduce the arbitrariness. Yet here too there was a snag: given two elements that belong together, it is not certain that a third will be intermediate. This could be certain only if there were a sure method of determining atomic weights, and this hardly existed at that time.

Nevertheless it was evident that there was some truth in the idea of triads even though it was far from the full truth.

7.2. Newlands's Law of Octaves

A later attempt to bring order into the chaotic assembly of elements established much closer relations between them. The object was no longer to find disjointed triads, but a coherent natural arrangement in a single series. This attempt became more promising after 1860, when a greater certainty and consensus about atomic weights was established (see chapter X, *Cleopatra's Nose*).

In 1862 A.E. Béguyer de Chancourtois (1819 - 1866) announced a 'Natural classification of the simple or radical bodies, entitled the Telluric Helix.' One of his main theses was that 'the properties of bodies are the properties of numbers.' He claimed to perceive 'at the same time approximations of the series of numerical characteristics to the series of musical sounds, and to that of the bands and rays of the spectrum.'

In 1864 J.A.R. Newlands (1837-1898) published a table of elements arranged in order of their equivalent weights. He gave each element a serial number (hydrogen 1, lithium 2, beryllium 3, etc.) and stated that 'the eighth element starting from a given one is a kind of repetition of the first, like the eighth note of an octave in music.' He remarked a year later that further on in the series the distances between similar elements frequently are twice as large: 'the numbers of analogous elements generally differ either by seven or by some multiple of seven: in other words, numbers of the same group stand to one another in the same relation as the extremities of one or more octaves in music.'[109] If we consider groups of cognate elements — e.g. N (nitrogen), P (phosphorus), As (arsenic), Sb (antimony), Bi (bismuth) — there are in the series of elements of ascending atomic weights seven elements between N and P, but fourteen between As and Sb. ('The numbers of analogous elements, when not consecutive, differ by seven or some multiple of seven.')[110] This system connected many triads, though Newlands supposed that many were still incomplete because parts of them were unknown or unrecognized.[111]

De Chancourtois's *vis tellurique* and Newlands's law of octaves met with little attention. Newlands was hardly taken seriously. When in 1866 he read a paper on his law to the London Chemical Society, one of the audience asked whether 'he had ever examined the elements according to the order of their initial letters,' and the society refused to publish the paper.[112] The mocking remark may be excused because it was followed by the serious criticism that in Newland's system such elements as iron, nickel and cobalt which clearly belong together, find themselves widely separated. Newlands himself rendered a serious reply.

7.3. The Periodic System

The arrangement of the elements which in the end found general acceptance was the Periodic System, put forward entirely independently, though at about the same time (1869), by D. Mendeléev (1834 - 1907) and Lothar Meyer (1830 - 1895). They were neither of them inspired by a belief in triads or octaves and at the time of the publication of their systems they were not acquainted with the work of de Chancourtois and Newlands, who, as Mendeléev recognized in his Faraday Lecture (1889), 'had made an approach to the Periodic Law and had discovered its germ.'[113]

The new systems were inspired by a belief of a more general character in the unity and analogy of nature, a belief in 'a general rule of law, which finds its formulation in the system of elements.' Mendeléev considered weight as the most fundamental property of matter; hence it was 'most natural' to seek some relation between, on the one hand the properties and the analogies of the elements (i.e. the similarity of their chemical properties), and on the other their one property that is 'measurable and indubitable, the atomic weight.'[114]

He then found that 'the elements, if arranged according to their atomic weights, show an evident periodicity of properties.'[115] As Lothar Meyer put it, 'The properties of the elements are largely periodic functions of their atomic weight,' this being shown most clearly by the lighter elements at the beginning of the series:

element	Li	Be	B	C	N	O	F
atomic weight	7	9	11	12	14	16	19
element	Na	Mg	Al	Si	P	S	Cl
atomic weight	23	24	27	28	31	32	35.5

With the list so arranged, each element is similar in properties to the one below it. Within each horizontal 'period' the valency to oxygen rises from 1 to 4 as the atomic weight increases (from Li to C, and again from Na to Si) and that to hydrogen decreases from 4 to 1 (from C to F, as also from Si to Cl).[116]

The result was thus a classification of the elements into 'groups' (as Mendeléev called them) or 'families' (Meyer) containing those which were analogous in chemical properties and in valency. Proceeding to heavier atomic weights, three groups — their members here arranged in horizontal sequence — stand out particularly clearly (see table). These groups, as Mendeléev remarked, 'reveal the essence of the matter.'

Classification of chemical elements in three groups according to Mendeléev					
	Elements with their atomic weights				
Halogens		F 19	Cl 35.5	Br 80	J 127
Alkali metals	(Li 7)	Na 23	K 39	Rb 85	Cs 133
Alkaline earth metals	(Be 9)	Mg 24	Ca 40	Sr 87	Ba 137

Mendeléev was careful to distinguish between the periodic *law* (formulated above), and the periodic *system* of elements of which the law is the foundation.[117]

The fact that the first periods strongly support the belief in periodicity was a needed encouragement, as further on in the system many difficulties arose. There are the 'rare earths' for which it was not easy to find a place in the system.[118] Then in order to maintain the law, Mendeléev had to correct some widely accepted atomic weights by doubling them (e.g. thorium and uranium). Like Newlands before them, the founders of 1869 had to leave open places for elements as yet unknown. If they had kept strictly to the ascending series of then known elements the system would have crumbled.

Although the *empirical* basis of the periodic table admitted the possibility of vacant spaces (for the inventory of facts is never final), its character of a *law* demanded the belief that these vacant places would be filled in due time. It is hardly surprising then that Mendeléev in particular assumed a somewhat ambiguous attitude when he had to divide his loyalties between practice and theory. In retrospect he attributed the lack of attention given to his predecessors, in their search for similar laws, to the fact that their contemporaries were interested in facts rather than in laws.[119] In 1891 he pointed out that neither de Chancourtois and Newlands nor Lothar Meyer ('who is considered today by many people as the founder of the periodic law'),

> ventured to predict the properties of undiscovered elements or to change 'accepted atomic weights' or to consider the periodic law as a new and certainly established law of nature, as I had done from the beginning (1869); therefore the *regularities* discovered by these investigators, which moreover were unknown to me, can only be considered as a preparation for the law.[120]

According to Mendeléev, a law can be confirmed only when conclusions are drawn from it which, in the absence of the law, would be neither expected nor thought possible, yet which prove to be justified by experiment. As soon as he had recognized the periodic law, he says, he drew logical conclusions from it which could allow its correctness to be tested. He had corrected the atomic weights of little-known elements (e.g. by doubling the atomic weight of uranium from 120 to 240), calculating them from those of the four 'atom analogues', i.e. by averaging the values of the atomic weights of the elements above and below them in the same family and those of the elements to the right and the left of them in the same period. He had predicted the properties of elements as yet unknown that should occupy places he had had to leave vacant in the system, and even several properties of their compounds. He did so in particular for the analogues of aluminium, boron and silicon, denoting them by the names eka-aluminium, eka-boron, and eka-silicon. He had not believed that he would live to see his predictions fulfilled and so it had been 'a great joy' to him when they were splendidly confirmed by the discovery respectively of gallium (1875), scandium (1879) and germanium (1886).[121]

Mendeléev was justifiably jubilant about these discoveries. Though he had criticized the reluctance of his contemporaries to accept a general law of periodicity — because they clung exclusively to facts that were available and had no faith in the facts still to be discovered — he now, as the facts confirmed the predictions deduced from his general law, more often emphasized the importance of such facts and even considered them indispensible. 'The recognition of a law ensues ... only when it has been confirmed by experiment, for the scientists should consider experiment as the highest jurisdiction for deciding about their ... opinions.' In this spirit Mendeléev even went so far as to describe de Boisbaudran, Nilson and Winkler (the discoverers of the three new elements) as 'the true founders of the periodic law' because they had proved 'its applicability to chemical reality.' Among those praised he also included people who, like Roscoe and Brauner, had experimentally verified the corrections Mendeléev had made to the atomic weights of some elements on the basis of their four 'analogues' in the system. It is evident that even when he first announced it in 1869, Mendeléev was already quite sure of the truth of his law. It seems, however, that for him there were two kinds of 'truth', for with apparent inconsistency he continues his retrospective review with the remark that '... the discovery of gallium, scandium and germanium showed the periodic law as a *new* truth, which makes it possible to see what is unseen and to know what is not yet known [italics RH].' The contrast here is between a truth founded upon many facts and contradicted by other 'facts' which were therefore considered untrue, and a more certain 'new' truth confirmed *a posteriori* by new facts.[122]

We see that however much a scientist may be sure of the truth of his theory, his conviction is strengthened when it turns out that it has rightly predicted new facts. In our time similarly, we see that however sure biologists may feel of the truth of Darwin's picture of evolutionary change, they all yearn for the discovery of missing links. The difference is that whereas Mendeléev lived to see the main gaps closed, the enormous increase of paleontological experience since Darwin's time, while it may have closed some gaps, has also revealed many more.

The question raised by these examples from history is: how can we know when deviations from a 'law' are caused by misunderstanding, lack of information and wrong observation and when they arise from the fact that the 'law' itself is not wholly correct? In Mendeléev's opinion one of the weaknesses of Lothar Meyer and Newlands was that they had failed to grasp the nettle when confronted with the dilemma: shall we alter the generally accepted values of atomic weights and adapt them to the propounded law or should we reject the law? Because they had done neither the one nor the other, they could announce only 'regularities' but not a 'law'.[123] Mendeléev criticized Meyer's statement (1869) that 'It would be rash to introduce a change of the atomic weights accepted up till now on the basis of such uncertain starting points.'[124] For example, Meyer, in order to get tellurium into the family to which it clearly belonged (that of oxygen, sulphur and selenium) placed it between antimony (Sb, atomic weight 122.1) and iodine (I, atomic weight 126.5), though its atomic weight (128) had been found to be higher than that of iodine. This was done on the basis of *facts*, *viz.* its chemical properties. However, on the basis of the theory or the law of periodicity he followed the value of the atomic weight with a question mark (Te 128?). In this way he maintained the natural classification, according to which tellurium is a member of the sulphur family and iodine is one of the halogens, while the question mark revealed his tentative attitude to the universality of the 'law' that the properties of the elements are periodic functions of their atomic weights, and his doubt as to whether the current value of the atomic weight of tellurium was incorrect. He ventured neither to deny the 'fact' in order to save the 'law', nor to assert the 'fact' remarking only that 'the atomic weight of tellurium ... probably has been found a bit too large.'[125]

Mendeléev, in his first table, had also placed tellurium between antimony and iodine and provided its atomic weight with a question mark (Te 128?). In the accompanying text, however, he made a firmer statement than Meyer: 'some atomic weights are expected to undergo a correction, e.g. tellurium *cannot* have the atomic weight 128 but 123-126' (italics RH). Meyer's probability is Mendeléev's certainty.[126]

In the same year, in Mendeléev's next publication, tellurium is still '128' or '128?', but he formulates his opinion more carefully: 'Only tellurium steps outside the series, but it may be quite possible that it has not been determined exactly and if we assume the atomic weight is 126-124 (*sic!*) instead of 128, the system becomes wholly precise.'[127] One of his final theses then runs as follows:

> 7. The atomic weight of an element may sometimes be corrected, if its analogues are known. Should then the atomic weight of tellurium be 123-126 and not 128?[128]

Finally, in his great publication of 1871, Mendeléev makes a decisive step by putting in the tables the sequence Sb 122 - Te 125? - J 127, elucidated by the text as follows: 'The atomic weight of tellurium must be, on the basis of the periodic law, greater than Sb = 122 and smaller than J = 127, i.e. it *must be* about 125 [italics RH].' Thus it turns out, that '125?' does not mean that this value in itself is uncertain (it is founded upon the periodic *law*, and in Mendeléev's opinion a law does not tolerate exceptions), but that the experimental determination of the atomic weight of this element is doubtful: 'A question mark after the atomic weight means that the available data about the atomic weight are open to doubt, in other words, that the equivalent of the element has not been exactly determined up till now,' and 'some atomic weights have been altered in the table in accordance with the periodic law; e.g. tellurium is shown — in accordance with the periodic law — as 125? and not 128, which Berzelius and others put for it.'[129]

So the only argument adduced against the correctness of the experimental determination of the atomic weight of tellurium is its incongruity with the periodic *law*: thus the 'fact' is here adapted to the 'law'.

This reveals an essential difference between Lothar Meyer and Mendeléev as regards the relation between theory and experiment. Mendeléev allows the theoretical demands to prevail over the available data. He calls the then accepted value of the atomic weight of tellurium 'an error', and asks for 'new experiments' because one can hardly assume that so large a deviation from the periodic law (128 - 125 = 3) could occur. According to Mendeléev, 'natural laws do not tolerate exceptions, and in this respect they distinguish themselves from rules and regularities.' The discrepancy between the atomic weight of tellurium *found* by experiment and that calculated from the atomic weights of the four adjacent elements was an exception and thus could not be true.[130]

The matter of the atomic weight of uranium did not raise the same problem. The question as to whether its atomic weight was 80 or 120 (or, as Mendeléev held, 240) involved such enormous differences that evidently what was as stake was a matter of different interpretations of the same fact and the periodic *system* might well be a help in untying the knot.

At the other extreme, a small difference between the expected atomic weight and that found by experiment could easily be attributed to experimental errors. Thus the fact that cobalt had a measured atomic weight slightly greater than that of nickel — though cobalt precedes nickel in the system — did not seem too disturbing.[131]

In the case of tellurium, however, the difference of three units was too large for an experimental error, and too small to be attributed to a mistaken interpretation of the chemical analysis. The discrepancy therefore remained a thorn in the flesh and Mendeléev, though convinced that, as the 'law' demanded, the atomic weight should be 125, very much wanted an experimental determination that would confirm this and discredit the accepted value of Berzelius and others. He seems to have been particularly pleased by B. Brauner's experiments on this subject; he tells (1891) how Brauner (1889) had come to the conclusion that the supposedly pure tellurium was not a simple body but a mixture from which a substance with the atomic weight 125 could be separated. This was the long-awaited grist to the mill of the St Petersburg chemist. He pointed out that the correction of the 'empirical data' about the atomic weight of tellurium which he had made on the basis of the periodic law had been confirmed 'by careful experiments' by Brauner, who found 'precisely the atomic weight predicted by the periodic law.' Brauner had thus done for atomic weights what de Boisbaudran, Nilson and Winkler had done for the vacant spaces in the periodic system: he had experimentally confirmed the Mendeléev predictions, and so finds himself praised, together with those other discoverers, as one of the founders of the periodic law.[132]

Strange founders, who instead of being first, come at the end![133]

The history of the periodic system clearly illustrates the dilemma of the scientist who, starting from facts, construes a hypothesis or tentative law, from which he predicts other facts to be tested by experiments. If the test does not yield the desired facts, the question is whether this is because the hypothesis (the posited 'law') was wrong or because the 'facts' are wrong. Should the fiction be adapted to the facts, or the facts to the fiction?

In the case of the atomic weights, tellurium refused to obey the 'law'; yet Mendeléev refused to accept the 'facts' and declared them to be necessarily erroneous. One wonders whether he did so with an uneasy conscience, for it was a great relief to him when (defective) experiments yielded data fitting in with his law. We now know that tellurium is indeed an exception. In Mendeléev's opinion one exception is sufficient to demote a law to no more than a rule. Nevertheless, the system built upon his law has increasingly turned out to be right!

If Mendeléev had kept strictly to the facts alone, he would not have left several vacant gaps in the series of the elements and probably neither the periodic law nor the periodic system would have emerged. As later discoveries showed, his (and Meyer's) refusal to keep to the facts alone was on this occasion more justified than in the matter of the irregular atomic weights. This latter problem could be evaded in two ways: by holding that the measurement of the atomic weight was incorrect or by postulating that the substance in question was a mixture and not a pure simple body. Both ways of escape were tried again and again but better analytical methods have unambiguously yielded 128 as the atomic weight of tellurium. The case of the vacant places is different in that we cannot know *a priori* whether more elements will be discovered in future: from the empirical point of view the possibility must be left open. The situation bears some similarity to the case of the hardness of diamond: when one has determined the hardness of diamond, the hardness is a fact; and the conclusion that diamond is the hardest substance yet known may also be a fact; but it remains an open question whether an even harder substance will be discovered in the future.

The Periodic Table of the elements became more and more accepted as an adequate natural classification. Several vacant spaces were filled by the discovery of new elements (e.g. hafnium and rhenium), mainly thanks to the fact that their properties were predicted from the Table. The difficulties with the rare earths (lanthanides) were gradually resolved.[134] The inert gases, discovered towards the end of the 19th century, fitted well into the system: after the lonely element hydrogen (atomic weight 1) follows the lightest inert gas helium (atomic weight 4); the next period runs from the alkali metal lithium (atomic weight 7) to the second inert gas, neon (atomic weight 20); then follows another period of eight elements (from the alkali metal sodium to the inert gas argon). If one takes a period as running from one inert gas to the next one, the two periods which follow are longer (18 elements), and that which contains the lanthanides has 32 elements. The 'octaves' have thus disappeared forever: the discovery of the inert gases meant that an element of the second period finds its next family member not in the eighth place further on but the ninth.

Though difficulties with the periodic system of the elements were solved one after another, the periodic law had to cope with another disharmony in the series of ascending atomic weights. The inert gas argon, which, from considerations of chemical properties, had to be placed before the alkali metal potassium (atomic weight 39), turned out to have a greater atomic weight (40). Moreover the more precise methods of chemical analysis finally made it certain that the true atomic weight of tellurium was about 128 (127.6) and not 125. If we hold Mendeléev to his dictum that 'a true law of nature is in advance of the facts, it guesses numbers ... etc.,' then we must judge that the periodic law, having guessed wrongly in several cases, cannot be a *true* law however great its success in most cases.[135]

It had eventually to be recognized that the 'law' was just a rule that had its exceptions and that a new law — 'the properties of the elements are a periodic function of their atomic numbers' — had to be accepted instead. Initially this said no more than that the properties of the elements were dependent on their place in the table — a place that was allotted because of their properties, in a system based on their properties. The atomic number had in this sense no physical meaning: it could not explain the properties upon which it was based. A physical interpretation of the atomic number was later given by the atomic theory of Rutherford: the nuclear charge of an atom is equal to its atomic number, and the character of an element's chemistry turns out to be determined by this nuclear charge.

So Mendeléev was wrong both in the statement in his first publication (on the relations of the properties to the atomic weights of the elements, 1869) that 'the value of the atomic weight determines the properties of the element'[136] and also in his triumphant declaration (1879) that the Periodic Law 'proclaims loudly that the nature of elements depends above all on their mass.'[137]

What matters most from the chemical point of view is not the weight of the atoms but their nuclear charge (or what amounts to the same thing, the number of electrons surrounding the nucleus). In 1913 it was discovered, for example, that the radio-active disintegration of thorium to radium finally yields products with the same valency and the same chemical properties as common lead, though they have a different atomic weight. It thus turned out that what could count as 'elements' from a chemical standpoint could actually be mixtures of atoms of different weight (Aston 1919). Tin, for example consists of twelve kinds of atoms with atomic weights varying from 112 to 124. Natural hydrogen not only contains atoms with the atomic weight 1, but also a small percentage (0.02%) of atoms twice as heavy (deuterium, Urey 1932). On account of their valency and other chemical properties, however, the 'same place' in the periodic system is allotted to them and they are aptly named 'isotopes'.

Mendeléev's belief in the 'universality of the law because it could not possibly be the result of chance' was thus far from justified. It was a piece of good luck that the series of the elements ordered according to ascending *mean* atomic weights of the natural isotopic mixtures is, with a few exceptions, the same as the series ordered according to ascending nuclear charges.

In contrast to Mendeléev's periodic law, the periodic *system* has stood fast. The nuclear charge of tellurium is one unit below that of iodine, and the nuclear charge of argon is one unit below that of potassium.

So, after all, the scientist's belief in simplicity has not been shaken. The revised periodic law ('the properties of the elements are a periodic function of their atomic numbers') is even simpler than the old one. In contrast to the atomic weights, all atomic numbers are integers and they increase in a straight series from 1 to over 90. Whereas the former efforts to find a mathematical law connecting the atomic weights within a period or within a group (or covering all of them) had only partially satisfactory results, according to the new law the nuclear charges of the atoms form a series of the utmost simplicity, an uninterrupted series of integers.

The older efforts were often based on the belief that the atomic weights of all other elements are multiples of that of hydrogen. They were frustrated because the deviation of several atomic weights (e.g. that of chlorine 35.5), from integers was too large to attribute to experimental errors. Hypotheses invoking a primary particle with the atomic weight 0.5 as the universal building block had to be abandoned when the results of more precise methods contradicted it.

In this matter of the step or jump from one element to the next, it is interesting that the periodic theory brought different kinds of satisfaction to different people. Though many scientists were pleased at the unification afforded by the coherent arrangement of much miscellaneous knowledge, Mendeléev's own perception was rather that the discontinuous steps within the system were a clear indication that the elements are wholly separate entities.

The existence of isotopes now explains to our satisfaction why no law connecting atomic weights could be found. On the other hand each isotope in itself proves by its atomic weight (which is practically an integer) that there is truth in the idea that all elementary atoms contain identical building blocks. As to the length of the periods, a most elegant numerical relation emerged, not from aesthetic ideas conceived *a priori*, but from facts recorded *a posteriori*. The periods contain 2, 8, 8, 18, 18, 32 elements, or, as Rydberg pointed out, $2x1^2$, $2x2^2$, $2x2^3$, $2x2^4$. Here nature, which had so long refused to show the secret of her harmonious relations, at last revealed a rule of unexpectedly greater harmony. Nothing could be more gratifying to lovers of numerical speculation. To the scientist an 'explanation' of this remarkable series within the framework of the quantum theory and the atomic theory of Bohr was given through W. Pauli's exclusion principle — an explanation which itself demands further explanation.

8. THE HARMONY OF LIGHT : BALMER TO BOHR

The step from harmony in sound to harmony in light seems, on the evidence of the history of science, to have been rather an easy one to take. We find Newton, whose theory implied more than the emission of particles of light, already drawing a parallel between the musical scale and the spectrum. Adherents of the undulatory theory of light made even more direct comparisons between waves of air and waves of aetherial matter.[138] In the 19th century physicists discovered that elements emit light with a spectrum which is discontinuous, consisting of bands and lines at wavelengths characteristic of each element. This created a challenge to find the *law* governing the emitted frequencies and the discontinuities.

The Swiss mathematician Johann Jacob Balmer (1825 - 1898) was encouraged by the physicist E. Hagenbach to seek a mathematical formula expressing a *simple* relationship between the spectral lines of a given element and also simple relations between the spectra of different elements. The fact that the line spectrum is discontinuous, in Balmer's opinion made such a search worthwhile: it might be compared with the discontinuity in the proportions of given elements in their various compounds (e.g. CO, CO_2, $-CO_3$) - that discontinuity which is the basis of Dalton's law of multiple proportions. Balmer was convinced that hydrogen — being the simplest element — would be most likely to open new ways of discovering the essence of matter by the study of its spectral lines.[139] This prophecy was indeed fulfilled almost three decades later in Bohr's work on the spectrum and atomic structure of hydrogen (1913).

To Balmer his search for numerical relations between the first four hydrogen lines did 'stir and captivate the attention — for the ratios of these wavelengths may be expressed in a surprisingly exact manner by small numbers.' The wavelength of the red line is to that of the violet line as 8 : 5; that of the red to the blue-green as 27 : 20; and that of the blue-green to the violet as 32 : 17. Balmer's first thought was that the vibrations of the various spectral lines of a substance might be, as it were, harmonics (*Obertöne*) of a specific fundamental tone (*Grundton*). Though all his efforts to find such a fundamental tone for hydrogen failed, Balmer's belief in simple numerical relations was sufficiently strong to encourage him to continue his search: 'there must be a simple formula, by whose aid the wavelengths of the four main hydrogen lines could be expressed.'[140]

It is important that Balmer had not only a tenacious faith that there *must* be a simple law, but also that he did not say *a priori* what form this law must take. This was something which the *facts* must reveal; they should decide which of the possible fictions was the right one. Balmer's search for a 'common factor' in the four wavelengths 'gradually' led to a formula by which they 'can be represented with striking precision.' The wavelengths may be obtained by multiplying the common factor (the *Grundzahl*, h = 3645.6×10^{-7} mm) successively by the coefficients 9/5, 4/3, 25/21, and 9/8. At first sight this does not seem a particularly regular series, but multiplication of the numerators and denominators of the second and fourth terms yields 9/5, 16/12, 25/21, 36/32. The numerators now form an ascending series of squares of integers and the denominators a series of squares diminished by 4:

$$\frac{3^2}{3^2-4}, \quad \frac{4^2}{4^2-4}, \quad \frac{5^2}{5^2-4}, \quad \frac{6^2}{6^2-4}$$

or

$$\frac{3^2}{3^2-2^2}, \quad \frac{4^2}{4^2-2^2}, \quad \frac{5^2}{5^2-2^2}, \quad \frac{6^2}{6^2-2^2}$$

General formula: $\frac{m^2}{m^2 - n^2}$ x h mm, where *m* and *n* are integers.[141]

These ratios agreed to one part in 40,000 with Ångström's data — a remarkable fact, for as A. Sommerfeld said, Ångström had measured these wavelengths with a degree of exactness surpassing that of astronomical observation.[142] Balmer made a very modest claim for his formula: 'only the facts themselves' can decide whether the formula will be valid for other lines of hydrogen; the exact determination of wavelengths, he insists, is the only thing required for finding the law. Sincere though was his respect for facts, it should not be overlooked that it was his *faith* in the harmony of nature that made him persist in seeking a simple law. Balmer, who did not himself perform a single physical experiment, could succeed where such a great experimenter as Ångström failed for lack of the necessary disposition.[143]

Balmer did not stop at the hydrogen lines. Believing as he did in the unity of nature, he tried to find a formula that could embrace all the elements. He thought he had found such a general formula:

$$\lambda_n = \frac{a\,(n+c)^2}{(n+c)^2 - b}, \quad \text{or} \quad \tau_n = A - \frac{B}{(n+c)^2}$$

For b = 4 and c = 0, this gives the hydrogen formula as a special case. The great simplicity of the general formula and its close relation to the hydrogen formula were to him warrants of its truth.

> Balmer put forward conjectures as to the physical meaning of the three constants of the formula, since he supposed them to have a material basis. This 'like so many other things in nature, presents us with riddles whose solution again and again exerts an irresistible charm.' He had no expectation of finding out these innermost secrets of nature. His firm belief in a mathematical plan behind the universe was counterbalanced by an equally strong awareness of the limitations of the human mind as compared with the divine Mind. The closing sentence of the article he wrote in 1896 expresses this. These 'arch-elementary relations' give to our contemplating mind the impression that there is an 'inexaustible wise contrivance of Nature, which with never-erring certitude performs functions which the thinking mind can but follow with difficulty and with humiliating imperfection.'[144]

Balmer enjoyed art, music and literature and everything 'beautiful and ideal'. With the Platonising scientists like Kepler he shared the belief that a divine mathematical architectonic plan was realized in nature. The romantic and aesthetic streak in his character had been stimulated more directly when, as a young man, he studied in Berlin and encountered the influence of Schelling. His interest in architecture is shown by the plans he devised for working class housing (1883) and in his activities for the preservation of ancient buildings in Basel.[145] In his description of a medieval church he pointed out the mathematical relations hidden in it. His predilection for mathematical speculation is most evident in his tract on *The Vision of the Temple by the Prophet Ezekiel* (1858), which according to the sub-title, was 'a short account and an architectonic explanation for adorers and students of the Word of God and for friends of religious art.' To him the temple, in all its proportions of numbers, measures and forms is the perfect, ideal exemplar, 'the crystallized image of the kingdom of God'; it possesses 'greatest symmetry and harmony.'[146]

It is not then by chance that he who as a young man tried to lay bare the significance of the mathematical proportion of the house of God, in his later years wanted to do the same for the architecture of the atoms. He was soberminded enough not to bring these two together in one system by regarding the temple as some kind of image of the cosmos, as had happened in pagan religions of nature, but he shared their conviction that, in the work of creation as well as in conveying a spiritual message to mankind, the Godhead geometrizes. The only allusion to physics in his earlier geometrical and exegetical dissertation is the following, which happens to foreshadow the topic of his later investigations, for it concerns the wavelength of light: the text of Ezekiel 40:3 'a man ... holding a cord of linen thread and a measuring-rod' brings out according to Balmer, how important measures are: 'the whole of natural Creation is so precisely ordered according to measure and number, that one knows that ... in one second 720 billion of vibrations of violet light take place.'[147] As Balmer's biographer says, 'he sought to comprehend everything by number and proportion;' the whole world, nature as well as art, was to him 'one unified harmony', and his whole life was a yearning to grasp 'harmonic relations in numbers.'[148] In the same way as he enjoyed finding proportions in biblical, classical and medieval architecture, he desired to find harmonic vibrations in the atoms. Without such a predisposition, it is probable that the mathematician Balmer, an outsider to physics, would not have persevered in his quest.

Yet Balmer had no tendency to impose ratios upon nature. He may have been influenced by *Naturphilosophie*, but he also stood firmly in the tradition of the Swiss Reformation, imbued with the Old Testament teaching that 'as the Heavens are higher than the earth,' so are God's thoughts higher than man's thoughts and ways.[149] Natural theology had for him a subordinate place. The garden of the temple symbolizes 'the revelation of God in nature,' which has been given to all people, but in his opinion a garden without a fertilising spring would become a desert. Nature alone cannot lead to the true knowledge of God, unless the fountain from the sanctuary — the revelation of God through the Holy Spirit — is joined with it.[150] This almost platonist lover of harmony and rational patterns in nature had a keen eye for the irrational (or non-rational) element in nature. He pointed out (1868) that not only is science often guided by preconceptions and prejudices, even in the greatest of its adepts like Tycho Brahe and Newton — but also that not all legitimate results of physics are 'clear and perspicuous to human reason.' Mathematics introduces unthinkable magnitudes (imaginary numbers) and obtains good results with their help. Physics does not understand what 'matter' is or what the atoms are. In spite of its great triumphs in the theories of light and gravitation — which led to great discoveries demonstrating the conformity of nature and theory — the undulatory theory is full of inner contradictions and gravitation is an enigma. In spite of advances in physiological science we do not understand why and how certain physical events cause the sensory experiences we undergo; that certain vibrations reach our consciousness as a sensation of sound is a sheer wonder. 'The whole world of sound, which springs forth from the jubilation of nature and brings the ear into captivity by its magic, is an incomprehensible wonder, closed, however, to that man who blocks his ears in order to check by cold reason the strange swarming of wave motions according to measure and number.' This verdict is worth listening to, the more so as it comes from a man so deeply convinced that the finding of measure and number in nature was also a discovering of wonder in it. It implies that a scientist should not only experience the wonder of his scientific results, but that he must also enjoy that 'naïve' feeling of wonder and beauty which he shares with the non-professionals.[151]

Balmer's formula is written today in a slightly different version:

$$\gamma = \frac{1}{\lambda} = R\left(\frac{1}{n^2} - \frac{1}{m^2}\right)$$

where γ is the frequency of the spectral line and R is the Rydberg-Ritz constant.
The factor R, which Balmer had expected to be specific for each element, turned out to have a universal character, so that the unity and harmony revealed by his formula are even greater than he dared hope. The various series for which n has values other than 2 have been found: when measured their lines turned out to match exactly the calculations based on Balmer's formula. For n = 1 (and m = 2, or 3, or 4, etc.) we have the Lyman series in the ultra-violet; for n = 3 (and m = 4, or 5, or 6, etc.) the Paschen series; for n = 4 (and m = 5, or 6 etc.) the Bracket series, for n = 5 (and m = 6, or 7, etc.) the Pfund series.

Balmer's faith in the possibility of explaining (as distinct from describing) the spectrographical data in terms of integers was eventually justified when in 1900 Max Planck announced his theory of the quantum of action, which introduced the idea that energy could be discontinuous: there can only be one, two or n energy quanta associated with a certain frequency of vibration, just as there can only be one, two or some other whole number of atoms of a certain element. Some of the older physicists were horrified by such a heresy. As late as 1926 Chwolson wrote that the 'old physics is in sharp contrast to the new one;' the quantum theory 'which came to the light at the same time as the new century,' has 'put its muddy stamp on almost all parts of the new physics of this century.'[152] It seems highly probable that Balmer himself, with his feeling for the element of apparent irrationality in nature, would have had a more positive opinion. He would probably have enjoyed the way in which another paradoxical combination of the 'rational' and the 'irrational' in physics, *viz.* the theory of Bohr, 'explained' Balmer's formula and led to a seminal advance in thinking about the structure of the atom.

Niels Bohr (1885-1962) recognized (1913) that if, in Balmer's formula, the wavelengths of a discontinuous atomic spectrum could be expressed in terms of integers, the emission of radiation by the atom must be a quantum phenomenon. But his dynamical 'planetary' model of the atom (founded on the static one of Rutherford) was extremely paradoxical. Negatively charged electrons, supposed to move in discrete orbits around a positive nucleus, were nevertheless not to radiate energy so long as they remained in an orbit. Here Bohr trespassed against one of the great rules which Newton had laid down for scientific method: 'thou shalt not deviate from the analogy of nature.'[153] Bohr deliberately abandoned the analogy of nature when, ignoring Maxwell's laws of electromagnetism, he postulated that light would be emitted only when electrons jumped from a higher to a lower energy level (as permitted by Planck's quantum conception).[154] In other respects the model was based on the laws of classical mechanics; the electrons being supposed to move around the nucleus like the planets around the sun, that is, in conformity with Kepler's laws. The electrons were attracted by the nucleus according to another law of classical physics, Coulomb's inverse square law of electrostatic force.

Bohr fully realized that there were great inconsistencies, and he regarded his theory only as a first step towards a total change of fundamental concepts (which indeed took place in due time). It was all the more astounding that on the basis of this *fiction* he was able to deduce not only the Balmer formula but also the numerical value of its constant (the Rydberg constant) both of which up till then had possessed a purely empirical character.

In 1919 Arnold Sommerfeld, in his *Atombau und Spectrallinien*, pointed out that ever since the discovery of spectral analysis nobody could doubt that 'the problem of the atom would be solved when men had learned to understand the language of the spectra.' Now, thanks to Bohr, 'what we hear today from the language of the spectra, is a real music of the spheres of the atom, a consonance of proportions of integers, an increasing order and harmony ...' But Planck's name also will be honoured in this connection: 'All laws of integers of spectral lines and atomistics finally emerge from the quantum theory. It is the mysterious Organon, on which nature plays the music of the spectra and according to whose rhythm she orders the structure of the atoms and nuclei.'[155] These words reveal an almost Keplerian enthusiasm, which was not to abate with the years. When much of the original theory had been revised Sommerfeld still spoke of the 'fundamental beauty' of the theory.[156]

In 1923 the adherents of the new physics saw no reason to share the morose feelings of Chwolson. A special volume of *Die Naturwissenschaften* was then published in commemoration of the tenth anniversary of Bohr's theory. Planck, the great initiator of the changes which had taken place in the intervening years, opened by pointing out that the great merit of Bohr's theory was its sticking to 'the main task of theory, that of adapting conceptions to facts and not the reverse.' A verdict the more impressive as Planck always hoped that the unity of physical theory might be restored in the spirit of classical physics. But he did not think the 'irrational' character of the theory too high a price to pay for its conformity to *facts*. He took that into the bargain and praised Bohr's 'sense of reality', 'his respect for facts, from which without doubt he drew the courage necessary for putting (over against a seemingly perfect and permanently established system of thought) a new one of an audacity unheard of.' Evidently Planck was struck more by the realism and the obedience to facts displayed in Bohr's theory than by its 'beauty', founded as it was at the cost of a loss of the unity of physical theory.[157]

To other physicists, however, it seemed as if Bohr had brought an increase of unity to the physical system. Max Born, for example, found it

> one of the most remarkable and most attractive results of Bohr's atomic theory ... that it represented the atoms as planetary systems on a small scale; ... the yearning for knowledge of the unity of the universal law found unexpected satisfaction when the study of the atomic world arrived at forms which once had been discovered by astronomy in the heavens.[158]

In Born's opinion the enduring pattern had been given by Kepler's laws, but something new had been added: by the application of the quantum theory certain 'stationary' orbits were sorted out from the continuous multiplicity of astronomical orbits. Born, too, recognized that the unity of cosmic events and laws had its limits and that probably the whole system of physical concepts would have to be revised.[159]

In the end, Bohr himself was less enthusiastic about this analogy. Though at first sight there is an 'extraordinary similarity of the image of the atom to the system of the planets,' there is, he stressed, an *essential* difference between them. The distances of the planets depend on the unknown historical development of the system; those in the atom are independent of its history, for the atoms are constant, just as their spectra are constant. Since it was by the introduction of discontinuity in the energy of the atom, and the theoretical deduction of Balmer's formula and calculation of its constant were reached *a priori*, this demonstrates that, in spite of the deviations from classical mechanics and electrodynamics, the quantum theory may be considered after all a 'natural reformation (*Umbildung*) of the fundamental concepts of classical electrodynamics.'[160]

Evidently Bohr considered Planck's quantum theory as more 'natural' than his own 'artificial' model of the atom based on it. That this was indeed his attitude is confirmed when he concludes that we have to be content with pictures that are but 'formal', and which therefore do not give an adequate image of the kind we are accustomed to expect in our usual representations.[161]

Since that time, physical theory has gone through a development which has seemed to lead it still further away from the 'analogy of nature', and in particular from the parallel between atoms and planetary systems. But the integers have remained, e.g. in that the atomic number, which represents the nuclear charge, is still a fundamental characteristic of the elements.

9. Conclusion

Our review has shown several examples in which the scientist's search for harmony was doomed to failure and others where it led to marvellous results through some degree of correspondence with the reality of nature. All these efforts were fundamentally inspired by a general belief, from the Pythagoreans to the present day, that there is order in nature, that it is a cosmos, not a chaos. In many scientists we have seen not only this general aim of discovering order in nature, but also such 'metaphysical' ideas as 'simplicity' and 'beauty' playing a role.

The discovery of order and simplicity does not, however, seem to be a matter of a guaranteed upward progress. Not so long ago Professor P.T. Matthews, in an article on 'Order out of sub-nuclear chaos', reminded his readers that each new advance in the field of particle physics has seemed 'like a step backwards, in that it brought ever-increasing complexity and confusion into a domain of physics where one had reason to expect the last word in precision and simplicity.'[162] From my own student days in the 1920s I can still remember the enthusiasm of one of the professors, who stated that the existence of protons and electrons as the building blocks of atoms showed that all things consist of two contrary constituents, and that therefore 'after all, old Hegel was right.' Yet it was not long before the neutron was discovered — followed by a host of other elementary particles. Nature, which had seemed to confirm our aesthetic expectations, now yielded so many kinds of particles that our belief in simplicity was frustrated. But it was not lost; for Professor Matthews was able to report that the tide was turning and that a coherent pattern was 'emerging to bring order and beauty into the sub-nuclear world.' He goes on to show how the present theory groups particles into families, how the members of each family can be arranged in patterns — hexagons and triangles — so that some kind of order is introduced 'into the puzzling assortment of particles that physicists have detected in their dissection of the atomic nucleus.' He then cites the story of the prediction of the 'omega-minus particle' (see figure 9) by Gell-Mann (1961): 'This is the type of situation ... which coordinates and clarifies a previously confused experimental situation; which is in agreement with the known facts, and which makes an absolutely precise prediction, specific to the new theory and appearing quite weird on the basis of previous ideas.' Matthews concluded that the announcement of the discovery of the missing particle might be 'expected any day.'[163] As we have already mentioned, it indeed came soon afterwards (see page 29).[164]

In this case, as in many others, (e.g. the classification of the elements, crystal forms, chemical compounds) the phenomena are at first of a bewildering variety. At a second stage science brings some order and harmony into the chaos. A third stage, however, reveals that with more precise observations the harmony, if not wholly spurious, is less perfect and the law less exact than had been supposed in our first enthusiasm — aroused on the basis of relatively coarse measurements. Though Romé de Lisle's law of the constant angles of ideal crystals and Boyle's law of ideal gases satisfy our sense of beauty, crystals turn out to have a 'mosaic' structure, and their lattice may be defective, and gases are 'real' not perfect. Despite this defiance on the part of Nature of our yearning for harmony, the

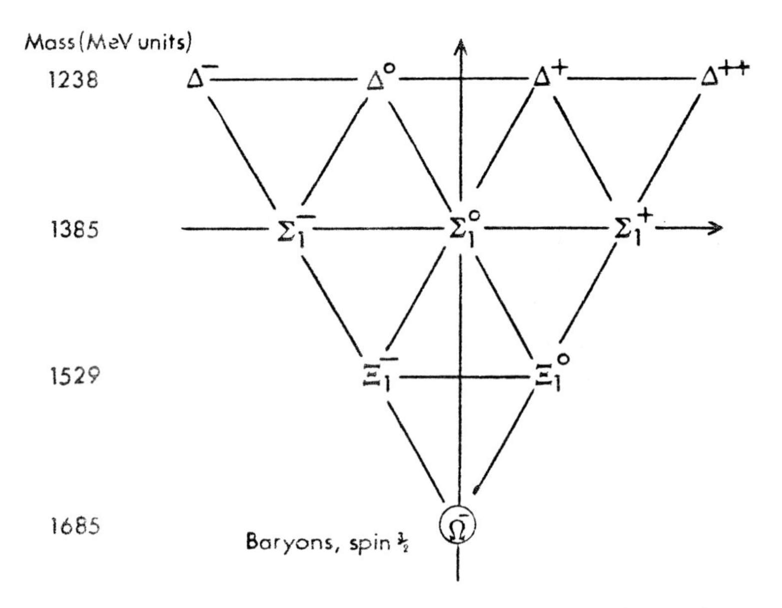

Fig. 9. Gell-Mann triangle: 'Plotting the nine known baryons of spin 3/2 suggests a Unitary-Symmetry ten-membered set with one particle missing — the omega-minus (ringed, at lowermost vertex). Its properties can be predicted but are very unusual. Its discovery will clinch the theory' (From: P.T. Matthews, 'Order out of Sub-Nuclear Chaos' in: *New Scientist,* 379 (20-2-1964), fig.3).

indefatigable scientists keep plodding on and do not rest until they have discovered a new law, which embraces the irregularities and transforms them into regularities and the former 'regularities' into 'limiting' cases. In this continual dialogue between man and nature the degree of 'simplicity' and 'harmony' of the picture is forever fluctuating — now increasing, now diminishing. Herein lies the unending excitement of science for its devotees.

Physical theory has assumed an increasingly abstract character (Heisenberg, Dirac, de Broglie, Schrödinger). It has become so 'mathematical' that Jeans could claim that 'the final truth about a phenomenon resides in the mathematical description of it ...; the Great Architect of the Universe now begins to appear as a pure mathematician'[165] (rather than the great mechanic of yesteryear). But there is one fundamental difference with respect to those older systems which proclaimed that nature consists of numbers or that nature is essentially geometrical. It is now generally recognized that mathematical deduction *in itself* does not yield the most satisfactory theoretical construction. In the sciences of nature, mathematics — notwithstanding its weighty and indispensible role in building theories — is but the handmaid, the means which, as Einstein, Planck and Bohr fully acknowledged, has finally to adapt itself to the observed *facts*, which are the beginning and the end of physics.

III. The Philosopher's Stone

1. Common Sense and Naïve Experience 79
2. Alchemy 83
3. The Sulphur-Mercury Theory 87
4. Corpuscular Theory 91
5. The Scientific Character of Alchemical Theory 92

1. Common Sense and Naïve Experience

Aristotle's ideas of Being and Change, of elements and compounds and of the structure of the universe had an overwhelming influence, even upon those, such as the Stoics and the Neoplatonists, who did not belong to his school of thought. In the Middle Ages all scholastic sects accepted his cosmology; later the vast majority of Renaissance philosophers, even though they were more inclined towards Platonism, would accept much of his natural philosophy and most of his astronomical system.

Thus the rise of modern science was also a struggle with Aristotelianism, particularly in astronomy (Copernicus), in mechanics (Galileo) and in chemistry (Jungius, Boyle). It must be recognized, however, that in the biological sciences the renewal was to a large extent a *revival* of Aristotle, and that in the humanistic disciplines (of metaphysics, logic and politics) his influence lasted until our own time.

Why then did Aristotle's natural philosophy have such a firm grip on most scholars from the fourth century BC until the 17th century AD and even later? One of the reasons might be that he constructed an all-embracing system, logically developed on the basis of immediate experience. Whereas the approach of Pythagoras and the Platonists to nature was in the first place mathematical and quantitative and, moreover, asked men to doubt the evidence of their own eyes, Aristotle started from *qualities* and events that are met with in everybody's daily experience. Next, having given the senses their due, the great master of logic hastens to erect upon them an impressive theoretical edifice — a building that satisfies reason. Thus, in Aristotle's methodology, it seems that both Reason and Experience get their fair share.

At first sight Aristotle's starting points seem self-evident; and once granted them, he invariably leads us by seemingly impeccable logic to theoretical conclusions which are as far from naïve observation and common sense as those of any of his rivals in Antiquity.

The Pythagoreans had baffled understanding with their doctrine that the world consists of numbers; the Platonists did the same by holding that true reality consists in invisible Ideas; and the Atomists explained away the world as we see it by reducing it to a multitude of never-seen particles, endowed with nothing but geometrical properties — particles, that is, without colour, smell or taste.

Aristotle, on the other hand, seemed to remain true to the facts of unanalysed direct observation, which are immediately recognized as familiar and conformable to one's own experience: the difficulty is only in the notional apparatus. In the realm of the phenomena of mechanics, for example, naïve observation will admit that there are 'light' bodies (air, smoke) which, when left to themselves, will move away from the earth, — and 'heavy' bodies (lead, iron, stone) which will fall downwards. Naïve observation admits also that the heavier a falling body is, the sooner it will reach its destination: a piece of lead will fall with a greater velocity than a feather. Again everyday experience teaches us that a very small force cannot move a large weight, and that a body, when moved by a constant force travels at constant velocity — as everybody knows who has ever pushed a cart.

From such simple facts Aristotle drew rather sophisticated conclusions: for example, the greater the resistance of a medium, the smaller is the velocity of a falling body. This loose formulation in everyday language is then taken as a 'law' in a precise sense, so that velocity is considered to be inversely proportional to the resistance or the density of the medium. It then follows that if the density of the medium were nil (that is if the space were void), the velocity would be infinitely great — which is an absurdity. But if, in such a case, the velocity were *not* infinitely great, then it would have to have a finite value which would be the same as occurs in a medium of some finite density: again an absurd conclusion. So the only way out is to conclude that a void is impossible.[1] Here it is human reason that decides what is possible in nature; and as so often happens, a mainly empirical start has led to rationalistic conclusions.

Unanalysed observation is also the basis of Aristotle's theory of chemical change. He does not reduce it to changing relations between abstract entities like numbers, points, or atoms, but to what can be seen and touched, i.e. to sensual qualities like colour, fluidity, taste. The changing of one substance into another is illustrated by the changing of wine into water: one drop of wine poured into ten thousand measures of water will completely lose its taste and colour. Evidently the great power of the large quantity of water completely overwhelms the qualities of the wine.[2]

The most general of the senses is the sense of touch, which must therefore help to characterize the most general bodies, *viz.* the elements. The most general properties and qualities felt by touching are, in Aristotle's opinion, hot, cold, moist and dry. The Greek feeling for symmetry and proportion then leads him to arrange these in four pairs of qualities: hot and dry (manifest in the element fire), hot and moist (air), cold and moist (water), and cold and dry (earth). Opposite qualities — hot and cold, moist and dry — cannot go together. The more active elements are air and fire; the more passive ones, earth and water.

Transformation of one element into another takes place when one of its qualities is annihilated by the opposite quality, e.g. when, under the influence of the heat of a fire the coldness of water (*cold* and moist) is overpowered, it is changed into air (*hot* and moist). If, however, the interaction of elements results in a state of equilibrium, a *tempering* of the several qualities takes place and a *compound*, with intermediate strengths of qualities, is formed. The preponderance of certain fundamental qualities in the compound may indicate which element has contributed most to its formation and would be yielded in the greatest quantity by its destruction. It is, however, energetically denied that the elements are actually (*in actu*) present in the compound (*mixtum*). They enter into the compound when it is 'generated' and they may emerge again at its 'corruption', yet they are only potentially and not *actually* (with their full being or essence) *in* the compound.

Here again direct observation, which shows the homogeneity of the substance, supports the theoretical conception; conversely, the theoretical conception interprets the observation: if the elements were really *in* the compound, a Lynceus - a modern physicist would say an X-ray analyser - would recognize that it is non-homogeneous and that there are particles of the four elements lying one beside the other in the compound. It would then not be a true compound but a mechanical mixture, a compound to the senses only (*mixtum ad sensum*).

The *organistic* character of this approach is evident. When elements and qualities are at *strife* with each other, either one of them wholly wins the victory over the others, or a balance is achieved, a mutual tempering toward the right 'temperatura' of qualities (a term still in use for the degree of heat). There is however, no mere addition of qualities.

The organistic character shows itself in the idea that the compound is something different from the sum of its parts. Whereas the properties of a mechanical mixture are the sum of the properties of its constituents, a living being is more than a mere aggregate of head, trunk and limbs; it is a 'whole'. Thus a true compound is something more than a mere addition of the four elements.

In the same way in 'mechanics', the resultant of two motions, e.g. projection and falling, was not perceived as amounting to a simple addition; one of them may completely overwhelm the other, or they may mutually weaken each other. In a mechanistic conception, on the other hand, the effects that each would have separately may be combined additively.

Now Aristotle's 'chemical' theory did not give rise to a chemical practice. The philosophers of the Aristotelian school neither decomposed substances in order to determine the proportions of the elements in them, nor did they synthesize compounds out of the elements. Why not? For three kinds of reasons:

Fig. 10. Philosopher's Stone. From: Jean-Conrad Barchusen, *Elementa Chymiae, quibus Subjuncta est Confectura Lapidis Philosophici Imaginibus Repraesentata.* Leyden 1718, plate 19, fig.78. According to Barchusen the alchemical plates at the end of his book were from a manuscript in a Benedictine monastery in Swabia (Partington, *A History of Chemistry,* p.701). Courtesy Bibliotheca Philosophica Hermetica, Amsterdam.

they *should* not do it (for religious reasons);
they *could* not do it (for philosophical reasons);
they *would* not do it (for social reasons).

1. They *should* not do it because trying to synthesize things produced by Nature — God's instrument — was regarded by pagan as well as by many Christian poets and scholars as impious arrogance.

2. They *could* not, for according to Aristotelian philosophy, all natural things are generated by similar things: the Ideas, or the Forms, or the 'seminal principles' of things *are* Nature, and human beings are not free to dispose them at will: Man can generate only men. It is impossible to produce a natural substance by means of human

art — that is by fabrication. By a mechanical procedure (in contrast to a natural one) only mechanical mixtures can be fabricated, in which the particles are contiguous but do not merge into a continuous whole; no really homogeneous compounds can ever be generated.

As mentioned above, the Aristotelian world view was organistic rather than mechanistic. An organistic world view is one that considers all things, living as well as non-living, after the model of an organism. Non-living things are treated as living things minus something (organisms minus life). An extreme form of this conception regards all things as animated ('hylozoïsm'). A mechanistic world view, by contrast, considers all things after the model of a machine; living things are like non-living things plus something (mechanisms plus life). In the extreme form it holds that living beings are mere machines. (In the 18th century La Mettrie; in the 17th century Descartes, who made, however, an exception for Man).

3. They *would* not do it, because chemical production implies manual work. Chemistry is a 'mechanical' art: a *pyrotechnia.* Mechanical arts or crafts, in particular those that require people to work with fire, like the art of a smith or a metallurgist, were deemed vile; fit rather for slaves than for free citizens, who should occupy themselves only with free (liberal) arts.

2. Alchemy

This brings us to our subject proper: the alchemists, who defied all these ancient prejudices, yet considered their work a divine art, the highest philosophy, and a work of the hands as well as of the mind. Alchemy claimed to change base metals into gold by means of a substance prepared in their laboratories, the so-called Philosophers' Stone.

In late Antiquity the techniques of 'colouring' ('tingeing') metals, (e.g. the change of red copper into a silverwhite metal by addition of a rather volatile substance, arsenic), led to the belief that it must be possible not only to give the right *colour* by means of a 'tincture' (dye), but also that it must be possible to induce the right density, ductility, etc in base metals so that they would turn into the perfect metal (gold). In terms of the hylozoïstic conceptions of the alchemists the base metals are 'ill'; they are short of metal force, of 'pneuma' or 'soul' and consequently, of all the qualities that depend upon it. More pneuma (in the form of a volatile, pneumatic substance) needs to be added in order to restore life. In this theory the four 'bodies' (*corpora*) copper, lead, iron and tin are contrasted with the three 'spirits', sulphur, mercury and arsenic. The ancient Egyptian alchemists described their procedures in the language of mystery religion: 'Osiris is dead; Osiris is a mummy; the cool grave covers his limbs and lets us see his face only.' The body of Osiris was restored to life by the divine water administered by a libation offering. In this ceremony sympathetic magic was supposed to help the revival — by means of the holy water of the Nile — of the grain that had been buried in the earth. In the same way the *hydor theion* (divine water, sulphur water) was supposed to give soul and colour to the dead or sick metals. In alchemical practice the soul was separated from certain 'bodies' by means of distillation, and after ascending to heaven (evaporation) it acquired a new, 'pneumatic' body by condensation (sublimation) in the head (*capitellum*) of the apparatus. A strong concentration of pneumatic spirit in a body, then, may yield a more-than-perfect medicine, which, when administered to the imperfect metals, will restore them to the perfection of silver and gold. In one of the most ancient alchemistic manuscripts we see the snake Ouroboros ('who bites his own tail'), the symbol of cosmic circulation, depicted together with a distilling apparatus in which on a microcosmic scale a similar process is taking place. A magic ring with a conjuration formula completes the picture (see figure 11).

Allegedly the inventor of this 'divine art' was Hermes Trismegistus (a Greek equivalent for the Egyptian god of wisdom, Thoth). The name survives in terms like 'hermetic art' and 'hermetically sealed.' Moses, being 'instructed in all the wisdom of the Egyptians', must have known this art and he was supposed to have taught it to his sister Miriam, to whom was attributed the invention of the water-bath, (used for gentle heating), the so-called 'bain-marie'. This Egyptian art (*chèmeia*; art from the land of Khem; black art) was further developed by Syrian and Persian scholars. The prefix 'al' (al-kimiya), introduced after the Arab conquest of the Middle East, implied no loss of continuity with the past: alchemy remained an art with the final aim of producing the Philosophers' Stone, the remedy for sick metals, or El-ixir, a remedy for sick people.

The medieval alchemists had no religious scruples about the legitimacy of their art. Provided one avoided black magic (incantations) and performed the Great Work in a spirit of devotion, God would favour a work that served to benefit mankind. As to the impossibility of transmutation of species, they did in general recognize this. They shared the organistic world view of their contemporaries; indeed, their hylozoïstic tendency made it even stronger than with the orthodox Aristotelians. But would not the fabrication of gold be a transmutation of species? No, it was argued, for the alchemist did not *make* gold; he just supported Nature in achieving her own aim: the production of gold, the perfect work of Nature. Moreover, as some alchemists pointed out, Aristotelian philosophy also admitted change of species, at least for plants and lower animals. Aristotle himself had recognized that living beings of a lower order did sometimes arise by spontaneous generation. The 13th century Spanish alchemist who wrote under the name of 'Geber' reminded the opponents of the Art that a worm turns into a fly.[3]

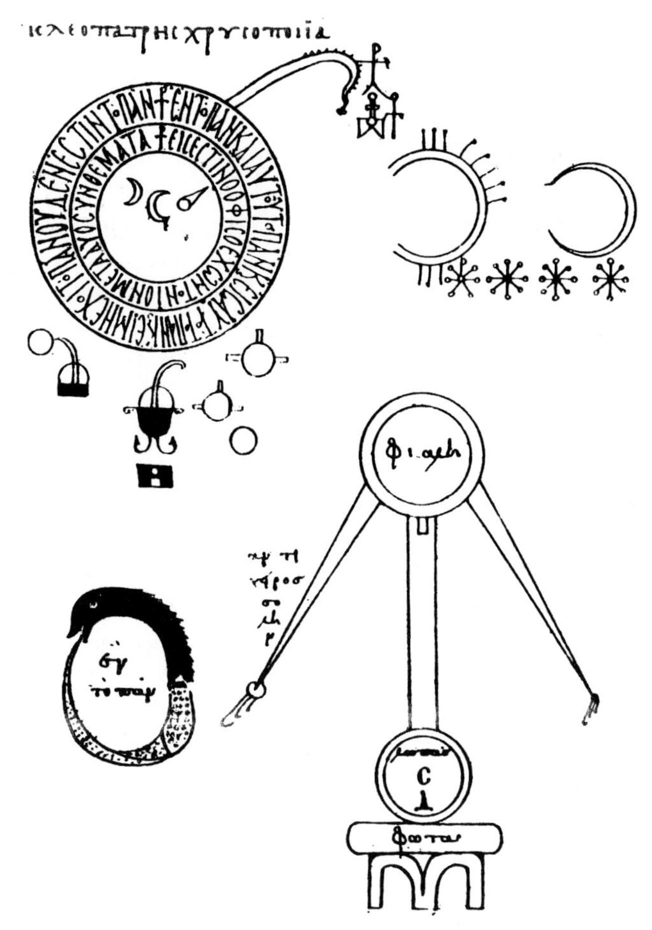

Fig. 11. Illustration from the 'Goldmaking Art of Cleopatra'. From: M. Berthelot, *Les Origines de l'Alchimie,* Paris, 1885.

As to the third objection: in spite of their procedures involving dirty and unhealthy manual work, the alchemists were proud to call themselves 'philosophers by fire'. Thus they defied prevalent prejudice, boldly asserting that a philosopher should use his hands. One of the worst things to say of a fellow alchemist — implying evidently that he had no laboratory experience — was that 'he had not made black his hands by coal.' It must have been extremely irksome for the School philosophers that these people, who at best cultivated a mechanical and not a liberal art, considered this art to be a *science* as well. A 15th century alchemist, Bernard of Trevisan, spoke (1453) of 'this worthy Art, this noble Science' which follows the true path chosen by Nature herself in her creation. He claimed that he worked 'by reliable experiments which I

Fig. 12. 'In the sweat of thy face shalt thou eat thy bread'

made with my hands and saw with my eyes', and thus succeeded four times in preparing the Philosophers' Stone.[4]

The position of the alchemists was not without its ambiguity. Although they interfered with nature, they maintained that they were doing no more than help Nature in her work of generating things — something which even Aristotle regarded as possible and legitimate.[5] They held a strongly organistic world view; yet inevitably they furthered mechanistic ideas, if only by the importance they attached to mechanical means (such as heating, dissolving, precipitating) and to the mechanician's art.

This ambiguity is evident also in their theories. In a thoroughly mechanistic chemistry the elementary constituents would continue their existence in their compounds without losing their own being, just as the letters of the alphabet are still present in all the thousands of words which they can form together. Moreover, on such a world view, some change may be effected without a strife of qualities, because constituents may 'hide' each other and hinder each other's activity.[6] Thus on the one hand one can stick to the idea that the elements in the compound show their activity by their qualities, yet on the other hand maintain that the qualities of the compound are not precisely those of the elements.

3. The Sulphur-Mercury Theory

According to the Arab alchemist who wrote under the name of Jabir-ibn-Hayyan (circa 9th - 10th century) all metals consist of two principles, sulphur and mercury. The duality, matter - form, female - male, passive - active, which played an important part in the Aristotelian philosophy of nature, was now applied to the four elements. The more 'active' elements (fire and air) constitute the so-called 'sulphur philosophorum', whereas the more passive ones (earth and water) form the 'mercury of the philosophers' (see figure 13).

The great difference from Aristotelian orthodoxy is that these two principles do *not* lose their identity when they unite to form their compounds (*viz.* metals and minerals), even though the metals are homogeneous to the eye: 'When sulphur and mercury unite to form one new substance, it is often thought that they are changed in essence and that a wholly new substance has been formed ... But they keep their own nature; their particles are finely divided and closely brought together, so that the product seems to be *one* to the eye. But if one could find a means to separate those particles from each other, it would turn out that they have preserved their natural Form and that nothing has been changed: *such* a transmutation is impossible for the philosophers.'[7]

The particles of the two principles are not absolutely simple; they are composed of the Aristotelian elements. In mercury, says Bartholomeus Anglicus (in his *De Proprietatibus Rerum*, circa 1240), earth and water are united 'in a strong and inseparable union'; they behave *as if* they were elementary.[8] That is: they are 'relative elements', which are so stable that they cannot be changed into one another. As the 13th century alchemist 'Geber' put it:

> Each of them is of the strongest composition and of uniform substance, and this because in them the parts of the earth are so strongly glued together with those of air, water and fire in their minimal parts, that none of them can leave the

Fig. 13. Allegorical representation of the two principles: the King standing on the sun, the Queen on the moon.

> other when dissolving takes place ... thus each of them is dissolved together with the other ...[9]

This is a corpuscular theory which rejects the absolute homogeneity believed in by the Aristotelians. The substantial Forms of the parts are still present under the rule of the substantial Form of the whole.[10] Consequently, though the whole compound is not merely the sum of its parts, the parts are still present in a certain mode. This shows a decidedly mechanistic trend of thought.

But the opposite, the organistic conception, is also implied in alchemical theory. In the treatise of Bernardus Trevisanus the two principles are called the 'two seeds', to be united in the mother womb (the vessel), which by its warmth heightens their virtues, so that the imperfect becomes perfect (see figure 14). These two seeds, however, are not made by Art, but provided by Nature (extracted from natural bodies):

Fig. 14. Allegorical representation of the generation of the Philosophers' stone. (From *Rosarium Philosophorum.* Francofurti: Jacobi 1550; John Read Collection, St Andrews University Library.)

> Art cannot create the seeds, nor the matter of the metals, but Nature first creates and next Art helps them. ... So put the red Man and the white Lady in the round chamber, heated by a small continual fire until the union takes place. Nature herself determines the spermatic natures and creates them; and afterwards Art, working on them, joins them together, thus following the purpose and intention of the natural spermatic virtue.[11]

There is here no question of *fabricating* gold, only of *generating* it naturally with human cooperation. Man accelerates the process of union by bringing together the male and female principles in the bridal chamber, and so the highest Metal, the Philosophers' Stone, is born. As Thomas Norton (1477) wrote:

> But two be Materialls, yet our Stone is one
> Betweene which two is such diversitie,
> ... Such as is found between Male and Female.[12]

In his *Faust* Goethe makes the same point in more colourful terms: 'There was a red Lion, a bold youngster, married to the Lily in the luke-warm bath; And then with an open fire, chased from one bridal chamber into another. Next, with motley colours, the young Queen appeared in the glass [vessel]'.[13] Thus the Philosophers' Stone is born as the result of a sexual union, which goes together with a change of colour: black, white, motley and red in sequence, or as they were poetically denoted: the raven, the swan, the peacock and the lion.

In another trend of thought it is emphasized that the base metals are on their way to become gold, because this is the final goal of Nature. But they remain unfinished or are hindered by circumstances, so that they contain impurities and lack thorough concoction, homogeneity and the right qualities. They are sick, and in need of the Medicine that cures their imperfection by its superabundance of 'virtues':

> By our Art we bring [the metals] one degree further than does Nature in the mines; for we make metals one degree more perfect, that is in the bloodlike redness ... the male body should be made more-than-perfect (*plusquamperfectum*) by the Art that follows Nature. Consequently by the more-than-perfect nature which it has acquired by Art, it can make perfect the imperfect metals.[14]

For this trend of thought the comparison with the development of a caterpillar into a butterfly was quite apt: in both cases the imperfect or the unfinished becomes perfect; in neither case is there a transmutation of one species into another.

> Thomas Norton claims that the 'growing' of metals takes place only underground, which is their 'natural place':
>
> > contrary place to nature causes strife,
> > as fishes out of water lose their life.
>
> Yet he remained much more within the pale of Aristotelian orthodoxy than did Trevisan with his far-going organistic theory. In Norton's opinion the metals grow out of their principles, but one should not speak of their 'seeds', as metals have no life:
>
> > Where Metalls be only Elementative,
> > Having noe seeds, nether feeling of life,
>
> they will not increase. The term 'grow', then, should not be taken too literally. The simile of fermentation, so popular with alchemists, is hardly applicable here. Most alchemists believed that the fermentative, self-multiplying power of the Stone was very high: 'I would change (*tingerem*) the sea into gold', said a 14th century adept, 'if it were mercury.'

Norton, more soberminded, stressed, in Aristotelian fashion, the strife of qualities which ensured that their right proportion might be attained:

> And soe of Alkimy the true foundation
> Is in *Composition* by wise graduation
> Of Heate and Colde, of Moist and Drye,
> Knowing other Qualities engendered thereby.

To Norton the transformation of base metals into gold implies the restoration of the right proportion of the elements. He complains that

> *Bacon* said that old Men did nothing hide
> But only Proportion, wherein was no guide.

But though he boasts of having prepared the Stone himself, he follows the example of the 'old Men' and hides the 'proportions', only advising his readers to follow Boethius, and to take care that the four elements have the right proportion of weights:

> Joyne them together also Aritmetically,
> By subtile Numbers proportionally.

He must have used the balance with the utmost care, judging by the picture of the laboratory which accompanies the manuscript of the *The Ordinall of Alkimy* (see Frontispiece).[15]

4. CORPUSCULAR THEORY

The most interesting aspect of alchemical theory is its qualitative character. The alchemists were keenly aware of the fact that ordinary brimstone and quicksilver are not the constituents of the metals. They considered them to be but the exemplars for the 'philosophical' sulphur and the 'philosophical' mercury. The philosophical sulphur is the hypothetical substratum of the qualities 'colour' and 'combustibility'; the philosophical mercury is the hypothetical cause of the typically metallic properties, fusibility and ductility. They are bearers of prominent qualities (qualities substantified) rather than concrete products of chemical analysis.

As metals show the qualities of these two principles, and yet appear to be homogeneous to our senses, this must be because the particles of these princples are very small and their union very close. This corpuscular aspect comes to the fore in the writings of 'Geber'. According to him, iron and copper can easily be calcinated because there are pores between their sulphur — and mercury — particles, so that fire can easily enter and convert the metal into ashes. As all metals consist of sulphur and mercury, their differences must be caused mainly by admixture of earthy particles, or by a loose texture and insufficient mixing of the particles, or by the difference of proportions of the two principles in the diverse metals. The transmutation of metals must therefore be achieved by purification, by closer union of the particles and by separation of the principles followed by re-union in the right proportion.[16] To Geber it is not in the first place a matter of giving the right *qualities*, but one of manipulating the material substances to which the fundamental qualities are attached.

In this way Geber goes in the direction of what might be termed a *qualitative corpuscular theory*. The practically elementary particles of sulphur and mercury have to 'explain' the qualities of their compounds *not* by their various shapes or motions (as was the case with ancient atomic theory), but by their being endowed with these qualities themselves. Thus the qualitative approach, which substantifies qualities, is the counterpart of the geometrical and mechanistic explanations which reduce sensible qualities to shape, size and motion of particles or connects them with numbers.

5. The Scientific Character of Alchemical Theory

One might ask: what is the value of such a theory? Wasn't it entirely wrong? A 19th century chemist voiced the opinion of most of his contemporaries, when in dealing with the sulphur-mercury theory, he exclaimed with indignation: 'How little effort was made to find the true chemical constituents, is eloquently testified by this theory.'[17] He saw only one ray of light in this medieval darkness: the alchemists were right at least in recognizing that cinnabar consists of sulphur and mercury (a praise hardly deserved, as they considered all 'minerals' to contain some sort of 'sulphur' and 'mercury'.) One is tempted to retort: How little effort has been made to find the true meaning of ancient theories, is eloquently testified by this verdict. If one makes a serious attempt to enter into the way of thought of that age, and into the factual knowledge available at that time, one cannot but conclude that the sulphur-mercury theory bears a truly *scientific* character.

There is much plausibility in reasoning on the following line: many substances have in common their combustibility; this common effect may have a common cause. Consequently, there must be a common material basis of combustibility in all combustible bodies. This general material basis may be called by the name of a wholly combustible substance, 'sulphur', which thus can serve as its symbol. Now iron is relatively combustible, so that it must contain much of that 'sulphur', and consequently little of the other ingredient of metals, 'mercury'. This is confirmed by its lack of two qualities that are considered in the (then) generally accepted definition of a metal as essential, *viz.* ductility and fusibility. (In the Middle Ages the founding of iron was not technically possible). Gold, on the other hand, is wholly incombustible (it does not 'rust') and thus contains little sulphur; hence it must contain much mercury. This conclusion is confirmed by its great ductility and its rather ready fusibility.[18] Just as a modern scientific theory is built upon a series of facts interpreted according to the prevailing framework of thought, so this medieval hypothesis about the composition of metals bases itself firmly upon facts, namely the qualities observed.

If a theory can be tested against a series of facts which are wholly independent of the first series of facts upon which it was based, and survive, its probability and its convincing power are considerably increased. This condition, to which we would submit a modern theory, is also fulfilled in the case of the sulphur-mercury theory. According to a generally accepted medieval belief (again supported by facts), substances that resemble each other will *attract* each other and mix together: oily or fatty substances are dissolved by oily liquids; watery liquids (like spirit of wine) are dissolved by other watery liquids (water). In this train of thought substances that have some kinship or affinity will combine most easily. The sulphur-mercury theory then turns out to be confirmed, because it is conformable not only to the principle that the same qualities must be caused by the same constituent, but also to the tenet that a common constituent urges two substances to attract each other. A substance, like iron, which contains little 'mercury', shows no affinity to quicksilver (which is extremely rich in 'mercury'). It is a well-known fact that quicksilver can be kept in iron vessels. Gold, however, which is rich in 'mercury', readily amalgamates with quicksilver: 'quicksilver is most friendly to other quicksilver, next to gold, and after this to silver.'[19]

On the other hand iron, which contains much 'sulphur', is easily united to sulphur (the reaction between finely powdered brimstone and iron filings is a well-known textbook experiment). Gold, which contains little 'sulphur' (it resists combustion), does not lose its lustre when exposed to sulphureous vapours.

In a 16th century exposition of Geber's work by Tauladanus, it is pointed out how easily mercury 'devours and gobbles up', gold, lead and tin and how difficult it is to combine it with iron. Some innate force urges bodies to unite with bodies of common origin by a bond of affinity and to 'enjoy to couple with them', whereas they have an 'antipathy' toward those with whom they have no affinity. The affinity is greater the more they share each other's nature: the more they share mercury's nature, the more mercury they contain. And as iron has a poor 'mercury' content in comparison with other metals and is rich in combustible 'sulphur' — and as quicksilver feels that iron contains little mercury and consists for a large part of combustible sulphur — it will reject iron which, as Geber (ch.64) says, 'it feels to be alien.' The more 'mercury' a metal contains, the more perfect it is; therefore the base metals, which contain little 'mercury', have to be 'restored to health' by an elixir derived from quicksilver.[20]

Here Tauladanus speaks about inorganic substances in biological terms. Yet the difference between living and non-living matter was fully recognized by most alchemists. In Norton's opinion the metals, in contrast to the animals, do not propagate themselves; they have neither seeds nor life: in non-living bodies the constituents are bound together more closely than in living beings.[21]

In alchemy, then, we meet with a wholly wrong theory which, nevertheless, shows the characteristics of a truly *scientific* theory. Firstly: it was adaptable to the generally prevailing natural philosophy. Although this did not admit the possibility of transmutation of species, the alchemists found a solution by maintaining that all metals belong to the same species and were all varieties or incomplete forms of it.

Secondly, it was in conformity with widely acknowledged facts. Substances seemingly not containing gold yielded gold in perceptible quantity after the processes of concentration involved in distillations, crystallizations, precipitations and cohobations. Moreover there were facts which seemed to prove that one metal could turn into another. The observation that iron, when immersed in certain Hungarian springs, seemingly turned into copper, was still accepted at its face value by some scholars in the 17th century.[22]

Finally, alchemy could claim to be built upon a series of facts, and to be supported by another series of facts independent of the first.

This may all be so, one might object, but the sulphur-mercury theory is simply untrue. Certainly, but that does not imply that it was not 'scientific'. Not all that is 'scientific' is necessarily true (and not all that is 'true' is 'scientific'!). Today we recognize that the starting-point was erroneous. It was a mistake to assume that the same quality (combustibility) always has the same material substratum, however plausible this may seem. Nature does not always follow our logic, though — to make things more confusing — it does indeed sometimes seem to do so. There is, for example, a common cause of acidity (the hydrogen ion). Sweetness, however, confronts us with an awkward situation: alcohol, with one hydroxyl-group (-OH) in its molecules, is not sweet; glycol, with two, and glycerol, with three, got their names because of their sweet taste; and glucose, with even more OH-groups, is 'grape-sugar'. Yet — quite unexpectedly — the sweetest of all, saccharine, has no OH-groups at all. Moreover, as to bitterness, there is not the slightest chemical relation between quinine and Epsom salt ('bitter salt'; magnesium sulphate). *Experience* alone has to teach us whether the principle 'similar properties indicate some similarity of composition' may be applied. So the objection is perfectly right, but we should realize that we are speaking thus with hindsight, after an age-long experience in chemistry by which we have had to learn which ways, though promising at first, turned into blind alleys.

In order to identify the constituents of a compound, chemistry follows two methods.

1. The properties of a compound are studied and then rational analysis is used to find an answer to the question; 'what are the constituents?' (In modern chemistry this procedure is followed, in principle, when determining the presence of elements or of groups of elements by means of spectral analysis).

2. The compound is dissected by chemical procedures, and the end-products of chemical analysis are considered to be the elementary constituents of the analyzed

substance. As long as the concept of element was that of an absolutely simple thing, this latter method was unsatisfactory, as there was no guarantee that better methods of analysis would not later split up a substance that was at first considered to be undecomposable. We can understand therefore why it seemed more satisfactory to find out which of the qualities show a certain permanency — that is, are common both to the elements and their several compounds — as these would indicate which substances entered into the compound and then remained there. To this way of thinking, combustibility indicated the presence of the 'combustible matter', the 'sulphur', in its compounds. This is a method which, fundamentally, is still employed in modern chemistry. The only — and important — difference is that the properties now chosen for identification are, from the 'qualitative' standpoint, common to *all* elements and their compounds, though with an 'intensity' or quantitative degree characteristic of each element. 'Carbon' (C) has the atomic weight 12, and oxygen (O) 16; and in carbon monoxide (CO) these numbers persist in the molecular weight 28 (12 + 16). Similarly, chemical compounds will manifest the spectral lines and the nuclear charges of their components.

We can see that the alchemists were already confronted with a difficult problem, which runs through the history of chemistry from ancient times up to the present day, *viz.* the need to distinguish between the concept of the *simple body* (the 'concrete' substance which is the end product of chemical analysis) and the rather abstract concept of a *chemical element*, which is an entity common to the simple body and the compounds made out of it.[23] The simple bodies graphite and diamond both yield carbonic acid gas upon complete combustion. Consequently, this product can hardly be called graphite oxide or diamond oxide: it is the oxide of something that is common to graphite and diamond: the combination of an 'abstract' substance 'carbon' with 'oxygen'. Graphite and diamond are different simple bodies: the former is black and very soft, the latter is colourless and extremely hard. But both of them contain one and the same chemical element, carbon, which is neither soft nor hard, but which has the atomic weight 12 and the 'order number' 8 in the Periodic System of the Elements. This system then is not a system of simple bodies, but a system of elements, as Mendeléev (1869) clearly recognized (see chapter II).

Just as it would have been wrong to identify 'philosophical sulphur' with the yellow brimstone found in nature, so it would now be wrong to identify the chemical (= 'philosophical') element 'sulphur' (atomic weight 32; atomic number 16) with the yellow stuff found in nature, or, for that matter with either rhombic or monoclinic sulphur, or with 'plastic' sulphur or any other of the concrete substances corresponding with it.[24]

The sulphur-mercury theory eventually succumbed because of its inconsistencies and ambiguities, and most of all, as a result of rapidly growing chemical experience. It became more and more evident that, for all its merits, it was not true to nature. In particular, in its more ambitious form as put forward by Paracelsus (1493 - 1541), it was vulnerable to attacks by the chemists themselves. Paracelsus extended the number of principles by adding *sal* as the principle of taste and solidity. According to him disintegration by heat of fire shows that *all* chemical bodies consist of a volatile part (mercury or spirit), an oily, combustible part (sulphur or soul) and a solid residue (*sal*), the body.[25] Again these were symbols of important properties bound to material bearers.

Paracelsus' theory was not restricted to metals and minerals; *all* substances were supposed to consist of these three principles. On the one hand they are the fictitious substrates of qualities; on the other hand they are the concrete substances, products of chemical decomposition. All three are fully present in the compounds, albeit invisibly as they are 'painted over with life' (*mit dem Leben übermalet*), and only when their life is taken away (e.g. by heating) do they come into the open. This way of expressing himself shows how attached Paracelsus was to a hylozoïstic world view.

Of course Paracelsus did not manage to separate all substances into three components. In order to keep to the number three he had to make an artificial separation into three fractions when distilling organic substances (in many cases he could as well have separated them into five fractions). And on the other hand it was impossible to separate stones or metals (which must also be compounds of the three principles) into three fractions. But Paracelsus was not at a loss. If a compound would not separate into three 'principles', he simply added some alien substance to it in order to 'open it up', and in next performing the procedure of separation he conveniently ignored the fact that he had added something to it earlier. So he could now state that he had obtained the three from the original body in all cases. In this way, for example, he 'extracted' a 'vitriol' ($CuSO_4$) from copper (Cu), after first adding vitriolic acid (H_2SO_4).[26]

> A second argument was the fact that Hermes Trismegistus said that all things consist of three: body, soul and spirit. Without saying so, he meant sal, sulphur and spirit. In the *Book of Lambspring* (which does not yet have the doctrine of three principles) nevertheless it is said 'two fishes swim in our sea'; the hermetic trichotomy thus competes with Aristotelian dichotomy (see figure 15).

Paracelsus' strong conviction of the threefold character of all things was connected with his belief that mercury, sulphur and sal were analogous to spirit, soul and body. These must be present in all species because God, who is Triune, made them.[27] Consequently, 'all art that seeks more than these is fundamentally false.'[28] Evidently, the result of his chemical dissection was fixed beforehand.

Fig. 15. 'Two fishes swimming in our sea' From: Lambsprinck (pseudonym of Lampert Spring), *Traité de la Pierre Philosophale* (translated from German).

Paracelsus' theory was not only utterly weak as to the chemical *facts*, it was also extremely weak in its chemical *fictions*. The alchemists had already felt obliged not to stick to one clearly defined 'sulphur philosophorum'; they had introduced a distinction between 'combustible' and 'incombustible' sulphur, and between sulphur that gave colour and sulphur that did not give colour. In this way the explanatory value of the theory was considerably weakened. As often happens when a theory meets with new difficulties and objections, it became involved in arbitrary complications and additions. Paracelsus went even further than the alchemists when he proclaimed that *every* substance has its own specific mercury, sulphur and *salt*. In this way the total number of mercuries, sulphurs and *salia* would be thrice that of the substances they had to account for. The explanatory value of the theory thus became practically nil: a scientific explanation should lead to a reduction rather than a multiplication of fundamental entities. If simplicity be not the mark of nature, it should at any rate be as much as possible the mark of the science of nature.

Small wonder then that Paracelsus' theory was demolished by the devastating critique of Robert Boyle, who in his *Sceptical Chymist* (1661) demonstrated that it was a fiction which evaporated into thin air when confronted by experimental facts.

The alchemists formed a far from homogeneous group. Some of them were pragmatic chemists searching for medicaments to cure sick metals or to restore the health of sick people. Their goal was the Philosophers' Stone or the Elixir of Life, and they claimed to work for the benefit of mankind. Some of them were soberminded scientists striving only after knowledge of the world. Most of them said that their aim could not be reached without a special divine grace. Others were mere impostors, who claimed to possess the secret of the Stone and who knew how to extort money from their royal protectors with a promise to satisfy their need for gold. Others again were magicians who wanted to rule over nature and over their fellow men — 'a great magician is a mighty god' (Marlowe) — and they used incantations to force the demonic powers of nature to help them achieve their work.

Why do they still fascinate us so strongly, even though we know that their work ended in failure? Probably because in many of us there is something of Doctor Faustus, who wanted to comprehend the incomprehensible and to achieve absolute power over nature. They often tell us that they had *almost* reached their goal, but at the last moment the vessel burst and all their labour was lost. They stubbornly refused to give up their search for the Absolute, just as we too, when our most beautiful systems crumble, start again building new ones on the ruins of the old.

Here again modern science has something in common with the alchemists. One problem solved immediately gives rise to new ones. Our thirst for knowledge is never fully quenched; the horizon we are chasing is ever receding. In our quest for Truth we too, seek for a Philosophers' Stone: each great discovery, each illuminating insight is a milestone on the path of Science, but it is not yet 'the Philosophers' Stone' for that is always the *next* milestone.

IV. THE UNDYING FIRE

QUALITATIVE THEORIES OF STAHL AND LAVOISIER

1. CORPUSCULARIAN THEORIES 99
2. PHLOGISTIC THEORY 100
3. LAVOISIER 'PHLOGISTONIST' 103
4. THE THEORY OF 'CALORIC' 105
5. PHLOGISTON RE-KINDLED 113

1. CORPUSCULARIAN THEORIES

After Paracelsus qualitative explanations in chemistry lost ground because of the development of corpuscularian theories. These theories, so warmly welcomed at the beginning of the 17th century that they were called 'the key to almost all natural science',[1] turned out however to be rather sterile. They interpreted the phenomena by using notions and terms of their own, but — except in some very limited fields — they could not predict new phenomena. In those cases, that where the compound was separated by heating into two other substances, it did seem that these already existed before the separation, in particular if it was possible to reunite them into the original compound. Ammonia [NH_3] and 'spirit of salt' [HCl] formed 'sal ammoniacum' [NH_4Cl], and the reverse could also be effected by heating the latter compound.[2] Similarly when Angelo Sala (1617) dissected copper vitriol [$CuSO_4 \cdot 5H_2O$] into copper-ash [CuO], vitriolic acid [SO_3] and water [H_2O] and synthesized it again from these constituents, the conclusion seemed justified that, in spite of the temporary loss of their qualities, small particles of copper and vitriolic 'spirit' were hidden in copper vitriol. The presence of finely divided copper in a solution of copper vitriol in water was proved by putting a piece of iron in the solution: copper then precipitated on the iron.

Robert Boyle (1627 - 1691), elaborating chemical corpuscular philosophy, brought together many similar examples, for a large part borrowed from his predecessors Sennert, Sala, Billich and Jungius. He assumed an unlimited number of kinds of particles differing in shape and size. In his *Sceptical Chymist* he successfully demolished the old qualitative theories of the four elements of Aristotle, and the three principles of Paracelsus. The old elements were discarded but he did not venture to introduce new ones; he gave plausible 'illustrations' of how things *could* happen in the world of minute particles behind the world of phenomena, but no phenomenon could be *deduced* from the hypotheses suggested (sometimes more than one 'conjecture' could be put forward for the same phenomenon). No comprehensive theory bound chemical facts together into a system that made predictions possible; the theory was too adaptable and too vague. The only solid conclusion was that chemical reactions took place between particles of unknown forms and size; it was indeed the theory of a 'sceptical' chemist.

Small wonder then that after Descartes and Boyle corpuscular theory resumed a more or less qualitative character: the particles were endowed with qualities other than the merely geometrical properties of shape and size (and motion and arrangement).

Georg Ernest Stahl (1660 - 1734) repudiated the speculations of the mechanistic philosophers. Though he, too, firmly believed that all substances consist of small particles (corpuscula, moleculae), he denied that we will ever know their size and form. Qualities may have their mechanical cause (particles of acid may have sharp points), but we cannot be sure of it; so we should not make any apodictical statement about size and form and we should not use them in our chemical interpretations. That is why he reverted to the more ancient qualitative theory, although in a corpuscularian sense, i.e. rejecting the Aristotelian conception of perfect homogeneity of compounds. In his opinion compounds are combinations of different corpuscles which retain their nature.[3]

2. Phlogistic Theory

Stahl wanted chemical theory to be built upon experience and experiments; it should be constructed *a posteriori* and not *a priori*.[4] In his 'phlogistic' system the two characteristics of the alchemical sulphur-mercury theory reappeared: prominent qualities — in particular combustibility (phlogiston = the combustible) — were hypostasized in material substances, and, secondly, chemical binding was explained by similarity of composition and qualities.

Without having isolated the combustibility-principle, Stahl simply stated that everybody knows that in base metals there is something that can make them burn[5] and that everybody can see that the burnt metals are 'reduced' to their metallicity by means of coal. Therefore some material inflammable principle, the indestructible basis of combustibility, must have been transferred from the coal to the metal ash, and — when the metal was burning — some material principle must have been yielded by it:[6]

$$\text{metal} \rightarrow \text{metal calx} + \text{phlogiston} \uparrow$$
$$\text{metal calx} + \text{phlogiston (coal)} \rightarrow \text{metal}$$

These burned metals can only recover their metallic character if inflammable matter is restored to them. In that case they will immediately have their former complete composition, fusibility, and ductility of which they had been deprived by the abstraction of their inflammable principle.[7]

When a metal is dissolved by an acid, its phlogiston is extracted first, which, in Stahlist conception, explains why a salt like copper vitriol — when heated — yields the metal calx and not the metal itself [$CuSO_4 \rightarrow CuO + SO_3$].

As Stahl's interpretation of combustion was the most conspicuous feature of his theory, it was called after the supposed material basis of this property. He said that nothing is more reasonable than to call this principle after the effects it causes in the compounds: 'the matter and principle of Fire, — not Fire itself — I began to call *phlogiston*, namely the basis of Fire (*primum Igniscibile*), inflammable, the *Principle* directly and eminently fit to receive Heat.'[8]

Though Stahl did not identify phlogiston with any known substance and could not isolate it, he believed that in particular soot[9] (which after combustion left practically no residue) was very rich in it. In his *Specimen Beccherianum* (1703) one of the main theses is: 'That phlogiston, that inflammable matter, is indeed something corporeal, and that it effectively enters into the composition of the metals.'[10]

Unlike the 'sulphur philosophicum' of the alchemists it was a definite substance and not a common name for a group of various 'sulphurs'. Though he recognized that there was some resemblance to the alchemical 'sulphur' — (he sometimes spoke of 'phlogiston or the sulphureous principle'[11]) — he emphasized that it was only the combustible part of common sulphur (brimstone) which, as he said, consists of an 'acid and phlogiston'.[12] It was precisely the demonstration that brimstone consists of phlogiston and vitriolic acid that was his most cherished proof of the truth of his theory.

He advanced it already in his first chemical work (*Zymotechnia*, 1697). Vitriolic acid [= sulphuric acid] was 'fixed' (made into a solid compound) with potash, so that potassium sulphate [K_2SO_4] was formed, and the product was fused with charcoal. The result was 'liver of sulphur' [potassium polysulphide], from the solution of which, — after adding an acid, — common sulphur precipitated. Evidently, the phlogiston of the charcoal had been combined with the vitriolic acid and thus sulphur had been formed.

Liver of sulphur can only be made directly by heating sulphur [S] with potash [K_2CO_3]. Consequently, liver of sulphur is formed from

sulphur + potash

as well as from

(vitriolic acid + potash) + phlogiston.

By subtracting 'potash' in both equations, it follows that:

sulphur = vitriolic acid + phlogiston.

This discovery was so important to Stahl that he proudly described it several times, mentioning even the date on which he made it (April 1698).[13]

The second tenet Stahl shared with the alchemists is 'like attracts like'.[14] Salt molecules, which consist of the principles 'earth' and 'water', when brought into contact with water, are attracted into it thanks to the aqueous principle they have in common with it, and thus a solution is formed.[15] Similarly 'what nobody would believe', an acid ('sal acidum') and 'our fiery principle' react to form a compound as 'earth with earth'.[16] Our 'phlogistic matter' (e.g. sulphur) has an aversion of water and therefore the connection with an acid cannot come from the side of the aquosity, but from the side of the 'earthy' substance which is part of the acid.[17]

Stahl's theory was less naive than that of the alchemists. He did not hold that the atoms or particles of phlogiston themselves possess the quality they explain, but rather that they are only the cause of it. Phlogiston is a colour-giving, not a coloured substance. Consequently, he assigned a role to mechanical explanations: in his opinion the 'formal' cause of colour is reflexion and refraction; the 'material' cause is phlogiston. This resembles Newton's conception of light particles which have no colour themselves but *cause* the effect that we experience.

Liquidity is wholly explained by Stahl in a mechanistic way, when it is attributed to a greater movement of particles; ice and water are said to be substantially the same thing.

The great merit of Stahl's theory was that it connected diverse phenomena that at first sight seem quite different: combustion (with flame and loss of weight), calcification or rusting (without flame) and even respiration were brought under the common heading of 'loss of phlogiston'. Moreover the relation between what is at present called 'oxidation' and its opposite 'reduction', was established. Where there is in fact addition of oxygen, Stahl assumed subtraction of phlogiston:

metal - phlogiston → metal calx; metal calx + phlogiston → metal

metal + oxygen → metal oxide; metal oxide - oxygen → metal

Similarly, Stahl pointed out that there is a decrease of phlogiston-content from sulphur to (what is now called) sulphureous acid to sulphuric acid. That is, the sequence of decrease of phlogiston-content is the same as the increase of oxygen-content in modern chemistry.

After the discovery of 'inflammable gaz' (hydrogen) by Cavendish in 1766, its discoverer took it to be phlogiston.

metal [i.e. metal calx + φ [=hydrogen]] + sulphuric acid
$\rightarrow$ a salt [i.e. metal calx + acid] + φ

metal calx + φ $\rightarrow$ metal

The water vapour yielded by the latter reaction was easily overlooked.

It is understandable that the phlogiston theory tenaciously withstood the attacks by the new chemistry of Lavoisier. Even chemists who by their experiments on gases (and in particular on oxygen) had contributed to its downfall, remained its faithful supporters. Combustion might, in their opinion, henceforth imply 'combination with oxygen', but it meant to them at the same time also 'yielding of phlogiston'. It was even an asset that the increase of weight when metals are calcinated found now an explanation by ascribing it to combination with oxygen: phlogiston could be held to be practically weightless, so that neither its addition could increase the weight, nor its loss decrease it. Hitherto there had been the rather embarrassing situation that the only way to account for the increase of weight at calcination of metals seemed to be to attribute a negative weight to phlogiston.

3. Lavoisier 'Phlogistonist'

To give up phlogiston seemed to imply abandoning the principle that common qualities have a common cause. Lavoisier himself, however, did not go so far; as on other topics (the choice of elements), he took here a rather inconsistent attitude. In a certain sense he, the great opponent of Stahl's theory, was himself a 'phlogistonist'. He could not withstand the attraction of the qualitative approach. *Oxygen* and its reactions form the core of his chemical theory, and the very name of this 'element' already bears witness to the qualitative aspect of the new theory: it means acid-former. Lavoisier traced oxygen in many acids and he thought that their common property of acidity was caused by a constituent they had in common, *viz.* the 'acidifying principle' (*principe acidifiant*; *principe oxygène*).[18] It should be realized that to him the acids were just oxides and did not contain hydrogen: by 'sulphuric acid' he meant sulphur oxide [SO_3 in modern notation], that is the anhydride of sulphuric acid [H_2SO_4 in modern notation].

Besides a constituent specific for each acid, there must be a common constituent for all of them, the common cause of acidity: 'We should therefore distinguish in every acid the acidifiable basis ... and the acidifying principle, that is: oxygen.'[19] In 1776 he declared that what had been at first but a rather probable conjecture, soon converted itself into a certainty, when he applied experiment to theory, 'and today I can advance with certainty that ... the purest part of the air enters into the composition of all acids without exception, and that this substance constitutes their acidity.'[20]

In spite of the fact that he could not dissect the acid of sea salt [*viz.* HCl!], he felt himself compelled by *analogical reasoning* to assume that oxygen was one of its components. Inevitably, this led him further astray, because he had thus to assume a hypothetical element, 'murium', of which the acid of sea salt was an oxide. This was a conclusion by the man who solemnly proclaimed that to himself he had set as a rigorous law, never to fill up the silence of the facts ('... la loi rigoureuse ... de ne jamais suppléer au silence des faits').[21] Here fiction, built upon an in itself legitimate analogical reasoning, led to the belief in spurious 'facts'.

This, however, was not yet the end. When 'acid of salt' is treated with an oxidizing substance [manganese peroxide] chlorine is formed. Lavoisier called this (in fact elementary!) substance oxy-muriatic acid. This pretendedly 'super-acid', however, shows no acidity at all, so that Lavoisier's conception here contradicted his own principles.

Yet, there came an occasion when Lavoisier saw that he was forced by hard facts to recognize that analogy may lead not only to important discoveries, but also to serious errors. Having found that carbon, phosphorus and sulphur, when burned form acids, 'analogy invincibly brought me to the conclusion that the combustion of 'inflammable air' [hydrogen], too, must produce an acid'. He found, however, that the reaction yielded nothing but water (1781).[22] Here Lavoisier was manifestly confronted with the truth of his own statement that 'imagination continually tries to lead us beyond truth; self-love and self-confidence ... invite us to derive conclusions which are not immediately derived from the facts; so that we in a certain sense are interested in seducing ourselves.'[23] 'The only means to avoid these aberrations is to suppress or at least to simplify as much as possible reasoning, which is of us and which alone may mislead us; and to put it continually to the test of experience; to keep to the facts which are but data (*données*) of nature and cannot deceive us ... Convinced of these truths, I have imposed upon myself the law to proceed always from the known to the unknown, not to deduce any consequence which does not derive immediately from experiments and observations.'[24]

Beautiful words, lofty intentions, inspired by the indestructible optimism of the 18th century Enlightenment. Lavoisier's own work contains many examples of 'invincible' logic and, in particular, of 'analogical reasoning' (which is also 'reasoning of us') which were a guide to new discoveries. On the other hand, there were also many occasions upon which they led him into error, for analogical reasoning can be a bane as well as a blessing to science (H. Davy).

> Lavoisier went still further in his qualitative conceptions. He explained the fact that, though lower metallic oxides show no acid character, sometimes the increase of oxygen content when they are oxidized to higher oxides, makes them into acids (e.g. first degree of oxigenation in 'oxide de tungstène' (*chaux de tungstène*) and second degree of oxigenation in 'acide tungstique').

> The red colour of the higher oxide of lead ['red lead', Pb_3O_4] and of mercuric oxide [HgO] and oxide of iron [Fe_2O_3] were attributed to their oxygen content, and this belief seemed to find support by the fact that 'air éminemment respirable' imparted 'in the same manner' a red colour to blood (1777).[25]

Though in general Lavoisier avoided the notion 'chemical affinity' as being too theoretical and too speculative, on some occasions he mentioned it. He does so in the Stahlist way: if two bodies easily unite with a third one, they must also have a mutual affinity and thus easily combine.[26] In his system this explains why metal oxides (basic oxides) easily combine with acid oxides (acid anhydrides) to form salts. In Lavoisier's opinion, when a metal seems to be dissolved by an acid, there is no direct reaction: first the metal is oxidized[27] and then the metal oxide combines with the acid; metals cannot immediately combine with acids: 'Combustible substances, being in general those with a great appetite for oxygen, must have affinity amongst each other, so that they tend to combine one with another: quae sunt eadem uni tertio, sunt eadem inter se.'[28]

In particular Lavoisier's theory of caloric clearly demonstrates a 'Stahlistic' way of thought, and it shows how right the 17th century philosopher Nicole Malebranche was when he said that 'prejudices are not abandoned like an old cloak.'

4. The Theory of 'Caloric'

Let us now follow Lavoisier in the general development of phlogiston into caloric matter and of the definition of combustion as loss of 'caloric' to that of union with the 'oxygenous principle'.

Lavoisier's work on gases ('pneumatic chemistry') was based upon what Joseph Black (Glasgow and Edinburgh) had begun earlier in the 18th century. This most able chemist and physician had discovered the relations between quicklime [CaO] and calcareous earth [chalk; $CaCO_3$]. The latter, according to Black, is the former on which a certain 'air' (gas) has been *fixed*. Black also discovered that this 'fixed air' [carbon dioxide] is a kind of acid, and that it is identical with the 'air' yielded when chalk is brought into contact with sulphuric acid or nitric acid.[29]

At first Lavoisier thought that the air 'fixed' in 'red lead' is the same as Black's air [CO_2] fixed in calcareous earth.[30] At that time it puzzled him that quicklime [CaO], which is 'calcareous earth minus fixed air' [$CaCO_3$ minus CO_2 produces CaO], has properties in common with metal calces, which are obtained by adding air [*viz.* oxygen!] to the metal.

In 1773 Lavoisier heated red lead with charcoal and found that an 'elastic fluid' was yielded [$Pb_3O_4 + 4C \rightarrow 3Pb + 4CO\uparrow$]. He then wondered whether the charcoal yields phlogiston to the metal calx (red lead) — as the Stahlists say — or, whether it enters into the elastic fluid.[31] He supposes then that each 'elastic fluid' (gas) may be composed of some solid or liquid substance together with a combustible principle, or perhaps even the pure 'matter of heat'. He also wonders whether the elastic state depends on this combination, whereas the substance that was fixed in the metal calx and thereby augmented *its weight*, was (according to this 'hypothesis') not an elastic fluid but the solid part of an elastic fluid, without its inflammable principle.[32] That is: according to Lavoisier's hypothesis a gas consists of some solid 'basis' plus 'matter of heat' or 'inflammable principle'.

This uncertainty, however, was not to last very long, for in that same year 1773 he heard from Priestley about his discovery of 'dephlogisticated air' [oxygen], so that the 'principle of inflammability' could no longer be identified with the 'principle of elasticity'. In a memoir presented to the Académie des Sciences in 1777 Lavoisier described an experiment that shows that atmospheric air consists of two 'elastic fluids', a 'mephitic air' and a 'respirable air'.[33] In the same year, in his memoir 'On combustion in general', he made an important step forward in the theory of combustion. He pointed out that *systems* in physics are nothing but instruments of knowledge, methods of approximation, which help to solve problems and which, when given the lie by experience, must be changed until they lead to the knowledge of the true laws of nature.[34] Consequently, he now proposes a new 'theory of combustion', or, so he adds, rather a 'hypothesis', which explains combustion (with flame), calcination (without flame) and the respiration of animals. He recognizes that in his earlier publication[35] he had little confidence in his own insight (that went against Stahl), but that now *facts* have been multiplied, and that he now writes 'in downright opposition to Stahl's theory'. (Yet, as we shall see, this 'downright opposition' boils down to a kind of phlogiston theory under a new name).

He puts forward some 'constant phenomena which appear to be laws from which Nature never deviates':[36]

1. Each combustion goes together with development of matter of fire or light. Just as with Stahl, combustion evidently is always accompanied by the yielding of some invisible matter. Even its name, 'matter of fire' (*matière du feu*), is just a translation of the Greek 'phlogiston'. According to Lavoisier this matter, however, is not yielded by the metal, for the combustion does not take place in vacuo. (This was a strong point: if the phlogiston were yielded by the metal, there would be no reason for the lack of combustion in a vessel devoid of air. The Stahlists could only answer that combustion required both, combination with oxygen as well as the yielding of phlogiston. In this way the difference between Lavoisier and his opponents was not so great).

2. Combustion takes place in the presence of only one peculiar species of air, which Priestley called 'dephlogisticated air' but which here is called 'pure air' (*air pur*).

3. Each combustion, then, goes together with the destruction or *decomposition* of 'pure air', whereas the substance that is burned increases its weight precisely in proportion to the decomposed air. That is: though combustion is loss of matter of fire [and this is also what Stahl said!, RH], this substance is [in opposition to what Stahl said!] yielded not by the sulphur or charcoal or metal or some other combustible body, but by the oxygen *gas*. Moreover, at the same time the basis of oxygen unites with the burning substance [to form oxides, RH]. In this stage of development of his theory Lavoisier still recognizes that combustion of non-metals (which then form acids of sulphur, phosphorus, etc.) and calcination of metals (which brings forth calces) are 'explained in a satisfactory way by the hypothesis of Stahl'.[37] But one has then to suppose with Stahl that there is matter of fire, or phlogiston, fixed *in* the metals (and in the sulphur or any other substance that Stahl considers 'combustible'), whereas he, Lavoisier, holds '*the opposite* hypothesis': the 'matter of fire or phlogiston' is not in those bodies that are called 'combustible'; therefore, if his own hypothesis is correct, 'the system of Stahl will be shaken in its foundations'.

The 'matter of fire', in Lavoisier's opinion, is a very subtle elastic fluid which seeks an equilibrium by freely flowing from one body to another, and which can enter into chemical combination; it is the solvent for many substances (when they are melting), as water is a solvent for salts and acids are solvents for metals. The bodies that combine with this elastic matter 'partly lose their properties and acquire *new ones*, by which they approach to matter of fire.'

We see here again the qualitative character of the new theory: bodies combining with the extremely elastic matter of fire become elastic themselves. As Lavoisier put it: 'every species of air is the result of some solid or liquid body with matter of fire or light: it is to this combination that aeriform fluids owe their elasticity, their specific lightness, their subtlety and all other qualities by which they approach matter of fire'.[38]

'Pure air' then is a compound of matter of fire: in this compound fire-matter is the 'solvent' and another substance — the 'basis of pure air' — is the dissolved body. When another substance — a metal for instance — is added, and the 'basis of pure air' has a greater affinity for the latter, then the matter of fire will be set free and fully resume its character, whereas phenomena of heat, flame or light accompany this reaction:[39]

[basis of pure air + fire matter] + metal → [basis of pure air + metal] + fire matter↑

Thus each gas — and in particular 'pure air' (which he later was to call oxygen gas (*gaz oxygène*) — is a combination of some basis with 'matter of fire'. When the basis of pure air unites with a metal, the product need not be volatile, for the basis itself was solid or fluid.

We should realize that this implies that oxygen gas ('pure air') is not the simple body 'oxygen' — which evidently has never been isolated — but a compound of elementary oxygen with fire matter.

All this leads him to the surprising conclusion that 'pure air ... is the true combustible body, and perhaps even the only one in nature'.[40] And that we need no longer assume a large amount of fire fixed in those bodies that 'people call combustible' in spite of their being solid (and thus without fire matter).

Now at this stage of development Lavoisier's theory may be the opposite of Stahl's in placing the phlogiston not in what we usually call combustible bodies but in the oxygen gas, — yet, his definition of combustion is wholly Stahlistic: the body that yields phlogiston is called 'combustible', and not the body combining with oxygen.

Lavoisier emphasizes that *all* gaseous (aeriform) bodies are compounds of matter of fire. The properties 'elasticity' and 'fluidity' are the marks of its presence in a compound: 'Matter of fire is the most subtle, most elastic and thinnest of all fluids and it must impart part of its qualities to substances with which it unites; just as solutions of salts in water preserve part of the qualities of water, so a solution by fire must preserve part of the qualities of fire.'[41]

According to Lavoisier, it now becomes evident why there is no combustion in vacuo, nor in an aeriform compound (a gas) in which the matter of fire has a very great affinity to the basis with which it is combined (nitrogen gas may contain much fire but it does not let it go easily).[42]

Moreover, 'it is now unnecessary to assume a large quantity of matter of fire in [solid, RH] substances that do not possess a single quality resembling that of matter of fire', says Lavoisier. Thus he underlines this argument against Stahl's theory by an argument of a typically qualitative (Stahlist) character.

And, finally, he reminds his readers that he is now under no obligation to maintain (as Stahl did) that 'bodies that increase in weight [as happens in calcination], lose part of their substance.'

Lavoisier ends this essay by saying that he attacks Stahl's theory without claiming to have substituted it by a 'strictly proven theory', but only by a hypothesis which 'seems to me more probable, more conformable to the laws of nature, less artificial and less self-contradictory.'[43]

This was true; nevertheless many difficulties remained. What to say of the reaction between iron and sulphur? [$Fe + S \rightarrow FeS$ + heat]. There is a violent development of heat, though both substances are solid and dense bodies. But Lavoisier easily overlooked such disturbing cases, as he hardly took notice of other compounds than those of oxygen. In his textbook *Traité élémentaire de Chimie* about 900 compounds are mentioned, of which 850 are oxygen compounds. His follower Gaspard Monge, however, pointed out that, since, according to Lavoisier's definition, each reaction yielding much heat, — for example the reaction of quicklime with water -, should be called a combustion, the reactions of metals with sulphur should also be called 'combustion'. Lavoisier mentioned Monge's critique without denying that it was fundamentally right![44] The 'Hollandish chemists', too, then rightly pointed out that reactions with sulphur as well as those with oxygen should be called 'combustion'.[45] In this way the analogy of sulphur with oxygen was for the first time established.

> The analogy was to be elaborated by Berzelius in his conception of sulpho-salts (K_3SbS_3) analogical to oxo-salts (K_3SbO_3). He also compared with them the chloro-salts, thus stating a parallel between oxygen, sulphur and chlorine as having all three a strongly electronegative character.[46]
>
> This was an important anticipation of the modern theory of oxidation, which, as we will point out below, has much in common with the phlogiston theory of Stahl.

Somewhat later (1783) Lavoisier made the final step in the development of his theory of combustion. His 'Réflexions sur le Phlogistique' was a frontal attack on phlogiston, which he had treated until then with reticence and even a certain respect.[47] He now points out that he starts from a simple principle, *viz.* that 'air pur' consists of a particular principle which is specific for it and which is its basis (the 'oxygenous principle'), and the principle of fire and heat. He declares that this enables him to explain all chemical phenomena in a satisfactory way without the help of phlogiston, so that it is 'infinitely probable that this principle does not exist, that it is a hypothetical being, a supposition made without good reason,' and he adds that 'good logic demands that we should not multiply entities without necessity'. 'Phenomena can be better accounted for without phlogiston than with phlogiston.' He calls it now 'an error fatal to chemistry, which has considerably retarded its progress by the bad manner of philosophizing it introduced therein.'[48] In his opinion the phlogisticists attributed qualities to the need of the moment: 'it is a vague principle ... which adapts itself to all explanations in which one wants to use it. Sometimes it has weight, sometimes not; sometimes it is free fire, at other times it is fire bound to the element earth, and it explains causticity and non-causticity, diaphaneity and opacity ... it is a veritable Proteus which changes each moment.'[49]

Bold language from a man who introduced 'caloric' (matter of heat) into chemistry, though it soon became evident that phenomena can be better explained without it! (He wrote with Laplace a treatise on heat, in which he stuck to heat as a substance, whereas in Laplace's chapters heat is treated as kinetic energy of molecules).[50]

Bold language also for a man who by his substantification of qualities and by his conception of combustion as a decomposition, interpreted the recently discovered phenomena of combustion 'in the bad manner of philosophizing' of Stahl.

At this stage Lavoisier began to create his own myth: that of the man who completely demolished the ancient edifice of chemistry. It is true that at this moment a great change was made in his theory: 'oxygen gas' was no longer called a 'combustible body' (in the way of Stahl, calling the body that loses a heat principle 'combustible'), but now in the traditional way (which again was the way of Stahl) sulphur (and phosphorus) are called 'combustible bodies'. A combustible body, according to his new definition, is that which has the property of decomposing 'vital air' (*air vital*; oxygen gas). A combustible body is 'that with which the oxygenous principle (*principe oxygène*) has more affinity than with matter of heat'.[51] In this new definition combustion consists in: oxygen abandoning caloric in order to unite itself with e.g. a metal; the emphasis is no longer on caloric matter abandoning oxygen.

Oxygen gas consists of a basis (*principe oxygène*; the cause of acidity) + caloric matter (*principe calorique*; the cause of elasticity):

Metal (element) + gaz oxygène [element O + element calorique]
→ metal oxide [MeO] + calorique↑

At long last Lavoisier then used the definition now generally connected with his name: combustion is no longer a losing of matter of heat and fire, but it is oxidation, a union with oxygen (the basis of the gas we usually denote by this name!). 'Calorique' now became the name for matter of heat and this element (which had been considered as such since Antiquity) was to live on until the very influential chemist Berzelius quietly omitted it in his table of elements. So, Lavoisier, intending to toll the knell of the old qualitative theory, in fact sang its swan song.

In 1791 Lavoisier declared with great satisfaction 'la révolution en chimie est faite',[52] and this statement has been almost universally taken for granted by posterity. The French chemist Adolphe Würtz started a book by the notorious statement: 'Chemistry is a French science; constituted by Lavoisier,'[53] and another French chemist even declared that 'Chemistry, like Minerva of yore, sprang forth fully equipped from the brain of an eminently French chemist, whose name was Lavoisier.'[54] We have seen that in fact the oxygen theory, far from having such a miraculous birth, was developed rapidly but gradually. Yet, it should be fully acknowledged that if we want to connect a name with the beginnings of 'modern' chemistry, it should be that of Lavoisier. He is the greatest of 18th century chemists: his theory of combustion, his work on chemical classification, his experiments on respiration, his systematic use of the balance, — they fully entitle him to this claim. But was his 'revolution' so revolutionary? By his rational mind and his dexterous hand he effected a thorough change in chemistry. But in the art of experimentation he owed much to the founders of 'pneumatic chemistry', Stephen Hales, Joseph Black, Joseph Priestley, Henry Cavendish and a host of smaller men, who had discovered the several 'airs' or gases and elaborated the equipment to collect and manipulate them. These were the indispensable conditions for the construction of the oxygen theory of combustion. It should be emphasized, however, that it was the lucid and logical spirit of Lavoisier that brought order in the chaos of experiments and conflicting hypotheses about the gases.

Stahl, unlike Lavoisier, did not pay much attention to quantitative experiments. But for our present topic the method of checking chemical reactions by means of the balance would have been of little use as long as gases could not be measured and their chemical functions were unclear and their differences hardly recognized.

Though the contents of Lavoisier's theory may have been widely different from Stahl's, they have a similar formal structure:

a) Lavoisier inserted in his table of elements (see figure 16) substances which — according to his own definition of elements — were simply limits of chemical analysis (without claiming them to be absolutely simple). But beside these he inserted some 'hypostases' (bearers) of qualities: matter of heat (*calorique*), matter of light (*lumière*). Stahl did the same: on the one hand he used the concept of concrete 'simple bodies' (i.e. undecomposed bodies), on the other hand he used the non-isolated principle of combustibility which was a substantified quality.[55] To Stahl, however, not the metals but their calces (= oxides) were simplest. That is, whereas Lavoisier considered lead, copper, sulphur as simple bodies, Stahl took as such lead oxide, copper oxide, sulphuric acid.

   To Lavoisier, sulphuric acid (anhydride) is sulphur + oxygen.
   To Stahl, sulphur = sulphuric acid (anhydride) + phlogiston.

   To Lavoisier, lead oxide = lead + principe oxygène.
   To Stahl, lead = lead calx + phlogiston.

b) To Stahl phlogiston is the bearer of combustibility; all combustible bodies owe their combustibility to it. To Lavoisier calorique is the bearer of elasticity (the gas-making principle; principe gazéfiant), whereas the 'principe oxygène' is the acidifying principle; together they form 'oxygen gas' (*gaz oxygène*). But the oxygen of his table of elements is not this gas, but the never-isolated 'principe oxygène', acidifying principle. All gaseous bodies owe their elasticity to the 'principe gazéfiant'; all acids owe their acidity to the 'principe acidifiant'. The phlogiston theory may have led the Stahlists into many errors, but the acidity theory led Lavoisier astray (misinterpretation of hydrochloric acid, hydrocyanic acid, chlorine).

192 DES SUBSTANCES SIMPLES.

TABLEAU DES SUBSTANCES SIMPLES.

	Noms nouveaux.	*Noms anciens correſpondans.*
Subſtances ſimples qui appartiennent aux trois règnes & qu'on peut regarder comme les élémens des corps.	Lumière.........	Lumière.
	Calorique........	Chaleur. Principe de la chaleur. Fluide igné. Feu. Matière du feu & de la chaleur.
	Oxygène.........	Air déphlogiſtiqué. Air empiréal. Air vital. Baſe de l'air vital.
	Azote............	Gaz phlogiſtiqué. Mofete. Baſe de la mofete.
	Hydrogène.......	Gaz inflammable. Baſe du gaz inflammable.
Subſtances ſimples non métalliques oxidables & acidifiables.	Soufre...........	Soufre.
	Phoſphore........	Phoſphore.
	Carbone..........	Charbon pur.
	Radical muriatique.	Inconnu.
	Radical fluorique .	Inconnu.
	Radical boracique..	Inconnu.
Subſtances ſimples métalliques oxidables & acidifiables.	Antimoine........	Antimoine.
	Argent...........	Argent.
	Arſenic..........	Arſenic.
	Biſmuth..........	Biſmuth.
	Cobolt...........	Cobolt.
	Cuivre...........	Cuivre.
	Etain............	Etain.
	Fer..............	Fer.
	Manganèſe.......	Manganèſe.
	Mercure.........	Mercure.
	Molybdène.......	Molybdène.
	Nickel...........	Nickel.
	Or...............	Or.
	Platine..........	Platine.
	Plomb...........	Plomb.
	Tungſtène.......	Tungſtène.
	Zinc.............	Zinc.
Subſtances ſimples ſalifiables terreuſes.	Chaux...........	Terre calcaire, chaux.
	Magnéſie.........	Magnéſie, baſe du ſel d'Epſom.
	Baryte...........	Barote, terre peſante.
	Alumine.........	Argile, terre de l'alun, baſe de l'alun.
	Silice...........	Terre ſiliceuſe, terre vitrifiable.

Fig. 16. Table of elements according to Lavoisier. From: Lavoisier, *Traité élémentaire de Chimie* I, Paris 1793, p.192.

c) To Stahl the body yielding phlogiston is the combustible body; to Lavoisier in the earlier stage of his theory the body yielding caloric matter is the combustible body.

d) To Stahl compounds containing a constituent in common, have a tendency to combine. To Lavoisier oxides of metals (basic oxides) and oxides of non-metals (acid oxides) form salts because of their common constituent.[56] Consequently, Lavoisier held that a metal is dissolved by an acid after first having been oxidized. (One should remember that in his opinion all acids have an oxidizing power). This belief that a salt consists of two oxides considerably hindered the development of chemistry in the following decades.[57]

5. PHLOGISTON RE-KINDLED

Understandably those who had only an eye for the *contents* of a theory firmly believed that with Lavoisier 'real' chemistry started. One of them, the geologist K.A. von Zittel, said in a presidential address to the Bavarian Academy of Sciences: 'Only — towards the end of the 18th century — after nitrogen, hydrogen and oxygen had been isolated as gaseous elements and Lavoisier had definitively discarded the phlogiston theory, does chemistry begin to enter into the circle of the sciences of nature.'[58]

Those scholars, however, who looked more to the logic and form of thought, recognized that the phlogiston theory had some truth in it, and that its mould had to be filled with different contents. In spite of its un-truth they did not deny to the old phlogiston theory a scientific character.[59]

In his 'Remarks on the Forces of Non-living Matter' (1842) Robert Mayer expressed a positive evaluation of the phlogiston theory, when holding that the phlogistonists gave to the heat developed when hydrogen gas was bound together with oxygen gas, the name of 'phlogiston': 'thus they made a great step forward, but they got confounded in a system of errors, by putting loss of phlogiston instead of loss of oxygen, so that hydrogen was "water minus phlogiston".'[60] It is worthy of remark that Mayer seems to mix up Lavoisier's theory of caloric with Stahl's phlogiston (which indeed were closely related in some aspects). A similar interpretation was put forward by Schroeder van der Kolk in an article on 'Mechanical Energy of Chemical Reactions'.[61] In his opinion the function of energy at chemical combination 'very often coincides with that of Stahl's phlogiston': it was held that carbon contained much phlogiston, and that the quantity of it was connected with the quantity of its heat of combustion. Carbon and oxygen together contain more energy than the carbonic acid they form: the difference is as it were heat of combustion: 'In this respect the phlogiston theory was wholly conformable to nature'. This evaluation was fully shared by G. Helm (1898).

At any rate the above quotations show that even after Lavoisier the 'matter of fire' left some traces in chemistry in that it lived on under the disguise of the heat made free at combustion, though this was just a form of 'energy' and not a chemical substance.

An even more telling appreciation that the matter of fire had not wholly died was given by G.N. Lewis's comparison between the electron and phlogiston (1926). He called the phlogiston theory a 'great step in chemical classification', and he pointed out its remarkable resemblance to the modern conception of oxidation and reduction. Oxidation is a loss and reduction an addition of electrons: $2K + O \rightarrow K_2^+ + O^-$. According to Lewis 'if they [the phlogistonists] had only thought to say: 'the substance burning gives up its phlogiston to, and then combines with, the oxygen of the air' the phlogiston theory would never have fallen into disrepute. Indeed it is curious now to note that not only their new classification but even their mechanism was essentially correct.'[62]

The structure of a theory is perhaps its most lasting feature. The 'Fire' kindled by the Ancients has not been extinguished. The search for quality-bearers is a logical one: similar effects point to causes having some similarity; a certain quality seems to demand a similar basis of the quality. Nature herself, however, will show us how far she conforms to our natural expectations.

> As to the tenet 'like dissolves like' (similia similibus solvuntur), this finds an interesting parallel in the theory put forward to explain the solubility of organic compounds with polar groups (-OH; -COOH) in water (W.D. Harkins a.o.; 1917).[63] In this theory it was supposed that the molecules with polar groups, when brought into contact with water (which itself shows polarity: H-O-H), will turn their non-polar hydrocarbon chains toward the air and the polar ends toward the water, which pulls them downward when they are poured on it. If the hydrocarbon chain is very short (ethanol C_2H_5OH; acetic acid CH_3COOH) the polar group is so strongly attracted by the water that the hydrocarbon chain is dragged into the water so that there is complete solubility. If the hydrocarbon chain is long (higher alkanols; fatty acids) the organic compound remains on the surface with its polar ends directed toward the water.

One might object to the arguments adduced in defence of the phlogiston theory that, after all, it was false. This may be granted, but a theory, however fictitious, which arranges facts in a rational system and which connects them in a way that is a mirror image of the 'true' relation, is more 'scientific' than a jumble of loose facts, however true they may be. The phlogiston theory was such a system: it established relations between phenomena of oxidation and reduction, between metals and their calces, between non-metals and acids. It is understandable that so many of its adherents stubbornly refused to give it up. As long as a theory offers a possibility of 'saving the phenomena' one has as much right to try to save the theory as to launch forth an alternative one. We similarly hung on to the old chemical theory of valency when it could not satisfactorily account for the properties of benzene, acetylene and carbon monoxide. We maintained the laws of fixed proportions (Proust) and multiple proportions (Dalton) in chemistry, though they forced us to invent highly phantastic silicic acids.

But, one might object, the existence of phlogiston could not be proved directly. That is true, but the Copernican theory was almost generally accepted long before the direct proofs — aberration (Bradley); parallax of fixed stars (Bessel) — had been found. Its more comprehensive character, in particular in its Kepler-Newton version, made it triumph over its rivals. Simplicity, whether ontologically true or only methodologically convenient, is highly valued in science. After the discovery of oxygen and the interpretation and classification of oxidation phenomena by Lavoisier, the phlogiston theory in its original form, after a valiant struggle, succumbed under the burden of its complicatedness. But it did not die completely, for there was *some* truth in it.

V. A Tunnel Through the Earth

1. The Place of Hell in Medieval Cosmology 119
2. The Solution of the Tunnel Problem According to the Via Antiqua 124
2.1. Walter of Burley 125
2.2. Swineshead on the Tunnel Problem 126
2.3. The Theory of Latitudes 128
3. The Impetus Theory 130
3.1. Impetus 131
3.2. Buridan on the Moving Heavens 131
3.3. Buridan on Falling Bodies 132
3.4. Buridan on the Tunnel Through the Earth 133
3.5. No Experimental Testing 135
4. Renaissance: the Italian Engineers 139
5. William Gilbert; Francis Bacon 141
6. The Mechanistic Philosophers 142
7. Isaac Newton 144

When the University of St Andrews was founded in 1410 its first rector was Lawrence of Lindores (circa 1437). All historians of the university mention that he was also 'Inquisitor of heretical pravity' — the main inquisitor of the kingdom of Scotland — and that, as such, one of his unattractive features was his zeal in bringing Lollards to the stake.[1] If the prior of St Mary's had had his way there would have been even more victims, for he wrote to Lawrence an admonitory letter (1418) richly larded with quotations from Scripture and from pious writers, chiding him for his laxity and ominously imputing to him a tendency to heresy himself.[2]

Some of Lawrence's biographers mention a more positive action: almost from the beginning (16 February 1417) he introduced at the university and maintained with a strong hand the teaching of the *via moderna*, the 'nominalist' philosophy.[3] In spite of opposition from the adherents of the *via antiqua* (the ancient way of Aristotle and Thomas Aquinas) he pushed it through with no less firmness than he had shown in persecuting the unorthodox. In Paris — where Lawrence and most of the other Regents had studied — at that time the 'new way' prevailed. From that centre of medieval learning the ideas of the great philosopher Jean Buridan had been spread all over Europe by his pupils and the pupils of his pupils, who elaborated or modified his commentaries on Aristotle.[4] The ideas of the Buridan school were still taught at Cracow university when Copernicus studied there in 1491,[5] and the writings of Lawrence of Lindores in particular became known throughout Middle and Eastern Europe.

In St Andrews, however, after Lindores's death the tables were turned; the old 'realistic' school gained influence and the principles which he had forcefully imposed upon his colleagues lost their monopoly. The adherents of the *via antiqua*, with help from outside the Faculty of Arts, obtained the freedom to teach the old way of Albert the Great and Thomas Aquinas, and this in spite of the fact that only one month earlier (13 October 1438) the Faculty had voted by 20 to 5 not to tolerate this.[6] In these circumstances we may be practically certain that the ideas of the Paris philosophers will give us reliable information as to the character of the physics taught at St Andrews university in the first decades after its formation.

The physical teaching of the several scholastic traditions had in common a predilection for philosophical disputation about imaginary situations; e.g. if the heavens were to stand still, could tow still be burned? Or: if a tunnel were bored through the centre of the earth and a heavy body dropped into it, what would happen to it? One might be tempted to dismiss such questions as inane, for although they may be imagined, they can never be realized. We should not, however, forget that in modern science too, purely imaginary problems play an important role in so-called thought-experiments. These consist of imaginary events and situations, and lead to deductions based on axioms, postulates and hypotheses. In modern science the conclusions are, as a rule, tested by experiment, but very often they are just an idealized component of real physical events; something that cannot be isolated in its pure form, like the motion of 'a body on which no force is exerted' or 'motion on a perfect frictionless plane' which parades in theoretical mechanics.

Fictitious though they may be, thought-experiments have a valid function in science. As we shall see, even that of the stone falling through a hole piercing the earth has a legitimate place in the development of scientific thought.

Within the framework of the medieval world view, this particular abstraction is removed not only from physical reality but also from a commonly held theological belief *viz.* that the centre of the earth is occupied by the infernal fire. That this belief never interfered with the working out of the 'philosophical exercise' of the Tunnel shows us how far philosophers could go in separating scientific from theological issues.

1. THE PLACE OF HELL IN MEDIEVAL COSMOLOGY

Whereas St Augustine was rather reticent about the character and place of hell, declaring that nobody knows what kind of fire it is and in which part of the world it is located, later authors (like Hugh of St Victor, circa 1135) were more precise. And in the 13th century Thomas Aquinas gave (as we shall see) rather detailed information on this topic. In physical matters (whatever some of his modern adherents may claim)[7] he showed, in general, little originality: he simply repeated the opinions of Aristotle, trying to make them more consistent and to make some corrections to the peripatetic system. In the case of the location of hell, however, he could not borrow any idea from the natural philosophy of Aristotle whose cosmology paid no attention to the underworld. For this reason the topics of 'heaven' and 'hell' were hardly touched in scholastic books on natural philosophy, these being practically always paraphrases or commentaries on Aristotle's works on natural philosophy.

For the 'heaven of the blessed' it was not in fact too difficult for the learned to find a place in the cosmic system inherited from the Greeks. They located it beyond the outermost (i.e. the 9th or 10th) sphere of the Aristotelian cosmos. By this philosophically harmless addition the Aristotelian world picture was not disturbed, as can be seen from the diagrams of the universe inserted in the astronomical textbooks of the German astronomer Petrus Apianus (1535), see figure 17, and the Flemish botanist Rembert Dodoens (1548), see figure 18.[8]

Such a combination of the knowledge obtained through 'reason and experience' with knowledge received by special revelation — as if they were on the same level — was, however, unusual. In general the medieval scholars did not mix the natural with the supernatural.

It is remarkable that, in contrast to the abode of the blessed, hell was allocated no place in the astronomical textbooks. They show an earth that is solid to the core and is enveloped by subsequent shells of water, air and fire. The same attitude is shown when the problem of the stone falling through a tunnel piercing the earth is considered. In a scientific discussion one could evidently be permitted some abstraction from religious belief.

The historiography of cosmology seems largely to have overlooked the question as to why the heaven of religion could so easily be inserted into the Aristotelian cosmos, whereas its negative counterpart was ignored. One reason may be that the God who dwells beyond the outermost material sphere of the universe, where there is no motion but eternal quietude, could with some good will be accepted in place of Aristotle's Prime Mover (or rather prime moving cause, *primum movens*) which is the cause of the motion of the outermost sphere (the *primum mobile*). The Prime

Schema huius præmissæ diuisionis Sphærarum.

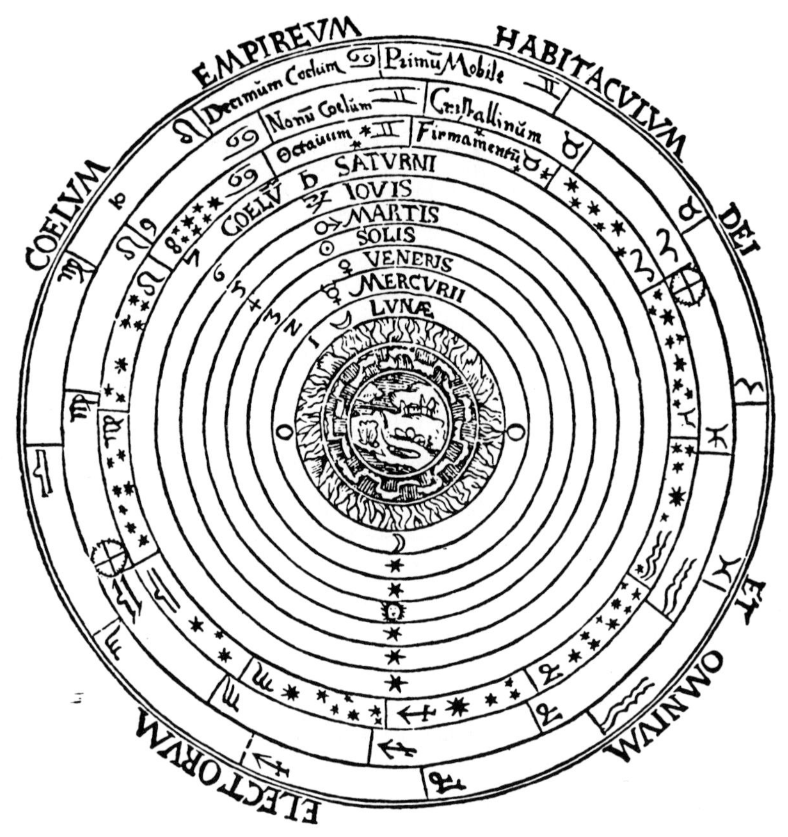

Fig. 17. The Aristotelian system of the universe according to Petrus Apianus' *Cosmographia* (1539). The abode of the blessed is beyond the moving spheres.

Mover (for which there is no fit place within the limits of the universe) is the First Cause, and in itself passive: the heavenly spheres are in unchangeable circular and uniform motion(s) because they are striving as much as possible to imitate the prime Mover.[9]

The God of the Judeo-Christian tradition, by contrast, is ever *active* in upholding the world He created by his Word. But once the (rather strange) identification of the prime mover with the God of Israel was accepted, it seemed a matter of course to place the abode of God and his elect beyond the outermost sphere.[10]

In a 'natural' system based on Aristotelian principles it was, on the contrary, impossible to allot a place to hell-fire at the centre of the earth. In agreement with Aristotle, philosophical works always declared that the element 'earth', which is

REMBERTI DODONAEI

Cœlum Empyreum. Beatorum ſedes & habitaculum.

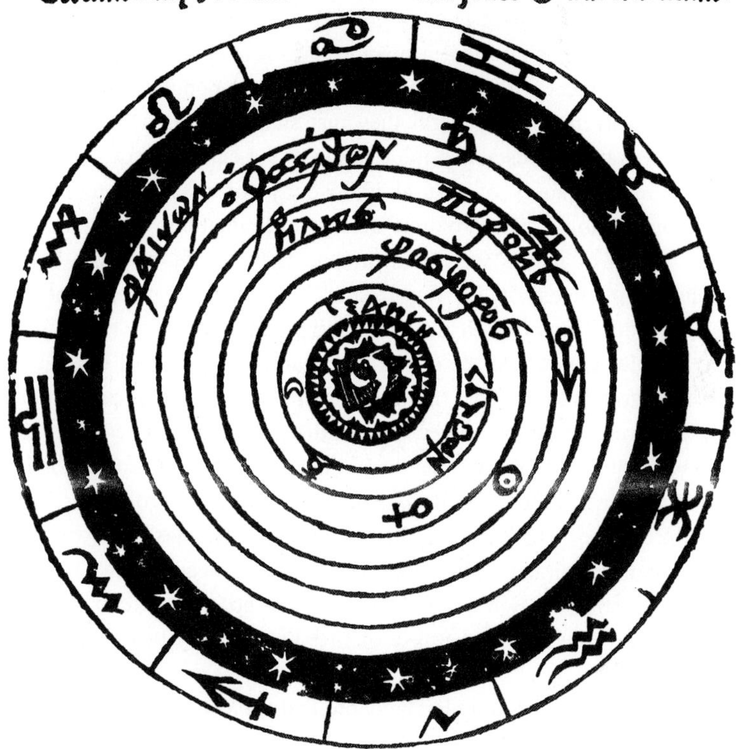

Fig. 18. The Aristotelian system of the universe according to Rembertus Dodonaeus' *Cosmographia* (1548). 'The abode and dwelling place of the blessed' lie beyond the 9th sphere.

absolutely heavy, extends itself to the very centre of the universe (which is the centre of the terrestrial globe) so that there could be no room here for anything else. Moreover, the element 'fire', which is absolutely light, has its 'natural' place as far away as possible from the centre of the universe (that is, just below the sphere of the moon). How then could hell-fire be in the centre of the earth — a most unnatural situation?

Whereas an immaterial supernatural heaven outside the material universe was neither according to nature nor against nature — it was beyond nature — a permanent fire below the earth was flatly against nature. Though the lighter elements, fire and air, temporarily may be mingled with earth and water upon or below the earth's surface, they will always tend to move upwards to join the spheres of air and fire.

In spite of these philosophical objections Thomas Aquinas concluded, after discussing various opinions and uncertainties about the place of hell, that it is best to stick to Scripture which says that it is under the earth.[11] This became the general opinion and we will meet with it in several more or less popular expositions of the structure of the universe.[12] In general they just stated this as a fact, but Thomas went further. He wanted to establish the relation to the current Aristotelian physics, for he considered the study of natural philosophy (of Aristotle) to be of great importance for theology.[13] For to him events and situations fitting in with the Aristotelian system of nature were *natural* and those deviating from it are thus recognizable as miracles. It is, thus, natural philosophy that enables us to distinguish the natural from the supernatural and miraculous.

Aquinas mentions as one of the objections against locating hell in the centre of the universe Aristotle's saying that 'nothing violent and accidental can be eternal.'[14] But if this fire were in hell for ever, it would be there not under constraint (by 'violence') but naturally. The opponents of the subterranean location for hell-fire say that 'under the earth there can be fire only by violence; ergo the fire of hell is not under the earth.'[15] In face of this conclusion Thomas now resorts to a supernatural, endless miracle. This fire, so he says, is accumulated there by the order of divine justice, though *according to nature* no element can stay for ever outside its natural place (i.e. in the case of fire, immediately below the orb of the moon).[16]

From the above it becomes understandable that *philosophers* writing about the *nature* of things (i.e. physics) avoided the problem of hell and that in most diagrams of the Aristotelian universe this topic was ignored.

For a picture of the cosmos with hell in the centre we therefore have to look to some popular work whose author was not sophisticated enough to heed the methodological distinction between philosophical and theological truths, or who just wanted to present a *complete* picture of the universe without caring for such niceties when addressing the unlearned. Such a work was that of the Flemish Brother Gheraert, 'Natuurkunde van het Geheel-al' (Natural science of the universe) a didactic poem written in Ghent in the 14th century.[17] In the centre of the diagram of the world (see figure 19), hell is indicated by the legend 'dit is die helle'. It is enveloped by the crust of the earth; the 9th sphere is 'the kingdom of heaven wherein are all 9 choirs of Powers'. The text says: 'Now I will tell you where hell happens to be. One can prove by Scripture that it is nowhere else than in the middle of the earth, which surely is the centre [of the universe].'[18]

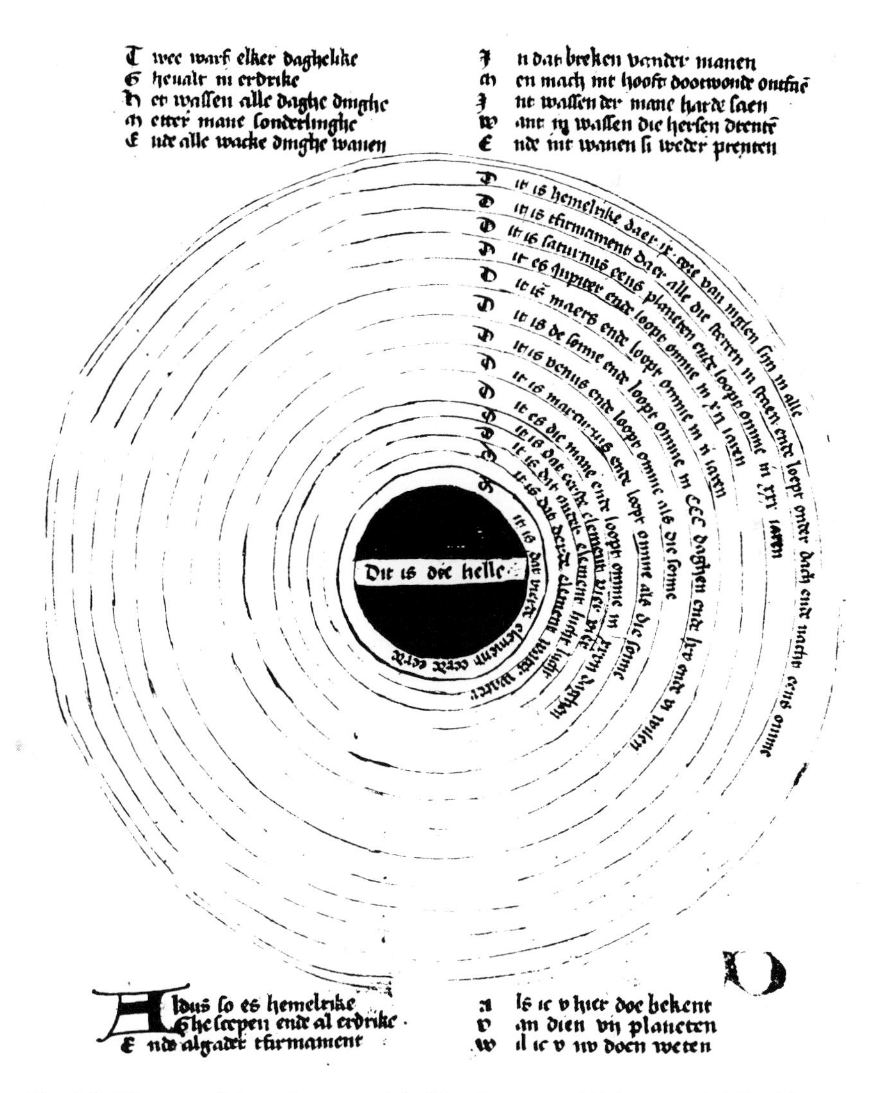

Fig. 19. Diagram of the universe with hell in the centre (14th century). The 9th sphere is 'The Kingdom of heaven wherein are all 9 choirs of Powers.'

In the popular encyclopaedia of Brunetto Latini (1230 - 1294) 'Li Livres dou Tresor', the author says that a stone dropped in a hole piercing the earth stops at the lowest part of the earth, that is, at its centre 'which is called abyss, where the hell is'.[19]

Similarly an early 16th century Portuguese navigator, D. João de Castro (1538) simply defines the semi-diameter of the earth as the distance 'from here to the centre of the world and to the middle of the hell of the damned'.[20] Castro was no professional scholar and he wrote for seafarers, so one can hardly expect great subtlety in his work. But earlier in the same work he had stated that a leaden ball, if dropped into a tunnel through the earth, would stand still in the centre of the earth and remain there quietly and at ease.[21] He overlooked the objection that the fire there would certainly melt the lead; but to such a remark he almost certainly would have answered that he just chose lead as an example of a heavy body and that a stone would have served the same purpose; that, moreover, he did not speak 'physically' (as it was only a thought-experiment, in which abstraction of some qualities is allowable); and that he intended least of all to speak 'theologically'.

Professional philosophers, in speaking about the same problem, did not wish to fall into inconsistencies: they ignored the problem of hell not so much because they forgot it but rather because in a philosophical discussion they could leave it alone.

At first sight it may seem that Erasmus was an exception. He had received a philosophical training at the university of Paris and so was familiar with the problem of the tunnel. When discussing it in one of his dialogues he says that a stone would finally arrive at the centre, but that lead would arrive in a molten state.[22] This, however, is a purely physical statement which has nothing to do with going through a fire. Aristotle had mentioned that leaden balls used as missiles melt and even catch fire.[23] Aristotle's 'Commentator', the Arab philosopher Averroës had said that just 'as it is a natural property of gravity to move things downwards ... so it is also a natural property of local motion to heat things.' The 14th century nominalist Buridan agreed with these statements.[24] With a small variation he quoted Aristotle as saying 'that motion heats so much that as a result sometimes leaden arrows will melt.'[25] Erasmus evidently associated the tunnel problem with the received ideas on the rise of temperature of bodies moving swiftly in air.

2. The Solution of the Tunnel Problem According to the Via Antiqua

The question as to what would happen to a stone dropped into a hole pierced through the earth must have been raised at an early date. Plutarch (circa 50 - 120 AD), who did not believe that the earth is a sphere, mocked those who held that a bar weighing a thousand talents, dropped into such a hole would stop at the centre of the earth.[26]

According to the Aristotelians the centre of the earth coincides with the centre of the universe and all heavy bodies have a natural desire to be as close as possible to that point, which is their natural place. Therefore a stone dropped in a hole bored through the centre of the earth will fall with an increasing velocity and suddenly halt when it has reached its goal, the centre of the world. The 'natural' motion of the falling body is a means of reaching its fullness of being, its 'Form' — and to that fullness belongs also being-in-its-natural-place. When this is attained, the urge to further motion ceases. Such an answer was given by the Arab philosopher al-Khwazimi (circa 1100), by Adelard of Bath (12th cent.) and by Vincent of Beauvais (13th cent.).[27] Among the more conservative philosophers it lived on until the end of the 16th century (Alessandro Piccolomini, 1508 - 1578).

In a Portuguese Nautical guide of 1509 it is said that a body that is heavy in the highest degree, i.e. consisting of the element 'earth', *yearns* after the Earth, and *enjoys* reaching it and then stops moving.[28] These expressions show how the 'organistic' spirit (see above, Chapter III) of the Aristotelian philosophy had penetrated into purely practical works. The old school thus solved the problem of the heavy body falling into a hole bored through the earth in a typically organistic way. The falling body, however great the velocity it may have acquired, suddenly stops when it has reached its goal, just as a living being would do: a horse its stable or a bird its nest.

It was realized, however, that not all parts of the falling body can be at the same place, and here the organistic conception that the whole body is something more than a mere addition of its parts, and that the latter are subject to the whole, must lead to a deduction different from a more mechanistic and mathematical conception which considered a body as a collection of its parts.[29]

2.1. Walter of Burley

Walter of Burley (14th cent.) considered what would happen if a clod of earth were dropped in the hole. Once it had reached the centre, half of it would gradually pass the centre and would then be going upwards in the direction of the heavens. This means that it would have a *violent* motion; against its nature. Burley's answer is that if a part of the earth is detached from the whole, it will move as long as it meets with no obstacle, towards that centre of the earth (which is also the centre of the universe) . But if it is *united* with the rest of the earth, it may be at rest outside the centre of the world, without violence being done to it. For in that case it is at rest not by itself but thanks to the whole to which it belongs and which is at rest.[30] In the case of a clod of earth falling through the tunnel, this means that the part of the body that has passed the centre of the universe goes on to cooperate in bringing the whole body to that centre.

> The tenet that the whole is more than the sum of its parts seemed, even in purely mechanical phenomena, to be supported by direct experience. Aristotle refers to the philosopher Zeno who had said that a bushel of millet, when falling to the earth causes a movement of the air (a sound), so that in similar circumstances one grain of millet will give rise to a proportionally smaller sound. Aristotle, however, — quite apart from the argument that too small a 'motor' (moving cause) cannot cause a motion, opposes Zeno's tenet by the argument that such a grain, as long as it is a part of the bushel, does not have

the same effect as if it were independent: it now has only a potential effect, and as a part of the whole bushel it need not contribute a proportional part to the movement of the air caused by the whole bushel. The total effect is due to the bushel as a whole and not to its separate parts simply added together.[31]

Within the Aristotelian framework this was not a well-chosen example, for a bushel of millet is not a unity under the impress of its own Form, but only a heap or mechanical mixture of separate grains which keep their individuality; it is not integrated into a composite natural body (in the way that parts of an animal are integrated to compose the whole animal). But Aristotle's intention is evident: the effect of the totality is different from the sum of the separate effects of its constituent parts.

2.2. Swineshead on the Tunnel Problem

The conflict between this organistic approach and the opposing mechanistic-mathematical one is clearly illustrated by Richard Swineshead's treatment of the tunnel problem. This Oxford philosopher (circa 1340), better known under the name of Suisseth, was surnamed the Calculator. He and his followers, the *calculatores*, applied mathematics to philosophical problems. They expressed the intensity of a property (e.g. velocity of motion) by a number, and then tried to find out how it would change in relation to time and distance.[32]

Imagine an ideal rod — perfectly homogeneous and of only one dimension (length), so that its centre of gravity is precisely in the middle. Let it be dropped lengthwise down in a tunnel in which there is no resistance of the air and let its foremost end have reached the centre of the earth. The centre of gravity (m), which wants to coincide with the centre of the whole (C) then still has to travel half the length of the rod. After the foremost end has reached the centre, those parts of the rod which have passed it will henceforth be going upwards with an unnatural, 'violent', motion in the direction of the other side of the world. Thereby they exert a retarding force on the part that is still descending.

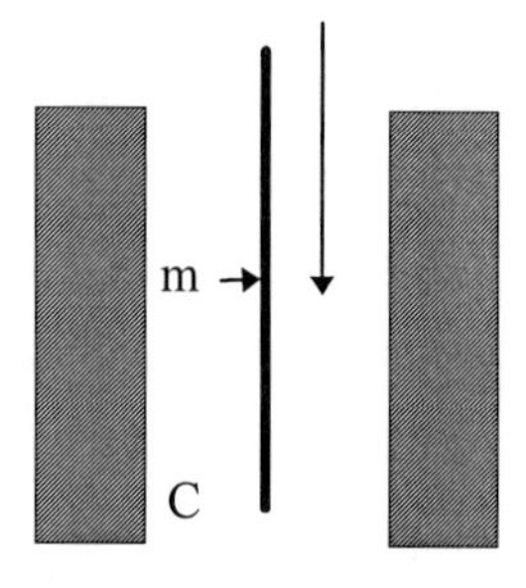

ideal rod dropped in a tunnel

At this point Suisseth makes use of Bradwardine's law of the relation between motive force, resistance and distance covered by the moving body for which the rationale ran as follows:

> Aristotle's fundamental law of mechanics had an awkward weak spot which prevented its being generally applicable. In his opinion the distance covered in a certain time by a moving body (i.e. the speed or velocity) is proportional to the applied force (F) and inversely proportional to the resistance (R). Anachronistically we might express this as $v \infty F/R$. If this were generally true, a force, no matter how small, would still cause a very slow motion. Since this is not what generally happens, the Oxford philosopher Thomas Bradwardine (1328) enunciated another law, which avoided this difficulty. He assumed that when starting from a certain value of Force/Resistance, in order to double the velocity, it is necessary to square the quotient F/R. In order to triple the velocity the value of the original quotient should be cubed, etc, etc. The velocity thus increases arithmetically as the ratio of force to resistance increases geometrically:
>
> v_1 if Force/Resistance is (F_1/R_1);
>
> $2v_1$ if Force/Resistance is $(F_1/R_1)^2$ (where $F>R$).[33]

Suisseth then demonstrates that, when the distance, e, from the falling rod's centre of gravity to the centre of the whole has been halved, the velocity will have been more than halved. At the next halving of the distance the velocity will be reduced to less than half of the second velocity. The motion becomes slower and slower and the conclusion is that the centre of the rod will not have reached the world's centre in a finite time.

e ...... v_1;
½e ...... v_2 (where $v_2 < \frac{1}{2}v_1$);
¼e ...... v_3 (where $v_3 < \frac{1}{2}v_2$ and $\ll \frac{1}{4}v_1$).

This result was reached (without using formulae) by quite a feat of mathematical acuteness and by far-going physical abstractions. An ideal law of falling bodies is posited; an ideal rod assumed; all disturbing factors — like the resistance of the air — are neglected; the only resistance is that of the part of the rod hindering the downward motion. Most important of all, the rod is treated in a mechanistic and mathematical way, as if it were just the sum of its parts. The growing force exerted by the part that has passed the earth's centre increasingly resists the decreasing driving force of the part that still wants to proceed towards that centre.

The above is just a mathematical 'description' of a fictitious event undergone by a fictitious body. In this respect Suisseth proceeded like Newton, or any modern theoretical physicist: a mathematically formulated working hypothesis is put forward and its physical consequences are found by mathematical deduction. There is, however, an essential difference: whereas the modern physicist will next try to test the conclusion by some experience of physical reality (in all its contingency and, maybe, unexpectedness), Suisseth immediately rejects the conclusion of his mathematical deduction because it does not stand the test of Aristotelian natural philosophy. He does not ask himself whether the absurdity of the conclusion might be a consequence of some incorrectness in Bradwardine's dynamical hypothesis. At the end of the treatise the mechanistico-mathematical hypothesis gives way to the organistic physics of Aristotle. He then concludes that even the part of the rod that has already passed by the centre of the world cooperates to satisfy the 'appetitus' of the whole (namely its yearning for letting its centre of gravity coincide with the world centre). The purpose of each part is wholly subsumed to the aim of the whole body, which wants to find its 'natural place': The part on one side of the centre neither desires to resist nor actually does resist the descent of either the whole body or the other part of it; on the contrary, since it is a part of the whole, it desires (*appetit*) that the middle of the whole body should be in the middle of the world.[34] Consequently Suisseth rejects the result of his own mathematical analysis as an absurdity.

2.3. The Theory of Latitudes

Alongside the calculators' method of mathematizing qualities in an arithmetical way, some Paris philosophers who followed Nicole Oresme (circa 1323 - 1382) developed a geometrical method. Graphs were drawn in which successive distances were plotted on the abscissa as 'longitudes', with the corresponding intensities of qualities (instantaneous velocity included) measured on the ordinate as 'latitudes'. A quality that does not alter with time (e.g. a uniform velocity) is then represented by a graph enclosing an oblong.

A 'uniform difform' quality (e.g. a uniformly accelerated motion) is represented by a triangle if there is no initial value; if the quality has an initial value it is represented by a trapezium. In this way the qualities are geometrized: Oresme speaks of 'square qualities' and 'semicircular qualities' etc.[35] In kinematics successive instantaneous velocities are represented by the latitudes at the times represented by the longitudes. Irregular motions are represented by irregular figures; their surface, determined by the lines of longitude and latitude, is the 'measure' (*mensura*) or 'quantity of the quality' (*quantitas qualitatis*) which, in the case of local motion is intuitively identified with the distance covered. When the motion is uniformly accelerated, this quantity is equal to that of a uniform motion with the average velocity (a constant latitude equal to the instantaneous latitude at the middle moment).

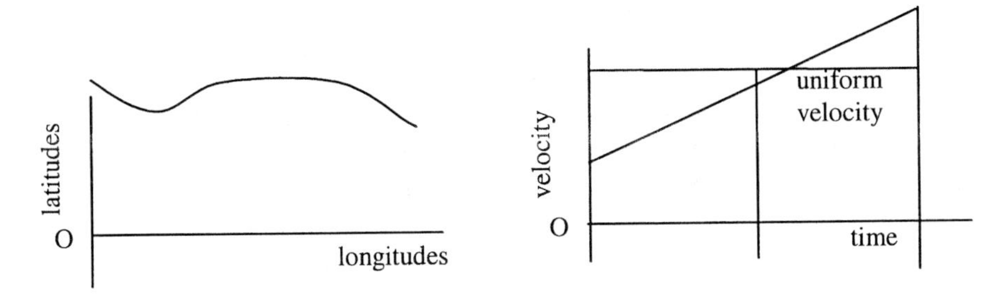

Fig. 20. Graphs representing the relation between qualities and distances, according to the geometrical method of Oresme. The distances are plotted on the abscissa (horizontal axis) as 'longitudes'; the corresponding intensities of qualities (including instantaneous velocity) are measured on the ordinate (vertical axis) as 'latitudes'.

Such graphs were also used in Lawrence of Lindores's writings.[36] They were considered to be more than merely abstract figures depicting the way the intensity of a quality changes when longitude is increasing; they had a concrete physical significance: expressions like 'a triangular quality' are meant literally. Oresme says that the geometrical form of the measure *determines* the behaviour of the quality it represents in the same way as the shapes of the atoms of the ancient atomistic theory. According to the atomists, penetrating, sharp-pointed particles cause an acrid taste; similarly in the theory of latitudes, the sharper the triangle of the 'measure', the more active its effect: hence the expression 'a sharp or acrid smell, taste or cold' for properties that operate with small, forceful shocks.[37] Something similar is the case with passive qualities: a porous body, which has alternately much and little matter along a given direction, can easily be influenced because of its lack of homogeneity. In the same way a body in which contrary qualities alternate at short intervals is more liable to alteration. A medicament will have a wholesome influence on the heart if some quality of it has an intensity whose figure resembles the configuration of the heart.[38]

However odd all this may seem to the modern scientist, it was quite a respectable effort towards mathematization of the world picture; it undermined the current organistic conception of nature, as the essences or substantial Forms were reduced to configurations of qualities.[39]

Chemical phenomena, too, were subjected to this geometrization of nature. Heinrich von Langenstein (1325-1397) held that the formation of a substance is determined not only by the proportion of its constituents, but also by the intensity and the configuration of the heat applied during its preparation (an opinion which, by the way, is quite familiar to the modern chemist). There are an infinite number of configurations; consequently it is possible, in principle, to arrive at new substances which have not yet been found in Nature. It appears that Langenstein does not make an essential distinction between the rise of new Forms by *fabrication* of artificial ('synthetic') chemical compounds and the origination of natural compounds produced by 'generation'. Moreover, in his opinion, more things and events are possible with God than those so far encountered; Man cannot conclude by the light of nature alone whether or not a wonderful fact ensues from the powers of nature. In holding this opinion Langenstein assumed a position opposite to that of Aquinas, who considered natural philosophy to be a means of distinguishing natural events from supernatural ones. It is wholly in agreement with Langenstein's and Oresme's standpoint that many 'magical' effects (e.g. of precious stones) should be attributed to configurations as yet unknown.[40]

Buridan had taught that God's will could have created a different world and that this too would have been 'natural'. In other words, God does not follow Nature or Absolute Reason or Eternal Ideas or Forms; rather 'Nature' is what He wills, whether or not we may deem this 'natural' or 'rational'. Similar ideas led Langenstein to the belief that in principle a higher philosophy, a philosophy based on knowledge of the configurations of the substances and their processes of production, could perfect the vulgar philosophy. This philosophy must be based on experience (also of rare events) and experiments in medicine, alchemy and astronomy.[41]

In principle this way of thinking had great possibilities; it gave some hope that problems whose solution was barred by the doctrines of Ideas and Forms might in future be solved. New substances, produced artificially, might be equivalent to products of nature. These *possibilities*, however, did not lead to practical consequences. For there are infinitely many proportions of the same constituents, and an infinite number of procedures for uniting them. But the theory does not give the slightest indication as to how a certain compound should be formed from the elements, for neither the right proportion nor the right treatment (e.g. for making gold) is known. This was also the weak spot of ancient atomism: the infinity of possible geometrical shapes, sizes, arrangements and motions of the atoms did not offer the slightest hold for scientific planning, e.g. in chemistry.

3. The Impetus Theory

The *via moderna*, introduced by William of Ockham (circa 1290 - 1349 or 1350) was characterized by its denial of metaphysical reality to eternal entities like Ideas, Forms and Species. Real existence, so they said, does not pertain to *the* lion but to this or that individual lion. They demanded a parsimony of notions; we should not introduce more notions (fictitious substances) than strictly necessary: 'entia non sunt multiplicanda praeter necessitatem.' Connected with this tenet was their emphasis on the contingency of nature (which has no need to conform to the expectations of human reason) and on 'voluntarism' in theology: God's *will* decides about the existence of things and the occurrence of events — whether *we* deem it 'rational' or 'natural' is irrelevant .

It is not by chance that even moderate nominalists like Buridan and Oresme, when making bold statements deviating from Aristotle's opinions, appeal to the authority of the bishop of Paris, Etienne Tempier. Tempier in 1277 had condemned 219 theses (stemming largely from the school of Averroës) that set restrictions on God's power. The bishop stressed that things that seem impossible to reason are possible with God. Implicitly he thus admitted the possibility of natural 'laws' that are not conformable to Aristotle's philosophy.

3.1. Impetus

Aristotle had laid down the axiom that 'everything moving is moved by something else'. In his opinion the motion of an arrow is sustained by the air which has been set in motion by the projector at the same time as the projectile.[42] Buridan and his followers, however, adhered to the doctrine put forward by one of Aristotle's commentators, Joannes Philoponos (6th cent.), that the moving cause resides in the moving body itself.[43] This 'impetus' was put into the projectile by the hand or instrument that set the body in motion. As Lawrence of Lindores put it: 'The projectiles are moved by a certain quality, which is called 'impetus', which the projecting body impresses into the projected body. This is evident because one cannot see by what else it would be moved ... Positing this 'impetus' we save all [phenomena].'[44]

Though the adherents of the impetus theory remained Aristotelian enough to make an essential distinction between heavenly motions (by nature circular) and terrestrial motions (of all sorts) and among the latter between natural motions (rectilinear) and violent motions (of all sorts), they unified the theory of motion in that in all cases they used the concept of impetus to explain the motions.

3.2. Buridan on the Moving Heavens

According to Buridan the heavenly spheres must once have received an impetus by which, without diminution, they perpetually continue their 'natural' rotatory motion.[45] He therefore deemed it unnecessary to assume that there are 'Intelligences' (as Averroës called them) or angels (as Thomas Aquinas said) who keep them moving. If Scripture had mentioned them, he would gladly have admitted them in spite of their philosophical superfluousness. But this not being the case he piously declares: 'I do not say this assertively, but I shall ask the reverend theologians to teach me how these things happen.'[46] He must have written this tongue in cheek, for he could have asked for their 'instruction' before writing down his rejection of these moving beings. At any rate it is evident that he was inclined to the idea of conservation of impetus in cases where no resisting forces interfered.

The concept of 'impetus' has sometimes been considered by modern authors as equivalent to 'inertia' and to the concept of 'impulse' or quantity of motion. It has indeed some affinity with both of these (quite different) concepts; yet it should not be identified with either of them. In Buridan's example of the spinning top (see below, page 202) or the rotating heavens it may be a kind of inertia (though not for rectilinear movement!) The fact that the impetus of a moving body is considered proportional to its weight and to its velocity brings it close to the notion of impulse or momentum. But there is always a risk in identifying ancient concepts too strongly with modern notions.[47] Ancient concepts only gradually developed into modern ones, and in that process have been split into different concepts thus assuming less vague contours. To take an example from a different field, Aristotle's concept of a chemical compound, as we saw in chapter III, is widely different from Dalton's, which in its turn is different from the modern one.

Even more remarkable is the comparison Buridan made between the rotatory motion of the heavens and the rotation of a grindstone. If there were no resistance because of 'gravity' (*gravitas*), the grindstone, once set in motion, would turn perpetually because of its impetus.[48] Having abstracted the resisting forces, Buridan here simply ignores the fact that in Aristotelian physics a rotatory motion is 'natural' only for the heavenly element, whereas in terrestrial things (which by nature move rectilinearly) it can only be 'violent' and thus temporary. By the making of the grindstone analogy two barriers are thus (hypothetically) taken away: that between heavenly and terrestrial motions and that between natural and violent motions. To put it otherwise, the phenomenon of the rotating grindstone is dissected into its constituent parts: the forces resisting the motion are imagined to be eliminated, so that the remaining (violent) terrestrial motion would resemble the natural heavenly motion.

Buridan's disciples did not go so far. Nicole Oresme held that the impetus (quite apart from external resistance and the tendency to other motions) will eventually spend itself. This would also be the case for the heavenly motions, were it not that they are kept going by the force of the Intelligences or angels which has just the right proportion to resistance to ensure a uniform velocity.[49]

3.3. Buridan on Falling Bodies

In Buridan's opinion, a projectile moved by the impetus imparted to it by the projector, will gradually lose this impetus by the resistance of the air and by the contrary inclination to another motion (as in the case of a body thrown upwards). 'This impetus is a thing (in itself) of a permanent nature,... an innate quality moving the body into which it has been impressed' ... 'it is decreased, destroyed or impeded by resistance or a contrary inclination.'[50] When the impetus is destroyed, gravity moves the body down to its natural place. Evidently the tendency to fall weakens the impetus even before the downward motion has set in.

This impetus theory also explains, for Buridan, why the natural motion of a heavy body downwards becomes continually faster: in the beginning only gravity moves it, but in moving it impresses an impetus in the heavy body which, in conjunction with gravity, makes the motion faster; and the faster it becomes, the more intense the impetus becomes.[51] So, according to Buridan, the gravity will remain the same and 'principal' cause, but the other moving cause, the impetus grows with the velocity and might therefore be called 'accidental gravity'.

3.4. Buridan on the Tunnel Through the Earth

In Buridan's works the problem of the tunnel through the earth turns up when he discusses the possibility of a vacuum. He argues that heavy bodies, like the elements earth and water, when dropped in the hole, would go to the centre; but that if these were not present, even air and fire would descend into that tunnel rather than leave a vacuum.[52]

This is a matter of the hierarchy of causes. Air and fire are 'light' elements, which, according to Aristotelian doctrine, ought by their nature to move upwards. However a space devoid of any material substance would be even more contrary to the order of nature than a downward motion of these light elements. Similarly water, which (being a 'heavy' element) by its nature ought to move downwards (i.e. to the centre of the universe with which the earth's centre coincides) will move upwards in a pump in order to prevent the formation of a vacuum.

The idea of a tunnel through the earth then, was known to Buridan. He did not, so far as I know, deal expressly with the question as to what would happen to a stone dropped into it, but there is a revealing passage in his commentary on Aristotle's *Physica*, which gives some indication of how he might have solved it. He posits that a ball thrown to the ground is compressed by the impetus of its motion, so that, next, it returns to its spherical shape and elevates itself, moved by a new impetus in the opposite direction. Similarly, the string of a zither after percussion remains a long time in vibration, for being bent in one direction and returning swiftly to its straight position it then is forced by the impetus to pass it by in the contrary direction, whereupon it returns again — and so on repeatedly. For a similar reason a bell, after the ringer (*pulsans*) has ceased pulling, swings for a long time alternately in one direction and the opposite one and cannot easily be brought to rest.[53] This latter example is fundamentally the same as that of a swinging pendulum and it is precisely this phenomenon that Buridan's pupils Nicole Oresme and Albert of Saxony later used as an analogy in their solution of the tunnel problem.

The answer of the adherents of the impetus theory to the tunnel question is thus quite profoundly different from that of the Aristotelians of the old school. Nicole Oresme (*Traité du Ciel et du Monde*, 1377) says that if we imagine the earth pierced by a hole towards the other end ('where the antipodes would be if the earth were inhabited everywhere') and we dropped a stone through this hole, it would descend and pass beyond the earth's centre and go upwards over a certain distance to the other side. It would then return and pass beyond the centre to this side, afterwards falling back again and going beyond the centre but not so far as before. It would go and come back several times, the reflex motion being smaller each time and finally it would come to rest at the centre of the earth. These motions would be caused by the impetus (*impetuosité*) which it acquires by the growth (*cressance*) of the velocity (*isneleté*) of its motion: 'And one can understand this easily from something we can see by our senses', for if a heavy object is hung on a long cord and we push it sideways it will move backward and forward, making several oscillations until it finally comes to rest with the cord in a perpendicular direction, that is, as near as possible to the earth's centre.[54]

In another passage Oresme had pointed out that impetus is not the same as weight or heaviness, for it can push a heavy body upwards beyond the earth's centre: 'it is an accidental and acquired quality'. He then makes also the comparison with 'a heavy object hanging from a beam by a long cord.'[55] In an earlier work he uses the comparison of the 'many reflexions' of the stone falling through a tunnel with a bouncing ball (cf. Buridan!) and with the swinging beam of a balance.[56]

The imaginary oscillating movement is also compared with the *observed* phenomenon of a swinging pendulum: in both cases the oscillations gradually decrease and the final standstill occurs at a point as close as possible to the body's 'natural' place. This comparison with a pendulum mars the perfection of the image of an ideal thought-experiment: just as the concrete pendulum comes to a halt, so the oscillating stone will finally cease moving. But in the case of the pendulum there is the resistance of the air and this disturbing influence might perhaps be the only cause of the decline of the impetus. According to Oresme, however, there would be an 'exhaustion' of the impetus in any case. In spite of his frequent and severe criticisms of Aristotle, he is Aristotelian enough to hold that the impetus is spent, in the way that the 'force' of a living being gets exhausted. The example of the pendulum may have supported his conclusion that there is a final standstill, but it was already implied in his theory.

Another philosopher of the Buridan school, Albert of Saxony, solved the tunnel problem in the same way: and he, too, was of the opinion that the oscillations would come to an end because the impetus would be spent.[57]

The *via moderna* took an important step in the direction of a mechanistic world view. The comparison with the pendulum makes the tunnel problem a more 'mechanical' one than it was with the orthodox Aristotelians. In their opinion the stone yearns for the centre of the world and it suddenly stops when its goal has been attained, like a horse running to the stable and halting at the trough, that is, like a living being that has reached its aim. With Oresme, on the other hand, the falling stone behaves like inanimate beings — a pendulum, a balance — which oscillate around their final resting place. His doctrine of the final exhaustion of the impetus, however, shows that there are 'organistic' traits left.

3.5. No Experimental Testing

It is notable that none of this talk of calculations, latitudes and impetus led to experimental measurements in order to test the conclusions of the reasoning. In his considerations on dynamics Thomas Bradwardine suggested several alternative laws relating force, resistance and velocity for a moving body, but he chose that which he held to be the correct one on the ground of subtle reasonings, not because of a decisive experience.[58] Oresme, in his kinematics, dealt with several kinds of motion (uniform, uniformly difform, etc.) but he did not enter into any concrete cases (with the exception of uniform circular motion of heavenly bodies). The relation between velocity, time and distance in the case of a uniformly difform motion (a motion with constant acceleration) was correctly deduced, but no one held that the motion of a freely falling body (which was known to be accelerated) might be an example of it, let alone that it might be tested by experiment.[59] It seems that the medieval philosophers were hardly interested in putting their theories to the test of anything more than coarse experience.

There were some profound reasons why they did not attribute a high value to measurement. They realized that their mathematical speculations dealt with idealized events: such as motion in a vacuum (which, in their opinion, could not exist in reality) and motions without friction. Only in the heavens could 'ideal' motions take place and so become liable to exact measurement. On an imperfect earth, however, no mathematical laws could be verified. Jean Buridan makes this evident by saying that although we suppose the moving forces to be constant, in nature this is never the case:

> Consequently these rules [of the relations between velocity, force and resistance] will seldom or never be found to have their effect [in reality, RH] ... Yet these rules are true under some conditions. And truly, if these conditions

> which we supposed in the rules are fulfilled, then it would happen as the rules posit. Therefore, it should not be said, that these rules are useless and fictitious (*inutiles ac fictitie*), for though those conditions could not be fulfilled by natural forces, yet it is quite possible that they could be fulfilled by the divine Power. Moreover, these rules are very useful, for they tend to inculcate a truth, that increase of the proportion of motor [= force, RH] to moveable [= resistance, RH] generally increases the velocity of the motion...[60]

So the hypothetical law of Bradwardine leads to a quantitatively rough approximation to phenomena. In this trend of thought the idealization of concrete reality is not based on an analysis of reality. In contrast to the modern scientist, Buridan does not say that the ideal phenomenon is an underlying and constituent part of a phenomenon in the world in which we live. Rather platonically, (what we call) the 'real phenomenon' is a crude analogy of the real one which may be realized somewhere at some time by the divine Omnipotence, but is not met with in common experience. It reminds us again of the decree of bishop Tempier (1277) who, while admitting that a vacuum does not exist in nature, nevertheless condemned the thesis that God cannot create a vacuum. One step further would be to postulate that perhaps a vacuum *does* exist in nature and to make an effort to realize it experimentally. He who has gone so far might hope that by art a situation might be realized in which even the law of bodies falling without meeting any resistance could (more or less precisely) be verified experimentally.

The strength of reasoning of the medieval philosophers was at the same time their great weakness: they wanted an absolute precision — a precision which apart from astronomical and perhaps some statical phenomena was far beyond their reach. Kinematical problems (measurement of velocity) let alone dynamical ones (measurement of force) were recognized to be blurred by disturbing influences; and even if those philosophers had been interested in the elimination of these factors in order to create a 'pure' phenomenon, the performance of such experiments would have been beyond their capacity.

Besides the fact that in nature we do not meet with 'pure' phenomena, there was still another reason that impeded these philosophers from tackling the procedure of verification, a reason not regarding nature herself but the investigator of nature. Measurements are never completely precise. Oresme pointed out that because we never know whether two things are absolutely equal, by our science we can never acquire an *exact* knowledge of proportions.[61] An approximate measurement, a slight error, was inadmissible to these punctilious logicians and metaphysicians. Their demand for absolute exactness was thus another impediment in the path of a mathematical-empirical science of nature.

The modern scientist, when considering the development of his discipline, tends to *trace back* selected aspects of problems still alive today. Consequently one could easily forget that 'latitude of forms' and 'calculations' were applied not only in kinematics and dynamics but also to all manner of things lying outside the sphere of mechanics — such as ethical problems — some of which have afterwards turned out to be hardly suitable for mathematical treatment. Consequently our forebears get praised for tackling the problems of the intensity of force and velocity and heat, but have been severely criticized and even ridiculed for having tried to submit 'virtue' and 'sin' to their calculations. We easily forget that not so long ago school reports expressed the degree of 'diligence' and 'behaviour' of the pupil by numbers from 1 to 10. *A priori* there is nothing against efforts to mathematize any realm of life whatsoever: in the long run only experience will show which of these efforts are futile and which subjects lend themselves to such a treatment.

The theory of the intensification and remission of qualities has borne fruit first of all in mechanics; but other 'qualities' have also shown themselves to be liable to quantification. The intensity of heat, for example, is expressed by 'temperature' (measurable by the thermometer), while the quantity of heat is measured by the calorimeter. It is only with hindsight, however, that we know that this quality, perceived by the sense of 'touch' is quantifiable; we have not been so successful with the causes of what the senses of taste and smell experience. Moreover we should realize that within the framework of Aristotelian thought it is quite as plausible to apply the same procedures to the intensification of sinfulness as to that of velocity. To Aristotle *all* change is motion (from a potential to an actual state); why then should *local* motion be singled out, and why then should not a change, a motion, from good to better be treated in the same way? Both 'good' and 'quick' are 'qualities' and both are liable to quantification. Consequently it could be deemed possible to have divergent and convergent series of 'merits' in mathematical ethics as well as in mathematical physics. And as we do not deem it strange that such 'mathematical' speculations in ethics did not lead to practical measurements, we should not wonder too much why kinematic speculations did not go together with measurements. Perhaps the above will help us to form a milder judgement about the fact that the *Calculatores*, who considered all qualities to be measurable (even those that cannot be measured), nevertheless did not measure any of them (not even those that are measurable).

The more the subtlety of their speculations grew and the more hairsplitting their discussions became, the farther the scholastic philosophers drifted away from reality. One gains the impression that for many Calculators their problems served only as a 'whetstone of wit', a mathematical game not really concerned with physical reality but performed only 'par esbattement'. In this way, even so, they contributed to the development of mathematics and to a better insight into epistemological and methodological problems regarding the relation between experience and reason and mathematics and paved the way for the further mathematization of mechanics in the 17th century by scholars who had been educated in their particular school.

> At the end of the 15th century the calculations lost credit. Many Renaissance scholars preferred concrete reality to abstractions, in particular when their own mathematical interest was slight, or restricted to numerological speculations. It is understandable that humanists who had just discovered the abundant joy of life should abhor the bloodless speculations of late scholasticism. Luis Vives (1492-1540) may have had in mind such problems as that of the stone falling in the tunnel through the earth when he wrote that 'they dispute about cases that never occur and even that never *can* occur in nature. They expatiate on foolish subtleties, which they call 'calculationes'. The young people who are educated with these silly questions do not know anything about plants and animals and they have not the slightest experience of the things of nature.'[62] Another humanist, Francisco Pico della Mirandola, spoke deprecatingly about the 'rubbish of Swineshead' (*quisquiliae Suiceticae*).[63]

> Renaissance philosophers with a mathematical interest, however, assumed a more positive attitude, and this was even more true of the still numerous adherents of the old methods of education. Consequently the work of Suisseth was printed and commented upon even in the 16th century. Hieronymo Cardano (1501-1576), himself a great mathematician, when enumerating the twelve greatest thinkers that had lived up to his own time, put Suisseth in the third place, after Archimedes and Aristotle and together with Euclid. The Averroistic and Aristotelian diehard Julius Caesar Scaliger, who denied practically every point Cardano advocated, for once agreed with him: Suisseth's intellect was 'beyond the measure' of human ingenuity.[64] The impetus theory, which had become acceptable not only to the nominalists but also to the 'realists' of the old school, was accepted also by Scaliger who even managed to trace the origin of this non-Aristotelian theory in the writings of the *praeceptor veritatis, dux noster, dictator noster*, though of course this required an exegetical *tour de force*.[65]

> Thus through an uninterrupted tradition, all Renaissance scholars and scientists were familiar with the impetus theory in some version or other and the thought-experiment of the tunnel continued to fascinate all parties.

The tunnel problem turns up even in the writings of the humanist Erasmus who — though highly critical of the School philosophy — had absorbed some of its teaching as if it were a matter of course. In one of his *Colloquia* a certain Alphius repeats the belief that a stone dropped in a hole piercing the earth would come to rest in the centre because this is the place of rest for all heavy bodies. His partner in the dialogue, Curio, answers that a natural movement that meets with no obstacles in the tunnel would go faster and faster, so that the stone or the piece of lead would go past the centre, and the backward motion would then be a violent one. Alphius agrees that the stone would go past the centre, and that it would then slow down and finally return to the centre in the way that a stone thrown high by force returns to the earth. Curio then adds that upon returning by natural motion the stone in the tunnel will again be carried past the centre by the impetus produced and thus never come to rest; but Alphius denies this: in the end, he insists, it will stop after having gone back and forth until equilibrium is reached.[66]

Evidently the solution implicit in Buridan's conception was too subtle and the result dictated by 'common sense' was more acceptable.

4. RENAISSANCE: THE ITALIAN ENGINEERS

Several 16th century Italian engineers accepted the theory of a self-expending impetus. In their treatment of the tunnel problem these practical men attributed high value to the analogy with the swinging pendulum. Their conclusion is always that, just as the swinging pendulum after many oscillations will stand still, so the oscillating stone will come to rest at the earth's centre when the impetus is wholly spent.[67]

In the first edition of his *Nova Scientia* (1537) the engineer Niccolò Tartaglia (1499-1535) still advanced the older theory. In his opinion a falling body moves more swiftly the more distant its starting point from the earth's centre and the closer it is to its final aim. He compares this with a pilgrim coming home from a distant place: the farther away he has been and the nearer he approaches his own place, the more he hastens his pace. Evidently, in Tartaglia's reasoning (as with so many people before him) the attractive force of the central point of the earth has a considerable accelerating effect.[68]

In a later edition, however, Tartaglia disapproves of those who would let the heavy body stop immediately at the centre of the earth:

> ...but I say that this is not true; instead, by the great speed ... it would be forced to pass by with very violent motion,... toward the sky of our subterranean hemisphere and thereafter return by natural motion; and so it would continue for a time, and then finally it would stop at the centre.'

Another Italian engineer, John Baptist Benedetti (1530 - 1590) — who was on many issues a declared opponent of Aristotle — adhered to the impetus theory. In his main work (1585) he dealt with the oscillating movement of a body falling in a tunnel through the earth. In his opinion Maurolyco and Tartaglia were right, whereas Alessandro Piccolomini (who held to the old theory) was wrong: the former two 'so far outshine Piccolomino in these sciences as the sun outshines the other stars'. Benedetti elaborated the comparison with a pendulum: imagine a cord to which a weight is appended, and let the end be fixed to the outermost heaven at the zenith. If the falling stone would oscillate over a path equal to the diameter of the earth, the arc it describes would practically be a straight line. Let the amplitude over which it oscillates be equal to the diameter of the earth; now the stone (which is an earth's radius distant from the earth's centre) would act in the same way as if it were oscillating at the end of the cord.[69]

Benedetti, like Tartaglia, held that the force impressed upon the stone would diminish, so that it would finally come to rest at the earth's centre. There is then no essential difference between his standpoint and that of Oresme, though these practical men lay the emphasis on the parallel with the concrete phenomenon of the pendulum.

Even Galileo Galilei in his *Dialogue on the Two Chief World Systems* (1632) does not seem to have gone much further. In the second chapter 'Salviati' says: 'if there were a well, that did pass through and beyond the centre [of the earth] yet would not a clod of earth pass beyond it, unless inasmuch as being transported by its *impetus*, it should pass the same to return thither again, and in the end there [come] to rest'. Further on, when speaking about a pendulum (swinging under ideal circumstances), Sagredo, the intelligent partner of Salviati, asks himself 'whether the vibrations might not perpetuate themselves' and adds: 'and I believe that they might, if it were possible to remove the impediment of the air, which resisting penetration, does to some small degree retard and impede the motion of the pendulum...' Salviati (who speaks for Galileo) does not *wholly* agree, but he says that he proposed the observation of this pendulum, in order that his partner should understand that the impetus acquired in the descending arc — where the motion is natural — is of itself able to drive the ball with a violent motion as far on the other side in the ascending arc, if all external impediments are removed. He then goes on: 'Hence it seems to me ... that I might easily be induced to believe, that if the terrestrial globe were bored through the centre, a cannon ball descending through that well would acquire by the time that it came to the centre, such an impulse of velocity that, having passed beyond the centre, it would spring upwards the other way, as great a space as that was wherewith it had descended ... and the time spent in this second motion of ascent, would, I believe be equal to the time of descent...' Though Galileo does not expressly say so, this implies that in 'ideal' circumstances the oscillation would never end, for after the first passage, the situation is exactly the same as at the start. It is remarkable that here the concrete example of the pendulum is first extrapolated to the ideal situation, and that the imagined ideal experiment of the stone falling through a perforated earth then follows.[70]

It is interesting that Galileo held that the parts of the chain or thread of the pendulum nearer to the fixed point want to make more frequent vibrations; they do not permit the lower parts of the chain to swing as far as they would naturally do; so that by continually detracting from the vibrations of the plummet, they finally make it cease to move.[71] To Galileo the analogy became inapplicable where the plummet of the pendulum differed from the falling stone, in that it was not free but linked to the chain or cord.

It should be stressed that in spite of the adherence of the outstanding Renaissance engineers to the model of the swinging pendulum, among less sophisticated authors the more ancient conception survived to the 17th century.[72]

5. WILLIAM GILBERT; FRANCIS BACON

William Gilbert (1540 - 1603), the author of the famous work *On the Magnet* (1600), went much farther in the direction of an organistic world view than even Aristotle: his conceptions could be characterized as 'hylozoïstic'.

He was a strong opponent of the newly emerging mechanistic conception of the universe. Even the current astronomy, in which the planets were conceived as fixed upon epicycles (which were concrete shells) irritated him. He complained that 'wheeling mechanicians make almighty God constitute the whole fabric of the world mechanically so that wheels (*rotae*) must be added to the primary heavenly bodies which move in circles ...' Even Aristotle was wrong in denying a soul to the earth, which is thus 'degraded to a level lower than the worm she herself brings forth.'[73]

To Gilbert, as to Copernicus the motion of a falling body is caused by its desire to join the larger body to which it belongs. This implies that lunar matter would always fall to the moon, and terrestrial matter to the earth. Only those bodies do flow together at the earth which had gone out from it and therefore gravity is 'the inclination towards its source; to the earth go those which had gone out from the earth.' Accordingly Aristotelians — who held that a point or place (*locus*) attracts heavy bodies — also incurred Gilbert's severe criticism. He deprecatingly named them 'locastri'. Attraction into union is caused by 'sympathy' and not by the yearning for a certain place. A point cannot exert attractive force.[74]

For the solution of the tunnel problem, Gilbert therefore followed neither the *via antiqua* nor the *via moderna*. Both held that the final resting place of the falling body is a 'centre': both were 'locastri'. The stone falling down a tunnel through the earth will not acquire an oscillating movement; it will not pass through the centre, but on entering the tunnel it will immediately start to *lose* velocity and eventually adhere to the wall of the tunnel. A magnet would come to rest even sooner, because of its even greater affinity with the Great Magnet Earth. But both would make haste to join the 'Communis Mater' and come to rest at a moderate depth.[75]

Francis Bacon (1561-1626), not otherwise sympathetic toward Copernicus and Gilbert, shows affinity with them in his opinion about the cause of fall. He opines that falling bodies will weigh less in the mine than on the surface of the earth, as 'the appetite for union of dense bodies with the earth ... is more dull ..., because the body has in part attained its nature when it is some depth in the earth.'[76]

These latter words show how much 'Aristotelianism' continued to influence even those who thought they had emancipated themselves from it. It is part of the nature of a body, in the Aristotelian conception, that it is in its natural place; and the nearer it is to its natural place, the closer it is to the fulness of its realization. Now to the Aristotelians the natural place of a heavy body is a point (the centre of the universe) and to Bacon it is the mother body (the earth). The physical contents thus are different, but the metaphysical form had remained.

According to Bacon the opinion of the Ancients that bodies move to a *point* or place is 'vanity'.[77] Elsewhere he calls the supposition that a place 'has any power', 'silly and childish': 'Therefore philosophers do but trifle when they say that if the earth were bored through, heavy bodies would stop on reaching the centre. Certainly it would be a wonderful and efficacious sort of nothing, or mathematical point which could act on bodies, or for which bodies would have desire.' This desire of ascending and descending depends, according to Bacon, either on the configuration of the body in motion or on its sympathy or consent with some other body.[78]

6. The Mechanistic Philosophers

With the rejection of the idea that a *point* (the centre of the universe or the centre of the earth) has some physical power, the problem arises as to whether bodies falling below the earth's surface follow the same law as those above it. In this respect Gilbert's conception, though hardly mathematical, was of great importance. This was henceforth a problem that kept resurfacing.

The French universal scholar, P. Marin Mersenne, who had an extensive correspondence with many contemporaries (Descartes, Beeckman, Gassendi, etc.) could not definitively make up his mind about bodies falling below the surface of the earth. In his *Harmonie universelle* (1636) in connection with the question whether falling bodies always increase their velocity or will reach a 'point of equality' from which they will thereafter fall with a uniform velocity, he concludes that human reason, *without experiments*, cannot solve this difficulty, 'though many people imagine that the earth attracts heavy bodies, and that they fall more swiftly towards the surface of the earth than when they are lower, between the surface and the centre'. For, when they fall through air towards the surface, the whole earth attracts them, but when a falling body is beneath the surface all parts of the earth that are above it will try to pull it back. Consequently the velocity will gradually diminish, 'until being arrived at the centre, it would not be able to pass by it, because the two hemispheres of the Earth will then pull equally to the one and to the other side'. But Mersenne adds that we do not know whether bodies descend only because they are *attracted* or because they have some weight in themselves independent of their attraction; therefore we cannot draw any conclusion that would satisfy judicious minds, for experiments one could make in the deepest pits and mines do not perceptibly diminish the velocity, and the earth has no holes penetrating deeply enough for this purpose.[79]

Mersenne is right, that as long as the ball has not reached the centre, the attraction downwards is greater than that upwards, so that there remains a surplus of attractive force causing a (gradually diminishing) acceleration downwards: not the velocity but the acceleration is then dwindling to zero (when the centre is reached), whereas beyond the centre the velocity diminishes and the acceleration is negative (the *retardation* increases until the velocity is zero). But during the accelerated motion *towards* the centre the velocity once acquired will be maintained, so that the body will have its maximum velocity at the moment when it is passing the centre (i.e. when the acceleration is nil). Why then does Mersenne hold that, under the given conditions, the velocity diminishes? Because he here accepts the main error of ancient mechanics, according to which a constant moving force causes a constant *velocity* (and not, as Newtonian mechanics implies, a constant acceleration). Consequently with the attractive force becoming zero at the centre, the velocity, too, must be zero according to Mersenne.

On the other hand Mersenne is also under the influence of Galileo's *Dialogues*, and in the next part of his work he discusses these problems along lines indicated by the Italian 'mathematician'. Falling bodies, he says, augment their velocity in the direction of the earth's centre as the square of the time, 'for the weight steadily adds new motion in every moment of the fall.'[80] *If* the weight augments the velocity during the fall as the square of the time, and if we now imagine that there is an opening through the earth, says Mersenne, a cannon ball will certainly descend from the surface of the earth to its centre in 19 minutes and 56 seconds, and it will arrive at the centre with a velocity of almost two miles in the last second of its fall.[81] If, next, the ball were to pass beyond the centre, its velocity would decrease in the same way as it increased during the fall, so that 'it would cover almost the same distance in climbing', and it 'would balance perpetually going hence and thence, as if it were attached to the cord of a pendulum.'[82] Evidently the word 'perpetually' should not be taken too precisely, for he adds that all its descents and ascents will end in 326 days and 15 hours. The swinging of a ball on a cord of 3 feet lasts for one whole hour; and as, in a fall to the centre of the earth, there is no hindrance by a cord, so the ball will go on oscillating for at least a year before resting at the centre.[83] It is not clear, however, what are his reasons for expecting it to stop.

One of Mersenne's correspondents, Isaac Beeckman, had mathematically deduced the law of falling bodies as implying that the distances covered are as the squares of the times, if equal minimal velocities are added in equal minimal times. His physical, dynamical, hypothesis was that the gravitational force 'pulls with little jerks' ('sy treckt met cleyne hurtkens')[84]. Also whatever velocity has been already acquired, will continue: 'what once moves, goes on moving if it is not impeded from doing so' ('Wat eens beweegt, beweegt altoos, soo 't niet belet en wort.')[85] It should be noticed that Galileo went no further than a kinematical treatment of the problem.

Mersenne could not decide whether the cause of the fall is the body's weight (i.e. an inner downward tendency) or an attraction by something outside the body).[86] But he felt the necessity of making absolute measurements. However faulty these might be, his empiricist inclination on this issue was stronger than that of Galileo.

Thirdly the comparison with the oscillations of a pendulum led him (like Galileo) to the conclusion that the swinging will not last for ever (but 'almost' perpetually), though the freely moving ball in the tunnel would go on for a much longer time.

It is remarkable that in this line of thought those who, like Galileo, rejected the idea that a point exerts any force, did conceive of the falling body, even when below the surface of the earth, as if it were attracted by a point. One might object that the centre of gravity has every right to replace the sum of all points of a globe, but this is no longer obviously the case *within* the earth, and precisely this latter case presented grave difficulties to mathematizers.[87]

7. Isaac Newton

Newton's theory was highly applicable to the problem of the tunnel, though it seems that he never explicitly mentioned it. Newton fully recognized that a *point* had no attractive power, but pointed out that 'mathematically' it may be *as if* it exerts attraction. In the first book of his *Principia*, he says he was 'treating the attractions of bodies towards an immovable centre, though very probably there is no such thing existent in nature. For attractions are made towards bodies...'[88] Newton treated bodies as being the sum of their parts: 'We are to compute the attractions of the bodies by assigning to each of their particles its proper force, and then finding the sum of them all.'[89]

He calculated what would be the effects of attraction according to the inverse proportionality to the distance, and that to the inverse cube of the distance, etc. But to gain agreement with concrete astronomical phenomena (the radii and periods of planetary orbits) he found that gravitational force had to be taken as proportional to the inverse square of the distance between bodies.

Assuming then that attractions between particles operate according to an inverse square law, he found that for a particle lying outside a uniform spherical shell, the force exerted on it is directed towards the (vacant) centre of the shell and is inversely proportional to the *square* of the particle's distance from that centre.[90]

If, however, the particle lies inside a uniform empty shell the force exerted on it by all the particles of the shell is zero, i.e. as it descends the tunnel the layers of the earth above it exert no net gravitational force. A solid sphere can be thought of as made up of uniform concentric shells. So the combined effect of Newton's two calculations is that the force exerted on a particle within the sphere is directed towards the centre and is *directly* proportional to its distance from the centre. 'If to the several parts of a given sphere there tend equal centripetal forces decreasing as the square of the distances from the points, I say, that a corpuscle placed within a sphere is attracted by a force proportional to its distance from the centre.'[91] This, though Newton does not say so, implies that the motion of a particle through an imaginary tunnel piercing a uniform earth (ignoring the effects of resistance) would be a permanent oscillation.

Thus we see that, just like the *Calculatores*, Newton started from imaginary cases, and computed the consequences; but quite unlike them he proceeded to *test* which of the imagined laws was the one prevailing in nature. In all his calculations he thought in a mechanistic way. This does not imply that he put forward a definitive mechanical model — let alone a mathematically manageable mechanical model — however much he would have liked to do so (see section VIII.5). But the principle that the effect of the whole is the sum of the effects wrought by the parts is a mechanistic principle and lay at the basis of his system, even if its mechanism remained obscure.

VI. 'And The Sun Stood Still'

1. Copernicus 148
2. The 'Impiety' of Copernicus' System 149
3. Man's Place in the Universe 153
4. The Absurdity of the Earth's Daily Motion 156
5. Physical Arguments for the Earth's Rotation 158
6. The Medical Analogy Between a Falling Body and a Sick Body 161
7. The Structure of the Universe 162
8. The Central Place of the Sun 167
9. The 'Novelty' of Copernicus' System 170
10. Tycho Brahe, Galileo and Kepler 175
11. The Paradoxical Character of the 'New Astronomy' 177
12. Towards the Final Triumph 179
13. Copernicus' Theory: 'True' or Only 'Scientific'? 180

Copernicus' great work *De Revolutionibus Orbium Caelestium* was published in 1543. Today it is widely accepted that this book, *On the Revolutions of the Heavenly Orbs*, marks the beginning of the 'scientific revolution': the Aristotelian Goliath was slain by the solitary Canon in remote Poland; a science based on observation triumphed over clerical dogmatism; the Earth was removed from its central position, and its human inhabitants thereby lost the privileged position accorded them by the Bible — whose authority in its turn was severely damaged. In short, with Copernicus fact triumphed over faith and fiction ...

The Copernicus commemoration of 1973 revived this quaint mixture of truths, half-truths and non-truths in many semi-popular articles. Copernicus was depicted as a 'progressive' thinker and even as a 'revolutionary character'.[1]

In the present chapter we shall try to gauge how far these claims fit the historical facts — and how far they belong to hagiographical fiction based on the belief that the development of science is a continuous and glorious progression from a dark and dim past toward the light of the present day. When we have brought alive in our imagination the situation of 16th century science, we shall, I think, recognize that however much we can now see, with hindsight, that Copernicus' theory *was* a new beginning in astronomy, the matter must have looked rather differently to his contemporaries. The scholarly literature of the second half of the 16th century gives, in fact, very little support to the idea that a 'revolution' was taking place. Copernicus' work was duly admired by the astronomers for the mathematical skill displayed in it. But a much greater commotion was caused by the then recent *geographical* discoveries and later by the rise of a new star in 1572.

1. COPERNICUS

Nicolaus Copernicus (1473-1543) belonged to a wealthy burgher family in Prussia. After his studies in Cracow he prepared himself for future ecclesiastical office by following courses in Canon Law and Medicine in Bologna and Padua. On his return to Poland he lived in Frombork (Frauenburg) as a canon of the semi-independent bishopric of Warmia (Ermland). He functioned there as an administrator, practiced medicine for his colleagues and wrote a treatise on monetary reform. He had the humanist predilection for the writings of the great Greek and Roman philosophers. In religious matters he took a moderate position, similar to that of his best friend Tiedemann Giese (who became bishop of Kulm (Chelmno) in 1538).

During his long stay in Italy Copernicus acquired a special interest in astronomy, a discipline important at that time for ecclesiastics as well as for physicians. At the end of the dedicatory letter of his main work he reminded Pope Paul III that the fifth Lateran Council (1512 - 1517) had not been able to solve the problem of the lengths of the year and of the month, both of which are important in establishing the church calendar, and that the president of the papal commission for calendar reform, Paul of Middelburg (bishop of Fossombrone), had urged him then to occupy himself with those problems and with that of the motions of the Sun and Moon.[2]

These astronomical studies led Copernicus to a new system of the cosmos, which he expounded in a small tract (*Commentariolus,* circa 1514), manuscript copies of which circulated among his acquaintances. The first *printed* account of Copernicus' ideas appeared in 1540 in Danzig (Gdansk). It was written by his only direct pupil, Georg Joachim Rheticus (1514-1574), a young Wittenberg professor of mathematics who worked under Copernicus at Frauenburg from 1539 to 1541. When he left Poland Rheticus took with him the manuscript of Copernicus' *magnum opus* and offered it to the Nuremberg publisher Johannes Petreius, a friend of his protector Melanchthon. The printing was finished under the supervision of one of the local ministers, Andreas Osiander.

In the preface to his great work Copernicus says that he withheld its publication not just for the Horatian nine years but for more than three times nine years.[3] He delayed it because he expected opposition to his theory from scholastic philosophers and theologians on account of:

a) its alleged impiety, b) its absurdity, and c) its novelty,[4] and it was thus only after long hesitation that he yielded to the pressure of his friend Tiedemann Giese [and of his pupil G.J. Rheticus] and let the manuscript be published. He added a letter from the Cardinal Nicolas of Schönberg (1536), who had also urged him to publish the work.[5]

In rough outline we might say that in Copernicus' system the Earth and the Sun have exchanged places: not the Earth but the Sun is standing still in the centre of the universe, and it is the Earth which goes around the Sun in an annual motion. Secondly, not the Heavens but the Earth rotates around its axis in twenty four hours. And thirdly, the Earth has a 'third motion', a slow rotation of its axis round the pole of the ecliptic which explains the slow motion of the equinoctial points (see figures 21 and 22).

We will now consider how Copernicus defended this thesis against the anticipated opposition.

2. The 'Impiety' of Copernicus' System

Most theologians accepted the Aristotelian and Ptolemaic world systems because, in them, the Earth stands still and it is the Sun which has a daily and an annual motion, which is conformable not only to immediate observation, but also to the letter of the Bible.

Copernicus himself hardly enters into the problem of biblical interpretation, but in one passage he speaks in a deprecatory tone about people who might adduce the authority of the Bible against his theory. In his opinion they distort texts from Scripture for their own ends: 'their judgment I despise as stupidly audacious'. He calls them 'logic-killers' (*mataiologoi*) and compares them with Lactantius — 'a famous author, but a poor mathematician, who spoke about the shape of the earth in a childish way when mocking those who hold that the earth is spherical.'[6] The reference to Lactantius is pointed. This 4th century theologian had denied the existence of the antipodes and so had become a particular laughing stock of the learned since the oceanic navigations of the 15th and 16th centuries had directly shown that the Southern hemisphere is inhabited and that the Earth is circumnavigable. Copernicus is thus making clear that by 'mataiologoi' he means those interpreters of the Bible who keep to the letter when it refers to things belonging to 'natural history'. He evidently felt that if literalism led to absurdities with regard to geography, this could also be the case when astronomical issues were at stake. A non-Ptolemaic geography had triumphed in spite of literalistic interpretations of the biblical text; a non-Ptolemaic astronomy should not be condemned on arguments based on such interpretations.

SECTIO POSTERIOR. 205

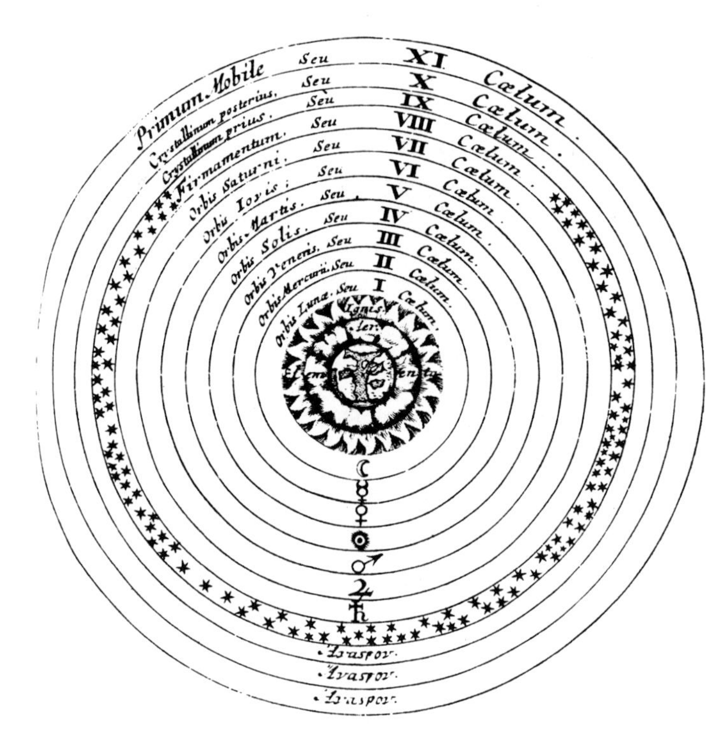

Quemadmodum verò Motus communis & proprius Stellarum per Orbes hosce peragatur, suprà tradidimus. Ipsum ordi-
IX. Apparentiis non satisfacit.
Cc 3

Fig. 21. The Ptolemaic system of the Universe (from: Joannes Luyts, *Astronomica Institutio*. Trajecti ad Rhenum: Halma 1692, p. 205).

Copernicus himself did not enter further into exegetical problems. His pupil Rheticus, however, wrote a treatise maintaining that the new system was not incompatible with Holy Scripture. Unfortunately this tract was lost for four hundred years and no particulars as to its contents were known, until (in 1972) I had the good luck to discover and identify an anonymously written text as Rheticus' missing defence of Copernicus' orthodoxy.[7]

In it Rheticus followed St Augustine, who had dealt with the literalists of his own time who were combatting Greek physics because it did not agree with the letter of the Bible. Against them the church fathers had maintained that Holy Scripture is intended only for teaching what is necessary for salvation. In scientific matters it does not speak apodictically but adapts itself to the common way of

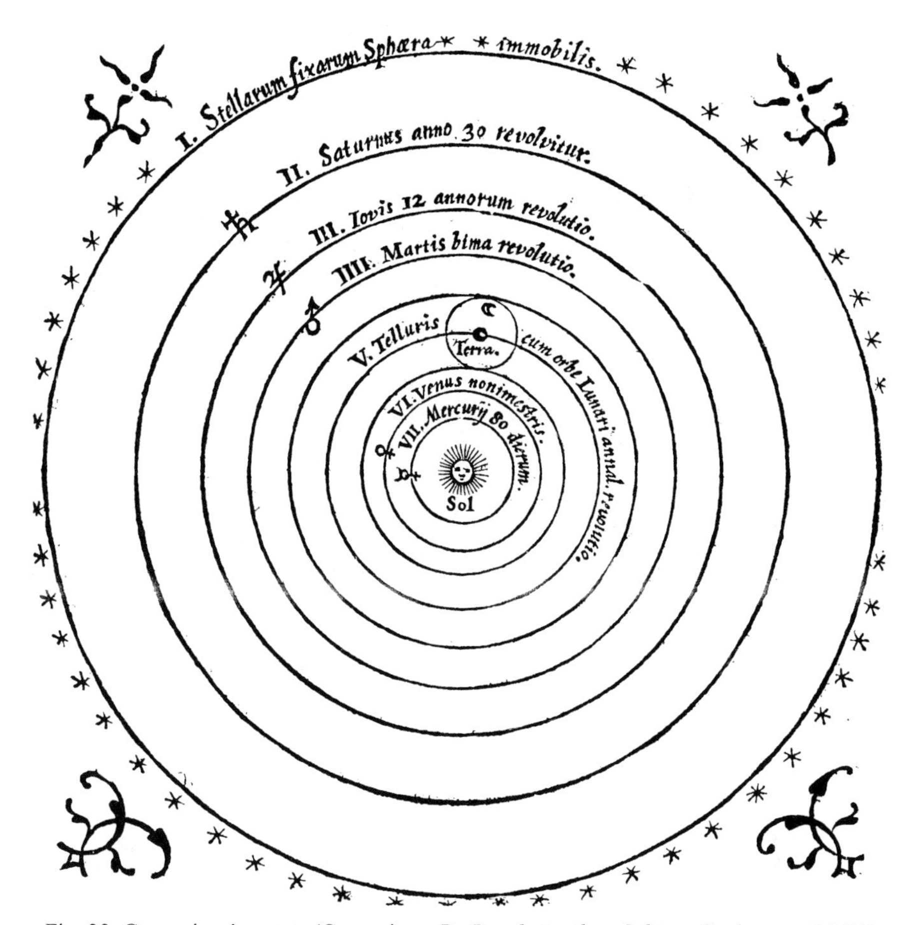

Fig. 22. Copernicus' system (Copernicus, *De Revolutionibus Orbium Caelestium* (1543)).

speech or the vulgar opinion. This principle of 'accommodation' to current opinions and way of speaking was now applied by Rheticus to biblical passages referring to the 'motion' of the Sun and the 'stability' of the Earth. He pointed out, for example, that Scripture does not mention the recently discovered New World, and thus evidently has no intention to give scientific information.

Within the Christian Church there has always been discussion as to whether the Bible speaks apodictically about things of nature. In his commentaries on the Book of Genesis and on the Psalms, Calvin frequently applied the principle of accommodation. More emphatically than any other commentator he pointed out that current (i.e. Ptolemaic) astronomy — which he accepted as scientific truth — seems to clash with the letter of the Bible, in which the Holy Spirit, in matters not touching doctrine and morals adapts Himself to the common way of speech and even to common error.[8] Calvin suggested, for example, that the 'waters above the expanse' probably means clouds and not a vast mass of water above a heavenly roof. The whole idea of a solid 'expanse' he considered an adaptation to naïve observation, for astronomy shows us that there are several heavenly spheres. It is probable that he did not know about Copernicus but he was familiar with the suggestion of the earth's daily rotation and mentioned it in one of his sermons *not* as something offending Holy Scripture but as an example of the absurdity into which some philosophers had fallen.[9]

Later on, the accommodation principle as put forward by Calvin was frequently appealed to by science-cultivating ministers of the Reformed Churches — like Philips van Lansbergen in Zeeland and John Wilkins in England. In the same way that Calvin believed that the Ptolemaic system might be true although it is not conformable to the letter of the Bible, these scholars concluded that the Copernican picture might be true in spite of its conflict with a literal interpretation of some biblical texts. In England and in the Netherlands Republic there was thus a predilection for Copernicanism amongst Reformed scholars. As to Holland this was remarked on by Galileo and by the French priest Pierre Gassend, and also — though with disapproval — by the Flemish priest Libertus Fromondus. They even seemed to believe that *all* Calvinists were Copernicans, and Froidmont went so far as to speak of the 'Calvino-Copernican system'. As late as the beginning of the 18th century the Italian Jesuit Thomas Ceva (1640 - 1727) asked himself why the Arctic wind turns the Earth around in Holland and England, whereas in other countries it is standing still.[10] Of course this was a generalisation, used also by Galileo to urge his Roman Catholic opponents to take the wind out of the sails of the heretics — and by Fromondus to discredit the theory in the eyes of his fellow believers. In fact, the 'freedom of philosophizing' in those two countries made it possible there for both the conservative and Copernican standpoints to be simultaneously respectable. The influential Reformed theologian Gisbert Voet (1588 - 1676), for example, was violently opposed to Copernicanism. For Reformed scholars there was evidently room for a free option.

On the other hand those Roman Catholics who sympathized with Copernicanism had to content themselves, after the condemnation of Galileo (1616, 1633), with the Tychonian system (see page 175).

Some adherents of the accommodation principle considerably weakened their position by trying to detect in Holy Scripture allusions to the motions of the Earth. Rheticus, who so vigorously defended the principle of accommodation with many references to St Augustine, was nevertheless of the opinion that Scripture, though 'under a veil', speaks about the daily, the annual and even the 'third' motion of the earth. While denying that Biblical passages speaking about 'the foundations of the earth' say anything in favour of the fixity of the earth, he interpreted them as referring to the centres of the diverse circular motions of the earth.[11]

Galileo went even further in this direction when he interpreted 'and the sun stood still in the midst of the heavens' as referring to the axial rotation of the sun, which he had recently discovered. That is, instead of an interpretation of *nature* dependent upon theological prejudices, he put forward an interpretation of *Scripture* dependent on the most recently acquired scientific knowledge. The Puritan minister (and future bishop) John Wilkins, showed a greater sense of balance and consistency, repudiating Galileo's practice of discovering modern science in the Bible as soon as this seemed advantageous to his own standpoint.[12]

3. MAN'S PLACE IN THE UNIVERSE

A supposed connection between man's rank in the hierarchy of created beings and the 'importance' of the place of his terrestrial abode has occupied many writers on the Copernican controversy. They attribute the opposition to the new system to the annoyance caused by the displacement of man's dwelling place from the centre of the universe. As the Bible deals with the relation between the Creator and His creatures, and as mankind, to whom was given the place of honour among the creatures, was 'deposed from it' by Copernicus, this caused — so they say — a violent opposition from the theologians.[13]

At first sight this seems quite plausible, and therefore the proponents of this explanation deem it unnecessary to adduce solid proofs for their contention. Yet it is one of those 'clever' constructions by moderns who think they know better than our ancestors themselves what were their real motives. Following the same method of interpreting the past we could just as easily and with as little foundation 'demonstrate' that the theologians were *not* predisposed to *reject* the Copernican system, but rather to acclaim it, on the grounds that it presents a Cosmos that is yet another proof of the divine economy. Jesus Christ, the Son of God, was born not — paradoxically — in imperial Rome but in a minor province, and then not in the capital, Jerusalem, but in the humble village of Bethlehem, and then not in a stately home but in a stable, and not to a high class family but as a carpenter's son. Now what would better fit with all these circumstances than that he should be born, not in the centre of the universe but on one of several planets at a considerable distance from the centre of the universe? An edifying thought, brought forth by the same kind of gratuitous phantasy as the other one. The fact is, however, that no solid proof has been given that *doctrinal* ideas played any significant role either in the rejection or the acceptance of Copernicanism.

In historical interpretation we have to keep to facts, and it is a fact that Copernicus expected (and afterwards got) theological opposition only on the ground of *exegetical* objections and not because of his changing man's place in the universe. From Rheticus onwards, those who defended his theory as compatible with Christian religion dealt only with the exegetical problems and never with matters of doctrine (apart from the accusation that a non-literalist interpretation of certain biblical texts was in itself 'unorthodox'). Man's dwelling place was, if anything, exalted by being put on an equal footing with other heavenly bodies. In the geocentric, Aristotelian system the Earth might be at the centre of the universe, but it was inferior to the heavenly bodies in that while they enjoyed unchanging and unending perfection of light and motion, the Earth was liable to change, darkness and decay.

The Copernican system, then, met with exegetical rather than dogmatic objections, for the idea that man's abode was not at the centre of the universe did not change anything in religious belief. In John Donne's *Ignatius his Conclave* (1611) Loyola says to Copernicus:

> what new theory have you invented, by which our Lucifer gets anything? What cares hee whether the earth travell, or stand still? Hath your raising up of the earth into heaven brought men to that confidence, that they should build towers or threaten God againe? Do not men believe; do they not live just as they did before? ... Besides ... those opinions of yours may very well be true.[14]

Obviously there is here no trace of the idea that man would feel humiliated by being ejected from the centre. Quite the reverse: he might perhaps feel exalted because of his abode's being raised into the heavens. Yet there was no reason for pride either, for as Donne wrote to Sir Henry Goodere in 1615: 'Copernicanism in the Mathematique, hath carried earth further up, from the stupid Center; and yet not honoured it, nor advantaged it, because for the necessity of appearances, it hath carried heaven so much higher from it.'[15]

Galileo was yet more positive about the enhancement of the earth by the new system. In the first chapter of his *Dialogues on the Two Chief World Systems* (1632) he pointed out to his peripatetic opponent that the Copernican system had raised the Earth to the rank of a heavenly body, by considering it as one of the planets: 'as for the Earth, we strive to ennoble and perfect it, whilst we make it like to the Coelestial Bodies, and as it were place it in Heaven, whence your Philosophers have exiled it.'[16]

It is similarly said today that Copernicus took away from man the safety of his quiet abode and made it an insignificant point in the universe. Ptolemaic astronomy, however, already considered the Earth as a mere point in a vast universe. The most popular astronomical textbook up to the 16th century, Sacrobosco's *Sphaera*, corroborated this conception. So Copernicus made no essential change in this respect. He emphasized that the Earth is *practically* in the centre of the universe. Precisely the fact that was adduced against the assumption that the Earth moves around the centre, *viz.* that no parallax of the fixed stars had been observed, was to him a proof that it is close to the centre, i.e. that the heavens are immensely great compared to the Earth's orbit: 'though the Earth is not in the centre of the universe, nevertheless its distance from it is immeasurably small in comparison with the sphere of the fixed stars.'[17] Or, to put it in the words of one of the earliest Copernicans, the English astronomer and military engineer Thomas Digges (1545 - 1595): 'But that Orbis Magnus [the Earth's orbit, RH] beinge ... but as a point in respect of the immensity of that immovable heaven, we may easily consider what little portion of God's frame our Elementare corruptible worlde is, but never sufficiently be able to admire the immensity of the Rest. Especially of the fixed Orbe garnished with lights innumerable and reaching up in Sphaericall altitude without ende.'[18]

Copernicus, Kepler and Galileo, though holding the universe to be immense, still clung to the idea that it is closed. With Kepler in particular aesthetic motives here played a role. He was — which was quite unusual with him — rather violent in his opposition to the idea that the universe was of infinite extent: for such a thing would destroy the beautiful harmony of the world.[19]

To Thomas Digges, however, the opening up of the heaven of the fixed stars was not too bold a thought. In his 'A perfit description of the Caelestiall Orbes' (1576) he was the first astronomer to assume an infinite universe. He was a staunch Puritan, 'zealous in religion', as the term ran, but evidently his religious convictions were not in the least disturbed by his scientific belief that the fixed stars are disposed throughout endless outer space. The geocentric astronomers (e.g. Petrus Apianus and Rembert Dodoens) had placed the 'heaven of the blessed' *beyond* the sphere of the fixed stars and eventual 'empty' spheres.[20] Digges had no difficulty in placing the 'habitacle for the elect' *among* the fixed stars (see figure 23).[21]

Man owed his high rank in creation to his being made in the image of God; the position of his dwelling place had nothing to do with this honourable stature. Indeed if it were true that the *central position* conferred dignity, Hell would have been the place most esteemed by the adherents of the geocentric system. For it, so they believed, was located in the centre of the Earth, i.e. nearest of all to the centre of the universe. The position of mankind in the universe was however much more a moral than a cosmological problem, and so played no serious role in the Copernican debate.

That the universe has been made for the *sake* of man was a tenet of Cicero's natural theology.[22] It was a generally accepted belief in the Christian Middle Ages too (see chapter 1 on Ramundus Sabundus) and it was firmly held by the Copernican Johannes Kepler.[23] Similarly Copernicus himself held that the wonderful construction of the world has been created 'for our sake (*propter nos*) by the best Maker, who works according to the most exact rules.'[24]

A perfit description of the Cælestiall Orbes,
according to the most auncient doctrine of the Pythagoreans. &c.

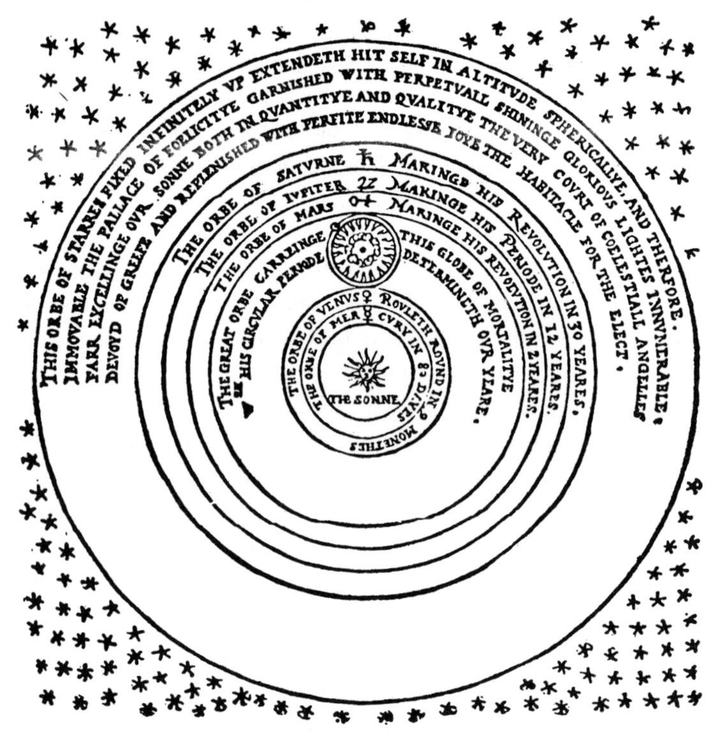

Fig. 23. The 'Pythagorean' [= Copernican] system of the universe according to Thomas Digges's *A Perfit Description of the Caelestiall Orbes,* London 1576. The last 'orbe' is no orb but the infinite abode of the blessed and the fixed stars.

4. The Absurdity of the Earth's Daily Motion

The possibility of a daily rotation of the Earth within a fixed outer heaven had been discussed by both Aristotle and the medieval philosophers. Some of them considered it to be a philosophically respectable standpoint, even though they finally decided against it for philosophical (Buridan) or for theological reasons (Oresme).

Probably the most weighty reason for rejecting the earth's rotation was the argument that, apart from seeing the moving skies, one does not *feel* any sensation of being on a merry-go-round. As so often Aristotle's 'philosophical' demonstrations on this issue were largely a sophisticated justification of what his 'common sense' and naïve observation had taught him. Later, the 'common sense' of educated people was further moulded by Aristotelian indoctrination in the schools: what has been inculcated into the mind in early youth and is believed by everybody, automatically becomes through long familiarity 'natural' and 'logical'.

Copernicus' fear that his work would meet with scorn because of its 'novelty and absurdity' was thus well-grounded.[25] Even a great admirer of his technical achievements, the Utrecht scholar Albert van Leeuwen (1543-1614), could not believe (1578) that Copernicus had meant seriously the idea of the Earth's rotation: 'the great man was only making a joke.'[26] Other scholars had taken the idea more seriously: Nicole Oresme (1377) adduced strong arguments in favour of it, though he finally declared that he had done so only by way of intellectual exercise (*par esbattement*).[27]

A less good humoured dismissal than that of Oresme was delivered by the Scottish humanist George Buchanan (1506-1582). He considered the notion of the Earth's rotation to be against good sense and ascribable only to sheer ignorance. In his didactic poem *Sphaera* (1581) he wrote: 'No matter how clearly reason will demonstrate these motions of the eighth sphere, i.e. of the fixed stars — ignorance, sunk in darkness, has not yet ceased to howl unabatedly, daring to damage heaven by assuring that it is at rest, and turning the immovable earth in quick motion.' ... 'Yet, they to whom nature gave the right sense of sight, see every day ... that Phoebus rises gradually from the sea in the new morning.'[28] With this Calvinist, as with Calvin himself, not the Bible but 'common sense' was adduced against the earth's rotation; it is an absurdity rather than an impiety.

On the other hand there were current arguments — some with their own 'common sense' appeal — which could be laid in the opposite scale. The 'mathematical' argument of the relativity of motion was recognized as valid even by opponents of the daily motion of the earth. If the earth has a certain motion we, who are living upon it, will see this motion as an 'appearance' in all things outside the earth.[29] Mariners out at sea have the impression that their ship is standing still and that all things around them are moving.[30] Metaphysical argument could also be drawn upon. On the principle of the 'simplicity of nature' it could be seen as more probable that the Earth should turn around in twenty four hours than that all the heavenly spheres should have that motion. Another argument was that motion being less 'noble' than rest, it is more appropriate that the noble heaven should be at rest while the vile Earth is in very rapid motion[31] (an argument that had been adduced by Oresme as pleading in favour of the Earth's daily rotation). Copernicus himself was no despiser of such metaphysical arguments and indeed added others; one was that it was more fitting to the order of the universe that the outer heaven (which 'contains' and provides the place for all things) should stand still, while the Earth and all other things 'that are contained' and to which a place within the universe is allocated, should move.[32]

A further argument put forward by Copernicus was that the earth is spherical and that the motion appropriate to a spherical body is circular motion: 'Why should we hesitate to allow the Earth a motion congruous with her nature, instead of assuming that the whole world, whose limits we do not know, is in movement?'[33] All this is conformable to the ancient belief that Nature does nothing in vain and that what can be done by fewer causes will not be done by more causes.

5. Physical Arguments for the Earth's Rotation

Today, of course, it is the *physical* arguments which Copernicus marshalled against Aristotle's in favour of a stationary Earth, that seem important. Quite a few of these were not entirely original, as they had been put forward in the 14th century by Nicole Oresme, and some of them also by Buridan whose ideas were being taught at Cracow university when Copernicus was a student there.[34] So Copernicus (and Galileo in his wake) shared with Oresme the arguments:

a) that *all* natural motions are circular and each element, when in its natural place, has a circular motion,[35]
b) that if a part of an elementary body is separated from the main body to which it belongs, it will try to go back to it by a rectilinear motion,[36]
c) that gravity is a tendency to *return to the main body* rather than to return towards a *point* (the centre of the universe), and that each thing wants to reunite with the body to which it belongs, (lunar matter will join the moon, terrestrial matter seeks the Earth),[37]
d) that circular motion belongs to wholes and rectilinear motion is an added property of separated parts,[38] and
e) that the movement of a falling body may be (Oresme) or is (Copernicus) a combination of the rotation of the Earth with the rectilinear motion of the fall.[39]

It should be stressed, however, that Oresme's cosmology was geocentric and that, unlike Copernicus, he did not base his argumentation on the firm conviction that the Earth is indeed in motion.

In the first outline of his system (*Commentariolus,* 1514) Copernicus had already posited that not the Earth but the Sun is in the centre of the world and that the Earth turns round on its axis in twenty four hours and that the observed annual motion of the Sun is in fact an annual motion of the Earth around the Sun.[40] But it is only in his main work of 1543 that his own *physical* principles on which he could defend the *daily rotation* against Aristotle are mentioned.[41]

The defence of these which he puts forward shows, however, that there is more affinity between Copernicus and Aristotle than appears at first sight. Copernicus' anti-Aristotelian physics is embedded in an Aristotelian metaphysics. Both Copernicus and Aristotle had a thoroughly teleological, organistic world view. All natural events occur, according to them, either to maintain or to attain their final aim: the full realization of the 'Forms' in their right place and their full essential being in the natural order.

According to Aristotle there is an essential difference between sublunar and heavenly elements, the former having a natural tendency either to seek by a rectilinear motion the centre of the universe (as do earth and water), or to approach as nearly as possible to the lunar sphere (as do fire and air). The rectilinear motions of the four elements are their *natural* motions. The heavens consist of a fifth element which is neither heavy nor light and the natural motion of which is *circular* and uniform, without beginning or end.

With Copernicus the novel departure is that he assumes that the Earth, too, is a heavenly body. *Its* natural motion therefore is also circular, uniform and everlasting. Consequently *all* natural motions in the universe are circular. From this anti-Aristotelian tenet, he draws in the following way a conclusion that still fits in with Aristotle's way of reasoning. Both Aristotle and Copernicus recognize that when a body is rotating with a large velocity, a disintegration will take place unless the rotation is a *natural* motion of the rotating body. But Copernicus points out that only objects to which force or violence are applied get broken up, but that things which are caused by nature are in the right conditions and are kept in their best organisation: 'Therefore Ptolemy had no reason to fear that the Earth and all things on the Earth could be scattered in a revolution occurring by the efficacy of nature, which is greatly different from that of art or from that which can result from the genius of man.'[42] Copernicus then wonders why his opponent does not feel anxiety about the universe instead, whose movement must of necessity be of greater velocity the greater the heavens are than the Earth? A rhetorical question, to which the answer would be analogous to that given by Copernicus when his opponents said that an Earth in daily rotation would fly apart: to the Aristotelian the daily rotation of the heavens was a *natural* one in the same way that to Copernicus the daily rotation of the earth was a natural one: so there was no reason for a centrifugal disintegration in the one case any more than in the other. Copernicus' cosmology, we see, may be anti-Aristotelian *in specie*, but it is Aristotelian *in genere*.

Both Aristotle and Copernicus held that all natural things seek by their nature to attain the completeness of their 'Form' (nature; essence) and one aspect of their perfection is that they should be in their 'natural or proper place.' According to Aristotle a piece of 'earth' (e.g. a stone) falls down by a natural action; 'motion towards its proper place is for each thing motion towards its proper Form.'[43] Similarly, Copernicus says that nothing is so repugnant to the order of the whole and to the 'Form' of the world, as 'being-out-of-one's-place.'[44] Both Aristotle and Copernicus thus explain the motion of a falling body by its desire to reach its natural place. To Aristotle this is the centre of the universe: a piece of earth falls towards the main body of the Earth not because it is attracted by the Earth, but because it wants to come as near as possible to the centre of the universe. As Aristotle explicitly put it: 'If the Earth were removed to where the Moon is now, separate parts of it would not move towards the whole (Earth) but towards the place where the Earth is now.' It is only because the Earth is now in the centre that we may say that the falling body is an example of the old rule 'like moves to like', for this is not strictly true. Thus did Aristotle in his time combat the 'ancients' and probably also his own teacher Plato, who had said that similar things try to come together: earth seeks earth, fire seeks to unite with fire.[45]

Copernicus, on this matter, wholly agrees with the older view: a fragment of earth falls toward the earth in a rectilinear motion because it wants to be in its 'natural place' (and thus far Aristotle would agree), but its natural place is not an abstract point, but the main body to which this piece of earth belongs — that is, the moving planet Earth, wherever it may be: 'I myself think that gravity or heaviness is nothing but a certain appetency implanted in the parts by the divine providence of the Maker of the universe, in order that they may unite with one another in their oneness and wholeness and come together in the form of a globe. It is believable that this desire is also present in the Sun, the Moon and the other shining planets so that through their affinity they remain in the spherical figure.'[46]

Copernicus, like Aristotle, spoke in a non-mechanistic way about the desire (*appetitus*) of the falling body to reach the 'natural place'. To him this implied reunion with similar parts to regain a whole. It would be going too far to regard this, as has sometimes been done, as a step towards Newton's general gravitation, for it implies that terrestrial matter seeks the Earth and lunar matter seeks the Moon. Each of them has its own specific matter which is attracted by the spherical body from which it had been separated.

Both Aristotle and Copernicus held that a body can have only one *natural* motion. To Aristotle the only natural motion of terrestrial bodies is rectilinear and of heavenly bodies is circular. To Copernicus (and Galileo after him), the Earth being a heavenly body, its only fully natural motions are circular. This is seen in the case of a fragment of earth when separated from the 'mother body'; it continues in its circular motion, but, if it is a falling body, it has a second motion combined with the first.[47]

This view does *not* imply, however, that all circular motion on the earth is natural. All earthly matter shares the 'natural' circular motion of the planet Earth, whether it is in it, on it, or above it, and that is why no disintegrating centrifugal motion is produced at the earth's surface by the Earth's rotation. But a circular motion caused by violence and human art shows a centrifugal effect. This may be the reason why Copernicus could not refer to *experiments* in order to support the Earth's rotation; he strictly kept to a fundamental difference between natural and violent motion.

6. The Medical Analogy Between a Falling Body and a Sick Body

Copernicus says that rectilinear motion is *added* to those bodies which are away from their natural place. This is quite Aristotelian, except that for Copernicus the 'natural place' is the main body to which they belong — as we have seen, a quite un-Aristotelian tenet. According to Copernicus nothing is more repugnant to the order and the Form of the universe than that something should be outside its natural place, so 'the rectilinear motion (*motus rectus*) occurs only to bodies that do not feel well (*non recte se habentibus*) and which are not perfect according to their nature, being separated from their whole and having abandoned the unity thereof.'[48]

This passage contains a play on words which I only lately recognized.[49] When reading so stately a work we must realize that not an oracle but a human being is speaking; yet we hardly expect a play upon words to be hidden in this so serious prose. The pun must have given some pleasure to Copernicus the *physician*. 'Right' (*rectus*) may mean 'rectilinear' (when speaking about motion) and 'well' (when speaking about the state of health). 'Recte se habere' is thus 'to feel all right', 'to be in good condition'. Copernicus had already used this phrase at the beginning of the same chapter when discussing an argument attributed to Ptolemy who (according to him) had said that if the Earth rapidly and violently whirled around, it would fragment.[50] Copernicus countered that this would be valid only if the Earth's rotation were a violent, unnatural motion caused by human 'art' and device. But the rotation of the Earth being a *natural* motion and since 'things that occur by nature are in the right condition and are preserved in their best composition', 'all terrestrial things in a revolution caused by the working of nature — which is quite different from that of art' — are not liable to dispersion.[51]

Copernicus then continues his medical comparison by concluding: 'as therefore circular motion belongs to wholes, but the [additional] rectilinear motion to the parts, we may say that circular motion remains with the rectilinear motion as does "animal" with "sick".'[52] That is to say that just as the sick animal remains an animal in spite of the abnormality of its sickness, so any part of the planet Earth shares its circular motion, even when in the abnormal state of being separated from the whole. The separated part is 'sick' — in an imperfect condition — because it is detached from the mother body, and now it tries to regain its state of health by means of a rectilinear local motion, just as the sick animal regains its normal state by the 'motion' of healing.

Now in Aristotelian philosophy (which on this point Copernicus follows), the motion of healing is not a wholly 'natural' motion, as would be for example the 'motion' of an embryo towards a complete animal, or the growth of a young animal into an adult. In the latter cases the motion starts from a situation that has nothing abnormal in it, whereas 'sickness' is *not* a perfectly natural situation. On the other hand, the process of healing is not an unnatural ('violent') one either, for it is a motion from an unnatural state to the natural perfection of the Form. Copernicus has now carefully pointed out that the rectilinear motion of the falling body is analogous to the process of healing. On the one hand it is *not* a perfectly natural motion as is the circular motion of the planet and its parts; it differs widely from it and manifests this by its rapidly increasing velocity which is far from uniform (*uniformis et aequalis*); thus, like a violent motion, it is changeable.[53] On the other hand, it is not an unnatural motion either, for it is a movement towards its natural destination — the perfection of the Form by the reunion with the main body, the Earth from which it had been amputated.

Copernicus' comparison of the 'rectilinear motion of a piece of earth towards its natural place' with the process of healing of the sick is fully Aristotelian. The physician Copernicus could find it in the *On the Heavens* of the physician Aristotle. The Philosopher says there that fire moves upwards and earth downwards *in the same way* as the curable, when moved and changed *qua* curable, progresses towards health. It is, says Aristotle, one of the cases of movement from potentiality to actuality (the attaining of the place as well as the attaining of quantity or qualities proper to the actual, perfect state). Just as earth and fire move to their own place unless something prevents them, so too the curable, when there is nothing to hinder, enters into the motion towards health.[54]

Medieval commentators on Aristotle, such as Averroës and Buridan, fully understood this simile: Buridan says that 'health (*sanitas*) is to the curable (*sanabile*) and the healing (*sanatio*) as "place below" is to the heavy body and its motion. Health is the goal of sanation and the "formal perfection" of the curable.'[55] When studying in Cracow, Copernicus had been steeped in Aristotelian doctrine, so it is not surprising that when he came to write his *De Revolutionibus* he made use of this same conception of the parallel between the motions of healing of sick bodies and the falling of heavy bodies. So yet again we see that however strongly some of his ideas might differ from the special cosmological tenets of Aristotle, they were still fitted into the old ways of thinking about nature. Because of this common ground his ideas might be seen by some of his opponents as far from 'absurd'.

7. The Structure of the Universe

The idea of an annual revolution of the earth was closely connected with Copernicus' conception of the structure of the universe. In the dedication of his work to Pope Paul III he pointed out that the Aristotelian theory of homocentric spheres did not satisfy the phenomena (e.g. the varying apparent size of the planets), and that the Ptolemaic theory of eccentrics and epicycles — though fit for computing the apparent positions of the heavenly bodies — was in conflict with the first principles of astronomy, which demanded strict uniformity of the celestial circular motions.[56]

No less than the Ancients Copernicus stuck tenaciously to the dogma of the uniformity of celestial motions, and in the first outline of the theory in the *Commentariolus* this lack of uniformity (which Ptolemy had tried to remedy by introducing 'equants'*) seemed to be his only reason for rejecting Ptolemy's theory. The equant was in Copernicus' opinion 'not sufficiently conformable to reason,' as it clashed with the principle that all motions in the heavens must be uniform.[57] He therefore wanted to find 'a more rational system of circles' to reduce to uniformity the apparent inequality of planetary motions, 'as perfect motion requires.'

In the dedicatory letter of *De Revolutionibus*, when emphasizing the lack of uniformity and harmony in Ptolemy's *structure* of the world, he said that the adherents of the system of eccentrics and epicycles could not discover 'the main theory', i.e. the 'Form' (*forma*: shape) of the universe and the correct 'right proportion (*symmetria*) of its parts.' The thought led on to a comparison of these astronomers with painters of the human body who take its various parts — hands, feet, head — from different individuals: these parts may be beautiful in themselves, but not in respect to one and the same body; consequently they do not correspond to one another, 'so that a monster rather than a man is composed from them.'[58]

It was indeed a normal practice and in accord with Renaissance theories of Art for painters and sculptors to borrow the several parts of the body from various individuals with respectively the most perfect shape of head, or the finest hands, and so on. But the condition was that the 'right proportions' were observed for reducing hands, feet, trunk and so on to the same scale so that a perfectly harmonious picture or sculpture might result. These 'right proportions' had been laid down in classical Antiquity, not only for the human body but also for architecture. There were parallel rules for the *macrocosm* (i.e. the universe) and for the *microcosm* (the human body) and it was held that works of art should follow them as closely as possible. The great architect in Roman Antiquity, Vitruvius, based the proportions of a temple on those of the human body. The planning of a temple depended, according to him, on 'symmetry', that is, on 'proportions' or analogy: 'It must have an exact proportionality (*ratio*) after the example of the members of a well-shaped human body.'[59] He pointed out that 'proportion' requires a fixed module, and that nature has so planned the proportions of a perfect human body 'that the face, from the chin to the top of the forehead is a tenth part ... the middle of the breast to the crown is a fourth part,' and so on, and that the ancient

* 'Equant'; a point outside the centre of the orbit, with respect to which the motion along the circle is uniform. Truly uniform motion implies a constant velocity with respect to the orbital circle.

painters and sculptors had attained great distinction by following these proportional measurements.[60]

Renaissance architects, sculptors and painters, like Luca Pacioli (*De Divina Proportione,* 1509), L.B. Alberti (*De Re Aedificatoria*, 1452), Da Vinci and Dürer (*Vier Bücher von menschlicher Proportion*, 1528) followed precepts that guaranteed that the mathematical principles of harmony would be recognized by the mind *via* the senses. Beauty in painting, sculpture, architecture and music thus found an objectivisation.

Copernicus shared these aesthetical and metaphysical convictions. With the Ancients he held that the universe was the most beautiful temple (*templum pulcherrimum)* to the divine work (*fabrica*) of the Most High.[61] He claimed to have found the most sure plan of the great artifice of the world, built for us by the best and most orderly Maker.[62] As he saw it, the current astronomy did not at all satisfy the demand for a realistic, and thus beautiful, system of the universe. It reduced the irregular motions of each of the five planets to a combination of (roughly speaking) two regular circular motions (deferent and epicycle), the relative dimensions of these two circles being chosen in such a way that the observed, apparently irregular, motions resulted from them. There was, however, no connection between the motions of the several planets; the possibility of alternative combinations was left open provided only that the visible result should be the same. It was, in Copernicus' eyes, an incoherent system, which might serve practical ends, but was not founded upon the solid ground of reality.

The way out was for him the heliocentric system in which the Earth and the sun changed places: the stationary sun took over the place of the (fixed) Earth of the old system, and the Earth now performed a yearly revolution along the course run by the sun in the old system. In the preface to his great work Copernicus proudly declared: 'I discovered by the help of frequent and long observation that if the movements of the other wandering stars are correlated with the circular movement of the Earth, and are computed for the revolution of each planet, not only do all their phenomena follow from that, but also this correlation binds together so closely the order and magnitude of all the planets and of the heavens themselves, that nothing can be shifted in any part of these without bringing into confusion the remaining parts and the universe as a whole.'[63] The key to the secret of the proportions of the heavenly spheres, the fixed module lying at the basis of the plan of the beautiful temple, thus was found. It was proven to be a structure of the highest beauty and therefore must be objective truth: the picture of the universe was no longer a *monstrum*.

Copernicus had recognized that the sun stands still and that its apparent course is a projection of the Earth's annual revolution and that, similarly, this annual revolution is also contained in the observed complicated motions of the wandering stars: 'the stoppings, retrogressions and progressions of the planets are not their own,' but are a consequence of the movement of the Earth; they borrow the appearances of this movement from the Earth.[64] If one of the Ptolemaic circles is conceived to be equal to the Earth's orbit round the sun, the other circle represents the planet's own orbit, whose relative magnitude can then be calculated in agreement with the phenomena. The 'common measure' — the fixed standard of comparison that was lacking in Ptolemy's system — was now discovered. Thus it became evident that 'the world has a wonderful commensurability (*symmetria*) and a sure bond of harmony (*harmoniae nexus*) of the motion and dimensions of the planetary orbits, such as cannot be found in any other way.'[65] Copernicus left no doubt that he had discovered the *true* system of the world; 'the ratio of order in which their bodies succeed one another and the harmony of the whole world teaches us their truth, if only we would look at the thing, so to say, with both eyes.'[66] Copernicus' Platonist spirit felt that the Great Architect could only make the most perfect building.

Besides its simple overall elegance the grand design had consequences in *detail* that had a confirmatory 'ring' to them. Copernicus remarked on the fact that, in contrast to the other planets, Venus and Mercury are always confined to parts of the sky close to the sun — they do not have the full range of 'angular elongation' from the sun — and that this is a sufficient reason for assuming that they are companions of the sun and do not revolve around the Earth. The orbits of Mars, Jupiter and Saturn, on the other hand, he considered to be out beyond the Earth's orbit. This overall picture accounted satisfactorily for several observed phenomena for which the Ptolemaic system had been at a loss to provide an explanation. Copernicus remarked that when Mars, Jupiter and Saturn have their *rising* at around the time of sunset they are at their brightest and largest. This is explained by the realisation that when they are opposite to the sun in the sky, the Earth is on a part of its orbit between them and the sun and so is much nearer to them than when they *set* in the evening. In the latter case the sun is between them and the Earth, so they are on the far side of their orbits and hence at their greatest distances from the Earth.[67] In the Ptolemaic system variations of distance (and so of size) were to be expected as the planets traversed their epicycles, but it could not account for these systematic correlations with the position of the sun. Nor could it get the expected range of planet size to tally with the facts of observation: it expected that the apparent magnitude of Venus should increase by about 36 times in the course of a revolution!

The order of the spheres is illustrated by a diagram of great simplicity (see figure 22, page 151). The sun, precisely in the middle, is surrounded by the homocentric spheres of the planets and the outermost, stationary, sphere of the fixed stars.[68] Copernicus suggested that his system is much simpler than the geocentric with its 'almost infinite multitude of spheres', and he claimed that it follows 'the wisdom of nature which takes the greatest care not to have produced anything superfluous or useless.'[69]

We shall see below that this claim was not strictly justified, though as to 'simplicity' his system had some advantages. For instance, he held, as Aristotle and Ptolemy had done, that the more distant from the centre the longer the period of revolution of a heavenly sphere. In the geocentric system, it is embarrassing that the outermost sphere turns out to have the fastest movement, its revolution being performed in twenty four hours. With Copernicus its velocity is almost zero, as becomes a sphere that is beyond that of the slowly moving planet Saturn. Furthermore, in the then current system, the planets moved from West to East. In Copernicus' system *all* heavenly spheres move in the same direction.[70]

Copernicus' explanation of the precession of the equinoxes was another simplification. Other astronomers held that the equinoctial points (i.e. the points where the ecliptic and the celestial equator intersect) are slowly moving. This motion was attributed to the sphere of the fixed stars. But this eighth sphere already had the daily rotation which it imparts to all the lower spheres. According to the principle that each heavenly sphere should have a characteristic motion of its own (distinct from those it receives from the higher spheres), a ninth sphere had been introduced which was held to be the originator of the 'precession of the equinoxes'. This sphere makes a revolution every 49,000 years, and the eighth sphere (of the fixed stars) thus has two motions; the daily rotation and that of the precession. Medieval astronomers added yet another motion; a periodic oscillatory movement of the equinoxes (trepidation), which was superimposed on their uniform progression. This led to the introduction of a tenth sphere, so that the 8th had three motions: the daily rotation, the precession and the trepidation. Copernicus now simplified the system by eliminating the 9th, 10th (and even 11th) spheres by transferring their motions to the earth's axis.[71] The 'empty' spheres (which contained no heavenly bodies) became superfluous.

Copernicus' explanation of the precession of the equinoxes was connected with the so-called 'third motion' of the Earth, to which he gave the name the 'motion of declination'.[72] He explained the fact that the earth's axis points in the same direction during its annual motion by assuming that it is describing the surface of a cone around the pole of the ecliptic in a direction contrary to that of the annual motion. Its period, however, is slightly less than a year so that the axis of the earth does not regain exactly the same position at the end of the year and the points of intersection of the ecliptic with the equator make a slow movement (precession).

8. THE CENTRAL PLACE OF THE SUN

In describing his system — which seems 'almost inconceivable and quite contrary to the opinion of the multitude'[73] — Copernicus starts at the periphery with the fixed stars and proceeds inwards *via* the spheres of Saturn, Jupiter, Mars, the Earth (with the moon), Venus and Mercury, to end the description with the jubilant passage about the central place of the sun.

> In the middle of all is the sun. For who could place this lamp of the most beautiful temple in another and better place than this, whence he can illuminate all things at the same time? As a matter of fact some, not unfelicitously, call him 'the lantern'; others the 'mind' and still others the 'leader' of the world. Trismegistus calls him a 'visible god'; Sophocles' Electra 'that which sees all things'. And so the sun, as if resting on a royal throne, governs the family of the stars which wheels around him.[74]

We may feel some puzzlement that in this triumphal statement of heliocentric belief an appeal is made to the authority of adherents of a *geocentric* world structure. But this was possible because the phrase 'in the middle' of the planets (*in medio vero omnium*) is an ambiguous expression which Copernicus interpreted to his own advantage. It can refer to the centre of the universe, certainly, but it can equally well mean the middle member halfway in a series, e.g. of planets. It expresses heliocentrism if it is used to denote a place in the centre of the universe which is occupied by the sun and around which all planetary spheres are arranged. But it can alternatively be understood in a geocentric sense, for it may equally well signify that there are as many planetary spheres beyond that of the sun as there are within it. The honorific titles 'mind', 'leader', 'visible god' do not in themselves indicate a local but rather a hierarchical centrality and both of the above positions in space could be seen as befitting the highest-ranking planet. So it is not clear from these quotations alone which choice these ancient authorities favoured. But we do know that in the writings attributed to 'Hermes Trismegistus' — as in those of Aristotle — it is held that there are as many planets beyond the solar sphere (namely Mars, Jupiter and Saturn) as below it (Venus, Mercury, the moon) and that the Earth occupies the *centre* of the universe.

Copernicus cannot have failed to be fully aware of the ambiguity of the terms 'the centre' and 'the mean': it had been clearly pointed out by Aristotle in his polemics against the Pythagoreans, whom Copernicus so much loved. They located Fire in the centre of the universe because it was esteemed more honourable than earth and thus should occupy the most honourable place. But, says Aristotle, we should rather suppose *that* to be true of the universe which is true of animals, namely that the centre of the animal and the centre of its body are not the same thing.[75] That is to say that in the macrocosm as in the microcosm the part that is most important and honourable does *not* need to occupy the geometrical centre.

Clearly, in Aristotle's opinion the most eminent and honourable part does not in either case occupy the geometrical centre. In the animal body the heart holds the place of primacy and governance; it is the source of the blood and the heat and the sensations. Nevertheless, its position is not in the mathematical centre: that is reckoned to be in the vicinity of the navel.[76]

A similar distinction had been made by the Roman architect Vitruvius and by the medieval philosopher Nicole Oresme. Vitruvius, after having expounded the system of proportions between parts of the human body which bring about its perfect 'symmetry', concludes that 'the navel is naturally the centre of the body'. 'For if a man is lying on his back with hands and feet outspread, and the centre of a circle is placed on his navel, his fingers and toes will be touched by the circumference...'; also a square will be found circumscribed, 'for if we measure from the sole of the foot to the top of the head', this distance will be equal to that between the ends of the outstretched hands.[77] We have already remarked that Vitruvius recognized a certain analogy of the mathematical proportions of the temple and those of the human body, and as the latter was a microcosm there had to be a similar analogy between the temple and the macrocosm or universe.

In his commentary (circa 1375) on Aristotle's *On the Heavens*, Nicole Oresme agreed with the Philosopher that 'mean' or middle has more than one signification, and that the 'mean' of a magnitude or of a body, and the mean or middle of a substance or a nature are wrongly identified by the Pythagoreans. For 'the heart is the middle or centre of the animal and occupies the noblest place, but the middle of its body is situated about the navel. And therefore there is a middle of a quantity or magnitude — like the centre of a sphere or of some other body — but another centre is that which nature selects as the best place.'[78] It could not have been said more clearly: the most honourable place, because of its 'nature', qualities and function need not coincide with the geometrical centre. The 'heart' of a thing need not always be the centrally situated component (though it may sometimes coincide with it).

The Roman orator Cicero had had similar things to say on the sun both as to its position and its function. In 'Scipio's Dream' he held that the sun is 'almost midway' between the outermost heaven of the fixed stars and the Earth,[79] and he called him 'the leader, chief and ruler of the other lights, the mind and principle of the universe.'[80] His commentator Macrobius pointed out that Cicero gives the sun the middle position (on the fourth of the spheres), whereas Plato says that it is just above the moon. But Cicero did not simply say 'the middle position', but '*almost* the middle position' and he did so deliberately, for in his system 'the sun holds the middle position in respect to number but not to space': the sun is further removed from the top than from the bottom of the spheres.[81]

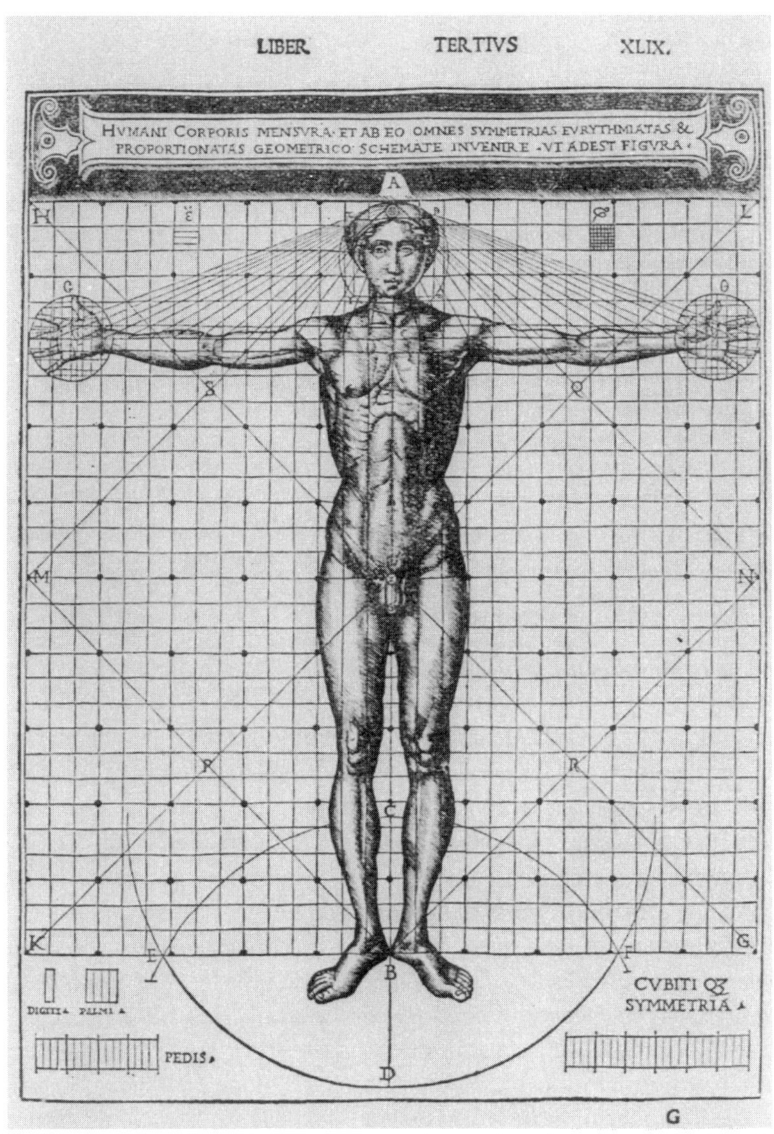

Fig. 24. Vitruvius' Humani corpori mensura (proportion of a man as he ought to be). From: Vitruvius, *De Architectura*. Ed. Como 1521. Courtesy Antiquariaat FORUM BV, 't Goy.

Of course Macrobius fully agreed with Cicero's calling the sun the 'leader, chief and regulator of the other lights, the mind and moderator of the universe.'[82] The latter appellation refers, in his opinion, to the fact that all things recur in the sky under fixed laws, and for the same reason the philosophers call him 'the heart of the sky', 'for we see that all things in the divine plan are accomplished through it': the sun is the source of ethereal fire

> and, as fire is always involved in motion and perpetual agitation, 'the sun's function in the ether is the same as the heart's function in the animal.'[83]

It was not only those who accepted Aristotle's world picture who held that the middle position of the sun does *not* mean that it occupies the geometrical centre of the world. Actually, on this matter the Pythagoreans were of the same opinion.

> We saw above that, according to Aristotle, the Pythagoreans placed the sun halfway between the centre and the outermost sphere. The neoplatonist Calcidius, in his commentary on Plato's *Timaeus*, relates that *some* Pythagoreans accept the (usual) sequence, and put the sphere of the moon closest to the Earth, and next Mercury, Venus, the Sun, Mars, Jupiter, Saturn and finally the fixed stars: 'among the planets the sun is placed in the middle, and even held to occupy the eminent position of the heart in all living beings.'[84]

In short, to both the Aristotelians and the Pythagoreans the sun had the function of leader and mover of the planets and both held that it was the most noble part of the cosmos, and they both accorded to the sun the non-central position which the heart has in the human body.

The geometrical ambiguity of the term 'middle', however, enabled Copernicus to blur the difference between the heliocentric theory and the quasi-geocentric world picture of the Pythagoreans. He directed attention to the fact that the sun is the 'leader' or king in the metaphysical sense in *his* system as well as in those of the Pythagoreans, Hermeticists and Aristotelians. It seems therefore that, having compared the perfections of the structure of the macrocosm with that of the microcosm, he cleverly did not call attention to the geometrical incongruity for he *avoided* calling the sun the 'heart' of the universe.

Others, however, were less cautious and precise. They called the sun the heart of the universe because of its having the *function* of a heart in their macrocosm, in spite of its having the geometrical *position* of the navel. Thus we find in the early 17th century that the fervent Copernican, Johannes Kepler was calling the sun 'the heart of the world' and spoke of 'the heart, the origin of life'.[85] William Harvey too — who had a great admiration for Aristotle — said that the heart is 'the beginning of life; the sun of the macrocosm, even as the sun in his turn might well be designated the heart of the world.'[86]

9. The 'Novelty' of Copernicus' System

The Renaissance was a time of revival of the ancient wisdom of the Romans, the Greeks and even of the 'Egyptians', a revival, however, that in fact brought much that was new. The word 'novelty,' when used by an author of the Renaissance period, was thus ambivalent: used in a deprecatory sense it denoted 'new-fangled ideas', but in a laudatory sense it might stand for some ancient truth, now restored and even further developed. So the introduction of a 'novelty' which could be proved to be indeed something very ancient — more ancient than the current systems or even than those of the Hellenistic age, was particularly applauded.

Copernicus' astronomical hypothesis included several things that were indeed entirely new and he did not conceal this, for they were improvements — in that they kept more strictly to the ancient rules of the discipline. But he claimed also that, although in general his system was new in that it differed from the current system, it was nevertheless in fact ancient. Consequently there was no inconsistency in his proudly quoting from a letter of Cardinal Nicolas of Schönberg (1536) asking him to publish his 'new system of the world' (*novam Mundi rationem*), while on the other hand, making efforts to heighten the respectability of his system by 'proving' that it had its roots in remotest antiquity.[87]

He had become dissatisfied wih the current system because Ptolemy had not been able to keep strictly to the ancient prescription of 'saving the phenomena' by reducing the irregular courses of the planets to combinations of circular and *uniform* motions. Ptolemy had referred the uniformity of some circular motions to a point (*punctum aequans*) which was not the centre of the circle. He had thus taken liberties with the notion of uniformity, which meant 'uniformity as seen from the centre of the circle along which the motion takes place.' When Copernicus began to be 'annoyed' by this and by other weaknesses of Ptolemaic astronomy he started to seek for a remedy in better measurements and in alternative theories by studying the books of the Ancients in order to see whether they had advanced anything better.[88] And thus he found first in Cicero that Hicetas had held that the earth is moving, and afterwards in Plutarch that other ancient astronomers were of the same opinion. He quoted from Plutarch that 'Philolaus the Pythagorean says that the earth moves round the Fire with an obliquely circular motion, like the sun and the moon' and that Herakleides of Pontus and Ecphantus the Pythagorean did not give the earth any locomotion but let it turn around its centre like a wheel.[89] Further on, Copernicus referred to 'the Pythagoreans Herakleides and Ecphantus, and Hicetas of Syracuse who made the Earth revolve in the centre of the world'. He added that 'nearly all' believed that the Earth is in the centre of the world, and if they did not do so, they deemed the distance of the Earth from the centre so small that it is as nothing to the distance from the sphere of the fixed stars, though it is considerable in relation to the orbits of the Sun and the planets. Finally he said it would not be surprising if someone attributed some other movement to the Earth in addition to the daily revolution, and noted that indeed Philolaus the Pythagorean is supposed to have held that it moves in a circle and wanders in some other movements and is one of the planets.[90]

Whereas Copernicus up till this point had quoted Pythagoreans who either let the earth make a daily rotation on its axis while remaining in the centre of the universe, or, like Philolaus, let the earth revolve around the central Fire in twenty four hours, he now referred to a Philolaus who appears to be silent about the central Fire, but who stressed that the earth is a planet. That is, though he had made evident up till now that the Pythagoreans did favour a daily rotation or (in the case of Philolaus) a daily revolution of an *Earth* which is at or near the centre, he now seems to hint that Philolaus tended towards the heliocentric system. At any rate all these references served to prove that belief in the movement of the Earth was no novelty.

As to the *structure*, which was the second point on which he was 'annoyed', he discovered at least partial support for the heliocentric system in the compilator Martianus Capella and some other Latin writers who held that Venus and Mercury turn around the sun (though in a geocentric universe).[91] Finally, as we have seen, in his laudation of the sun he recruited the most ancient 'Egyptian', Hermes Trismegistus, as joining him in praise of the 'centrally' placed sun, but remained silent about Hermes' geocentric conception.

Up till now Copernicus had not explicitly claimed that the whole theory was based on Pythagorean principles: and one might feel uncertain whether this was indeed the case. At the end of book I, chapter 11, however, there was originally a passage which has been obliterated in the manuscript and consequently did not appear in the printed work. He said there that, though the course of the sun and the moon could admittedly be demonstrated by supposing a stationary Earth, this is impossible for the other planets, and that it is possible that for this and similar reasons Philolaus held the conception of a moving Earth and that, according to some people, Aristarchus of Samos was of the same opinion.[92]

Now it was generally known (*via* Archimedes' *Sand Reckoning*) that Aristarchus (and not Philolaus!) had put forward a heliocentric system. So by placing Philolaus in the same category, Copernicus showed clearly that he disagreed with Aristotle's critique of the Pythagorean system.[93]

Copernicus had already identified himself obliquely with the Pythagoreans when writing in the preface to his main work that, when considering how absurd his theory of the Earth's motion would seem to those who had been educated in the tradition of many centuries, he hesitated as to whether he should bring his ideas to the light, 'or whether it would not be better to follow the example of the Pythagoreans and certain others who used to hand down the mysteries of their philosophy not in writing but personally and only to their relatives and friends — as is evident from the letter of Lysis to Hipparchus.'[94]

By referring to 'a secret tradition', Renaissance philosophers could and did project their own ideas into really or supposedly very ancient philosophies. Since, for example, tradition said that Plato had visited Philolaus, the *Timaeus* was interpreted as being in favour of the heliocentric structure so that Plato, too, became one of the protagonists of this conception of the universe. One could even go further back to the Egyptians, Pythagoras having been said to have met the 'Egyptian' Hermes Trismegistus.

The result was that Copernicus' system has been dubbed the 'Pythagorean' system not only by those who were proud of adhering to the 'most ancient' astronomy, but also by the much more numerous opponents of it, for whom the *old* thing was what their direct ancestors had believed and to whom a recently revived theory — even if it were indeed older — seemed to be a novelty.[95]

The first book of *De Revolutionibus* had expounded the heliocentric structure of the universe and the motions of the Earth in broad outlines. It had put forward physical arguments for the daily rotation, which were convincing if one accepted the physical principles that had been excogitated on behalf of the new system; it had expounded some metaphysical points and had presented a strong argument for the annual revolution of the earth (the 'common measure').

The subsequent five books had to tackle the task of Astronomy in the (then) proper sense: computation of planetary motions in conformity with the observed phenomena. Most astronomers did not demand that an astronomical theory such as Ptolemy's should be physically true. One physical condition, however, had to be fulfilled even by such a 'mathematical' theory, *viz.* that it strictly kept to the rule that only circular and uniform motions should be used. Ptolemy had violated this condition by acknowledging the equant and thus committed a cardinal sin not only in the eyes of those who demanded that an astronomical theory should be physically true, but also of those who required only useful fictions. Consequently Copernicus won the praise of all astronomers when he managed to discard the 'equants' from his theory.

There was, however, a 'second inequality' in the observed planetary motions, which was caused by the *fact* that in nature the motions of the Sun, Moon and five planets are not perfectly uniform and circular. Copernicus, therefore, like Ptolemy, had to resort to eccentrics and epicycles to account for them in detail.

Having read the first Book with its lofty harmonious unfolding of the structure of the universe and the motions of the heavenly bodies, it may be disappointing to have to plod in the other five through complicated diagrams and computations similar to those in Ptolemy's Astronomy. So it was not quite fair of Copernicus to claim in his first Book that it is 'much more easy to grant' *his* system 'than to unhinge the understanding by an almost infinite multitude of spheres — as those who keep the earth at the centre of the world are forced to do.'[96]

He compared the *idealized* diagram of the heliocentric cosmos not with the idealized simple Aristotelian cosmos, but with the very complicated system of

Ptolemy's *practical* astronomy. If we compare the picture of his system with analogous diagrams of the homocentric spheres of the current cosmography, we see that, apart from the moon (which in Copernicus' system is a satellite of the planet Earth), and the exchange of places of Earth and Sun, the arrangement and the degree of 'simplicity' in both idealized systems is practically the same (see figures 21 and 22).[97]

The multitude of Ptolemy's circles should be compared with the multitude of circles used by Copernicus in the Books II to VI of his major work. The suggestive influence of the system of homocentric spheres as depicted in Book I, however, had the effect that many (though not all) modern writers have accepted the claim of the much greater simplicity of Copernicus' astronomical system.[98] Nevertheless Copernicus' system was, even in the eyes of contemporaries who did not share all his tenets, a great achievement, because he *restored* the uniformity of the heavenly motions and discarded from the then current astronomy some irksome non-circular motions. He interpreted the oscillatory motion of trepidation (which alternately accelerated and retarded the precession of the equinoxes) and the other 'libration' (the periods of increase and decrease of the angle between the equator and the ecliptic) as combinations each of two circular uniform motions[99] and thus derived non-circular motions from uniform circular ones and at the same time transferred these motions from the 'empty' outer heavenly spheres of the current system to the axis of the earth, as he had done in his explanation of precession.

His removal of the equants and his explanation of the librations were *novelties*, but they were generally considered as a great improvement, not only by the few who accepted his hypothesis as physically true, but also by the majority who clung to the geocentric system. However great their objections against the heliocentric structure of the universe, they warmly applauded his restoration of the ancient rule of the *art* of astronomy. The young astronomer Albert van Leeuwen wrote a treatise in which he promised to revise astronomy 'according to the doctrine of Copernicus'. He enthusiastically welcomed (1578) Copernicus' explanation of precession and trepidation as slow movements of the earth's poles. Yet, as to the daily and yearly revolutions of the earth, he thought, as we have already noted (page 157), that 'the great man just made a joke.' He could not believe that such a great scholar could seriously mean what he wrote about these motions.[100]

Copernicus, then, was generally considered as a great astronomer, the greatest since Ptolemy: If one did not require physical truth from an astronomical theory, it was a great feat that the rules of the astronomical 'art' — the circularity and uniformity of motion — had been restored. It is true that soon afterwards trepidation was recognized to be a spurious 'phenomenon', and that the earth's 'third motion' (the motion of declination) was discarded as superfluous, the earth's annual revolution being conceived as a 'circular translation'. Yet the *fact* remained that Copernicus had improved astronomical computations and discarded the equants, and thus could be considered a 'second Ptolemy'. When in 1617 Nicholas Mulerius (Groningen) — who did not accept the earth's mobility — started a series of astronomical 'classics', he began with an edition of Copernicus' work.[101]

10. Tycho Brahe, Galileo and Kepler

The rise of a new star in 1572 and Tycho's demonstration that this phenomenon took place in the supra-lunar region of the fixed stars, together with his similar proof that the comet of 1577 was not a 'meteoric', sublunar phenomenon, were perhaps more damaging to Aristotelian physics and astronomy than Copernicus' 'fictions'. Here were phenomena of *change* in the sphere of the fixed stars: the Aristotelian dogma of the immutability of the heavenly regions was overthrown not by arguments but by 'facts.'

In Tycho's geo-heliocentric system (see figure 25) the stationary Earth occupies the centre of the universe. The moon and the sun revolve around her, while the other planets revolve around the sun as in Copernicus' system. This system had the astronomical advantages of the Copernican system without the 'absurdities' of the daily and annual motion of the Earth. Tycho abandoned the solid heavenly spheres: the planets move freely and the orbit of Mars cuts through that of the sun. His system met with wide approval, in particular after 1616 from those who wished to obey the decree against Galileo's theory of the earth's motion, several Jesuits included.

In spite of Copernicus' able advocacy, then, his system in its entirety found only a few adherents during the 16th century. To most people the motions of the Earth remained too 'absurd', even as fictions. Copernicus' devices for maintaining the uniformity and circularity of heavenly motions could be transferred to some geocentric system and so Copernicus' system remained to most astronomers an alternative mathematical fiction, but not a physical truth.[102]

In the 16th century the controversy remained largely one of arguments against arguments. Whereas Ptolemy's geography had been abandoned because hard facts bore witness against it, there were no incontrovertible facts for demonstrating the truth of the systems that claimed to replace his astronomy.

NOVA MVNDANI SYSTEMATIS HYPOTYPOSIS ab Authore nuper adinuenta, qua tum vetus illa Ptolemaica redundantia & inconcinnitas, tum etiam recens Coperniana in motu Terræ Phyſica abſurditas, excluduntur, omniaq; Apparentiis Cœleſtibus aptiſsime correſpondent.

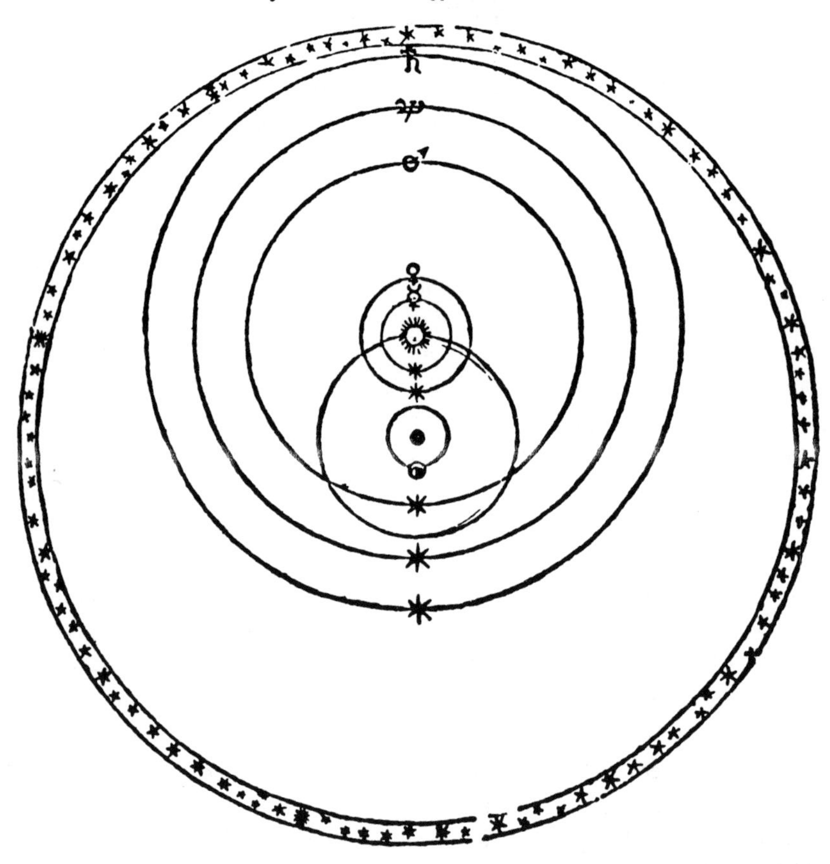

Fig. 25. Tycho Brahe's Diagram of the universe from the second issue of his *Progymnasmata* (1610). The diagram first appeared in Tycho's *De Mundi Aetheri Recentioribus Phaenomenis* (1588).

It was not until the beginning of the 17th century that the balance gradually tipped in favour of Copernicanism when Galileo, with the help of the recently invented telescope, discovered sunspots, the satellites of Jupiter (1609) and the phases of Venus. The Jupiter satellites showed that a planet could have 'moons', which made it plausible that the Earth having a moon could also be a planet. Within the geocentric frame it was impossible that Venus should go through all phases, but within the Copernican system this was to be expected. (This latter phenomenon, however, fitted equally well into the Tychonian system). These observed *facts* turned Galileo from a lukewarm adherent of the Copernican system into one of its most zealous apologists.[103]

Though the young Kepler was already in 1596 a fervent Copernican on grounds we would now call metaphysical, his discovery of the general laws of planetary motion (1609, 1619) was mainly based on Tycho Brahe's very accurate measurements of the motion of the planet Mars. Kepler found that the planetary orbits are not circles but ellipses in which the sun occupies one of the foci. That is to say, that Copernicus' removal of the equants became obsolete, for the planets were no longer thought to move with uniform velocity in circular orbits and the whole idea of epicycles became superfluous as one single orbit was sufficient to describe the course of a heavenly body.

Through Kepler and Galileo hard facts were thus adduced in support of the 'Pythagorean' system.

11. The Paradoxical Character of the 'New Astronomy'

Even so there remained reservations. Nicholas Mulerius declared in 1611 that the Pythagorean system, 'rejected by Aristotle, has recently been restored by outstanding mathematicians.' Nevertheless, where the topic of the immobility of the earth was concerned, he himself intended to 'stick to the best philosophers and to Ptolemy and to Holy Scripture.'[104] Yet that his heart was with the Polish astronomer, was clearly demonstrated some years afterwards by the fact that he put *De Revolutionibus* first on the list of astronomical classics he wanted to edit with elaborate commentaries for the benefit of his students (1617). The great stumbling-block for him with the Copernican system was the enormous extension of the universe it demanded in order to explain the fact that no yearly parallax of the fixed stars could be observed.[105] He called it a 'paradox'[106]: it is bold, 'not to say odd', but it is indispensible to the system, for 'if it is not admitted the hypothesis cannot be maintained.'[107] Though Mulerius did not accept the Copernican system as physically true, it was to him the best 'astronomical' system: he was astonished that in order to account for the phenomena, one had to resort to it. In his opinion this clearly demonstrated 'the weakness of the human mind, which cannot explain God's wonderful handiwork in the artifice of the world, without meeting the absurd.'[108]

Mulerius was not the only 17th century scholar who, while sympathising with the Copernican system, yet felt some misgivings because of its 'absurdity'. The Puritan-Anglican clergyman Henry Gellibrand, who was professor of astronomy at

Gresham College in London, when visiting his Netherlands-Reformed colleague Philip van Lansbergen in Zeeland, found that 'that Great Astronomer ... did most seriously confirm that he should never be dissuaded from that Truth.' Lansbergen was a staunch believer in the Copernican system, but Gellibrand remained more sceptical: 'that which he was pleased to stile a Truth, I should readily receive as an Hypothesis, and so be easily led on to the consideration of the imbecility of Man's understanding, as not able rightly to conceive of this admirable opifice of God or frame of the world, without falling foul of so great an absurdity.'[109]

There is nothing strange in such an attitude: it will occur when a 'new' science, built on principles strongly deviating from the familiar ones, is introduced. And, in spite of the strenuous efforts to demonstrate its antiquity, Copernicus' system was a 'novelty' in that its principles deviated from those inculcated into the minds of his contemporaries in the schools as well as by 'common sense'. When new conceptions are in process of overthrowing the old dogmas to which the scholar has been so much accustomed since early youth that they seem 'reasonable' to him, then these new conceptions may seem illogical and even absurd and he can only hope that one day all will be clarified. The next generation, acquainted with the new doctrine since its schooldays, will take it for granted and easily admit its reasonability. The older generation meanwhile will be inclined to use the new theory in a 'mathematical' way, hoping that the future will bring a more acceptable interpretation.

One can think of later instances when new conceptions demanded much of the imagination. When in the 18th century Newtonianism replaced the generally accepted Cartesianism (with its 'clear and distinct' notions), the hue and cry was raised against this irrational system which was said to re-introduce the 'occult qualities' of scholasticism. When Augustin Fresnel, who had conceived of 'light' as consisting of longitudinal vibrations of an extremely thin fluid, felt himself obliged (1816 and 1821) under the pressure of the polarisation phenomena, to admit the *transverse* character of the waves, he himself at first regarded this as an inevitable absurdity: it implied the existence of an ethereal fluid that possessed the elastic properties of a solid body (i.e. deformability), as well as those of an ideal thin fluid.[110] When Isaac Beekman at the beginning of the 17th century in attempting to replace an explanation of physical phenomena in terms of 'substantial forms' (which he considered chimerical) by one in terms of the shape, arrangement and motion of the smallest particles of matter, found that he had to attribute to these atoms perfect hardness as well as perfect elasticity, he recognized the paradoxical character of this concept. But the atoms introduced by him were at least imaginable concrete objects, whereas the nebulous Forms were to him just empty words, purely imaginary things. Later atomists like Christiaan Huygens, however, hardly paid attention to the paradoxical character of the collision of atoms.[111]

And after Planck (1900) had formed his quantum theory, a renowned Russian physicist — though as willing to use it as Gellibrand had been to use the Copernican theory — stated sadly that this theory, 'which sees the light of the world at the same time as the new century, has impressed its turbid stamp on almost all departments of the new physics of the new century.'[112] He found Maxwell's electromagnetic theory, Bohr's atomic theory and Planck's quantum theory unavoidable and fruitful for the moment but also a retrogression because of their lack of perspicuity so that they are far from the aim science has set before itself. Nevertheless he reluctantly admitted that it may be useful (*ein Nutzen*) when with the help of an incomprehensible hypothesis the explanation of a very large number of incomprehensible facts is attained: 'the many incomprehensibilities are, so to say, replaced by the one.'[113]

12. Towards the Final Triumph

In the first half of the 17th century quite a few scholars did not make a definitive choice between Ptolemy, Tycho and Copernicus, though like Isaac Beekman they for the moment preferred Copernicus. Pascal, who was no astronomer, was against the condemnation of Copernicanism, but it was the freedom of science — not Copernicanism in itself — that he was advocating, for he did not feel able to decide which of the three had the true system. At that time, however, Copernicanism was in the ascendancy, so much so that by 1657 Christiaan Huygens could declare that practically all astronomers accepted Copernicus' 'divinely invented' system, except those who were a bit slow-witted or 'under the superstition imposed by merely human authority.'[114]

Towards the end of the 17th century Newton in his *Principia* at long last managed to insert the Keplerian version of Copernicus' astronomy into the framework of a new mechanics (in which his own law of general gravitation and Huygens's law of centrifugal force played a prominent role). The new astronomy — now well over a century old — was given a solid physical basis which united a great many terrestrial and cosmic phenomena into a self-consistent system.

Scientific tenets do not always have to wait for direct proofs in order to be generally approved: when they fit into a system of well-established other tenets, this is already accepted as a proof of their 'truth'. When Alfred Wegener (1880 - 1930) put forward in 1912 his theory of continental drift, there were several strong arguments for it, but the results of direct measurements (on the suggestion that the continents were still drifting apart) were unconvincing and a satisfactory mechanism for the drift could not (as yet) be put forward. The theory was looked upon by the majority of geologists with no less scepticism than Copernicus' by 16th century astronomers and it was equally ridiculed. Even one of its (critical) adherents, the Edinburgh geologist Arthur Holmes (1890 - 1965), confessed in 1953: 'I have never succeeded in freeing myself from a nagging prejudice against continental drift; in my geological bones, so to speak, I feel the hypothesis as a fantastic one.' Yet, he added, 'there are weighty reasons enforcing us to recognize it as true.'[115] And indeed, in the 1960's oceanic research and the new paleomagnetic methods led to the development of 'plate tectonics' and the general acceptance of continental drift as a theory. It had been geophysicists who had provided the main arguments *against* Wegener's theory, but finally geophysics produced the most convincing proofs *for* it.

For Copernicus' theory the period between its general acceptance (circa 1687) and its confirmation by direct proofs, lasted even longer. The first direct proof by measurements of the Earth's daily rotation was given by Benzenberg (1802) by experiments on falling bodies which were found to come down slightly to the east of the vertical, and afterwards by Foucault's pendulum experiments (1851). The annual motion of the earth found confirmation in the aberration of light (Bradley 1726) and the parallax of the fixed stars (Bessel 1838).

13. Copernicus' Theory: 'True' or Only 'Scientific'?

It is noteworthy that *our* contemporaries, when praising Copernicus, refer to Book I of *De Revolutionibus*, and are silent about Books II - VI, whereas by 16th century scholars — with a few exceptions — Book I was hardly taken seriously and Books II - VI were reckoned to be the greatest feats in astronomy since Ptolemy. Both parties, then, agree that he was one of the great men of science, though they do so for quite different reasons. Even within Book I our contemporaries are carefully selective: the heliocentric structure and the daily and yearly motions of the earth are extracted for praise but Copernicus' 'third motion' ('declination') and his cosmophysics, being outdated, are usually ignored. Modern astronomy is based on Kepler's rather than on Copernicus' theory, but as Kepler borrowed from Copernicus the heliocentric system and never tired of repeating that his theory was based on Copernicus' work, modern astronomy is 'Copernican' in the wider sense.

The first Book of *De Revolutionibus* fits into the historical development of 'true' astronomy (the series Copernicus - Kepler - Huygens - Newton). Books II - VI, however, have lost their scientific significance, because they purport to save the scientific 'fiction' that all heavenly phenomena can be explained by combinations of circular, uniform motions.

From the standpoint of most mid-16th century scholars, however, Book I clashed with the generally accepted philosophy of nature and with common sense by introducing the motion of the Earth. Some astronomers therefore recognized with gratitude the discovery that either the deferent or the main epicycle of a planet could be taken as equal to the sun's orbit and thus lead to Tycho's system which still maintained the immobility of the earth.

Books II - VI, on the other hand, nowadays considered obsolete, were by all astronomers considered fully 'scientific', as they accounted for the astronomical inequalities without violating the basic principles of uniformity and circularity of heavenly motions — principles that were as fundamental as conservation of energy in modern physics.

But — one might object — as a matter of *fact* the planets have no uniform, circular motion and that great feat of the interpretation of trepidation concerns an event that does not exist. How can the bringing of spurious 'facts' into line with a general prejudice, and the saving of scientific 'fictions' be called 'scientific'? Here we may repeat what we said in our account of the phlogiston theory; 'in order to deserve the epithet "scientific" a theory need not be objectively true; it is enough that it satisfies the conditions we may demand of a scientific theory (self-consistency, conformity with observation, etc.).' Copernicus' theory 'explained' even non-existent 'facts', but *it is more 'scientific' to talk sense about things that do not exist than to talk nonsense about things that do*. In the latter case there is no science but a combination of bare facts and bad reasoning; in the former case there has been scientific reasoning, though on the basis of wrong 'facts' or false axioms. *Not all that is 'scientific' is true.* The belief that *we* have reached the top, that our science is the definitive truth and that the way that led to our present day science is the only scientific way, makes us evaluate the events and theories of the past with the measuring rod of the present. If, however, we assume a really 'historical' viewpoint, we imagine ourselves to be in the past, knowing the facts, sharing the fictions and opinions of our predecessors, and in our imagination following their methodological principles. So, when trepidation is a 'fact' to everybody, and a scholar succeeds in inserting it into the prevailing scientific system, he deserves praise for this 'scientific' achievement.

But does science not strive after discovering the *truth* of nature? Of course it does. But this does not imply that so long as it has not found the truth, or mistakes wrong tenets for the truth, it is 'unscientific'. It goes without saying, however, that when the problem of the trepidation of the equinoxes was 'solved' because it was found to be non-existent this was a real advance in science. We only want to stress that the dialogue between Nature and the natural scientist is remarkable in that when — as sometimes happens — the part of 'Nature' is played by the scientist himself projecting *his* answer to his questions onto Nature, then Nature has the last word by passively refusing to behave as we would like or expect.

VII. Thinking with the Hands

1. Experimentation ... 183
2. Types of Experiment ... 186
3. The Interplay of Fact and Fiction in Experiment ... 187
3.1. Ptolemy ... 188
3.2. Haüy ... 189
3.3. Dalton ... 189
3.4. Thought Experiments ... 190
4. The Emergence of Experiments ... 195
4.1. Petrus Peregrinus ... 196
4.2. Nicolas Cusanus ... 197
4.3. João de Castro ... 198
4.4. De Orta and Magnetism ... 200
4.5. Possible but Unperformed Experiments for Testing Aristotle ... 201
4.6. Magnetism Again ... 202
4.7. Isaac Beeckman: Scholar-Artisan ... 204
5. Experimental Science after 1600 ... 205
5.1. Galileo ... 205
5.2. Increasing Cooperation between Head and Hand ... 214
6. Conservation Laws ... 220
6.1. The Law of Conservation of Mass ... 220
6.2. Conservation of Energy ... 225
7. Conclusion: Head, Hands and Instruments ... 228

1. Experimentation

An 'Experiment' is a deliberate act which aims at knowledge of nature and power over nature. In medieval Latin, however, 'experimentatum est' can also refer to a mere rather passive observation; in modern French 'expérience' is a word for experience in general as well as for experimentation.

The scientific experiment is performed to discover rules prevalent in Nature ('laws') or causes of natural phenomena ('explanations'). This knowledge is used to get power over nature as well as to quench the thirst for knowledge for its own sake. The former originally implied that 'experimentum' also meant a magical act. A manuscript from the Hellenistic period, the 'Goldmaking Art of Cleopatra' has a picture of a distillation apparatus standing on a (sacred) tripod, in conjunction with a circle with a conjuration formula, while the snake Ouroboros (which bites its own tail) symbolizes the circular course of nature (see figure 11 on page 85).

In vain we look for experimentation in the ancient works on *physics*. This was a contemplative science which occupied itself with the nature, the essence, of things and did so without the help of manual work. The fabrication and manipulation of tools and the contriving of apparatus was part of 'mechanics'.

'Mechanics', the discipline teaching the making of instruments for experimentations as well as technological products in general, is a name derived from the word 'mechane', which means a 'clever device'. 'Mechanics' then is the cunning by which man outreaches nature, for example in moving a great weight by a small force with the help of a lever, or in moving water upwards, against its natural tendency downwards, by creating a space devoid of air as in a siphon. The ancient engineers who excogitated instruments and machines were aware of the origin of their art in magic. They considered their art as going against nature by using their superior knowledge of nature; theirs was a rationalized magic. Aristotle and, several centuries later, Pappus declared that engineers were formerly called 'miracle workers' (*thaumatourgoi*) and their problems were, as the author of 'mechanica' of 'Aristotle' puts it, neither identical with physical problems, nor entirely separate from them. The Alexandrian engineers made useful machines (a fire engine) but also toys to amuse (*automata*). Occasionally they encouraged a belief in their magical powers, e.g. by devising a machine which opened the temple doors that hid the god, when the fire on the altar had been lit (see figure 26).

Mechanics belonged, with optics and hydraulics, to the mechanical arts in the wider sense. The theoretical part of it might be held in a certain esteem, but not so the practical side which required use of the hands and so bore a 'servile' character. Plutarch tells us that Archimedes, the genius in mathematics and theoretical mechanics, took to making machines only when the fatherland was in peril. The great philosophers (Plato, Aristotle, Cicero, Seneca) considered manual work as beneath the dignity of a philosopher — and of a free citizen in general — because of its 'vile' and 'servile' character. Military engineers were looked upon with more favour because of their function in defence of the state, and the more mathematical of the mechanical sciences — astronomy and optics — were also sometimes cultivated by scholars who could not be reckoned as belonging to the engineering trade.

Most reprehensible, however, in the eyes of most ancient, medieval and humanist scholars were the arts that used fire: the trade of the smith, mining (metallurgy), and alchemy. The great medieval philosopher Albertus Magnus, though one of the most

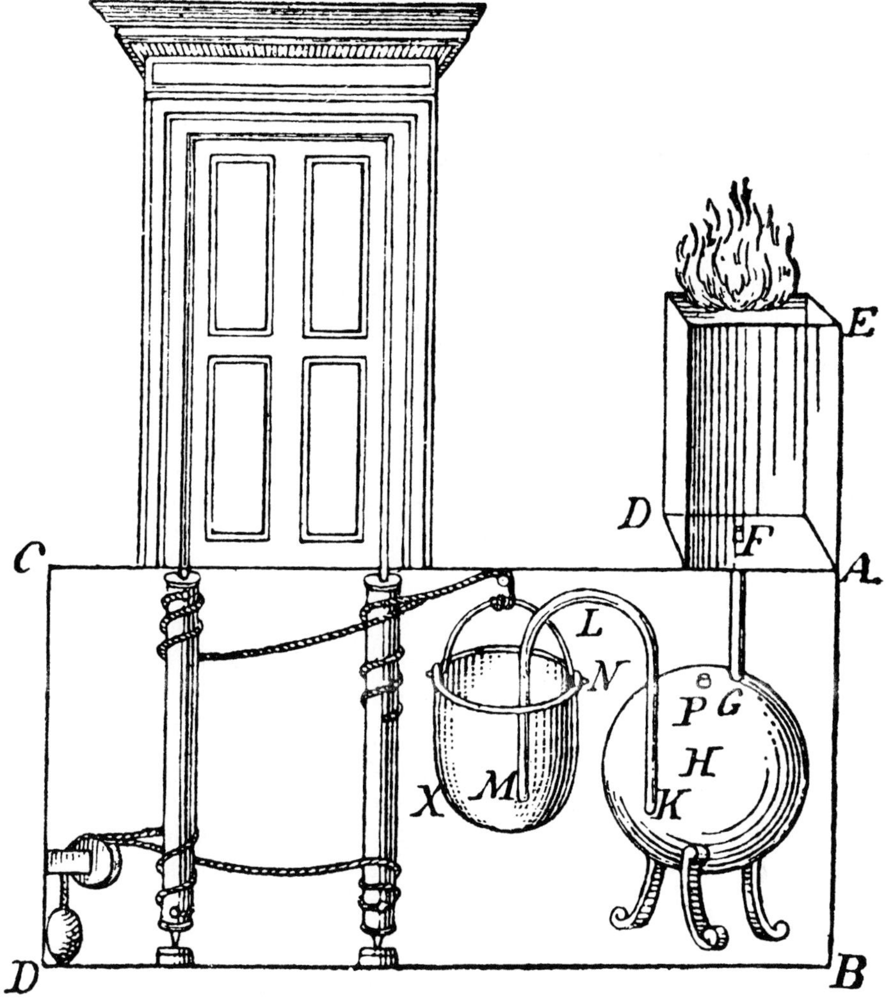

Fig. 26. Hero's mechanism for temple doors. From: F. Dannemann, *Die Naturwissenschaften in ihrer Entwicklung und in ihrem Zusammenhang*. Leipzig ^{2}1920, p. 195.

science-minded of his class, dismissed the arguments of the Arab pharmacist Ibn-Ostrul-Oshul with the remark that he was but 'a mechanician, not a philosopher'.

The idea that manual work might bring forth 'science' seemed unacceptable to most scholars. The 16th century Portuguese humanist poet Sá de Miranda stingingly expressed this when he said of the alchemists, 'they make science from gusts of wind'. On the other hand, some engineers found recognition from the ruling powers, e.g. Vitruvius, but then they generally stressed that they did not perform the manual work themselves but excogitated it. Vitruvius was an '*architekton*', a master builder, and not a simple 'tekton' (builder).

Of course there are exceptions; at any rate most of the experimentators themselves during Antiquity and the Middle Ages considered their work — though not contemplative thinking — as a product of the hands guided by thinking. The alchemists, doubly defiantly, called their mechanical art '*philosophy* by *fire*' and vaunted themselves of having 'blackened their hands by coal'.

Years afterwards the philosopher Roger Bacon was to declare that he had learned much from Petrus Peregrinus, 'the master experimenter' who performed his experiments himself, by the work of his hands (*manuum industria*). In similar vein the French philosopher Jean Bodin declared that the peoples of North-Western Europe in particular excelled in 'thinking by the hands'.

2. Types of Experiment

As we shall see below, it took a long time before it was generally accepted that 'thinking with the hands' — which after all is just a metaphor for letting the thoughts be followed by action — is an intrinsic part of natural science (or as it was called: natural 'philosophy'). Yet every type of experimentation has been applied to some extent since Antiquity. Let us consider some of them.

a) Some experiments could be denoted as 'heuristic'; they serve to discover new phenomena. In particular when a new instrument has been invented or a new substance has been discovered these burgeon. The invention of the telescope led to the discovery of many phenomena in the heavens: the satellites of Jupiter, the phases of Venus. The invention of the microscope opened a new world of hitherto invisible things: small insects, the structure of plants, little crystals. Though one might consider these as mere observations, they led the scientists to further research. The construction of air pumps by Otto Guericke, and the much improved versions of Boyle and Hooke, led to experiments *in vacuo*, many of which were made simply to find out what happens when certain things are put in a space devoid of air. After the fabrication of powerful burning glasses tests were made of how various substances behave when submitted to heat engendered by light.
b) In most cases, particularly in mechanics, a 'law' is posited or a theory is developed and afterwards experiments are used to test its validity or see whether phenomena predicted by the theory are realized. In these cases the experiment does not serve to *discover* a law but rather to *verify* a law or a relation deduced beforehand.

If possible, experimentation involves *measurement*, particularly in what we might refer to as 'testing' experiments. As an example of a mathematical law discovered in an empirical way we might consider the discovery of the law of constancy of angles of crystals by Carangeot and Romé de Lisle (1783); the dihedral angles between similar faces of crystals of the same crystal species are the same for all individuals of that species. One of a multitude of examples of the hypothetico-deductive sort is the experimental confirmation of the law of falling bodies, tested by rolling balls down an inclined plane (Galileo). Or, if we prefer examples in which there is no such interference with nature, there is the testing of the laws of refraction of light, and also the direct measurement of the declination of the magnetic needle in various localities on the same meridian in the (vain) hope that this might be characteristic of the meridian.

c) In many cases the experimentator's work is not only planned observation but also an interference with the 'natural' course of nature ('nature left to herself'), in order to elicit from her answers which she does not give 'voluntarily'. The 'natural' phenomenon then is often regarded as some ideal main event combined with 'disturbing' circumstances which have to be discarded as much as possible. The fundamental laws of theoretical mechanics with its frictionless motions in a non-resistant medium are approximately seen at work by creating empty space with the help of sometimes complicated machinery: the more the disturbing factors are eliminated the more the fundamental phenomena and laws come to the fore and finally an ideal event is approached asymptotically. In such experiments the tacit assumption is that the total visible phenomenon is the sum of its parts. A mechanistic world view may further such an attitude, though it must be recognized that the experimentator's pragmatist attitude can live with inner contradictions, provided they lead to practical results. William Gilbert (1600), who held an organistic world view, yet was a zealous and successful experimentator.

d) There is another kind of experimentation, which consists in making *models* that claim to imitate natural phenomena that are not accessible to direct observation and cannot be manipulated unless the scale is diminished (or enlarged). Such model experiments will be dealt with in chapter IX, 'Works of Nature, Works of Art'.

In a certain sense *all* experiments are works of art: the experimentator is a 'philosopher', but also an 'artificer'. Even theoretical physics finds its justification in experiments.

3. THE INTERPLAY OF FACT AND FICTION IN EXPERIMENT

The experimenter who performs measurements in order to check a mathematically formulated hypothesis or evokes 'desired' phenomena that would logically ensue if this hypothesis were true, is often tempted to adapt the results of his measurement to the expected values or to overlook cases in which the expected phenomena do not turn up. Scientific experimentation may collect ‘facts’, but these facts require critical consideration, and a correct observation can often be followed by faulty interpretation. The temptation to adapt data of measurement to the expected ‘law’ or to adapt the observed ‘fact’ to the preconceived hypothesis may thus lead to conflict between fact and fiction.

3.1. Ptolemy

An early example of adaptation of experimental data to theoretical fictions seems to occur in the *Optics* of Ptolemy (circa 150 AD). His measurements of the reflection and refraction of light belong to the outstanding contributions to ancient experimental science.

The device he used for his experiments on refraction is fundamentally the same as that depicted in modern textbooks. He found that at the transition of light from a less dense medium (air) to a denser one (water) there is refraction towards the perpendicular to the water surface, whereas a refraction away from the perpendicular ensues when the light moves in the opposite direction. Some of his data are quite correct, which shows that he did make the measurements involved. When light passes from air to water, for example, he found that

when the angles of incidence α were successively	0°	10°	20°	30°	40°	50°	...,
the angles of refraction β were	0°	8°	15° 30'	22°30'	29°	35°	...,
	[	][	][	][	][	][	][
so that their difference decreases:	8°	7½°	7°	6½°	6°	5°	...

That is: the differences between the refraction angles decrease from 8°, to 7½°, to 7°, etc. and subsequent differences differ from each other for each interval of 10° of the incident angle by precisely 30'.

When measuring the refraction of light passing from air into *glass*, the values of β were not the same as in the transition from air into water, but (allegedly) here too all final ‘differences of differences’ were 30'.

This ‘law’ is not conformable to the facts, though some of his measurements were correct. So it has been suggested that, once the idea of the *simple* law had entered his mind, he adapted other data to the hypothetical law.

There are many instances of experimentators who cling to their preconceptions, even when better experimental methods and more precise instruments give the lie to them.

3.2. Haüy

R. J. Haüy (1784), as we have seen in chapter II, launched a theory of crystal structure, and determined the interfacial angles of crystals using a contact goniometer. A certain angle in calc spar crystals was determined as 45° and this datum became one of the cornerstones of his theory. This value was found to be incorrect — particularly after the invention of the reflection goniometer at the beginning of the 19th century made more exact measurements possible. The angle was not 45° but 45°23'. Haüy (mistakenly) believed that the new value would undermine his whole theory and tenaciously clung to 45° as the correct value. He refused to use the more precise new instrument (precision 1') by which his error was detected. Similarly he maintained that certain dihedral angles of the calc spar rhombohedron were equal to their counterparts in the iron spar rhombohedron, *viz.* at 104°28'. Wollaston, measuring them with the reflective goniometer, however, found respectively 105°5' and 107°. So great a difference was easily detectable even by the contact goniometer whose precision (according to Haüy himself) was about 30', or even (with particularly perfect crystal surfaces) 20'. As Haüy's conception of mineralogical species did not admit precise equality of forms for similar crystals of different species, he whittled away even so great a discrepancy.

3.3. Dalton

Something similar took place when Dalton was defending his law of multiple proportions against Bostock, who had said that chemical analysis did not confirm this hypothesis. Dalton answered that Bostock himself had demonstrated its truth when he found that a certain quantity of lead could combine with different quantities of acetic acid that were as 24 : 49, that is [in Dalton's opinion] as 1 : 2.[1]

Thus the correctness of simple multiples was proved! On the other hand, when Dalton's explanation of chemical combination was endangered by Gay-Lussac's *experimental* law according to which gases combine with each other in volumes which are to each other as simple integers, Dalton emphasized that experiments do not support Gay-Lussac's law as he had found that the volumes of hydrogen and oxygen gas that combine to form water are as 1.98 : 1 and *not* as 2 : 1.[2] Even Dalton's best friend, W. Henry, pointed out that Dalton's instruments were 'incapable of affording accurate results' and that Dalton did not appear to advantage in his contest with Gay-Lussac. He recognized Dalton's 'vast inferiority in experimental chemistry' in comparison with the man 'whose beautiful law of volumes' he never assented to.[3]

It is evident that Dalton did not blame Bostock for the incorrectness of his experiment but he blamed him for not taking 24 : 49 as being 1 : 2. This would have fitted better into the idea he, Dalton, had of the simplicity of nature. But when simplicity of the ratio of the volumes in a reaction between gases clashed with that of theoretical pre-conceptions, he criticised Gay-Lussac for interpreting 1.98 : 1 as 2 : 1.

How do we make the decision in the choice between keeping to the data of measurement or of 'correcting' them slightly into conformity with the dictates of our theory? If a critical examination and alteration of the apparatus and continued agreement between different observers do not increase the deviation from the 'simple' relation, but rather support it, the decision is not too difficult. Therefore when Haüy excused the small deviation of his measurement from the 'limit' value of 45° (half of 90°, which is half of 180°), he had no right to appeal to the fact that Coulomb's law of mutual attraction or repulsion of electric charge ($f = k\, ee'/d^2$) was generally accepted though not exactly supported by measurements.[4] For improvement of measuring methods and repeated testing by different observers did not bring his result closer to 45°, whereas the measurements upon which Coulomb's law was based, came ever closer to the 'law'. An absolute agreement between theory and experiment, however, cannot be expected; for no experimental apparatus and no experimenter is absolutely precise.

3.4. Thought Experiments

Most real experiments first pass through the stage of being *thought experiments*. These are not experiments in the strict sense as they refer to imaginary events and imaginary instruments excogitated by the human mind. Hypotheses about the causes of phenomena or about the laws according to which they work in reality are put forward, for example in theoretical mechanics, but in their ideal form they cannot be realized.

Some thought experiments cannot be realized even in an imperfect or quasi-perfect way. They are deductions within the framework of an Aristotelian or a mechanistic philosophy of nature. The problem of the stone dropped into a tunnel through the globe (see chapter V) had to remain mathematics without measurement and physics without experiment. But the comparison of the stone's motion with vibrations known from daily life, the swinging pendulum or the bouncing ball, connected it with the reality of events which — conversely — could in principle be idealized by divesting them of disturbing circumstances.

The 'wreath of spheres' by which the Flemish engineer Simon Stevin (1548 - 1620) deduced the law of the inclined plane is a pure thought experiment. An endless chain, a 'wreath', of equal spheres is placed around a vertical triangle (see figure 27). The apparent weight of a sphere (i.e. the component of the weight along the plane) will be inversely proportional to the length of the side. When AB = 2BC the number of spheres on AB is twice that on BC, but the total apparent weight is

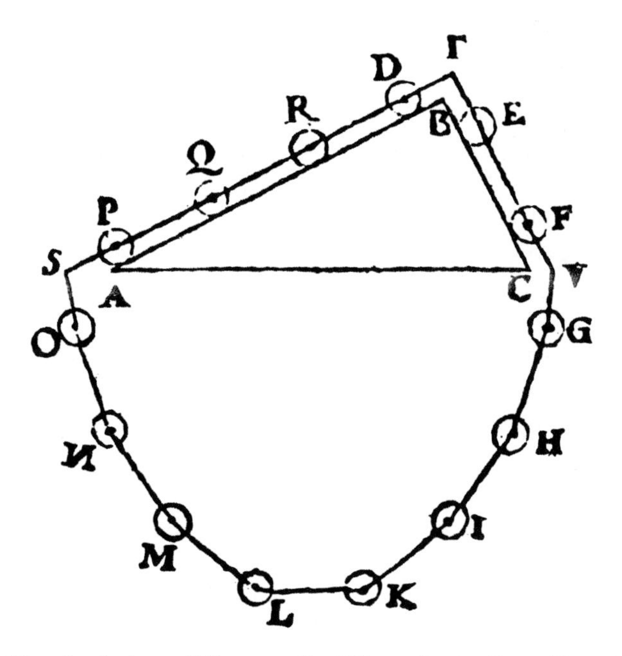

Fig. 27. Wreath of spheres ('Clootcrans') of Simon Stevin. From: Simon Stevin, *The Principal Works* I (Amsterdam 1955), p.176.

the same on the two sides. For imagine that the apparent weight on the left was greater: there would be movement but the overall situation would not change. Consequently the spheres would perform a perpetual motion, *which is absurd.* Therefore the wreath will not even start moving. Suppose now the spheres on each side replaced by single bodies, respectively of the weight of the spheres they replace. In that case equilibrium will be maintained when the ratio of the respective weights of the two bodies is equal to the ratio of the lengths of the sides.[5]

Stevin was so pleased with his demonstration that he used the wreath ('clootcrans') as the vignette on the title pages of most of his works, adding the legend: 'wonder is no wonder' ('wonder en is gheen wonder'). He thus reminds his readers of the fact that mechanics is a *natural* 'magic'. His admirer Isaac Beeckman generalized this statement: 'Philosophy proceeds from wonder to non-wonder; theology from non-wonder to wonder', thus saying that while the duty of science is to rationalize seemingly miraculous things, that of religion and metaphysical contemplation in general is to discover the wonderful and transcendental character of common things which we think we comprehend because we are accustomed to them.[6]

A third kind of thought experiment could have been put to the test, though it was thought superfluous to do so because the conclusions seemed to need no further confirmation.

Aristotle's tenet that the air in some way pushes forward a projected body after it has left the hand or a ballistic machine was 'disproved' by William of Ockham (14th century): if two arrows are shot respectively from A to B and from B to A, they pass each other halfway. According to Aristotelian theory this would imply an absurdity as at the meeting point the same air would be pushing them in opposite directions.[7]

Benedetti's demonstration (1553) of the law stating that heavy bodies fall at the same rate as lighter ones of the same material is a typical thought experiment that is meant to *dis*prove a current tenet. According to the Aristotelians a heavy body falls faster than a lighter one. If this were true, Benedetti remarks, and the two bodies are connected together, the heavy body would be retarded by the lighter one, and the lighter one would be accelerated by the heavier. The result would be a velocity intermediate between those of the separate bodies. But the two bodies, when connected, may be considered as one body, heavier than either of the components and thus should fall *faster* than either of them. The Aristotelian theory is thus self-contradictory and the only way to avoid the absurdity is to attribute the same velocity to all falling bodies, heavy and light (see below, section 5.1 on Galileo).

It should be pointed out that Galileo and Descartes had a tendency to skip real experiments when their result appeared to be a foregone conclusion.

Finally there are thought experiments which are presented as having been executed, though in fact their results were taken for granted because they had been described by earlier scholars of repute and seemed acceptable to later writers. In many such cases it is certain — notwithstanding the details given in their description — that they were never performed: if they had been, the falseness of their conclusions would easily have come to light.

We should remind ourselves that in general the ancient and medieval scholars were not rationalists *in principle* though they were often so in practice. Aristotle and his followers declared observation and experience to be the beginning and the end of scientific endeavour. Aristotle was a keen observer of animals and their way of life. When dealing with the generation of bees he recognized that the facts had not been sufficiently ascertained and that — when this will be done in the future — then we shall have to believe the direct evidence of the senses (the

WISCONSTIGE
GEDACHTENISSEN,
Inhoudende t'ghene daer hem
in gheoeffent heeft

DEN DOORLVCHTICHSTEN
Hoochgheboren Vorst ende Heere, MAVRITS Prince van
Oraengien, Grave van Nassau, Catzenellenbogen, Vianden, Moers &c.
Marckgraef vander Vere, ende Vlissinghen &c. Heere der Stadt Grave
ende S'landts van Cuyc, St. Vyt, Daesburch &c. Gouverneur van
Gelderlant, Hollant, Zeelant, Westvrieslant, Zutphen
Vtrecht, Overyssel &c. Opperste Veltheer vande
vereenichde Nederlanden, Admirael
generael vander Zee &c.

Beschreven deur SIMON STEVIN *van Brugghe.*

TOT LEYDEN,
Inde Druckerye van Ian Bouvvensz.

Int Iaer CIƆ IƆ CVIII.

Fig. 28. Title page of Simon Stevin's *Wisconstige Gedachtenissen* ('Mathematical Memoirs'). Leiden: Bouwensz. 1608 ([1]1605).

phenomenon) more than reasonings. We owe credit to theories provided their results agree with the phenomena.[8]

Similarly that eminent medieval scholar Jean Buridan (14th century) declared that each general thesis in natural science has to proceed by 'experimental induction' from many particular instances. He then refers to Aristotle who had said that without the memory and experience of the senses we could not know that fire is hot; and that, by a similar 'experimental induction', we know that there does not exist any empty space as everywhere we find some natural body: air, water or something else.[9] Buridan then gives examples of 'mathematical experiences' (*experientie mathematice,* that is in this case, 'physical experiments') which prove that there is no vacuum: we cannot separate the boards of a pair of bellows if all the holes are closed; even twenty horses, ten on each side, would not be able to pull them apart. They part only if something is broken or perforated so that air can enter. Similarly when one end of a siphon (*calamus*) is put into wine and the other in the mouth, and air is sucked out, the wine will ascend and it will not fall — although it is heavy — as long as a finger is put on the other end. These cases deal with experiments that had been, or could have been, performed since long.

Buridan's pupil Marsilius van Inghen declared likewise that natural science is founded upon experience and he too makes no distinction between immediate observation and contrived experiment. When describing the experiment with the pair of bellows he augments the number of horses to a hundred, which is almost a proof that he does not claim to have witnessed the trial. He mentions also an experiment in which a candle is burning under an inverted vessel that is standing in water: the water will ascend. He describes experiments with a siphon and a clepsydra for which he acknowledges his source as a treatise 'de inani et vacuo'.[10]

Most of the 'experiments' mentioned by medieval philosophers are borrowed from the ancient Alexandrian engineers (Ktesibios, Philo, Hero) whose works, mainly via Arab translations penetrated into the West after the Crusades.[11]

These philosophers often referred to phenomena known from daily life but where they dealt with more sophisticated experiments, which in fact they *could* have performed themselves, they preferred to repeat what they had read in the works of their predecessors, mainly those from Antiquity.

In Antiquity some people who held that seemingly homogeneous bodies are interspersed with vacuoles, believed that a pot full of ashes can absorb as much water as the empty vessel would have received. Clearly, they said, the water particles fit into the vacuoles. Aristotle accepted this as a fact, but he regarded it as a proof of his own (opposite) theory according to which homogeneous bodies are absolutely continuous, so that the same quantity of water could adequately fill the smaller as well as the larger space.[12] On his authority Buridan (borrowing the story from Averroës) accepted the 'fact', but, as so often happens, the story had grown more wonderful in the repetition: he declared that the pot with ashes can hold even more water than the empty one. [13]

Another famous story that, in spite of its being false, lived on from early Antiquity far into the 16th century is that Pythagoras had discovered that pairs of hammers of weights as 1 : 2, 2 : 3 and 3 : 4 upon striking an anvil produce respectively an octave, fifth and fourth. Pythagoras was said to have found also that strings of lengths 1 : 2 etc. produce an octave etc. (which is correct) and that strings of *equal* lengths produce the octave, etc., when under tensions that are as 1 : 2, etc. Thus by a false analogy the law valid for the lengths of strings was transferred to one concerning weights of hammers and tensions of strings (see also chapter II). According to Macrobius (circa 410 AD) 'Pythagoras ascertained with his eyes and hands what he had been searching for in his mind.'[14]

Even Cusanus (end of 15th century) took the truth of the story for granted and Franchino Gaffurio (1492) gave it a sense of concreteness by a picture (see figure 3 on page 30), in which the strings are seen loaded with weights marked 12 and 8 (for producing a fifth) and 12 and 6 (for producing an octave). Such a picture must have helped to convince people of the reality of the experiment, in the same way that the pictures of basilisks, dragons and unicorns in bestiaries made these legendary animals more 'real' than the rhinoceros about which at best vague memories existed. In 1547 Arnaldus Fabricius in his address at the opening of the academic year at the *Collégio das Artes* in Coimbra told his audience that if the weights stretching two similar strings are as 1 : 2 an octave is brought forth 'as has been taught by experience.'[15]

Yet it would have been quite easy to perform the real experiment, which would have shown that not two but four times the weight is needed to produce this effect and that for the fifth the weights must be as 9 : 4 and not 3 : 2. It lasted more than a thousand years before the falsity of this 'experimental' law was exposed. Vincenzo Galilei (circa 1520 - 1591) actually made the experiment and pointed out that there were 'two false opinions of which men have been persuaded by various writings, and which I myself shared until I ascertained the truth by means of experiment, the teacher of all things.'[16]

4. THE EMERGENCE OF EXPERIMENTS

In the above we have met with pure thought experiments (which are not experiments in the strict sense) and with pseudo-experiments which are not experiments either. This does not imply that no original research and no real experiments were performed during the Middle Ages.

4.1. Petrus Peregrinus

Magnetism, which had not been dealt with by the Alexandrian school or the Romans, was tackled in a 'Letter on the Magnet' (1269) by Peter of Maricourt (Petrus Peregrinus). It is a masterpiece of exposition of the properties of the lodestone. The author declares that besides the 'mathematical' also the 'experimental' method should be applied, for by manual skill (*manuum industria*) errors may be corrected which would never have been discovered if natural philosophy and mathematics were the only means at our disposal. The Franciscan friar Roger Bacon, who had the reputation of being a magician, greatly admired Peregrinus. In his opinion he is skillful in theory and practice of all technical work, 'a master of experiments' (*dominus experimentorum*) knowing natural things by experience in medicine and alchemy, expert in military matters, agriculture, geodesy and also the experiments of the sorcerers (*vetulares*) and the 'illusions' of the jugglers (*joculatores*). He worked three years to fabricate a burning mirror. Bacon thus stressed that Peregrinus not only occupied himself with the *liberal* arts (which are considered as befitting a philosopher) but also with the mechanical arts (which because of their 'servile' character were below the dignity of the philosopher). Indeed he considered the latter to be necessary for a natural philosopher and held that an experimenter should not be ashamed of learning from carpenters and smiths and of visiting their workshops in order to acquire their manual skills. (He thus anticipated and in some respects even surpassed some 16th century innovators: Vives, Paracelsus, Ramus). Bacon's panegyric shows also how close natural science and natural magic remained. As for Bacon himself, he claimed for himself feats that went beyond the limits of natural magic and verged on sorcery.

Peregrinus described several compass needles; he distinguished their two poles, also magnetic attraction and repulsion, and gave prescriptions for magnetization by contact. He put forward the hypothesis that a globular magnet must follow the daily motion of the heaven, but he also acknowledged that he had not succeeded in confirming this by experiment.

Bacon's enthusiasm was justified: more than two centuries were to pass before experiments of the same quality were made on magnetism — the oceanic navigations having made it a hot topic. In general these later writers made reference to their predecessor.

For the time being magnetism remained the subject of fabulous experiments. The philosopher Thomas Bradwardine (1328) mentioned an alleged 'phenomenon' which 'the common people will deem wonderful, namely that it is as easy to lift a magnet with a piece of iron adhering to it (whether underneath it, on top of it or enclosed within it) as to lift the magnet alone, without the iron.' According to Bradwardine the iron does not resist the raising of the magnet, but it moves of itself (*per se*) with the magnet, so the same weight will be found whether the iron is attached to the magnet or not.[17]

The story is repeated by Henry of Langenstein (1325-1397) as a 'result of experience', though he admits that he has not seen it himself, but that it is true 'as some people say'. Evidently he did not take the trouble of a simple verification by 'manuum industria'.[18]

Further, several medieval philosophers were acquainted (mainly via the 'Arabs') with the works on mechanics and hydrostatics of late Antiquity. Some of them (Witelo, Grosseteste, Dietrich von Freiberg), in the wake of their Greek and 'Arab' predecessors (Ptolemy, Alhazen), contributed to the development of theoretical knowledge of optics, a 'clean' appendix of the mathematical liberal arts. Though they deemed experimental confirmation of their tenets desirable, in general they paid only lip service to it, contenting themselves with citing authors who supported their preconceptions.

4.2. Nicolas Cusanus

We have pointed out before that real experiments were made especially by 'engineers' and artisans of the 16th century, and that they found increasing recognition by many scholars of that epoch. An anticipation of this cooperation of scholars and practitioners is to be found in Nicholas Cusanus' treatise *De Staticis Experimentis* (1450) in which an unlettered practitioner ('Ydiota') is interviewed by a scholar ('Crator') about devices for measuring various properties of bodies by means of a balance. For example the humidity of the atmosphere is to be measured using a wad of wool put on the scale.[19] To the Crator's question as to how the forces (*virtutes*) may be weighed by some artifice, Ydiota answers:

> I hold that the force of a magnet could be weighed if one puts on one scale the magnet and on the other iron until there is equilibrium. Next the magnet is replaced by an equal weight while the magnet itself is held above the iron which then apparently becomes less heavy. The weight on the other scale is diminished until equilibrium is reestablished; it may be assumed that the force of the magnet equals the diminution of the weight of the iron.[20]

It seems doubtful whether these measurements were anything more than mere thought experiments, for the Ydiota adds that in a similar way the force by which diamond *impedes* the action of a magnet may be measured. Similarly when the Crator asks whether the artificer can determine how much 'mercury' and 'sulphur' are contained in metals, the reply is that this may be rather precisely (*propinque*) found from their weights.[21]

The problem whether such substances do exist in metals does not interest him (probably he is thinking of an analogy with Archimedes' famous experiment with

Hiero's crown of gold and silver). The present problem remains purely theoretical as he could not know the densities of the fictitious chemical principles.

The truth of the story delivered by Macrobius and Boethius, about Pythagoreans who determined the weights of two bells or two hammers as being to each other as 1 : 2 when they produce an octave, Cusanus takes for granted, which confirms our impression that his problems and their solutions are largely borrowed from literature or from personal fantasy. Moreover the results of measurements are never given; whereas those people who really did perform measurements were inclined to communicate the numerical data they obtained. Ptolemy mentioned, as we have seen, the angles of incidence and refraction of light; al-Biruni mentioned the relative weights of various substances and all astronomers indicated the positions of the celestial bodies by numerical data.

It must be recognized, however, that Cusanus never claimed to have made the measurements. The answers are given in a conditional tense: 'if you were to do this, you would find that.' After all, Cusanus wrote the questions of the Crator as well as the answers of the craftsman. One might easily forget this as sometimes rather precise details of the experimental devices are given and it is modestly said that the measurements will not be absolutely precise: 'Many things can be known by very probable conjectures.'[22] The weight is 'nearly' (*propinque*) proportional to the force. But such statements need to spring from the modesty of the practitioner who knows that all measurement is imperfect because of the inevitable imperfection of man-made instruments and of human observation. What we see here is rather that the platonizing philosopher Cusanus holds that 'nothing in this world can reach precision.'[23]

Nevertheless, though most (if not all) of Cusanus' experiments were thought experiments, the advantage over similar 'experiments' by the *Calculatores* (see chapter V) is that they have the semblance of concreteness by their description of experimental devices and by the frequent 'use' of such a familiar instrument as the balance. It was also a step forward when the *philosopher* Cusanus let the artisan say that 'without the balance nothing certain could be done,' which is an acknowledgement that not calculation, like that of the Calculatores, but practical measurement is the best way of establishing mathematical relations and thus objectifying them. It is tacitly recognized that valuable knowledge may be acquired via measurement, that is by work of the hands; and that it is not the philosopher but the unlettered artisan — the intelligent worker with manual skill though without philosophical and literary training — who has the answers.

4.3. João de Castro

Concrete results of measurements will be found with those people who *needed* precise measurements: astronomers, geodetians, navigators, architects. With them imprecision could lead to disastrous consequences. Moreover they were not — like mathematizing philosophers — content with relative values but wanted *absolute* values.

A particularly sophisticated method of measuring and a discussion of possible errors is found in the journal of the maritime voyages (1538 - 1541) of the Portuguese admiral D. João de Castro which show what was possible in this field in the early 16th century. He was well aware that accuracy of the data he collected concerning the situation of rocks, the depths of bays and harbours, the direction of winds and currents was of great importance for the safety of those who would come after him. He clearly saw that the results may seem capricious because of

a) human errors (defects caused by the observer's lack of skill in manipulating instruments, erroneous calculations caused by lack of mathematical knowledge);
b) defects of the instruments, and
c) the disturbing influence of the environment.

Castro's main task was to determine the latitude and longitude of the places he visited. The method for measuring the latitude was borrowed (after some adaptations) from astronomy; it was based on measuring the height of the sun. Longitude, however, gave insurmountable trouble because no clock was reliable when submitted to the rocking of the ship. It was generally believed that the magnetic compass needle would solve the problem. It was known that the needle did not point precisely to the geographical north pole: its indications deviated from the meridian. Castro was given the task of finding out

a) whether there is an 'agonic' meridian (for which the 'variation' (declination) is zero) running through the Canary Islands,
b) whether on all meridians the declination is a constant, and
c) whether the declination changes equally when going east or west from the null meridian.

One after the other he found these hypothetical laws to be false. 'From these operations it is clear that the variation of the needle is not as the difference of the meridians.'[24] The negative results did not discourage Castro. He was convinced that some law (regra, rule) was hidden in nature which 'up till now has not come to my knowledge.'[25]

Castro let his measurements be checked by educated people (the ship's doctor) as well as uneducated ones (the caulker, sailors etc.). Moreover he checked the measurements of latitude by an instrument especially devised by the astronomer Pedro Nunes.

On several occasions he met with capricious deviations of the needle and then tried to find their cause by really analytical experiments, having first ascertained that neither personal errors nor defects of the instrument were at stake. Thus he discovered that abnormal behaviour of the needle was caused by iron objects on the ship. And the fact of great changes of declination over a short distance along the same meridian was ascribed to the influence of the rocks of the coast. After many experiments from the land and on the ship he concluded that 'these rocks are of the kind and nature of the magnet; their matter and composition is ferruginous and through their cause they attract the iron of the needle making it deviate from its natural direction.'[26]

Castro's meticulous measurements and precise descriptions of phenomena (e.g. of the basalt columns near Bombay) had little influence, mainly because of the secrecy maintained around the voyages of discovery. Rock magnetism as the cause of anomalies of magnetic declination did not become generally known before the end of the 18th century (Humboldt 1796).[27]

The work of Castro shows what science *could* be in about 1500 if it was not interacting with 'philosophical' circles. Castro himself, however, considered that his work, though useful and necessary, belonged to the 'lower parts of the mathematical sciences' and not to 'physics'. However, his opinions in philosophy of nature which may be found in his *Tratado da Sphaera per Perguntas e Repistas a Modo de Dialogo* are medieval in a particularly conservative way.[28] That work does not contain even a trace of the *via moderna* of the Nominalists. His work shows some analogy with the Manueline style in architecture, which was basically gothic but with a very realistic ornamentation based on the recent navigations and the objects they had brought to light.[29]

4.4. De Orta and Magnetism

At that time there was in Goa a 'philosopher' who likewise showed a 'modern' approach to experimental science, namely the apothecary Garcia de Orta (1504 - 1570). He actually carried out the experiment of the magnet weighed together with a piece of iron and found that earlier reports were false: 'we together with some persons have already experienced the opposite ... and be not amazed for people do not test all matters'.[30]

He put it mildly: the habit of describing experiments that were never performed lived on for further decades. Orta also proved that the current beliefs that diamond attracts iron and impedes the attraction of the lodestone by iron was 'a great phantasy ... for I showed the opposite before many people ... and I will make the experiment before you if you wish it.'[31]

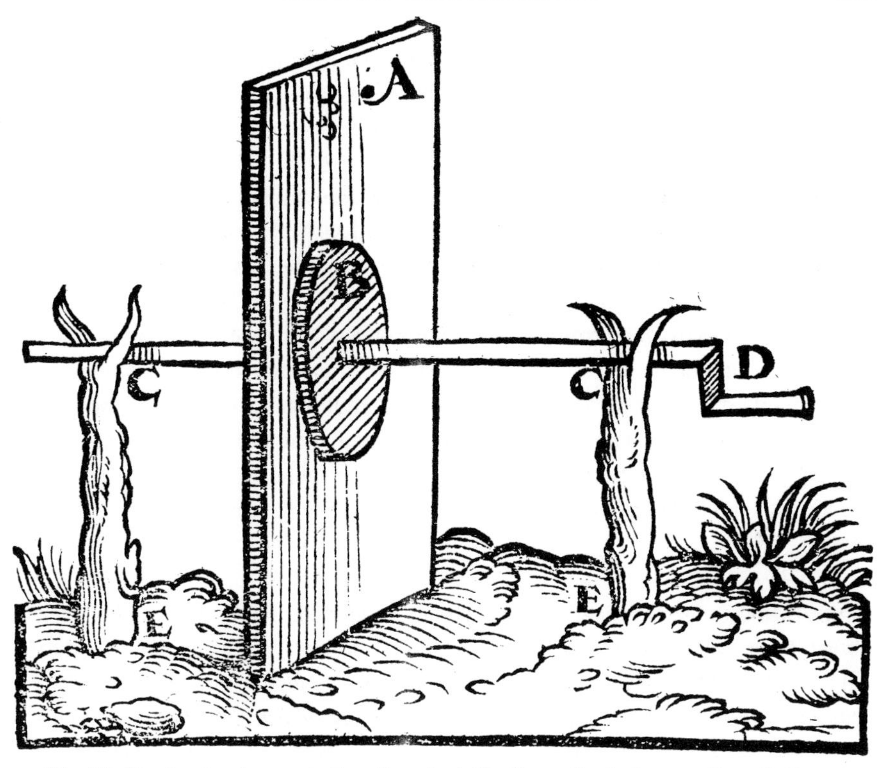

Fig. 29. Device for demonstrating 'impetus' (Scaliger, *Exotericarum Exercitationum* 1557).

4.5. Possible but Unperformed Experiments for Testing Aristotle

Castro and Orta were exceptionally realistic, free from fables and not afraid of using their own hands in order to find truth about nature. Half a century later a great contrast to them is still seen e.g. in the case of the Italian innovators in 'philosophy', Telesio (1586) and Patrizzi (1587) who maintained, against the prevalent opinion, that a void can be made by experiment. They adduced as a matter of experience that, when a vessel full of water is hermetically sealed and the temperature is lowered so that the water freezes, this will contract so that an empty space will be produced above it. Against them the notoriously conservative Jesuits of Coimbra staunchly defended the Aristotelian doctrine. They too resorted to experience, which teaches us that when the water in the vessel is strongly cooled either the water will not congeal (for nature will rather prevent this normal event from taking place than permit a void to be formed) or the water will freeze, thus diminishing its volume, whereupon the vessel will break up in order to prevent the occurrence of a vacuum.[32] Both parties spoke as if the experiment had really been carried out; both took for granted that water, like practically all other substances, contracts when solidifying. (This reveals that neither of them had in fact performed the experiment). Both claimed that the experiment confirmed their theory, respectively that a vacuum can exist and that it does not exist. They each adapted the fictitious experiment to their preconceived theory and spoke as if diametrically opposite effects ensued.

Some authors of the second half of the 16h century indiscriminately mixed genuine experiments with fictitious ones. The philosopher J.C. Scaliger depicted (1557) an apparatus for proving that a body once set in motion keeps on moving not by the air but by the impetus given to it. A circular disk is sawn out of a wooden board and fixed to an axle (with a handle) perpendicularly through its centre (see figure 29). The disk, once set in motion 'you will manifestly see' go on moving though the gap between the disk and the surrounding board is so small that the air therein could not have sufficient force to move the disk 'into that imagined motion'.

In the 14th century Jean Buridan, when suggesting a similar device, had chosen an object from an artisan's workshop (a 'smith's wheel') as an example, whereas Scaliger seems to have constructed a special apparatus for demonstrating a scientific thesis. Yet in spite of his giving a picture and describing its fabrication, he does not assert that he made it or that he ever saw it himself, but only that 'you will see the motion.'[33]

4.6. Magnetism Again

In the meantime 'mathematical' practitioners (engineers, shipwrights, gunners, navigators) were getting more self-confidence and gaining larger influence. This was the more so as the number of 'philosophers' with technological and experimental skill was increasing. In particular the literature on navigation (magnetism), gunnery and mining vastly increased: both scholars and outstanding artisans contributed to it. Their works contained discoveries of a truly experimental character. Robert Norman in his *The Newe Attractive* (1581) described experiments on magnetism which he had performed himself, among which were some on the inclination (dip) of the magnetic needle. The discovery of the latter phenomenon had been anticipated in 1544 by the Nuremberg clergyman Georg Hartmann.

Norman claimed that his experiments on magnetism were founded not on conjecture and imagination, but only on experience and demonstration; experiments, so he said, are 'Reason's finger' pointing to the truth.

Fig. 30. The fabrication of magnetized iron, indicating *Septentrio* (North) and *Auster* (South). From: William Gilbert, *De Magnete* (1600).

The greatest expression of the new approach was William Gilbert's (1540 - 1603) monumental *New Physiology on the Magnet, Magnetic Bodies and the Large Magnet Earth* (1600).[34] In the title it is claimed that this new physics has been 'demonstrated by arguments as well as experiments.' The book is a systematic collection of experiments made by his predecessors and by himself. Gilbert promises to unveil the secrets of nature by experiments that are apparent to the senses and can, as it were, be pointed out by the finger and not by probable conjectures and quotations from the Philosophers. He did not expect that 'the lettered idiots and Sophists' who are addicted to the opinions of other philosophers would listen to him, and therefore he addressed his *new* way of philosophizing to those 'who truly philosophize' and who look for a science not from the books but from the things themselves.[35]

It is a frontal attack on bookish physics, and this time not by an artisan or engineer but by a scholar (physician). His 'new physics' is written in the language of the learned but approaches the simple style and method of the 'unlearned'. It had a large part in persuading the scholars to join forces with the practitioners and to transform physics into an experimental science. The picture it shows of the fabrication of magnetized iron is one of the oldest pictures showing how much physical experimentation owed to the trades (see figure 30). Whereas the chemists (alchemists, pharmacists, metallurgists) had long since, thanks to a long tradition, had their special laboratories, the 'physicists' in the then current sense had been spectators. The real experiments were performed by 'practical mathematics' which was exercised in mechanical workshops.

4.7. Isaac Beeckman: Scholar-Artisan

An interesting example of this transition is Isaac Beeckman (1587 - 1637), who after finishing his theological studies in Leiden let himself be apprenticed to his father, a chandler and manufacturer of water conduits. Having established himself independently, he used the equipment of his trade for experiments on heat and on hydrology. He joined measurement to his experimentation. He is one of the examples that show mathematization penetrating further into experimental physics from the beginning of the 17th century. Beeckman's diaries, which were discovered by C. de Waard, contain simple observations, experiments checking hypotheses, experiments for measuring causal relations already established, model experiments, and pure thought experiments.[36] As an accomplished scholar as well as intelligent craftsman he assiduously cultivated the art of 'thinking with the hands'. This skill was clearly manifested when he tested the Aristotelian tenet that water can be transformed into air. In various ways he tried to discard any air dissolved in the liquid or left in the vessel.[37] Descartes wrote that he had never before met with any scholar who so closely united mathematics with physics (1619). Evidently this is a reference to experiments in which measurement played an important role. In 1614 - 15 Beeckman determined the rate of flow of water (through an opening in the bottom of a vessel) as a function of the height of the water column, and he found (correctly) that it is proportional. The numerical data, however, show that they were considered sufficiently precise if they did not flatly contradict this (expected) simple relation.[38] His experiments on the relation between volume and pressure of a certain amount of air yielded data that are far from inverse proportionality: the comparison was found to be as the 2/3 power of the volume.[39]

A pure thought experiment, his deduction of the law of falling bodies (1614, 1618) is remarkable not only because it is correct, but also because it bears a dynamic character (this in contrast to Galileo's purely kinematic approach). He assumed the law of inertia ('what once moves, moves always, if it is not hindered', 1613 - 14) and a regular increase of the velocity of the falling body by a gravitational force that 'pulls with little jerks.'[40]

5. EXPERIMENTAL SCIENCE AFTER 1600

Most scholars who employed the experimental method to obtain knowledge of nature were at the same time keenly interested in practical applications of their knowledge. The gap between the 'philosopher', the 'engineer' and the skillful artisan was considerably narrowed. Unfortunately much outstanding work in the pure and applied sciences was not published at the time. Yet Castro, Leonardo da Vinci, Thomas Harriot and Beeckman exerted influence on their contemporaries without public acknowledgement: Beeckman's diaries, for instance, were seen by Descartes in 1628 and 1629, by Gassendi in 1629 and Mersenne in 1630 — three fellow mechanists.

Even if this were not so, these people deserve attention, because they show how far one *could* come at that time. Together with those who divulged their results in books they demonstrate how much the attitude towards experimental science changed within their own lifetime. For example, in 1613 and 1614 Beeckman still shared the widespread belief that just as fat occupies more space when melted than when cold, so water too occupies more space as a liquid than when frozen to ice.[41] By the next year, however, he says the opposite — because ice floats on water, and an experiment shows that when a tube full of water is frozen 'the ice protrudes at the upper end.'[42]

Similarly Kepler in 1596, in his first work, related that 'as experience teaches' a magnet does not increase in weight when a piece of iron is clinging to it; but in the second edition of this same work (1622) he revokes his earlier statement by declaring that by 'manifest experiments' the weight of the magnet with the iron is equal to the sum of the weights of each separately.[43]

In sum, by about 1600, the process of the 'mechanization of the world picture' made a great leap forward in three respects: further mathematization, the rise of a mechanistic world view and the testing of theories by means of manual experiments ('thinking with the hands'). But though it was generally recognized at that time that in physics, as Gilbert said, proofs have to be borrowed from experiments that are apparent to the senses rather than from probable conjectures and quotations from Philosophers,[44] there were differences in the emphasis laid on each of these three characteristics.

5.1. Galileo

A case apart is Galileo Galilei (1564 - 1642), with whom, according to a widespread opinion, modern physics began. By his experimental work and his mathematical descriptions of phenomena he is said to have inaugurated the epoch of 'classical' modern science. In his own time his greatest fame was gained by his telescopic observations (the discovery of the satellites of Jupiter, the rotation of sunspots, the phases of Venus), his defence of the Copernican system and the 'discovery' of the law of falling bodies.

Because of their great importance for the further development of science, we will now concentrate our attention upon his main works *The Dialogues on the Two Main World Systems, the Ptolemaic and the Copernican* (1632) and the *Discourses and Demonstrations on Two New Sciences, Touching Mechanics and Local Motion* (1638).[45]

Galileo firmly believed that the Book of Nature is written in mathematical characters. He neither fell into numerological speculations (as did so many of his contemporaries) nor did he begin in an empiristic way by enouncing laws that were a direct result of experiment. His treatment of the law of falling bodies is a rational deduction from hypotheses which were stimulated by rather crude observations; indeed hypotheses and deductions are more characteristic of his work on local motion than the 'experiments' he performed to confirm *a posteriori* the laws which he had already found by his hypothetico-deductive method.

5.1.1. Falling Bodies

According to one of the many myths about Galileo, he dropped heavy and light bodies simultaneously from the leaning tower of Pisa and discovered that they touched the ground at the same moment. As Aristotle had stated that the velocity of freely falling bodies is inversely proportional to their weight, Galileo had thus demonstrated the falseness of the old doctrine.

The experiment allegedly made by Galileo was not new. Before 1586 Simon Stevin and Johan de Groot (burgomaster of Delft) had made a similar observation.[46] They simultaneously dropped two lead balls, one ten times as heavy as the other, from a height of 30 feet onto a wooden board and could not hear any difference in the times at which the balls struck the wood. Their experiment seems to have received as little attention as a similar one made by Philoponos in the 6th century. Other experiments made in Galileo's own time, however, showed a small lag between the times needed when bodies fell from a more considerable height.

What then about the experiments from the tower of Pisa allegedly made by Galileo? There *were* such experiments in Pisa, made not by Galileo but by an Aristotelian opponent, Giorgio Coresio (1612), who found that a smaller weight reached the ground a little later than a heavier one. In Galileo's *Discorsi*, when the Aristotelian Simplicio adduces this fact as a proof of the incorrectness of Galileo's theory, Salviati, the representative of Galileo's own opinion, does not deny the fact, but points out that this very small difference is very far from Aristotle's tenet that the times of fall are inversely proportional to the weights of the bodies. In Salviati's view, Simplicio should not try to hide behind a difference of a hair's breadth the Aristotelian error which is as thick as a ship's cable.[47]

Galileo held that *in vacuo* the heavier and the lighter bodies would descend in equal times, his best proof of this equality being a thought experiment which turns out to be identical with that given by Benedetti before him. If the bodies have equal weights but are of different densities, that which has the greater density falls a little faster, but experience shows that the thinner the medium through which they fall, the smaller the difference of their velocities, so that the difference would be nil if there were no resistance at all, *viz. in vacuo.*[48]

Observation shows that fall is an accelerated motion. Galileo assumes for this acceleration the simplest law, namely that it is a 'uniform acceleration'. Deduction then leads to the conclusion that in a uniformly accelerated motion 'the spaces passed in various times are to each other as the squares of the times',[49] that is, that in subsequent equal time intervals the 'spaces' passed by a falling body are to one another as the series of odd numbers 1 : 3 : 5 : 7 etc.

It was impossible to measure directly whether this law was correct but by a detour Galileo managed to perform an experiment supporting it. He rationally demonstrates that bodies rolling down an inclined plane will have a uniformly accelerated motion and that their final velocity upon reaching the ground will be the same as if they had been freely falling from the same height. He did not try to determine the absolute value, but like his medieval predecessors he formulated the laws in terms of proportions.

Consequently he had done enough by the mathematical demonstration that the proportions found for bodies that are moving more slowly along the inclined plane are identical with those deduced for free fall. Though in general Galileo was very vague about the way in which he performed his experiments for testing the laws and about the apparatus he used, on this occasion he described them with great precision. He emphatically declared that he had made many experiments to ascertain 'that the acceleration of naturally descending bodies occurs in the aforesaid proportion'.

On an inclined plane a perfectly straight groove about an inch wide was made. The groove was made smooth by glueing a polished piece of vellum onto it. A bronze ball, 'very hard round and smooth', was allowed to descend along it. The same experiment was repeated many times 'to assure ourselves of the time, and no, we never found any difference in them'. First, the ball was made to descend the whole groove, next a fourth part of its length, which took 'punctually half the time' of the first one. Trials were made with 2/3 and 3/4 of the length of the groove and 'by experiments repeated near a hundred times, we always found the spaces to be to one another as the square of the times'. When the experiments were repeated on planes with another inclination, the same proportions were retained. Time was measured by weighing the water emerging from a small hole in the bottom of a bucket full of water. The proportions of the weights denote the proportions of the times 'with such exactness that the trials being many times repeated, they never differed any considerable matter.'[50]

It sounds too good to be true: experiments repeated 'a hundred times', yet never yielding any difference in their results, even when time is measured with a water clock! Yet we need not go so far as his contemporary, Marin Mersenne, who was more of an empiricist than Galileo, and who did not at all find the claims of the Italian scholar confirmed. He stated that the ball will bounce when the slope is steep, whereas when the slope is slight the friction is relatively too great. 'I doubt whether Mr Galileo has performed the experiments of fall along an inclined plane ...; the proportion which he gives often contradicts experience.' [51]

It is noteworthy that Galileo only rarely, and then rather perfunctorily, mentions absolute numerical data. His mentality was that of an Archimedean theoretical mechanist rather than that of an engineer who measures exactly. One of the few occasions upon which he says something about numerical data is when he mentions that a freely falling canon ball of 100 pounds covers in 5 seconds 100 ells (a value that was about half the real one).[52]

He says that perhaps the result is not wholly exact, but that this is irrelevant for the purpose for which he made the measurement. Contingent values were not very important to him.

Though Galileo recognized that both reason and experience are indispensible in science, the balance tips in favour of reasoning. It is as if a precise experimental verification of his theoretical deductions is considered a work of supererogation.

5.1.2. The Earth's Motion and the Fall of Bodies

In the *Discourse* on the Motion of the Earth Simplicio compares the motion of a stone falling from the top of the mast of a ship with that of a stone falling from a tower. If a ship is standing still, the stone will fall to the foot of the mast, but (in Simplicio's opinion) when the ship is under sail the stone will come down as far from the foot of the mast as the ship has progressed during the time of fall. Consequently, from the fact that a stone dropped from a tower comes down at its base, we may infer that the earth is at rest.

Salviati then asks: 'have you ever tried the experiment on a moving ship?' and Simplicio answers: 'I have not, but reliable authors have done so'. Whereupon Salviati retorts that these authors probably remit this problem to *their* predecessors' authority; but that whoever shall make the experiment shall find that the result is quite contrary: the stone falls down at the same place on the ship whether it stands still or is moving. The same holding true for the earth as for the ship, we cannot conclude from the stone's coming perpendicularly down at the foot of a tower anything touching the rest or motion of the Earth.

Simplicio then surmises that Salviati himself has not delivered a proof of the motion of the earth, and yet does affirm it for true? In his response Salviati does not adduce an experiment but a *rational* proof; 'for it is necessary that it should'.[53] He drives Simplicio into a corner, asking him to *imagine* a flat surface 'as polished as a looking glass' and as hard as steel; and a perfectly spherical ball of a heavy and hard substance (e.g. brass), and moreover 'all external and accidental impediments taken away,' and all obstructions caused by the air removed. That is, he wants Simplicio to imagine the ideal situation of Archimedean theoretical mechanics. Salviati's further demonstration thus refers to a world — to use Galileo's own expression — 'observed, if not with the eye of the forehead, with that of the mind.'[54] If the plane were somewhat inclined and the ball laid upon it, it would start moving and continue with an accelerated motion, whereas on an ascending plane it would diminish its velocity. If the plane were neither ascending nor descending, the ball carefully placed upon it would remain still. If, however, it was given an impetus, it would move on with neither an accelerating nor retarding motion: in this case there is no cause of retardation or acceleration, and also no cause of rest: the ball would go on forever along the same plane if this plane remains always at the same distance from the centre of the earth. Similarly a ship moving in a calm sea would (if all obstacles external and accidental were removed) move on 'with the impulse once imparted, incessantly and uniformly'. [It will be noted, incidentally, that here the law of inertia is proclaimed for a horizontal motion, i.e. for a motion parallel to the surface of the terrestrial globe, a motion in a circle which has neither beginning nor end. This implies that the law is more or less valid for a rectilinear path if this is relatively short.] Similarly, the stone at the top of the mast is, together with the whole ship, carried along the circumference of a circle around the centre of the earth. Simplicio then agrees that, if Galileo's hypothesis is correct, the stone 'with a motion indelibly impressed upon it, follows the ship and will fall down at the same place as when the ship lies still.'

On the third day of the *Dialogues* Galileo points out that a knowledge of 'accidents' (mere phenomena) is necessary for the investigation of things, and as a praiseworthy example he refers to Gilbert's research on the properties of the magnet, which had to be ascertained by experience.[55] (He could have mentioned also his own discoveries ('observations' rather than 'experiments') made with the help of the new instrument, the telescope).

(The introduction of the telescope added data to 'natural history' no less important than the observations made, some decades afterwards, by Leeuwenhoek and others with the help of another instrument, the microscope).

Galileo's final aim, however, was the 'mathematization' of natural history, and therefore he wished that Gilbert had been more of a mathematician: in that case he would have been less resolute in accepting those reasons for demonstrations which he now addresses as the true causes — reasons which do not bind so fast as necessary and lasting conclusions would do.[56]

5.1.3 The Tides

The mathematization, however, is lacking in the last of Galileo's *Dialogues*, dealing with his theory of ebb and flow. These phenomena allegedly provided him with one of the three strongest arguments for the motions of the earth. The basis of his explanation of the tides is that each locality on the earth moves during a daily rotation for 12 hours in the same direction as the earth's annual motion and during the next 12 hours in an opposite direction. Consequently the mass of water on the earth's surface will assume a periodic deformation. The oscillations of the water mass lead to the action of secondary causes which manifest themselves in straits and almost closed seas, so that, e.g., in the Mediterranean the period will be 12 hours, not 6 hours, the length of the period mainly determined by the length of the seas in the direction East-West and by the depth of the sea.[57]

Galileo thus puts forward a theory which does not bear a mathematical character, but is rather a description with a plausible explanation *a priori* in each case. But it was a mechanistic theory and thus avoided bringing in the attractions by the moon which he abhorred as an absurd pseudo-solution of the problem.

One of Galileo's later admirers, E. Wohlwill, in an elaborate evaluation of this ingenious theory, pointed out that Galileo believed that he had given an incontrovertible proof of the motion of the earth, but that in reality he had put forward 'an error which only a genius can excogitate', a castle in the air built with the help of the new science.[58]

What interests us here, though, is not the quality or quantity of the data upon which his theory is based, but the fact that this is another case where Galileo thinks it is enough to say that he made the experiment about a theoretical tenet. The fourth *Dialogue* (1632) repeated mainly what he wrote in a little treatise of 1616.[59] Isaac Beeckman, who read a copy of it in 1630, was struck by the fact that Galileo there 'wrote that he had made an instrument by which this motion could be represented' but that he 'did not add its construction' ('fabricam vero non addit').[60] In the *Dialogues* Salviati (i.e. Galileo) again claimed to have given experimental proof of his theory of the tides: 'though it may seem impossible to some, that in artificial Machines and Vessels we should be able to experiment the effects of such an accident: yet it is not absolutely impossible to be done; and I have by me the model of an Engine in which the effect of these admirable commixtions of motions may be particularly observed.'[61] However, any expectation that at long last the secret of this almost impossible artifice may now be revealed is frustrated by the remark that for our present purpose 'that which you may have hitherto comprehended with your imagination may suffice.'[62]

And then the third party in the *Dialogue*, Sagredo, meekly declares that as the phenomenon is in a sense undreamed of and without parallel among motions, it is not hard to believe that effects may be claimed from theory which cannot be imitated by our artificial experiments. A strange conclusion, after having declared that he could make an experiment, Galileo immediately relieves himself of the need to show it by letting his supporter say that it would be impossible.

5.1.4. The Vacuum

According to the opinion current in Galileo's time, nature does not permit empty space: it has a 'horror vacui' and it will rather let a heavy fluid ascend without limit and against its nature than allow a vacuum to be brought into being. The pump makers, however, told Galileo that water in a suction pump cannot be raised higher than 18 Florentine ells (*braccia*).* Galileo therefore modified the doctrine of the 'horror vacui'; the 'resistance to vacuity' (*la repugnanza al vacuo*) holds up a standing column of water whose height is not limitless but attains at most 18 ells. Galileo says that in order to measure the 'force of vacuity' one has to take a 'continuous' substance whose parts lack any resistance to separation save that of vacuity, 'such as water'. To perform the experiment he 'has imagined a device', which he describes in great detail. A cylindrical vessel, filled with water upon which rests a piston, is turned upside down; the force of the vacuum is then measured by the weight needed to separate the piston from the water. An exceptionally elaborate description of the experiment is given; and even a picture is added (see figure 31).[63]

Galileo concludes that weighing the stopper and rod together with the contents of the bucket 'you will have the quantity of the force of the vacuity.'[64] However, in spite of the precise description of the apparatus (of which but a shortened version is given above) this experiment seems to have been performed only in thought and not in fact. No numerical results are given and it seems that the description of the apparatus and the experiment are meant as a suggestion to other people rather than as a report of a personal experience. E. Gerland and F. Traumüller (1899), though joining in the then general praise of Galileo for having 'shown the ways to the modern art of experimentation', pointed out that his 'mechanical' investigations are mainly theoretical and that the *Discorsi* do not yield much for the art of experimentation.[65] Consequently they could only say of the apparatus described above, that 'one is tempted to believe that the great investigator did not at all perform the experiment but put it forward only as a theoretical requirement.'[66]

* The Florentine ell is 0.5414 meters; the height of the water in the pump would thus be about 9.75m, corresponding to 67cm mercury. The maximum value is 10.34m., for an atmospheric pressure of 76cm of mercury.

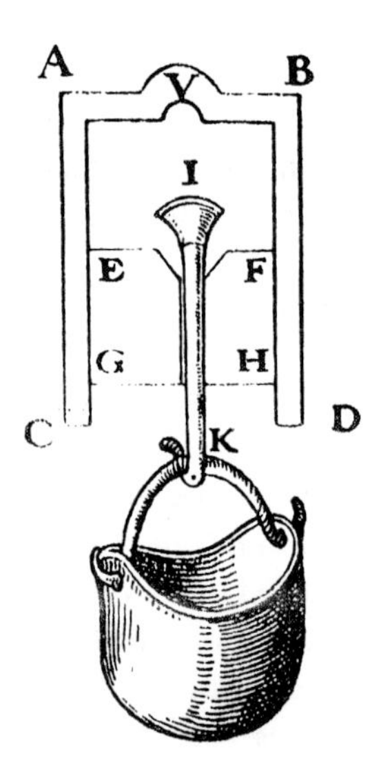

In a copper or glass cylindrical vessel ABCD fits a moveable wooden piston or stopper EFGH, through which can be moved an iron rod JK. The rod's end J is a cone that can fit precisely into a hollow of the stopper's surface EF. Some water is poured into the cylinder; the piston is placed in it and pressed down, holding the conical part of the piston some distance from the hollow, so that the air can go out (the hole in the wood is a little wider than the rod). The air is let out and the rod is pulled back so that its conical end J shuts the wooden piston. Next the whole apparatus is inverted so that the mouth is at the bottom; a bucket is fastened to the hook K. Sand is poured into the bucket until the superior part EF of the stopper forsakes the inferior part of the water.

Fig. 31. Illustration of cylindrical vessel used to measure the weight of vacuum. From: Galileo, *Discorsi,* First Day (*Opere* VIII, p.62).

Clearly this reluctant statement did not fit in with the then current image of Galileo as the founder of experimental science.

Having given this exceptionally detailed description of an experiment, which evidently was not intended to confirm the truth of his theory of limited 'horror vacui', but rather to determine its contingent numerical value, Galileo's fingers were not itching to perform it himself. His pupils occupied themselves with it, though on a different theoretical basis.

5.1.5. Galileo's Method

Galileo's hypothetico-deductive method contains several stages. After a direct, rather superficial observation of facts, a hypothesis (as simple as possible) is put forward; next a rational 'mathematical' (mechanistic) deduction leads to consequences that are capable of being tested by observation and experiment, and finally there is the verification of the conclusions by experiments. We have seen that the last stage does not have his full interest. It is evident that he, like Aristotle, starts from phenomena but that, in contrast to the Philosopher, he is not inclined to take phenomena at their face value, but subjects them to critical analysis to order them into a self-consistent mechanistic system. He could not enough admire Copernicus for the fact that, against the direct testimony of the senses, he maintained the motions of the earth.

Galileo had an astonishing talent in laying bare how the human mind is directed by the senses and by tradition. In doing that he used a Socratic method showing Simplicio 'in our wonted manner, that he has his answers at hand, though on first thoughts he does not discern them.'[67] He even used the trick of leading his opponent further along his erroneous path until he discovers that the end is a blind alley, a logical absurdity.

Galileo's greatest achievement is that (like Copernicus) he deduced 'laws' — e.g. in the case of projected bodies. He dissected the path of a cannon ball projected in a 'horizontal' direction into a uniform motion parallel to the surface of the globe and a uniformly accelerated motion toward the centre of the earth. It is only because of the relatively short distance covered that he allowed the horizontal component to be treated as if it were rectilinear.[68] By deduction Galileo determined the path of a projectile as being parabolic. He gave no experimental proof of this statement — it was enough that it was found by several reasons. Afterwards his followers in the *Accademia del Cimento* (1667) were satisfied with a rather crude agreement of their experimental results with the theory and they, too, found an easy explanation of any differences in 'disturbing circumstances' (friction and air resistance).

It was precisely one of Galileo's merits that he did not get entangled in the complicated phenomena of nature, but abstracted the kernel from them and thus could create a really 'intellectual' science. 'Things which no fixed mind can believe' are left out; the awkward question at what stage 'a hair's breadth' becomes 'a ship's cable', which tormented many empirical scientists, was left aside. His aim is to find conclusions 'abstracted from the impediments in order to make use of them in practice over those limitations that experience shall from time to time show us.'[69]

With Galileo, then, we meet with one of the greatest scientists, whose main instrument evidently was (to use his own words) the eye of his mind rather than the eye in his forehead. Though fully accepting that true physics must be founded on observation and be checked by experiment, he easily dodged the last part of the procedure or at least evaded revealing how he performed it.

Perhaps our criticisms of Galileo's experiments have given the impression that he is not one of the greatest physicists of all times. It should be emphasized that his mathematical treatment of the above problems shows that in the two masterpieces (the *Dialogues* and the *Discourses*) he is one of the founding fathers of classical-modern physics. But he himself would not claim that his glory lies in experimentation.

5.2. *Increasing Cooperation between Head and Hand*

During the 16th century the share of the engineers and of the chemists in scientific publications considerably increased. The upsurge of experimental research and the publication thereof by engineers and chemists, in particular in North-Western Europe struck the French philosopher Jean Bodin and — in his wake — the English geographer Nathaniel Carpenter who stated that 'their wits might seem to be in their hands.'[70]

We must think here in the first place of people who by their practical occupations rather than by theoretical meditation and mathematical deduction were, as it were, led by their hands into the construction of bridges, houses, canals, etc. and into the fabrication of metallurgical and chemical products. In particular the pharmacists and metallurgists stood in long traditions of using their hands in the development of instruments, in distillation and in smelting.

The great apostle of this cooperation between head and hand was Francis Bacon. From the artful discovery of 'natural laws' by means of experiments he expected a new world to come into being.

5.2.1. 'Experimental Chemistry'

By about 1600 many phenomena could easily be interpreted with the help of the then reviving 'corpuscularian philosophy'. As a result chemistry, so long despised by official scholars, was promoted from an 'art' to a part of natural 'science' by the German physician Daniel Sennert (1619) to be 'the key to natural philosophy'.

When speaking about alchemy we have already pointed out that Thomas Norton (1478) in his *Ordinal of Alchemy* states that:

> But mettall holdeth his holle Composicion
> When corrosive waters have made dissolucion.[71]

That is, the metal loses its typically metalline qualities (colour, ductility): it must have been divided into small particles, though without losing its 'being'. No theoretical preconception, no immediate sensory observation, but only the experimental result could lead to such an essentially un-Aristotelian conclusion. The metals disappear into the liquid; afterwards they emerge undiminished and unchanged from it; ergo they must have been in it in spite of the temporary loss of their characteristic properties. In a qualitative theory (like that of the Aristotelians, or even that of the alchemists) this was an absurdity. Yet the practice of the manual workers — alchemists, metallurgists — here overcame 'sound' philosophical thought.

Shortly after 1600 a similar, more or less corpuscularian interpretation was prominent with Angelo Sala, who pointed out that 'vitriols' crystallized from solutions of metals in sulphuric acid as well as nitric acid must contain these metals as invisible particles. In particular he combatted the concept of 'aurum potabile',

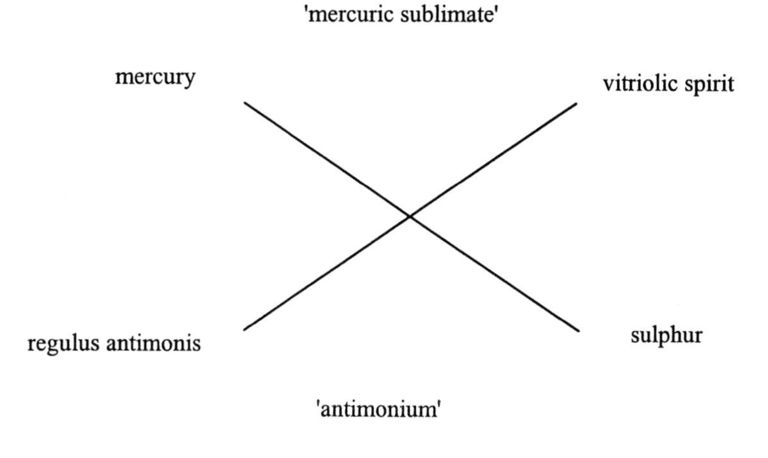

$[3HgCl_2 + Sb_2S_3 \text{ -- } 2SbCl_3 + 3HgS]$

Fig. 32. Béguin's interpretation of the reaction of 'antimonium' [Sb_2S_3] with mercury sublimate [$HgCl_2$]. Modern notation: $3HgCl_2 + Sb_2S_3 \rightarrow 2SbCl_3 + 3HgS$.

which was held to contain the medicinal properties of gold *extracted* from it by dissolving it in *aqua regia*. In Sala's opinion gold was in its totality in such a product, as it could easily be recovered. He then emphasized that it was only by observation by means of the eyes and the hands that he had been led to this conclusion.

A typical example of a new insight independent of any philosophical preconceptions is found with the apothecary Jean Béguin (circa 1615). His philosophy consists at best of vaguely Aristotelian ideas which could not lead to any discovery in chemistry. Yet he found something that comes close to a chemical equation. 'Antimonium' (at that time the name for sulphide of antimony [Sb_2S_3] which had already been recognized as a compound of elementary antimony (regulus antimonii) and sulphur), was heated together with mercury sublimate [$HgCl_2$] (product of the interaction of 'vitriolic spirit', an acid, with quicksilver). The results yielded by this process were 'butter of antimony' (antimonium oxychloride) and cinnabar [HgS] (then long known to be a product of 'common' quicksilver and 'common' brimstone). Béguin interpreted this reaction as an exchange between the four components (quicksilver, antimonium, sulphur and 'acid spirit') and summarized this in a diagram (see figure 32).

There was no reason to expect that there would be an interplay of these four components, and as they were considered to be themselves compound bodies, there was no reason to expect that they would function as constant component parts of other compound bodies. Evidently practice alone, aided by common sense interpretation of facts, may lead to solid scientific results. The saying that chemistry — or any other discipline — is a 'science' insofar as there is mathematics in it, is an exaggeration.

The phenomena referred to above may have furthered the acceptance of corpuscularian philosophy — but they were not based on it. Angelo Sala had no intention of attacking Aristotle's philosophy: but his experiments were grist to the mill of the atomistic anti-Aristotelian philosophers.

A striking result of chemistry in about 1600 was the discovery of the origin of 'metallic spirits'. Joseph de Chesne (Quercetanus) found that 'sugar of lead' (the product of the interaction of lead with vinegar, i.e., acetic acid), when heated *per se*, yields a combustible 'spirit'; 'the combustible spirit of lead' (spiritus ardens Saturni). A black residue remains at the bottom of the distillation vessel; when this is heated with 'soap' (i.e. with a reducing substance) the original quantity of lead may be recovered.

According to the then current conceptions, during the heating of sugar of lead the 'spirit' of the metal that was formerly contained in it escapes, while the dead body of lead (the 'caput mortuum') is left behind. The latter could then be 'restored to life' by heating it with coal or with soap.

A similar 'spirit', however, was obtained when the product of the dissolution of corals (mainly calcium carbonate) in vinegar was heated, so that a combustible coral spirit, 'spiritus ardens corallorum', escapes. This 'spirit', though practically identical with that obtained from lead acetate, was not recognized as such, not only because impurities caused differences of colour and smell but also because they *could not be* identical as long as they were believed to be respectively the 'spirit of lead' and the 'spirit' of corals.

In 1627 the German pharmacist and physician Anton Billich (Sala's son-in-law) put forward a different interpretation of these chemical phenomena. He pointed out that the black residue from sugar of lead could be dissolved in acetic acid and that then sugar of lead resulted again, which, when heated, yielded as much so-called 'spirit of lead' as the first action — and so on. As, in his opinion, lead would only *once* be able to give up its spirit, he concluded that neither the first nor the second spirit could be *the* spirit of lead. Consequently, because in both cases vinegar had to be added, the so-called 'spirit of lead' must be a 'changed vinegar'. Similarly the spirit of corals, too, must be just a 'changed vinegar', and these two spirits must be identical.

Thus his eyes were opened to the great resemblance of these two spirits so that he could easily attribute their difference to impurities — which can be discarded by repeated distillation so that in both cases pure acetone (propanon) is obtained.[72]

After these hopeful results, it is remarkable that experimental chemistry did not develop as fast as experimental physics. Robert Boyle, in the second half of the

17th century, heavily drew data from people like Libavius, Sala and Sennert. Why, then, did chemistry show so little progress? Probably because the gaseous substances, ('spirits', gases, vapours, airs) were hardly known. The technique for collecting, measuring and weighing them was not developed before the 18th century.

It should be noticed that the progress of chemistry owed much to the almost forgotten instrument makers and demonstrators who not only worked for the scholars but also earned a living by giving public lectures to inform and entertain the amateurs. They introduced auxiliaries facilitating chemical manipulation, e.g., the chemical stands and clamps, flexible (caoutchouc) tubes. The collecting of 'air' in an inverted jar filled with water and placed upside down in a 'trough' filled with water was introduced by one of these demonstrators (Moitrel d'Elément), and assiduously applied by the scholars (S. Hales 1727, J. Priestley, H. Cavendish, A.L. Lavoisier) who caused a revolution in chemistry by cultivating the new 'pneumatic chemistry'.

5.2.2. The Gas Law

The gas law We have seen that Galileo tended to use experiments not so much for discovering new laws of nature as to confirm 'rational' suppositions, theories and mathematical relations founded upon theory. A high degree of conformity between theory and experiments was not required.

A more empiricist approach characterizes the work of Robert Boyle (1626 - 1691). Following Francis Bacon, he regarded as his first task to contribute to the building of a new 'Natural History'. As a protagonist of corpuscularian theories in general he gave causal 'explanations' of observed phenomena, attributing them to the form, motion, size and arrangement of invisible particles. These did not lead to the deduction of other facts and were posited *a posteriori*. He had no way of checking their truth and thus often suggested several possible (or plausible) explanations for the same phenomenon, e.g. the elasticity of air.

With the help of his assistants Robert Hooke and Denis Papin he improved the air pump which Otto von Guericke had invented. He demonstrated how various phenomena (combustion, boiling, respiration, production of sound) changed when the air is removed or its pressure lowered. In his *Sceptical Chymist* he showed that the allegedly elementary bodies yielded no satisfactory explanation of chemical phenomena. His definition of true elements required absolute simplicity and he did not content himself with the 'limits' of practical chemical decomposition. His greatest merit for chemistry is that he proved that the four elements of Aristotle as well as the three or four principles of the chemists had no elementary character and had better be discarded from chemical explanations.

His empiricism ended in scepticism also in his treatment of the law by which his name is perpetuated. When Richard Townley suggested that there might be inverse proportionality between the volume and the pressure of a certain amount of air, he found experimental confirmation (at about the same time given also by Henry Power and by Lord Brouncker). He tested the hypothesis by increasing the volume of the air 32 times and also by compressing the volume to a quarter of its original value. He found that the law fitted better in the case of 'dilution' than in that of compression of the air. Having concluded that there was a 'suitable enough' confirmation of 'our hypothesis which supposes that the pressures and volumes are in inverse proportionality', the question was whether deviations from the law are due to deficiencies in the experiment (imperfection of calibration of the glass tubes, convexity of the surface of the mercury; the quicksilver might contain moisture and thus yield water vapour), or whether the ideal law does not obtain in nature?

Boyle remained uncertain, even so much so that he afterwards did not mention it among the many instances of 'proportions' in physics. Here again the cause of his uncertainty was the desire to keep strictly to the facts, and on the other hand, his awareness of the inevitable imperfection of experimental measurement. If he had deduced the law from a theory in which he firmly believed (like the law of the lever in mechanics) he could have attributed the deviation from it to experimental errors (at least within reasonable limits). He could have pointed out that the ideal law may exist in nature but that in the phenomena there is always the superposition of disturbing influences or the coincidence of several sub-phenomena. But in contrast to Galileo he wanted strictly to *verify* the law in an empirical way. At the same time he was hampered by being too conscious of the fact that measurements with the help of the human hand can never be absolutely perfect and thus can never give an absolute confirmation of a law. Empiricism, when not connected with either a certain *a priori* belief or a more or less positivistic attitude, must lead to scepticism. Small wonder then that Boyle never came back to the 'law' by which he got his place in the history of science.

A more resolute attitude was assumed by Robert Hooke. Having heard of Townley's suggestion,[73] he, too, found the law of the elasticity of air. The accuracy of his measurements was practically equal to that of Boyle, but his formulation was unwavering: 'from which Experiments we may safely conclude that the Elater of the Air is reciprocal to its extension or at least very near to it.'[74]

Hooke did not bother about absolute conformity. He even generalized the law when formulating it in an anagram (in 1665) followed by an overt statement (1678): 'The theory is very short, the power of any Spring is in the same proportion with the Tension thereof.' 'This is the Rule or Law of Nature.' '... the force or power thereof to restore itself to its natural is always proportionate to the Distance or space it is removed therefrom.'[75]

On that occasion he spoke of steel springs: perhaps Boyle's image of air as an aggregate of coiled springs led Hooke to the analogy with the law of elasticity of air. But analogy has to halt somewhere: a spring will not extend infinitely. Yet it was believed for a long time (though not by Hooke, 1661) that there is an end to the expansion of air: just like a steel spring its elasticity is 'spent'. On the other hand, neither air nor a spring can be compressed infinitely: the spring stops, compressed when the turnings touch each other, or in the case of air when its particles touch each other (Jacob Bernoulli, 1683). That is why Daniel Bernoulli in his correction of Boyle's law introduced the subtraction of the proper volume of the particles of air from the true volume: instead of pV it became p (V-b) is constant.

The history of the gas law exemplifies the fact that as long as experimental equipment is not very precise, a 'simple' law may be rather easily put forward. Improvement of the experimental auxiliaries diminishes the experimental errors and thus incontrovertibly reveals that the law is only approximately correct from a purely empirical standpoint. Further theoretical development then restores the law to its pristine simplicity by accounting for the 'deviations' due to natural causes. Not only the imperfections of Boyle's experiments, but also the fact that the law *cannot* be found as precisely valid prevented its verification as an 'absolute' truth. The proper volume of the particles of a gas and their mutual attraction need the transformation of Boyle's law into that of Van der Waals (1873).

5.2.3. The Third Factor

Pure empiricism and experimentalism may expand knowledge of nature but easily leads — as in Boyle's case — to scepticism. In particular when it goes together with a passionate and scrupulous yearning for truth it may hinder the progress of science by its being over-conscious of the imperfection of experimentation. Even for an experimentally inclined scientist like Pascal a conditional faith is most fruitful in creating a simple (preferably 'mathematically' formulated) theory. The greatest physicists of the late 17th century, Huygens and Newton, combined a hypothetico-deductive method in the Galilean spirit with a deep respect for empirical fact and a zealous application of experiments. The 'experimental philosophy' of the 17th century became a 'natural history' as well as philosophy of nature: it was a harmonious cooperation of head and hand.

'Reason and experience' are the foundations of science, yet there is a third, rather embarrassing factor for the scientist. Though it is easily overlooked, 'intuition' plays a role: his enigmatic faith in some tenet which is neither based on rational deduction, nor found by experimental induction. The great conservation laws, however much they may be supported by experience, were accepted *a priori* quite apart from their being 'proved' or 'disproved'.

6. Conservation Laws

6.1. *The Law of Conservation of Mass*

The laws of conservation of matter, of mass, of heat, of elements, of energy, they are all based on the faith that *something* in nature remains the same: the faith most radically held by the Eleatic philosophers of ancient Greece, who said that *all* things in nature remain the same because 'change' is irrational.

In particular the law of conservation of matter bears a wholly meta-physical, non-experimental character. We are wont to identify it with conservation of weight (or rather conservation of mass), which from the beginning of the 19th century to far into the 20th century has been considered as a matter of course: the quantity of matter is constant, the quantity of matter is proportional to its mass and thus there is conservation of mass. And at each place the weight of bodies being proportional to their mass, we have also conservation of weight. Strangely enough, in this same period, many physicists used the notion of the 'ether' (the vehicle of light) which was characterized as an 'imponderable matter'. The thesis that ponderable and imponderable matter cannot change into each other saved the 'law of conservation of weight'.

Conservation of matter is often confused with 'conservation of mass' and 'conservation of weight'. Even the great chemist D.L. Mendeléev's statements on this topic were far from impeccable. He stated that all practice of chemistry is based on the law of the 'eternity of matter', and that Lavoisier has 'demonstrated that it is impossible that matter could be created.'[76] But he also pointed out that the balance is 'imprecise', and thus recognized that experimental 'proof' of the absolute law is unattainable.[77] He spoke about the law as if it was based on experiments while at the same time recognizing that it *checks* all experiments and if observations do not agree with it, something must have escaped our notice.[78]

If such an outstanding and philosophically minded chemist (who, for example, so clearly distinguished the 'chemical element' from the 'simple body') mixes together an absolute dogma and an empirical law, one cannot be astonished to meet with the same confusion in many textbooks of chemistry.

We shall now try to show that:

1) 'conservation of matter' does not necessarily imply 'conservation of mass' or of 'weight';
2) conservation of mass is not identical with conservation of weight;
3) the conservation law was not a discovery of Lavoisier;
4) the law of conservation of the mass of matter has an aprioristic character.[79]

(1) The conservation of matter is based on faith and is not so cogent an *a priori* as it might seem at first sight. Aristotle supposed that a rather 'weighty' element

(water) could lose its weight by turning into a 'light' element (air). To him 'weight' was not the measure of the quantity of matter (which, it must be added, did never exist *per se*). Descartes considered 'extension' the measure of the quantity of matter; there is as much matter in a so-called vacuum of Torricelli as in the same space filled with mercury, or with water.

(2) On the other hand, some ancient philosophers did assume that *all* matter has weight and that under all qualitative transformations the constancy of its quantity is expressed by its constant weight. A second century author, Lucianus, relates a story about a philosopher who, when asked to determine the weight of smoke, answered: weigh the wood, burn it and subtract the weight of the ashes. A similar trend of thought had also been followed much earlier by the physician Erasistratos (280 BC) who put a bird in a cage on the scale of a balance and ascribed the gradual loss of weight of the animal to invisible effluvia. In these cases the *faith* in conservation of weight (and proportionality of weight and quantity of matter) seemed to make it possible to determine indirectly the weight of the supposed invisible matter that escaped. To Aristotelians, however, that conclusion would be unconvincing. They could say that some invisible 'light' substance had been imbibed by the bird from the environment. (A similar argument was advanced by some adherents of the phlogiston theory at the end of the 18th century: when a metal is calcinated the weight increases because phlogiston — supposed as having a negative weight — has escaped). As long as no direct data before and after the change are available the interpretation depends on the theoretical preconceptions.

(3) The introduction of the law of conservation of mass is usually (but wrongly) attributed to the French chemist A.L. Lavoisier (1743 - 1794). When enumerating his own merits in 1792 he did not mention the law of conservation of weight, though he was not the man to put his candle under a bushel.[80]

> It is even more unjustifiable to say that Lavoisier introduced the use of the balance in chemistry and as a result 'discovered' the law of conservation of weight. J.B. van Helmont (beginning of 17th century) demonstrated by weighing that as much sand enters into combination with an alkaline salt as can be regained from the product by adding an acid. Angelo Sala (1617) proved the presence of metals in vitriols by weighing the metal used for making the vitriol and recovering it by decomposition of the vitriol. Joseph Black (1756), whose experiments were the examples Lavoisier followed, heated a certain weight of calcium carbonate so that quicklime was formed, then slaked the quicklime by water, added a solution of sodium carbonate and found that the calcium carbonate precipitated from the solution had the same weight as that with which he started the series of manipulations. The reporters on the work of the young Lavoisier in 1773 praised him for 'submitting all his results to measurement, calculation and the balance, a rigorous method, which fortunately for the advance of chemistry, begins to become indispensible in the practice of this

> science.'[81] This was a statement by the adherents of the pre-Lavoisier phlogiston theory.

Lavoisier did not find the law of conservation of weight as a result of experiments; on the contrary he started from the metaphysical belief that nothing takes rise on its own account and — though an avowed empiricist ('I have put myself to the rigorous law never to fill up the silence of nature'[82]) — he never performed an experiment to check the truth of this law. How could he have done it with a balance which, loaded with 5-6 pounds (2 kilograms) had a sensitivity of 25 milligrams?[83]

In general his weighings were not very precise: he speaks about 135 to 136 grains of a 'martial aethiops' (iron oxide).[84] The numerical results of his experiments often are too good to be true: more modern and more precise methods would not yield such marvellous results as he mentions for his experiments on fermentation.

> In 1770 he proved by weighing that a closed vessel with water, when heated, does not absorb 'matter of fire', as the total weight remains the same. But he finds a precipitate in the water, which must be due to the glass of the vessel, for this lost a weight equal to that of the precipitate. Conclusion: water does not change into earth. Afterwards he shared Boerhaave's opinion that the weight of matter of fire is negligible.[85]

Lavoisier did not admit that the 'matter of heat' (caloric) can be transformed into more ponderable matter, for he assumed also a 'law of conservation of elements': each of his simple bodies keeps its weight throughout all chemical transformations, and consequently the same must be true for 'calorique'. This caloric matter may be almost weightless but its quantity is measurable (if not chemically bound) by a calorimeter. In checking the 'law of conservation of heat' he found that his calorimetric experiments confirmed his aprioristic law. That is to say: he confirmed it more or less, for on this issue his eagerness to find it confirmed was even greater than in the case of conservation of weight.[86]

As he recognized: 'when looking for the fundamentals of a new science, it is difficult not to start with rough approximations (*des à peu près*).'[87]

(4) Lavoisier's applications of the law of the conservation of mass were inspired by his faith in an incontrovertible truth. Thus in his early publications he did not weigh all the substances taking part in a reaction, but commonly used the law to weigh them indirectly. Most of his research touched the chemistry of gases (pneumatic chemistry) and the technique for measuring their quantities was only gradually developed. He took, for instance, the weight of nitric acid plus the weight of chalk [nitric acid + chalk → a product + gas]; subtracted the weight of the remains after these two had reacted, and concluded that the weight of the gas ('air') evolved is thus known.[88]

In his later experiments Lavoisier did, in addition, make direct measurements of the weight and volume (and specific weight) of gases; but the imprecision is so great that at best one might say that the experiments did not *contradict* the conservation of weight.[89] When dealing with the composition of water (1783) he wrote: 'nothing is annihilated in experiments ... the two gases ['airs', *viz.* hydrogen and oxygen], which are heavy bodies could not have disappeared.'[90] Hydrogen being non-metallic, he surmised that an acid would be formed by its combustion. Against expectations, however, he found only water as the product of the combustion: 'But as it is no less true in physics than in mathematics that a whole is equal to the sum of its parts [and the weight of 'caloric' is negligible] ... we have a right to conclude that the weight of this water is equal to that of the two gases that served to form it.'[91] In this case his strong belief in the conservation of mass led to a positive and lasting result.

The combustion of spirit of wine ('alcohol') yields water. Lavoisier was unable to collect this without loss, but 'it is easy to determine it by calculation', for 'we cannot doubt that the weight of the substances partaking in the process of combustion is the same before and after the reaction.'[92] There may be some difference because 'heat' escapes, but this has a negligible weight.[93] Yet Lavoisier recognized that, although this method is *sure,* it is less satisfactory than direct weighing of the water. He declared that he '*started* from the supposition that the weight of the substance is the same before and after reaction, which I regard as *evident* [italics RH].'[94]

It is *so* evident that in his main work, the textbook *Traité de Chimie*, the law of mass (or of matter) is not formulated at the beginning (as it is in all modern chemistry textbooks) but only when he deals with a formulation on page 141:

> Nothing is created, neither in the operations of art, nor in those of nature and one may posit as a fundamental tenet that in each operation there is an equal quantity of matter before and after; that the quality and quantity of the principles [i.e. chemical elements] are the same and that there are but changes, modifications. On this principle the whole art of making experiments in chemistry is founded.[95]

Here the 'conservation of matter' is tacitly conceived as implying the conservation of the weight of matter and moreover specified as the conservation of the weight of each of the various elements which together make up all 'matter'. In Lavoisier's publications the numerical values of the quantities of substances always agree 100% with the yields of reactions found for their component elements. This is too good to be true and he admitted this himself when he said that he 'corrected experience by calculation'.[96]

But what then of the practically weightless 'element' caloric (matter of heat)? It could not be checked by the balance, but here the recently invented calorimeter could be used. The 'law of conservation of heat' bears also an aprioristic character.[97] In checking it he had to permit himself an even greater liberty than in the case of the conservation of ponderable elements: he frankly avowed that 'it is difficult not to begin with a "more or less".' [98]

Whenever Lavoisier did not find conservation of weight in an experiment, the discrepancy was unhesitatingly attributed to an experimental error. Consequently Lavoisier did *not* discover conservation of weight by his experiments, but thanks to his belief in it he was able to improve his experiments. It was not the experiments that checked the law, but the law that checked the experiments. If the deviation from the law was deemed greater than the imprecision of the method of measuring, some substance must have escaped notice. Lavoisier was not very precise in his statements of experimental confirmation of the law. He often used expressions like 'about', 'more or less' and 'rather precisely' (*environs; plus ou moins; assez exactement*) when speaking about results he deemed to be positive. To him it was enough that experience did not flatly contradict a law already accepted beforehand. He says, for instance, that 100 grains of iron, when combined with oxygen increased in weight by 35 to 36 grains. The absorbed oxygen was measured 'rather exactly' as 70 cubic inches which weigh 'rather exactly' half a grain per cubic inch and this is 'exactly' equal to the increase of weight of the iron. That is: 'about' 70, multiplied by 'about' a half, yields 'exactly' 35 to 36.

The law of conservation of mass, then, is in its historical origin not an experimental law. Nor is it a necessary (compulsive) law of thought — Descartes, the apriorist *par excellence*, had not accepted it. It is a plausible proposition, put forward *a priori* (like the law of inertia). Man yearns for something constant (a real substance) in nature, a solid basis remaining steady behind all apparent change. But he must find out whether the right item satisfying the demand for constancy has been chosen.

As it functioned so satisfactorily, conservation of mass acquired the character of an experimental law, an absolute truth, confirmed empirically — with the reservation, however, 'as far as precision of measuring instruments reaches'.

Modern physics states that loss of heat (in a chemical reaction) implies a loss of weight (or rather of 'mass'). 'Conservation of matter' has historical priority over 'conservation of mass' but today conservation of mass has the priority in physical theory. Not only 'matter' but also 'energy' has mass. When four hydrogen nuclei form a helium nucleus the latter weighs less than the former; the 'mass defect' has to be ascribed to the great loss of energy yielded at this unification.

6.2. *Conservation of Energy*

6.2.1. Robert Mayer

The German physician Julius Robert Mayer (1814 - 1878) was struck by the fact that friction seems to cause a loss of motion and the generation of heat. This led him to the conception of conservation of energy which he put forward as a 'law': 'Nothing can arise out of nothing and no thing can become nothing'.

'Motion' and 'heat' must be different forms of one and the same indestructible 'force'. Qualitatively causes may change, but quantitively they remain the same: energy of fall (gravitational potential energy), energy of motion (kinetic energy) and energy of heat are transformed into one another without loss of quantity. A number of units of another kind (but of equivalent energy constant) takes its place.

Mayer did not state that if motion becomes heat, this implies that the essence of heat is 'movement'; the relation is one of quantity, but not of quality: 'One should beware of leaving the solid ground of the objective.'[99] In Mayer's view, little is known about matter; the existence of atoms is not proven, and we are ignorant about the nature of 'latent heat'.[100] In stating that heat is transformed into a mechanical effect, he is putting forward nothing but a fact; the transformation itself has not at all been explained. This positivistic statement is underscored when he adds: 'True science is satisfied with positive knowledge and is pleased to leave it to the poet and the 'Naturphilosoph' to try to solve eternal enigmas with the help of phantasy'.[101] Mayer was thoroughly convinced that he had kept to the facts alone: not only the atoms, but all imponderable matter was discarded: 'we want to banish the last remnants of the deities of Greece from the doctrine of nature'; nature in 'her simple truth' is greater and more glorious than any product of the human hand and than any illusions of the created spirit.[102]

This sharp distinction between the facts of the world created by God and the ideas created by His creatures seems to be in contrast with his aprioristic belief in the conservation of that great unknown: energy ('force'), and also with his Kantian belief that *our* notions about nature (insofar as they are not phantasy!) are *a priori* adapted to that world.

6.2.2. Determination of the Mechanical Equivalent of Heat

Mayer held that the determination of the quantitative equivalence between a certain unit of heat and a number of units of mechanical energy ('force'), the so-called 'mechanical equivalent of heat', did not require new measurements: only a thought experiment was needed. The values of the specific heat of air under constant pressure (c_p), and at constant volume (c_v), the expansion of air per degree rise of temperature at constant pressure, and the density of air were all known. Provided with these data, a thought experiment is imagined: 1 gram of air is heated at constant pressure (and with a resulting increase in volume) in an imaginary cylindrical vessel that is closed by a (weightless and frictionless) piston. To perform this, c_p calories are needed, whereas if the heating takes place at constant volume only c_v calories are needed ($c_v < c_p$). The difference ($c_p - c_v$) is assumed to be equivalent to the work done by the air's expanding (in the first case) against the constant external pressure. It is assumed that all the work done in the case of constant pressure was spent against the external pressure.[103] Mayer found (1845) for the mechanical equivalent of 1 kilocalorie of heat, 367 kilogram-meter, which was later (1874) replaced by 425 kilogram-meter.[†, 104]

Mayer first began to think about conservation of a 'force' after he made a voyage to Batavia (Dutch East Indies) as ship's physician, observing a rise in the temperature of the water when there was a storm at sea. The conservation principle was formulated (1842) by analogy with other conservation laws (e.g., conservation of mass). He considered it an application of 'logical rules' that from 'nothing' nothing arises, *causa aequat effectum.*[105] To him it certainly was no 'hypothesis' — a word that for him stood for a 'loose conjecture'. In his opinion, a thing that cannot arise out of nothing cannot be destroyed either; this has to be held for as long as no understandable fact proves the opposite.[106] That is to say, Mayer recognized that one fact against it would be proof of its nullity and that experiments did not absolutely *demonstrate* its truth. However much he was convinced of the truth of the law, the experimenter in him acknowledged that such a negative fact would overthrow his intuitive conviction. But Mayer did not believe that such a fact could ever turn up: in his view there is agreement between what is thought subjectively and what is true objectively; God, so he says (1869) has predestined an eternal harmony between these two.[107]

An over-bold thesis, which reminds one of Descartes' statement that God would not 'deceive' us — a statement which emboldened him to enounce his (erroneous) law of 'conservation of quantity of motion' which was linked up ontologically with God's immutability. This is a natural theology in which the analogy between God's and Man's thought is pushed to the extreme. Boyle said he wondered how Descartes knew the extent of God's immutability so precisely that he could make it an argument for his law of conservation of motion. A similar question could be put to Robert Mayer. The tradition of Boyle, Pascal, Newton, Boerhaave, which emphasized the ultimate incomprehensibility of God and His works could perhaps admit laws of this same type, but it would shrink back from tying up absolute Truth so closely with a scientific tenet based on human intuition.

[†] 1 kilocalorie is the heat required to raise the temperature of one kilogram of water through one centigrade degree; 1 kilogram-meter is the energy used in lifting a weight of 1 kilogram through one meter against gravity.

6.2.3. Joule: Direct Measurements of the Mechanical Equivalent of Heat

J.P. Joule (1818 - 1889), in contrast to Robert Mayer, made direct measurements of the mechanical equivalent of heat. From 1840 onwards he investigated the development of heat by electrical currents and found that it was proportional to the square of the value of the current (and to the resistance). He assumed that the quantity of heat developed in the electrical circuit is equivalent to the chemical heat that would be developed by the reaction taking place in the battery. At that time he was already convinced of the equivalence of heat and chemical, electrical and mechanical energy. Heat, in his opinion, had a mechanical character — a thesis also implied in his kinetic theory of gases (1848).

In 1845 - 49 he performed the famous, most direct experiments in which a paddle-wheel driven by falling weights stirred the water of a calorimeter. The work done by the weights was calculated and it was assumed that the friction between the scoops and the water yielded an equivalent amount of heat.

On metaphysical grounds Joule was *a priori* convinced of the truth of the conservation of energy. He wrote:

> We might reason, *a priori*, that such absolute destruction of living force cannot possibly take place, because it is manifestly absurd to suppose that the powers with which God has endowed matter can be destroyed any more than that they can be created by man's agency; but we are not left with this argument alone, decisive as it must be to every unprejudiced mind.[108]

However, was Joule himself so unprejudiced? He had posited that energy was indestructible, because God created it in the beginning in a definite quantity; consequently no experiments could change this. He said that he wanted to ask experiments whether there is a *fixed* relation between the mechanical work done and the heat yielded. In his first experiments he found that 1 British Thermal Unit is equivalent to varying numbers of foot pounds force ranging from 587 to 1040 ft lbf.[‡] Although this was later narrowed to 742 to 1040 ft lbf, clearly the conclusion from so wide a spread of results should have been that the relation is *not* constant. Joule admitted that there is a 'considerable' difference between some of the results 'but not, I think, greater than may be referred with propriety to mere errors of experiment'.[109]

In a postscript (August 1843) he decides for 770 ft lbf but adds 'I intend to repeat the experiments ... indeed I shall lose no time in repeating and extending these experiments, being satisfied that the grand agents of nature are, by the Creator's fiat, *indestructible*; and that wherever mechanical force is expended, an exact equivalent of heat is *always* obtained.[110] Joule did improve his experimental

‡ [*Editor's note (JCB):* expressed in SI units 1 BTU (British Thermal Unit) is equal to about 1055.06 J; and 1 ft lbf (foot pound weight) to 1.366 J.]

devices considerably and by 1849 he could rightly claim that he had found a constant value.[111]

Joule was a great experimenter, but this does not alter the fact that his law was founded upon a faith, which was a particular form of a conception that says that under all changes there is *something* that remains constant. The metaphysicist in him having first drawn his conclusions from the infallibility of the law of conservation of energy, the experimentalist in him did not rest until he had come to a convincing experimental confirmation *a posteriori*.

Joule speaks of God's *will* as governing the world, but though he started from a voluntaristic principle, he put forward essentially a Greek (Eleatic) doctrine of nature: 'Thus it is that order is maintained in the universe — nothing is deranged, nothing ever lost, but the entire machinery, complicated as it is, works smoothly and harmoniously. And though, as in the awful vision of Ezekiel, 'wheel may be in the middle of wheel', and everything may appear complicated and involved in the apparent confusion and intricacy of an almost endless variety of causes, effects, conversions, and arrangements, yet is the most perfect regularity preserved — the whole being governed by the sovereign will of God.'[112]

Joule presents us with a combination of Athens and Jerusalem in theology, and of apriorism and experimentalism in science. As to the latter, we find in him *faith* in a conservation principle, *fiction* of high quality in his kinetic theory of gases, and an accomplished *fact*-finding in his measurements in several fields of physics.

7. Conclusion: Head, Hands and Instruments

The general acceptance of *experimental* philosophy implied the recognition that the invention and the manipulation of instruments is essential for the development of science. This implies also a rise in status of the makers and inventors of these instruments. Not only were many scientists able instrumentalists, but also many instrument makers and other artisans turned out to be scholars of merit, and high expectations of their cooperation rose during the 17th century. The scientist is no longer an armchair scholar: he is also an 'architect'. The need for the cooperation of hand with head was generally recognized towards the end of the century; as Thomas Sprat put it: 'philosophy will then attain to perfection, when either the Mechanic labourers shall have Philosophical heads or the Philosophers shall have mechanical hands.'[113] Newton was an example of a happy combination of reasoning with experimentation. He too recognized that 'all knowledge begins and ends by the senses: We are certainly not to relinquish the evidence of Experiments for the sake of dreams and vain fictions of our own devising.'[114]

VIII. PHYSICAL AND MATHEMATICAL THEORIES

1. 'MATHEMATICAL HYPOTHESES' IN ASTRONOMY 230
2. KEPLER ON THE STATUS OF HYPOTHESES 232
3. DESCARTES'S THEORY OF THE REFRACTION OF LIGHT 235
4. 'HYPOTHESES' AS 'ILLUSTRATIONS' 240
5. NEWTON'S STATIC THEORY OF ELASTIC FLUIDS 242
6. THE UNDULATORY THEORY OF LIGHT 246
6.1. Thomas Young (1773 - 1829) 246
6.2. Augustin Fresnel (1788 - 1827) 247
7. HAÜY'S SUBTRACTIVE MOLECULES 249
8. GERHARDT'S RATIONAL CHEMICAL FORMULAE 254
9. THE CONTROVERSY OVER RESONANCE THEORY 257
10. CONCLUSION 263

Hypotheses and the theories built upon them are of two main types:
1) *they may be intended as constructs conformable with physical reality;*
2) *they may be only conventions which need not be true but which are useful in describing and predicting facts.*

Both types demand that hypotheses and theories should be as simple as possible; the latter for methodological reasons (theories should be as simple as possible); the former for ontological reasons (nature is believed to be simple and economical in its means). In the long run the two types of hypotheses and theories frequently converge.

Are theories mere artificial devices for bringing some order into the chaos of phenomena or should they be mirrors of objective reality? This is one of the perennial problems with which scientists have been confronted. A theory is an attempt to connect diverse phenomena by means of the smallest possible number of simple and fundamental causes and laws. But then the question arises: are these fundamental 'fictions' less real than the things we touch and see — simply aids helping us to recognize relations and affinities between things, and supporting our memory and imagination — or are they possibly more 'real' than the phenomena we observe by our fallible senses?[1] Are our models of atoms, molecules and crystals, while admittedly fictions bearing some analogy to what really is in nature and expressed unavoidably in human language and imagery, yet true images of certain aspects of physical reality — or are they only auxiliaries for memorizing known phenomena and predicting new ones?

The answer depends on one's metaphysical standpoint and on what one considers as the final aim of scientific theory: mastery over nature or knowledge of nature.

1. 'Mathematical Hypotheses' in Astronomy

Ancient astronomy was *primarily* the practical art of calculating the positions of the stars for purposes of navigation, of fixing the calendar and predicting events.

To the physicist-philosopher, however, it was more than that: since the time of Pythagoras (circa 530 BC) and Plato (429 - 348 BC) it had been unshakeable dogma that the apparently irregular motions of the heavenly bodies cannot be their real ones. It belongs to their divine nature (*physis*) to move along circular paths at a uniform (unchanging) rate. The task set to the astronomers by Plato was to 'save the appearances' (σωζειν τα φαινομενα; *salvare apparentias*).[2] That is, to reduce the seemingly irregular motions of the planets to a combination of uniform circular motions. A practical bonus of such a mathematical transformation was that it would enable the 'mathematician' to calculate the future course of a planet along its apparently irregular path. Plato's pupil Eudoxus (408 - 355 BC) produced a solution to the problem. He developed a system of the universe as consisting of homocentric spheres with the earth at their centre; each planet's motion, being determined by the motions of various spheres rotating along differently directed axes with different velocities, was apparently irregular while in every part circular. In an elaborate form this system was adopted by Aristotle and thus became for nearly 20 centuries part of the accepted *physical* system of the world.

In general later astronomers considered the homocentric spheres unsatisfactory and followed Hipparchus (190 - 120 BC) and Ptolemy (circa 150 BC) who had propounded a system in which each planet is fixed upon a circle (epicycle) that turns round a centre which moves along another circle (deferent), at or near whose centre the earth is situated. Though some astronomers tried to give a physical interpretation to this system, it was usually conceived of as but a convenient auxiliary for calculating the celestial motions rather than as an image of the real structure of the universe. As it was part of the *art* of astronomy and not a truth of the *science* of physics, it was often dubbed a 'mathematical' system, probably not so much because it applied mathematics (for this could be done also in the 'physical', realistically meant system of the homocentric spheres), but because it was used by the 'mathematicians' (i.e. astronomers) whose aim was calculation of the positions of the heavenly bodies for practical purposes.

There can be only one 'natural' system of the world, but the possibility has to be admitted that there is no such limit on the number of 'artificial' ones that can be invented to fit the facts. Hipparchus demonstrated that the apparent motion of the sun may be satisfactorily described either as the result of the rotation of a small epicycle moving along a large deferent circle (with the earth at its centre), or as motion on an eccentric (a circle whose centre is at a small distance from the earth). Ptolemy preferred the eccentric for the sun's motion because it seemed to him simpler and thus closer to physical reality.[3] Evidently even astronomers ('mathematicians') tried to keep their 'fictions' as closely as possible in agreement with physical reality. In the later chapters of *De Revolutionibus*, Copernicus in seeking to fit the facts more precisely, had to introduce epicycles, and he then gives reason to think that he too considered some of his hypotheses as only mathematical devices. He advanced, for example, several hypotheses for the motions of Mercury before finally accepting one possibility as being 'not less credible.'[4] The dedicatory letter as well as the whole of the first Book leaves no doubt that he wanted to find a system that was physically true. The astronomers' disagreement about hypothesis was precisely one of the reasons why he sought new ones which give a '*sure* foundation'.[5]

The Nuremberg pastor Andreas Osiander, however, to whom Rheticus had left the final stage of the editing of Copernicus' work, took the opposite viewpoint. In letters to Copernicus and Rheticus he advised that the motion of the earth be presented as a harmless mathematical artifice, in order to 'divert the opponents from the rigour of rejection toward the charm of investigation, and so become more benevolent, ... and join the author.'[6] He wrote to Copernicus that he always considered hypotheses not as articles of faith but only as foundations of calculation; if they are wrong this does not matter 'so long as they precisely reproduce the phenomena.'[7]

Osiander then prefixed a letter 'to the reader' of the work to appease likely opponents.[8] In it he said that the astronomer, after observing the motions of the planets, 'as he can in no way find their true causes or hypotheses, must invent and imagine (*fingere*) some arbitrary ones by which — if they are assumed — these motions may correctly be calculated according to the principles of mathematics (*geometria*), for the future as well as for the past'. And he added that 'it is not necessary indeed that these hypotheses be true, nor even that they be probable, but it is sufficient that they yield a calculation that agrees with the observation'. And further on: 'This art (of astronomy) is wholly and simply ignorant of the causes of the apparently irregular motions': when it enunciates hypotheses these are 'absolutely not intended to convince anybody that it is so in fact, but only that they may serve in the right way as the foundations of computations.' 'Since for the same motion different hypotheses present themselves — for the motion of the sun, the eccentric and the epicycle — the astronomer has a preference for whichever one is easiest to understand... while the philosopher for his part will rather ask which is more probable, but neither of these two can understand or transmit anything certain unless it should be revealed to him by God.'[9]

2. KEPLER ON THE STATUS OF HYPOTHESES

The staunchest defender of the realistic character of hypotheses was Johannes Kepler (1571-1630). The very title of his main work, *New Astronomy or Celestial Physics on the Motion of the Planet Mars* (1609), shows that to him the *art* of astronomy was part of the *science* of physics. In his *Apologia Tychonis contra Nicolaum Ursum* (1601) is a paragraph on 'what an astronomical hypothesis should be' in which he expounds his own ideas on hypotheses in opposition to Osiander and Ursus.[10] Kepler points out that the term was introduced by the first mathematicians, who assumed axioms, that is, postulates which are recognized immediately and intuitively ('by natural light') as true and generally accepted as such. When we move to science such hypotheses again form the foundations — on the same level as the facts obtained by observation.

It was from mathematics, according to Kepler, that the notion of hypothesis came into logic and the other sciences. In the practice of logic the term denotes the premisses from which the consequences are deduced by syllogisms. In the sciences of nature such hypotheses have to stand the test of experience: these hypotheses 'may be true or not true, but are supposed true in order that demonstration may show what would be the consequences if they were true.' All physical and astronomical theories, whether obtained by observation or in some other way, are the 'hypotheses' from which the observed phenomena as well as hitherto unobserved phenomena are 'demonstrated' by necessity.

Kepler's main thesis is that only *true* hypotheses lead to true consequences. When the adherents of the 'mathematical' hypothesis point out that it makes no difference whether the heaven of the fixed stars is said to make a revolution in 24 hours or whether, as Copernicus maintains, the heaven stands still and the earth rotates in 24 hours, Kepler adds that there is an infinity of other possibilities, and agrees that all these hypotheses account for the apparent daily rotation of the heaven of the fixed stars. The *real* hypothesis, which all these various suppositions have in common, is that there is a *separation of motion* between the earth and the heaven. In Kepler's opinion the systems of both Ptolemy and Copernicus explain the apparent motions of the planets, but Copernicus *also* explains why the outer planets are always in opposition to the sun when they are closest to the earth. Evidently, Kepler wants to say that the more facts that are explained by a theory, the closer it must be to the truth.

In contrast to the 'vulgar' astronomers, however, Kepler was not satisfied with a mere *description* of the *motions* of the planets. He remarked that even Ptolemy, who was often described as using hypotheses only for the sake of calculation, in fact introduced 'physical' hypotheses into his system, e.g. when he put forward the sequence of the planets. Kepler could have added that *all* astronomers, Copernicus and himself at that moment included, in that they admitted to astronomy only circular and uniform motions, were subscribing to an ancient position which had to do with the quasi-divine essence of the planets and thus bore a physical or rather metaphysical character.

To Kepler, in short, the task of the astronomer was not only to find the geometrical positions and the real motions of the planets, but also to discover *why* the structure of the universe is as it is and why the planets have the motions they perform. He wanted teleological hypotheses which would lay bare the divine plan of the universe. The solution he found was that in the main plan of the world the Godhead expressed His own triune Being: the sun, the sphere of the fixed stars and the space between them represent Father, Son and Holy Spirit.[11] This is a christianized version of Plato's tenet that the Maker of the universe made the world in his own image, as the Best One could make only the best.[12] Kepler was not the first Christian author to advocate the trichotomic character of the universe with similar arguments. Paracelsus (1493 - 1541) held that the world consists of three chemical principles, mercury, sulphur and salt, for 'the Trinity has spoken it', and he even extended the analogy to the three main parts of the earth, Europe, Asia and Africa.[13] This trend of thought was not, of course, genuinely Christian: the Bible tells us that *man* was made in the image of God but never makes a similar statement about any other created thing.

For the moving parts of the world Kepler found that the number and the relative distances of the planets were determined by the five regular solids (see figure 4).[14]

Kepler has often been criticized because of these 'vain' speculations. But it is to his metaphysical beliefs that we owe it that he discovered the fundamental astronomical laws that go under his name: the elliptic form of the planetary orbits, the law of equal areas and that of the 'simple' mathematical relations between their distances from the sun and the periods of the several planets.

The mathematical plan of the universe and the harmony of the heavenly motions were 'physical' hypotheses in that Kepler was convinced of their reality. So strong was Kepler's conviction that a plan, an 'archetype', needs a plan-maker, that he held (1621) that Aristotle, had he known the 'archetypal' causes that he, Kepler, had discovered, would have accepted God as their Architect, because 'archetypes' cannot 'work' themselves.[15]

Kepler's acceptance of God not only as the planning but also as a working cause did not, however, imply that he did not look for the instruments by which the divine Cause executes His plan. He sought, besides the mathematical description and the metaphysical cause, also the physical cause in the modern sense of the mechanism through which the phenomena occur. He supposed that the planets are moved in their orbits by an efficient cause, a 'magnetical' force emanating from the rotating sun, which pushes them on laterally as if by an arm.[16] Though this theory did not yield much in the way of results, it is remarkable because it shows clearly how Kepler, in spite of his tendency towards an organistic world view, nevertheless seriously considered the mechanistic alternative.

In contrast to such true astronomical hypotheses stand, in Kepler's opinion, the mathematical or, as he calls them, 'geometrical hypotheses' of 'vulgar astronomy'. These are devices for calculating the planets' positions in the sky for practical purposes; they are as it were scaffoldings which may be demolished after having served their end. They may thus prepare for the establishment of the true, physical-astronomical hypotheses (1601), and so Kepler did not wholly reject these 'small-money hypotheses'. When, for example, an astronomer says that the moon's orbit is an oval, this is an astronomical (realistic) hypothesis, but if he tells from which circles such an oval can arise, he uses 'geometrical' hypotheses. Similarly, Ptolemy's equant could be used as an auxiliary for calculation. Osiander however, Kepler remarks, caused confusion by applying what is valid only for 'geometrical' hypotheses (*viz.* mere usefulness) to all hypotheses, even to those of Copernicus.

Apart from making observations, the astronomer has the truly astronomical task of finding astronomical (realistic) hypotheses of such a kind that the apparent motions ensue from them and secondly the geometrical task of finding such geometrical hypotheses that the true motions can be calculated.[17] In his textbook (1618 - 1621) of Copernican Astronomy — or rather of 'Keplerian astronomy' — Kepler recapitulated that there are in astronomy: firstly the observations, next the astronomical hypotheses which go more deeply than the mere phenomena, and finally the 'causes' of the hypothesis i.e. 'physics'. This last by most people is deemed unnecessary for an astronomer, though it is the final aim of this part of 'philosophy'. For the astronomer has no licence to excogitate hypotheses without a reason, but has to penetrate to the causes and to confirm his astronomical hypotheses by physical or metaphysical principles and to present before our eyes the true form (*genuinam formam*) of the world. For this is 'the very Book of Nature in which God the Creator — though partially and by a certain kind of writing without words — has revealed and depicted His essence and His will towards man.'[18] Kepler felt himself to be a priest of this Book, which has first of all to be read and next rightly interpreted. All hypotheses, the metaphysical and meta-mathematical ones included, 'can be true or untrue' and have therefore to be tested by observations that must be critically examined before they can be recognized as 'facts'. The search for the facts and the 'true' hypotheses gave Kepler an immense intellectual and aesthetic pleasure: all Greek and Latin philosophers and all poets too, so he says, felt a divine rapture in investigating the works of God, and this not only for themselves but also for teaching them publicly.[19] In their company this most extrovert of the founders of modern science followed an inner urge to share his enjoyment with his fellow-men.

3. Descartes's Theory of the Refraction of Light

Descartes's main work, *Principia Philosophiae* (1644), contained a profusion of hypotheses which found wide approval in the 17th century, though some of the greatest scientists of that epoch — Huygens, Pascal and Newton (by mouth of Roger Cotes) — considered it a 'roman physique' i.e. a physics fairy tale.

His system of the world was allegedly deduced from 'certain germs of truth which God has laid in the mind' which give rise to the right ideas about the universe, the heavens, the earth, and also about the animals, minerals, etc. upon this earth.[20] Among the 'innate truths' belonged the conservation of momentum and the tenet that physical space and matter are identical.[21] These were for him axioms that lie at the basis of physical science and from them true deductions about the real world were possible. His law of conservation of momentum, according to which the absolute quantity of motion remains constant, led him to seven laws of collision. These, with one exception, Huygens was to prove on rational and experimental grounds to be false, the fundamental hypothesis being wrong: the constancy of momentum is valid only in the algebraic and not in the absolute sense.[22] But had Descartes lived to know of Huygens's demonstrations, they would not have disturbed him in the least, for he held that when the senses teach the opposite of what he had demonstrated, one should give more credit to the conclusion reached by reasoning.[23]

Descartes was a prolific generator of suggestions for mechanisms which might lie behind various phenomena. In order to explain the phenomena of magnetism, electricity, light, etc he invented specific particles endowed with forms and movements that 'explained' them: water particles, for example, were assumed to be flexible like eels, though when freezing these became rigid. But as Pascal remarked: it is easy to explain all phenomena if for each of them matter is invented that manifests its existence only in these same phenomena. Descartes, however, remained convinced of the correctness of his approach. He held his mechanistic phantasies and his 'innate' axioms to be physically true.[24]

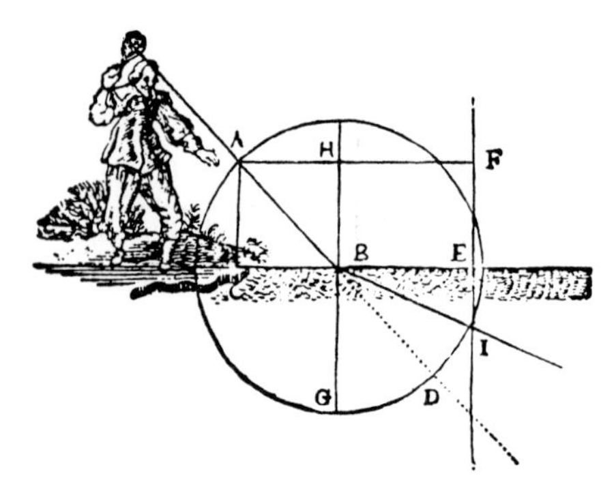

Fig. 33. Descartes's refraction law compared with a mechanical model I (From: Descartes, *La Dioptrique* (*Oeuvres* VI, p.98)).

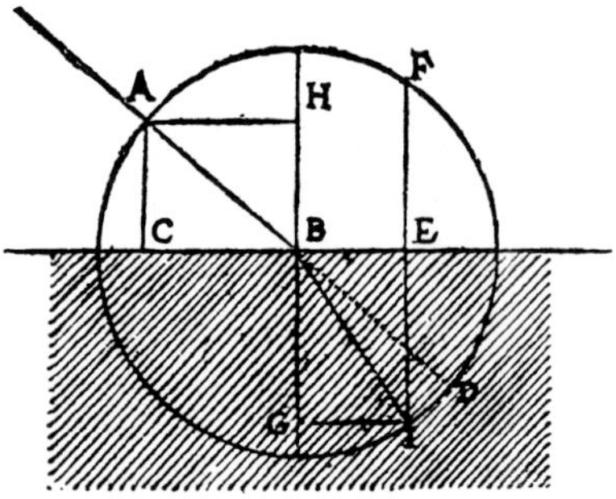

Fig. 34. Descartes's refraction law compared with a mechanical model II (From: Descartes, *La Dioptrique* (*Oeuvres* VI, p.100)).

This makes it the more surprising that he, who was never at a loss to invent a pretendedly 'physical' realistic hypothesis, resorted to a 'mathematical' hypothesis when in his *Dioptrique* (1637) he wanted to explain the sine law of refraction. He first declares that on this occasion he need not say 'what, in truth, the nature of light is', and that he therefore will use only some comparisons 'which help to conceive it in the way that seems easiest to me', in order to explain its properties known by observation, and then deduce other ones not so easily observed.[25] 'Herein I imitate the Astronomers, who, though almost all their suppositions are false or uncertain, nevertheless — as they are connected with various observations made by them — draw several very true and certain conclusions from them.'[26] Nevertheless, he emphasizes that in reality light comes instantaneously ('on the instant') from a source via a transparent medium to the eye — in the same way that the hand feels at once via a walking stick that the ground is touched.[27]

In order, however, to explain refraction Descartes next changes to a different mechanical model which introduces the idea of the 'velocity' of light. He makes a comparison with a ball that travels from A to B (figure 33) and touches a piece of cloth CBE which is so weak that the ball can break through, losing half its velocity. Let the ball after piercing the cloth travel on over a distance equal to AB. It will then arrive at some point on the circle AIDG in a time which is twice that needed for covering AB. Let the motion after the impact be resolved into a vertical and a horizontal component. The latter goes along the surface and thus will retain the former velocity and arrive at E (where BE is 2CB). So the ball arrives at I.

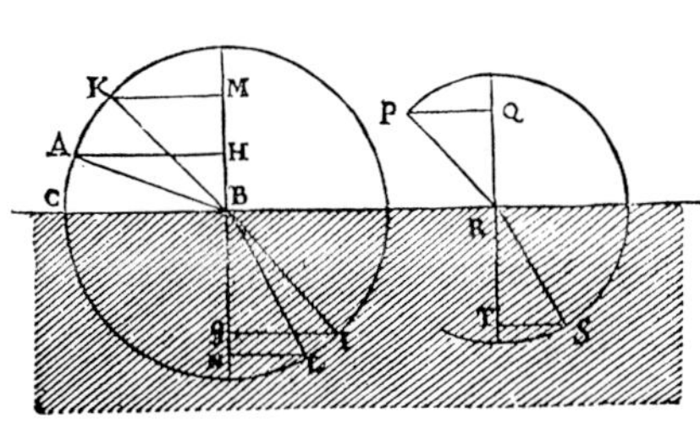

Fig. 35. Descartes's refraction law compared with a mechanical model III (From: Descartes, *La Dioptrique* (*Oeuvres* VI, p.102)).

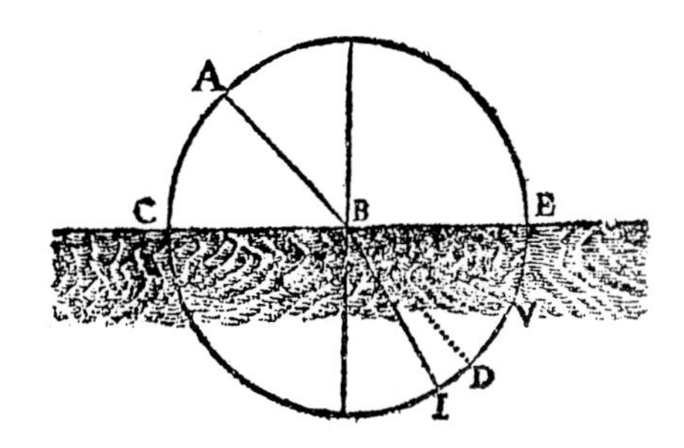

Fig. 36. Descartes's refraction law compared with a mechanical model IV (From: Descartes, *La Dioptrique* (*Oeuvres* VI, p.103)).

Now let the ball coming from A to B meet not cloth but water whose surface CBE also diminishes its velocity by ½. In this case progressing through water, it will arrive at I in twice the time needed for covering AB.[28]

Next, suppose that the ball on arrival at B is pushed on by a racquet CBE, 'which increases the force of its motion' in such a way that it can cover in the second medium the same distance in two time units for which it needed three time units in the preceding case. Again the velocity of the component parallel to CBE will be unchanged, so the distance BE will be to the distance CB as 2 : 3 when the ball has arrived at I, BI being equal to AB (see figure 34).[29]

If the action of light follows the same laws as the motions of the ball, so Descartes goes on, one may say that when the rays pass obliquely from one transparent body (air) to another (glass) which lets it pass more easily than the first, they will always become less inclined to the surface separating the two media on the side where the speed is greater. Descartes emphasizes that, though the ratio of the angles of incidence and refraction changes when the direction of the incident ray is changed, the ratio of AH to GI, or of KM to LN, or PQ to ST remains the same (see figure 35). Evidently this conclusion implies Snel's sine law ($\sin i/\sin r = n$).[30]

But a *ball* passing through air from A to B, upon touching the water surface CBE will go towards V, whereas light is known to go toward I (figure 36).[31]

Descartes therefore switches to yet another model. With an appeal to the *real* nature of light he explains why the behaviour of rays of light is the opposite of that of the ball. He recognizes that one might wonder why the rays of light are more inclined towards the refracting surface in air than in water, and more in water than in glass, this in contrast to a ball whose path is more inclined in water than in air (and which does not pass through the surface at all when meeting with glass). He then reminds us that light in reality is 'a certain motion or action received by very subtle matter which fills up the pores of the other bodies.' As a ball loses more of its motion when it impinges on a soft body than on a hard one and rolls less easily over tapestry than over a wooden table, so will the action of the subtle matter be more hindered by particles of air (which are 'soft' and loosely connected) than by those of water (which cling together) and more by those of water than by those of glass. The harder the little particles of a transparent body are, the more easily they allow the light to pass: 'for this light does not have to chase any of them away out of their place as a ball has to chase those of water away in order to find passage through them.'[32] Consequently light goes *faster* through water than through air.

A strange procedure: in his opinion light is in reality an instantaneously propagated pressure of subtle matter, but 'mathematically' it is convenient to compare it with the motion of a ball in spite of the disanalogy that light (in that 'mathematical' picture) would behave differently from a ball in that it passes less rapidly through air than through water. Next he abandons this rather awkward analogy and claims to have found the physical cause of the refraction of light by returning to its real nature and explaining the allegedly greater velocity of light in water than in air by comparing the resistance which the particles of 'subtle matter' or light meet with when passing through these two media.[33] On the other hand he had initially compared the transmission of light with the feeling of contact with the ground via a walking stick, both of which require no transmission time being an 'action', a pressure, rather than a 'motion'.[34]

He had declared that his 'explanation' of the refraction of light is but a 'mathematical' hypothesis, so one would expect that it would be to him merely a means of calculation or at best a mechanico-physical picture that could be true or not true. Quite the contrary, however: he claims to have found the *cause* of refraction (which implies that this was a physically true explanation).[35] To the critical remark that a 'mathematical' hypothesis cannot reveal a 'cause', he answers that an 'explanation' is not the same as a 'proof'.[36]

His deduction of the laws of reflection and refraction was avowedly based upon his experimental knowledge of them: but his hypothesis was in fact an 'explanation' solely of that phenomenon to which it had been adapted, and it did not lead to any further scientific discovery or insight.[37]

One cannot but wonder why Descartes chose such a tortuous way of justifying the law of refraction; *viz.* by employing a mechanical explanation that violated what he considered as an absolute truth, namely the 'fact' that light is transmitted in an instant. If this were not so, he had declared in 1634, he would be ready to recognize that he knew nothing of natural science. And if it were to be shown by experiment that there is a time lag between the emission of light and its perception by the eye at a large distance, his whole system would be radically destroyed.[38] By this rhetorical statement on the infinite speed of light Descartes unwittingly passed the death sentence on his own system of physics.[39] It was to be knocked down by experiments which showed that light is *not* transmitted instantaneously (Fizeau 1849) and that it does *not* travel more rapidly in water than in air, but precisely the reverse (Foucault, Fizeau 1850). Both his 'true' physical conception and his 'mathematical' explanation of it thus turned out to be untenable.

It is characteristic of Descartes that although he firmly believed in the truth of his own 'roman de la nature', he always kept a back door open. His proud statements that he had the absolute physical truth keep company with mitigating qualifications. His intellectual responses to the trial of Galileo are an example. Being both a convinced Copernican and a faithful Roman Catholic he was severely shocked by the condemnation of Galileo. He then found a way to maintain the motions of the earth and yet to obey the decree of Rome, by saying in his *Principia* of 1644 that motion is always understood in relation to the immediate surroundings and that the earth is in the centre of a whirlpool of celestial matter the lowest layer of which turns round with the same speed as the earth so that the latter is in fact standing still (see figure 37).[40]

He declared that 'it is hardly possible that causes from which all phenomena may clearly be deduced are not true'; it would be offending God to suppose that the causes of the effects we have found in nature are false.[41] But then follows a retreat when he says that those causes which he here proposes are to be considered 'as hypotheses only', and he goes still further by avowing that some of these hypotheses are certainly false.[42] There follows an example: I do not doubt that the world — sun, moon, earth, plants, etc. and Adam and Eve in their adult state — have been created in perfection in the beginning, as Christian religion demands us to believe, and natural reason absolutely convinces us of this truth. But we know their nature better when we imagine some very intelligible and simple principles from which all things of the visible world could have been produced, though we know that they have not been brought forth in this way.[43]

This odd combination of fideism and rationalism occurs repeatedly in his main work. One of its paragraphs is summarized in the statement: 'The falsehood of these suppositions does not impede that what is deduced from them (*viz.* the visible world) may be true and certain.'[44] At the beginning of Part IV ('On the Earth') he asserts that in order to understand the true nature of all things on Earth, one has to hold on to the false hypothesis about their origin because their nature is as if they have been brought forth in that way.[45] Towards the end of the *Principia Philosophiae*, having deduced from the principles and germs of truth that 'God had laid in his mind', the world and all that is in it, he declares again that all natural things could have been brought forth in the way he described, but that it is not certain that it is true. The very next paragraph, however, states that nevertheless his descriptions have moral certainty, and the following paragraph reaches the climax: 'we have a more than moral certainty that things are in nature as has been told in this book.' But then follows an anti-climax: all that has been said is readily submitted to the authority of the Church.[46]

Strange conclusion by a man who had said in the same work that to suppose that the causes he had found were wrong, would be an 'offence to God'![47]

Descartes was an adherent of two faiths: Roman-Catholicism and Cartesianism.[48] He did everything possible to defend the latter without getting into conflict with the former. In his young days he depicted himself as walking upon the scene in disguise.[49] May something similar not be the case with his declaration that his 'deduction' of the law of refraction was but a 'mathematical' hypothesis? If so, he still leaves us with the enigma of its being a turbid mixture of physical explanation and 'mathematical' device which hardly befits a genius in philosophy and geometry who attributed such great value to 'clarity and distinctness'.[50]

4. 'HYPOTHESES' AS 'ILLUSTRATIONS'

The distinction between convenient fictions and realistically intended suppositions that have the value of facts of nature or even articles of faith (physical axioms) never disappeared. In general the term 'hypothesis' was increasingly used of suppositions that had no claim to be truthful. It implied rather that the ideas were plausible — or possible — images of the causes behind visible things. The work of the Baconian empiricist ('experimental philosopher') Robert Boyle is full of such hypotheses. He 'explains' the elasticity of air, for example, by supposing that it consists of rapidly rotating coiled springs: on being heated these extend, because the centrifugal force increases, so causing expansion. He is also willing to consider a static image in which particles form a resilient mass like a pack of wool. And he admits that there might be still other possibilities. 'Hypothesis' has here completely lost its Platonic claim to be a basic physical truth: it can hardly be tested by experiment; it may well be erroneous. Such a conception was appropriate to a scholar who called himself the 'sceptical chymist.'

It should be noticed that to Boyle such hypotheses (he calls them 'conjectures') were just a support for the scientific imagination. As such they had already, before Boyle's time, rendered important service to chemical corpuscular theory. An instance is the image of copper particles as 'hidden in' crystals or solutions of

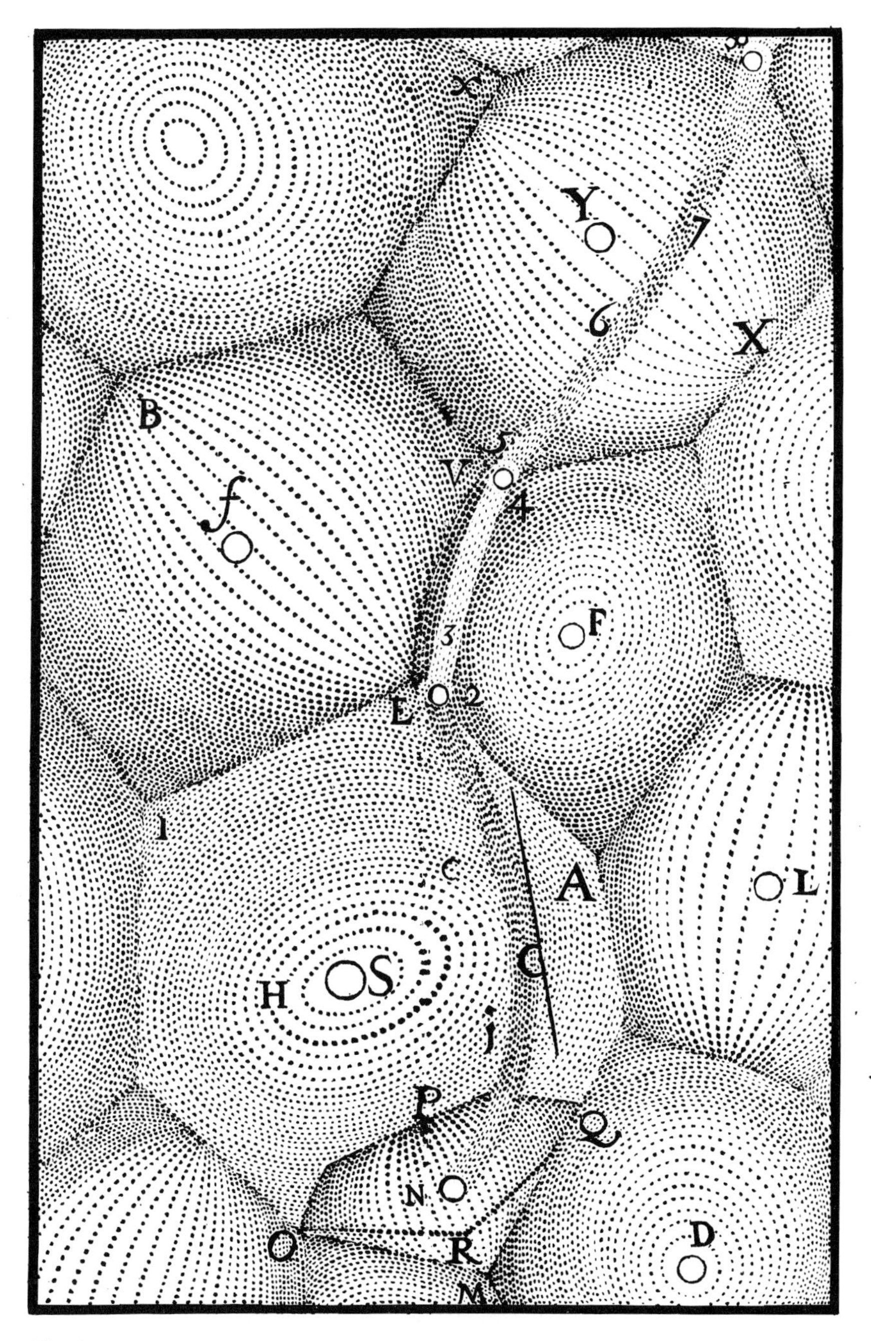

Fig. 37. Descartes's 'whirlpools of celestial matter'. Between the whirlpools of the sun (S) and those of the fixed stars (F, f, etc.), the orb of a comet is represented (1-7). (From: Descartes, *Principia Philosophiae* 1644, p.106).

copper vitriol rather than as 'potentially' (but not really) present in copper compounds. This way of accounting for the persistence of metals in their compounds satisfied a generation which asked for concrete images rather than the subtle distinctions of scholastic science.

For a long time conjectures of this kind remained fictions that were hardly ever used for explanation or for the deduction of mathematically formulated laws. Boyle's law was — insofar as it was accepted — an empirical law, not 'explained' by a mathematical deduction based on a hypothesis about the structure of gases. Kinetic theories of gases that derive the gas laws from the mechanical forces exerted by moving particles did not arise until the 18th century. It was then that in 1727 Leonhard Euler gave a physico-mathematical explanation of Boyle's law in terms of the hypothesis that air consists of small eddies of minute particles whose centrifugal force causes the pressure.[51] And in 1738 Daniel Bernoulli derived this same law from the elastic collision of air particles which in linear motion impinged on the wall of the vessel. This was essentially the same idea as that of Krönig (1856) and Clausius (1857) with whose names the modern kinetic theory of gases is usually associated.

An early example of a theory that not only had a physico-mathematical character but was at the same time a 'mathematical' theory in the ancient sense of being fictitious, was the static theory of elastic fluids advanced by Newton in 1687.

5. Newton's Static Theory of Elastic Fluids

When Newton wrote his *Mathematical Principles of Natural Philosophy* (1687) his purpose was to find what Aristotle had called the 'mathematical aspect of physical reality'. His emphatic declaration that it was only the mathematical aspect of the phenomena that he was considering did not, however, signify that he was dealing with fictions of the kind used by pre-Copernican mathematical astronomers. He tentatively put forward various mathematically formulated suppositions (e.g. about the attractive forces between bodies as a function of their mutual distance) in order to see what consequences would ensue. And whenever one of them led to conclusions which were conformable to measurements made in the real world, he considered he had discovered a 'mathematical aspect' of physical reality.

His demonstrations that bodies act on each other *as if* by forces directly proportional to their masses and inversely proportional to the square of their distance apart is the mathematical formulation of a physical reality and in this sense it is a *physical* law. But Newton made no decision as to the nature, the hidden mechanism, of these 'attractive' forces: it might be that the bodies are *pushed* towards each other, e.g. because there is a rarefaction of an unknown substance between them. Descartes would immediately have created the subtle fluid required;

but Newton, though he had something similar at the back of his mind, kept an eye on loose conjectures and spoke his famous words 'hypotheses non fingo' (I feign no hypotheses). Of course there were few things he loved better, but he put them as 'Queries' at the end of his *Opticks*. Perhaps an exception is his 'fits of easy reflection' and 'fits of easy transmission' which he introduced into the main body of the *Opticks* in order to account for the fact that part of the light is reflected and part refracted when meeting the surface of another medium, and also for explaining Newton's rings and the coloured fringes formed by thin plates. But he did so for 'those that are averse from assenting to any new discoveries, but such as they can explain by an Hypothesis.' 'But whether this Hypothesis be true or false I do not here consider. I content myself with the bare discovery, that the rays of light are by some cause or other alternately disposed to be reflected or refracted for many vicissitudes.'[52]

The verdict 'hypotheses non fingo' first appeared only in the second edition of the *Principia* (1713). In the first edition (1687) in bold letters there had paraded seven 'Hypotheses'. In the second edition some of these were renamed 'Principles of Reasoning' (i.e. methodological principles) and others 'Laws of Nature' (e.g. Kepler's third law). Only two remained and they really were sub-positions (one on absolute motion) and so quite different from the mechanistic conjectures which he had indeed kept out of both editions of the book in accordance with his statement towards its end.

A more notable exception to Newton's 'non fingo' is to be found in the *Principia* when he sought to deduce Boyle's law — of inverse proportionality of pressure and volume of a certain amount of an 'elastic fluid' — from an assumption that goes farther than a tentative hypothesis about the relation between attractive force and the mass and distance of material bodies.

The phenomena of attraction (centripetal forces) and the structure of the universe, which play such an important role in the *Principia*, are well-established as existent in nature. So Newton's 'mathematical' treatment of them amounts to finding out which of several hypothetical quantitative laws yields results which conform with observations of nature. Now the structure of an 'elastic fluid' (air) is not observed by the senses, and the forces at work between their parts are not observable. But the law that the air's pressure and volume are 'reciprocal' had been discovered by Richard Townley (Boyle's law). In order to deduce it Newton now wanted to find out from which hypothetical structure it could ensue. He assumed

a) that an 'elastic fluid' consists of particles which:

1) are at rest,
2) repel each other,
3) do so with a force that does not act further than the immediately neighbouring particles.

Next — pursuing his characteristic method for discovering the quantitative formulation of the laws at work in nature — he tried

b) to deduce the consequences if this repulsive force between the particles was inversely proportional to their distance apart — or, possibly, to the square or to the third power of this distance.

The result was that only on the supposition that this force is inversely proportional to the distance would Boyle's law ensue.[53]

Here Newton had gone much further in the Cartesian direction of feigning hypotheses about a hidden mechanism than he had in the case of the laws of friction and resistance of fluids or of centripetal forces, where he dealt with observable phenomena. The putting forward of the fiction as to the constitution of 'air' preceded, as we have seen, the variety of mathematical laws that were to be put to the test of experiment. Quite apart from making a choice from a variety of mathematical laws of repulsion (b), there was a hidden choice from a variety of mechanico-physical hypotheses about the constitution of elastic fluids (gases). There was, for example, the possibility of kinetic explanations. Newton was aware that there were such alternatives: in the Latin edition of the *Opticks,* the *Optice* (1706), he himself attributed the tendency of vapours to extend themselves to the fact that their particles (having overcome the mutual attraction they had in the liquid state) flee from each other so that they may occupy more than a thousand times their original space. This immense contraction and expansion would be hardly imaginable, he said, if the particles of air were 'feigned' (*fingantur*) as being elastic and ramified, or as being like coiled, flexible twigs, or any other cause, unless they have a 'repulsive force' by which they flee from each other.[54]

In the *Principia* he defended his somewhat gratuitous condition that the repulsive force does not reach farther than the direct neighbours, by comparing this with the force of a magnet which is almost terminated by an iron plate. He pointed out that if the particles did influence others more remote than those that lie next to them, a greater force would be required in the case of a large quantity of air than for a small quantity to produce an equal condensation, so that Boyle's law would not stand.[55] Why all these arguments if he believed that his hypothesis about the structure of a gas was a mere fiction and not a reality? The only thing that Newton did prove was that *if* air consists of particles satisfying the three conditions (a), then the intermolecular repulsive force between the particles (b) must be inversely proportional to their distance.

There is a distinct difference between the 'mathematical hypotheses' according to Osiander's conception (which were merely means of calculation and prediction, see also page 231) and Newton's mechanical picture of the causes behind Boyle's law.

There is also, however, a great similarity in that Newton's hypothesis (a) was only one among several other possible explanations (which he did not take into consideration). Newton realized this, for he said that he treated the problem 'mathematically' and that he left it to the 'philosophers' (physicists) to find out whether an elastic fluid *does* have the structure mentioned in the three conditions (a). It was not 'mathematical' in that it gave the mathematical aspect of a physical phenomenon (like the law of gravitation) but in that it was 'fictitious' and could become 'physically true' only after investigation.[56] The hypothesis being not (or not yet) verifiable, was as yet of little use in experimental science. But it was *not* in conflict with its author's physical world picture and herein it was less 'mathematical' than the hypothesis about refraction put forward by Descartes, which, had it been true, would have made one of the foundations of its author's physical system null and void.

The contrast of 'physical' and 'mathematical' hypotheses did not die after the 17th century. It became an ever returning issue not only of method but also of epistemology. If the physicist holds that human reason is able to grasp physical reality in an adequate way because he can re-think the thoughts of God, or because he has an affinity with Nature which enables him to penetrate into her secrets sympathetically, it follows that scientific hypotheses should be and can be conformable to physical reality. The belief in analogy between the human mind and the mind of God or the intrinsic logic of Nature raises the hope that there may be at least analogy between our picture of nature and the real world.

On the other hand, we may consider science as a description of phenomena according to an order we impose on Nature, an order that is largely determined by our senses and intellectual constitution. The world has the colour of the spectacles through which we look at it. From a positivist standpoint the chaos of phenomena becomes a cosmos by our imposing on it an order that largely depends on our way of classifiying them in an artificial system that we hope will run parallel with 'objective' nature. In practice most scientists will not take a firm standpoint on this issue. They put forward their hypotheses first as working hypotheses just to find out how workable they are. Sometimes they use them without deciding about their 'truth' but considering them as provisional 'truths' which will be abandoned when something more sure turns up. In this way many scientists of the 19th century applied the chemical atomic theory without believing in it.[57] The purists avoided the term 'atomic weight' and introduced that of 'equivalent weight', thus expressing the objective experience that only certain relative quantities of elements or multiples of them enter into chemical combination or substitution. Non-purists recognized that chemistry could be cultivated while acting *as if* the existence of atoms was a fact.

6. The Undulatory Theory of Light

At the beginning of the 19th century the undulatory theory of light became a serious rival to Newton's emission theory. By analogy with sound waves in air, light waves were considered to be periodic motions of particles along the direction of propagation of the rays, alternately direct and retrograde, with concomitant condensations and rarefactions. These longitudinal waves of light were supposed to be motions of a hypothetical ethereal matter.

The discovery of the polarization of light (Malus 1808) raised a difficulty. Although polarization could be easily explained on the assumption that the light waves had a transverse character like the vibrations of a violin string, this required a certain rigidity of the vibrating matter: an extremely subtle fluid such as the 'ether' was supposed to be, could, however, hardly sustain transverse vibrations. Transverse undulation was therefore considered an absurdity by great physicists like Laplace and Poisson.

6.1. Thomas Young (1773 - 1829)

Dr Thomas Young had occupied himself with the undulatory theory since 1800. In 1817, in order to explain interference phenomena, he assumed 'as a mathematical postulate in the undulatory theory, without attempting to demonstrate its physical foundation, that a transverse motion may be propagated in a direct line.' He considered this to be 'a tolerable illustration'. That is, he was using the term 'mathematical' here in its ancient sense: he called the motion the 'imaginary transverse motion'.[58] Its truth was problematical, for it was 'a tolerable *illustration*', a term used also by Boyle for his 'conjectures'. Young was anxious to explain polarization, 'without departing from the genuine doctrine of undulations'. Evidently he wanted to cling to the analogy with sound waves, as if longitudinal undulations were the truly physical ones: indeed they are the only *mechanically* transmitted waves that can be readily conceived.[59] Yet Young ventured the supposition that the transverse vibrations could be true 'even in a physical sense',[60] if they were 'minute beyond the power of imagination' and nevertheless effective on the senses. He recognizes that its 'inconceivable minuteness suggests a doubt as to the possibility of its producing any sensible effects: in a physical sense, it is almost an evanescent quantity, although not in a mathematical one.'[61] The minuteness may 'lessen the probability of the theory as a physical explanation of the facts, but it would not destroy its utility as a mathematical representation of them.' This utility, however, he acknowledged to be only really fruitful when the 'mathematical' representation could be made 'reducible to calculation'.[62]

Evidently 'mathematical' here has, besides the meaning of 'imaginary but useful', also that of lending itself to mathematical calculation. Yet if this were not possible, the theory remained 'even in a physical sense' for Young almost 'unavoidable', for he deemed it 'easier to imagine the powers of perceiving minute changes to be all but infinite, than to admit the portentous complication of machinery' which the doctrine of emission would demand in order to explain the phenomena of polarization and interference colours.[63]

Nevertheless, Young was not satisfied with the wholly 'mathematical' character of the transverse motion. Not being able to find a *physical* possibility for transverse motions of the ether (besides the preponderant longitudinal vibrations), he contented himself with the thought that the *physiological* efficiency of the posited transverse vibrations has to be incredibly great. He accepted such an assumption as being more probable than the cumbersome auxiliary hypotheses by which adherents of the emission theory had to account for polarization and other newly discovered optical phenomena. The methodological principle of *simplicity* here exerted pressure in the direction of ontological belief. But such considerations apart, it is evident that Young wished to use the concept of transverse vibrations, whether or not it answered to physical reality, because of its 'utility' as a 'mathematical postulate.'

6.2. Augustin Fresnel (1788 - 1827)

The French engineer A. Fresnel commenced his work on optics a decade later than Thomas Young. He recounts how, in September 1816 when working on the colours of thin crystal plates (chromatic polarization), he saw immediately that polarization could be explained 'in the most simple way' by transverse vibrations: but what would then become of the longitudinal oscillations?[64] At first, however, he, like Young, supposed that light involves transverse motions together with longitudinal ones, but in 1821 he decided that he should stick to transverse waves exclusively. And from that time onwards he viewed this as a theory that is physically true, because it has the main characteristics of such a theory: 'If a hypothesis is true, it must lead to the discovery of numerical relations connecting phenomena which are far distant from each other; if it is wrong, on the contrary, it may strictly represent the phenomena for which it had been imagined — as an empirical formula represents the measurements within whose limits it has been calculated — but it will not be able to unveil the secret knots which unify these phenomena with those of a different class.'[65]

While Fresnel was finding himself able to establish such relations between known phenomena and to discover highly surprising new phenomena predicted by his theory, the protagonists of the emission theory, such as Biot, were having to complicate their system by auxiliary hypotheses in order to account for each 'new' phenomenon. To explain polarization they endowed the particles of light with two different poles, while in order to explain 'chromatic polarization' they assumed that the particles that cause certain colours turn, to different degrees, about an axis coinciding with the direction of the rays (mobile polarization). In speaking of the adherents of the emission theory, Fresnel remarked that 'the system of emission has been presented as a kind of Proteus, who escapes objections by assuming all forms, by adopting all hypotheses it needs. The multiplicity of hypotheses is no probability in favour of a system'; if there are too many of them, they easily become contradictory — a remark reminiscent of Lavoisier's similar critique of phlogiston, which he called 'un véritable Protée' because of its assumption of properties needed at the moment.[66]

Fresnel pointed out that a good theory not only calculates forces when the laws of the phenomena are known, but also helps to find laws so complicated or strange that observation alone, aided by analogy, would never lead to their discovery. To solve these enigmas 'one ought to be guided by theoretical ideas that are based on a *true* hypothesis' [italics RH].[67] The undulatory theory has such a character; we owe to it the discovery of laws of optics which one would never have guessed, whereas the discoveries made by the adherents of the emission system — beginning with Newton — are 'the fruit of their observations rather than mathematical consequences deduced from his system.'[68]

In Scotland David Brewster, though he remained faithful to the emission theory, nevertheless recognized that the undulatory theory was one of the most fortunate 'fictions' human ingenuity ever invented to connect natural facts: if it is not true, it deserves to be so. Fresnel, however, was convinced that not only because of its fertility in discovering and connecting phenomena, but also because of its greater *simplicity*, the undulatory theory must be physically true. His arguments were similar to those of Kepler. Like the great astronomer he, too, had a firm *faith* in the simplicity of nature, a simplicity which was to him not just a methodological principle of economy, but also an ontological truth: 'nature loves simplicity', 'simplicity is the seal of truth'.[69] The antinomy between the thinness of the ether (as evidenced by its extreme penetrability by heavenly bodies) and its rigidity (as evidenced by its apparently admitting of transverse waves) pressed Fresnel to work on the theory of elasticity. A satisfactory solution was not forthcoming and Fresnel had to sacrifice for the time being the 'analogy of nature' on one point in order to gain more analogies on others. Sometimes fictions which seem less 'rational' have to be introduced for the sake of saving the phenomena and for arriving at a more comprehensive and more simple system. Evidently faith in the unity, harmony and simplicity of nature gained more by the theory that light is a transverse vibration than it lost by abandoning the analogy between the mechanics of a fictitious ether and the mechanics of ponderable matter.

There was a way out, *viz.* that of leaving mechanical models aside. In 1814 Joseph Fraunhofer (1787 - 1826), the discoverer of the black lines in the solar spectrum, pointed out that even those who do not adhere to the undulatory theory, must concede that the so-called 'wavelength' of light is a real magnitude as it says that in some respect one half of a certain interval is contrary to the other. This means that 'formally' a 'wave', a periodicity, has to be accepted, even if we cannot form a physical image of it.[70] The Newtonians could not deny this: had not the master himself introduced the periodic alternation of 'fits of easy reflection' and 'fits of easy transmission'?

A strict proof of the physical truth of the undulatory theory had not been given, though it had become more probable than its rival. When, however, Fizeau (1849) and Foucault (1850) gave experimental proof that the speed of light in water is less than that in air, the dispute seemed to be settled in favour of the undulatory theory, for (as we have seen) according to the emission theory light particles move faster in water than in air. This alternative, however, is valid only within the framework of the purely mechanistic world view that was accepted by both parties. But one might ask whether there are other possiblities that might account for the periodic character of the propagation of light and the relation of the velocities in various media. At any rate, Fresnel's theory had to give way to the electromagnetic theory of light (Maxwell 1861; Heinrich Hertz 1886, 1887), and Einstein's discovery of the photoelectric effect (1905) restored a kind of emission theory to the scene without ousting the electromagnetic undulation theory.

7. Haüy's Subtractive Molecules

The French priest René Just Haüy (1743 - 1822) constructed a crystal theory (1784) based on the specific 'primitive forms' of the various kinds of crystals. A primitive form was arrived at as the result of the cleavage of all the more complicated secondary forms that a certain crystalline substance could assume. The cleavage form gave Haüy indications for finding the form of the 'constituent molecules', or, as he later called them, the 'integrant molecules'. He distinguished three types: the parallelepiped (the simplest solid form limited by pairs of parallel faces), the trigonal prism (the simplest prismatic form) and the tetrahedron (the simplest pyramid). Tetrahedral molecules cannot form a continuum, so Haüy had either to accept octahedral vacuities between them in the crystals they build up or tetrahedral vacuities between octahedral molecules.[71] At any rate he rejected the idea that there might be two shapes of molecules in such a crystal (e.g. fluorspar) as this would be 'equally contrary to a reason based on the analogy drawn from the uniform structure of other crystals, and to the simplicity which all things compel us to recognize in the composition of natural bodies.'[72] The number of the faces for the above three crystal types is respectively 6, 5 and 4, so that in this respect too nature observes the greatest possible *simplicity.*[73]

Haüy approached crystallographic problems in two ways: by experience and by theory. As to the former his 'principle' is 'that things are deemed to be in themselves as they present themselves to our observation.'[74] Consequently the final results of mechanical division of the minerals, 'if they do not yield the form of the true integrant molecules of the crystals, they deserve the more to replace them in our conceptions as — when taking them as given — we succeed in faithfully representing the facts which nature offers us, and in establishing their connection and mutual dependence.'[75]

So, if we doubt whether the limits of division indicated by our experience (of the cleavage of crystals and striation of their faces) should be believed to be the final ones, it is at least *as if* the crystals consist of them, for these are data for a satisfactory (i.e. a simple) theory.

It is interesting to see how Haüy, who tenaciously defended the physical reality of his polyhedral integrant molecules, tried to convince the sceptics that, though they did not recognize them as physically true, they should at least acknowledge that they are 'mathematically' useful because of the simple means they provide for connecting so many phenomena. They should be accepted 'because with such a small spending of means we succeed in establishing a theory which embraces so many diverse results.' That is, if you do not choose for the ontological simplicity of nature herself, accept at least the methodological simplicity of the *science* of nature.[76]

Of course Haüy realized that mechanical division does not yield the integrant molecules directly. But the fact that salt crystals by cleavage easily split into a cubic primitive form or nucleus, which yields smaller cubes by further splitting, led him to the conclusion that the final units, too, must have the cubic form. This assumption, together with the law of decrescence* enabled him to deduce the secondary forms, whereas, inversely, from the secondary forms together with the law of decrescence he could calculate the primitive form. And though this may not be physically true and the continuation of the mechanical division 'in thought' ('en continuant par la pensée la division mécanique, jusqu'à sa limite') may not yield the molecules which Nature herself uses when building the crystals, they would at any rate be equivalent to them, for the results of the theory obtained by adopting them are so conformable to observation, that one does not go wrong in 'taking the molecules of the theory for those of nature.'[77] Haüy himself, however, firmly believed in the reality of his 'molécules intégrantes'. He held that Nature is simple: in the three types of integrant molecules we recognize 'the familiar device of nature: economy and simplicity in its means, richness and inexhausible variety in its

* See Chapter II on Haüy, pp.46-51.

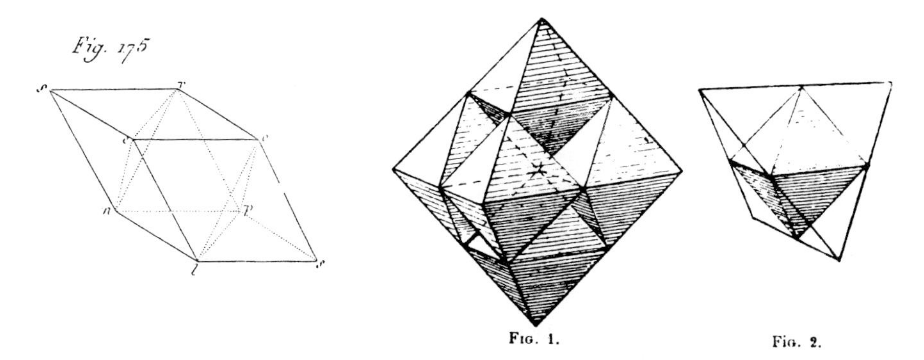

Fig. 38. Haüy's 'molécules soustractives': a) an octahedron with a tetrahedron on two opposite faces forms a parallellepiped; b) an octahedral crystal consisting of smaller octahedra and tetrahedron; c) a tetrahedral crystal consisting of smaller octahedra and tetrahedra.

effects.' He attributed it to the 'power and wisdom of God, who has created and leads Nature.'[78]

It seems that his belief in the ontological simplicity of Nature was even surpassed by his desire for methodological simplicity in the human systematization of its phenomena and his wish that those who did not share his belief would at least follow his method. This desire for methodological simplicity led to a 'mathematical' theory that was simpler than physical reality.

The trigonal prismatic and the tetrahedral molecules disturbed the unity of his original theory. Therefore in 1793 he put forward a way of reducing the theory of non-parallelepipedal molecules to that of the parallelepipeds.[79] Two triangular prisms with equilateral bases, when joined together may form a rhombic prism (i.e. a parallelepiped whose bases have angles of 60° and 120° and whose lateral faces are perpendicular to the bases). As to the tetrahedral molecules, Haüy pointed out that each octahedral vacuity (e.g. in a crystal of fluorspar; fluorite CaF_2) together with two tetrahedral molecules bordering opposite faces of the octahedral space, also form a parallelepiped. In this way, in our imagination *all* crystals can be considered as assemblages of parallelepipeds and the theory developed for this latter form of integrant molecules could also be applied to the two other forms. For calculation now all crystals could be considered as aggregates of units packed together in parallel positions so as to fill space. He called these units 'molécules soustractives' (see figure 38). Except in the case of parallelepipedal 'molécules intégrantes', these units were mathematical fictions introduced for the sake of simplification of the theory. Their abstract character was strongly emphasized:

> Calculation needs nothing but these parallelepipeds for reaching its goal and the sort of anatomy which these little parts undergo next, when one tries to ascend to the true form of the integrant molecules, is a matter of pure observation which stands outside the theory.[80]

The parallelepipedal 'subtractive molecule is the unit on which the theory operates; it has the double advantage of generalizing the theory and of simplifying its applications', and it evades the embarrassing question as to whether the ultimate products of chemical division are in fact the integrant molecules.[81]

The separation between the physical molecules — found by observation of the cleavage of crystals and striation of their faces, and not by theory — and the fictional molecules of the theory had now become complete. The former are no longer needed for reaching Haüy's final aim, the calculation of the geometrical crystal forms that do occur or can occur in nature. What matters in his theory now is not in the first place the physical character of the molecules but the mathematical relations between phenomena.

Yet, even the simplicity of his theory of subtractive molecules is not a pure fiction: it shows that each crystal may be conceived as an infinite three-dimensional repetition of the same geometrical units. With what contents these geometrical units are filled up is a matter of physics. Haüy himself had considered his 'molécules intégrantes' as solid polyhedra, mini-crystals that form the visible crystal. In some rival theories (with a less general scope) the crystals were considered as results of piling up of globular or ellipsoidal molecules (Huygens, Wollaston, Seeber, Prechtl, J.F. Daniell). L. Seeber (1824), who for physical reasons rejected Haüy's polyhedral 'molécules intégrantes', imagined globular molecules whose centre was in the centre of Haüy's 'molécules soustractives', so that the mathematical advantages of Haüy's theory were retained. H.J. Brooke, while following Haüy's method, went even further than Seeber when he wrote: 'The whole theory of molecules and decrements, is to be regarded as little else than a series of symbolic characters, by whose assistance we are enabled to investigate and to demonstrate with greater facility the relations between the primary and secondary forms of crystals.'[82]

Haüy's successor G. Delafosse (1843) reduced the molecule to a mathematical point occupying the centre of the former subtractive molecule. The cleavage along certain planes shows, according to him, that the molecules, considered as material points, are distributed over series of parallel planes. If there are two other cleavage directions, the molecules must be arranged in a uniform and symmetrical way and their centres of gravity must coincide with the intersection points of these planes.[83] This theory clearly contains the idea of 'space lattice'. When A. Bravais in 1848 found, by geometrical deduction, that in principle 14 types of space lattice are possible, a certain culmination point had been reached: Bravais had nothing against considering these points as 'molecules' or rather as the centres of the figure of these molecules.[84]

For the time being, however, most crystallographers followed the purely geometrical method of description of crystal forms which had been started, though not strictly quantified, by J.B.L. Romé de Lisle (in 1783), rather than the more explanatory, structural method of Haüy. The school of Weiss and Miller preferred that description which invoked three coordinate axes which were intersected by the several plane faces according to intercepts in rational proportions (which thus took over the function of the decrements in Haüy's theory) and in which the angles between the three axes and their unit lengths replaced the angles and the relative dimensions of the edges of the subtractive molecules. In this way what was left of Haüy's theory was a purely mathematical-formal description of crystals without reference to the underlying structure.

Yet the structure theory underwent a further development, though without much contact with mineralogy. The physicist L. Sohncke (1867) introduced the 65 'point systems', while the geologist E. von Fedorov (1885) and the mathematician A. Schönflies (1888) deduced the 230 space groups.[85] In mineralogy, however, they played a very subordinate role until, thanks to X-ray diffraction (M. von Laue 1912; W.H. Bragg and L. Bragg 1913) experimental verification showed the fundamental truth latent in the structure theories from Haüy to Schönflies, and an upsurge of experimental and theoretical research on crystal structure followed.

Haüy's 'molécules soustractives' (1793), Schönflies's 'Fundamentalbereiche' (1888) and Fedorov's 'parallelohedra' (1885) may be mere geometrical entities, but they have led to the conclusion that crystals are periodically three-dimensionally repetitive assemblages of certain units, which may be 'fictions', but nevertheless express 'physical' truths no less than was done by the transverse vibrations of the fictitious ether or the atoms of the chemical theories. Already before more solid proofs had been given of their physical existence (their 'Dinglichkeit'), they expressed the truth that something periodic, capable of measurement and mathematical treatment, was behind the phenomena.

'Mathematical' theories, originally intended as artificial devices for unifying the description of nature, and the 'physical' theories which often were bold hypotheses about the causes of phenomena, converged in the long run. In many cases what remained of value in the former merged with what was left of the latter, after crude analogies with objects that can be touched and seen had been discarded.

8. GERHARDT'S RATIONAL CHEMICAL FORMULAE

Haüy had left untouched the problem of the arrangement of the atoms within the molecules as being irrelevant for his purpose and, one might add, as beyond our ken. The rise of organic chemistry with its many compounds consisting of the same elements (carbon C; hydrogen H; oxygen O) made the tackling of this problem urgent, were it only because of the discovery of isomerism: i.e. that entirely different compounds, e.g. alcohol and ether, can have the same molecular weight and the same proportions of the component elements (C_2H_6O). The fact, however, that in alcohol one sixth of the hydrogen can be easily replaced by sodium (to form C_2H_5ONa), and that the oxygen together with one sixth of the hydrogen are easily replaced by chlorine (to give C_2H_5Cl) by the action of phosphorus pentachloride, clearly showed that one of the six hydrogen atoms occupied a special position, whereas in the isomeric ether all of them were indifferent to these agents. Tentatively one might express this by writing C_2H_5OH for alcohol and C_2H_6O or $(CH_3)_2O$ for dimethyl ether. Which formula was chosen depended on the reactions the chemist concerned thought most revealing: when alcohol is treated with concentrated sulphuric acid it yields ethene gas (C_2H_4). Evidently it has lost water, so one might be inclined to write its formula as $C_2H_4.H_2O$.

Therefore Charles Gerhardt (1816 - 1856) deemed 'the notation of the formulae a purely conventional matter' which nevertheless 'accustoms the mind to look upon them as absolute things, whereas it should only express mere relations' (1844).[86] It is, in his view, 'a generally prevailing prejudice that one can express by chemical formulae the molecular constitution of bodies, that is the true arrangement of their atoms' (1856).[87] Barium sulphate ($BaSO_4$) for example can be composed from and decomposed into baryte (BaO) and anhydrous sulphuric acid: so it could be conceived of as $BaO.SO_3$. But it can also be made from barium peroxide (BaO_2) and sulphurous acid: $BaO_2 + SO_2$, or by the combination of barium sulphide (BaS) and oxygen: $BaS + O_4$.[88] It has become usual to conceive of it as $BaSO_4$ because in many reactions of double decomposition this substance may be formed by barium's taking the place of another metal in sulphates (or (SO_4) taking the place of some other residue in another barium salt):

$$Ba(NO_3)_2 + Na_2SO_4 \quad BaSO_4 + 2\ NaNO_3.$$

In the same way one might choose to write ammonium compounds either as

$$NH_4.Cl \text{ and } NH_4.NO_3$$

or as

$$NH_3.HCl \text{ and } NH_3.HNO_3.$$[89]

In spite of his opinion that the true structure of a molecule is beyond our ken, Gerhardt certainly did not consider the choice of what he called the 'rational notations' an indifferent matter. He wanted the fiction that embraced most; as he put it: 'it is the better, the more analogies it awakens in the mind and the more fertile thoughts it suggests.'[90] But 'chemical formulae are not intended to represent the arrangement of the atoms, but have as their purpose to make evident in the simplest and most exact manner the relations which link the substances as regards their transformations.'[91] Now 'these relations we render more or less evident by images. We do not know what in reality goes on in the interior of the molecule when it is transformed.'[92]

Therefore, although Gerhardt is the father of a once famous 'theory of types', one would be much mistaken if one attributed to him leanings of a platonic-idealistic kind. The idea of 'type' is to him just a convenient means of classification, but it has no static character and no ontological value. Many bodies, of widely differing properties, may be derived from the type 'water':

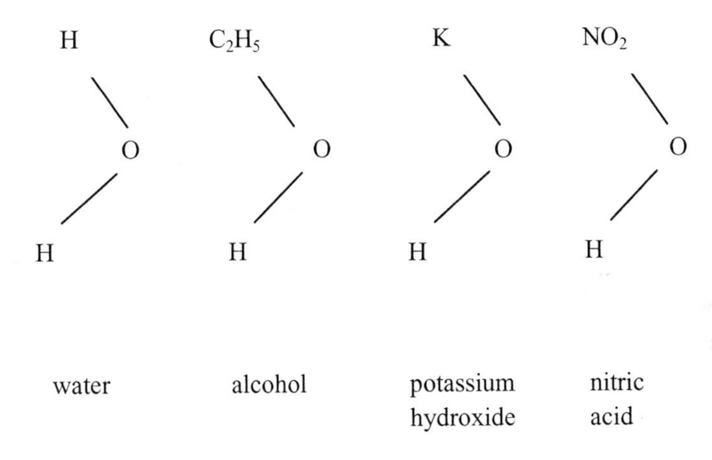

But in saying that a certain body is derived from the type 'water', or represents water in which hydrogen or oxygen is replaced by some other radical, 'I do not intend to express the manner in which the elements are arranged in the bodies to which this comparison is applied.'[93] Here again his own (very fertile) theory does not claim the status of a *physical* theory, but only that of a 'mathematical' theory in the ancient sense, a convenient way of grouping chemical substances.

In his theory of types he had found one that embraces most reactions of the substances classified. He pointed out that the difference between his theory and Dumas's theory of molecular types is that the latter 'refers to the supposed arrangement of the atoms in the bodies, an arrangement which, in my opinion, is inaccessible to experience.'[94]

Gerhardt had introduced four types:

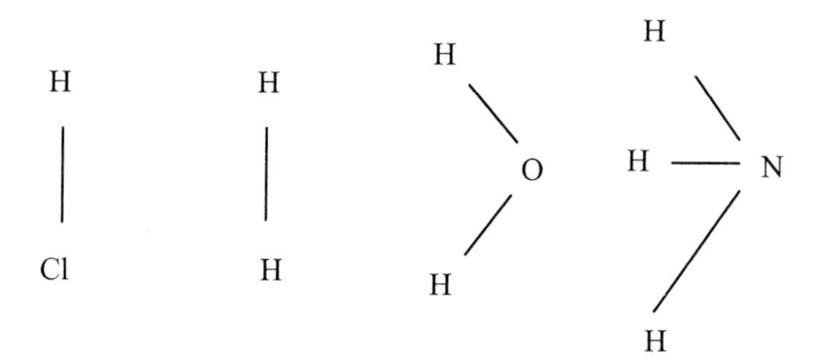

which he later changed to

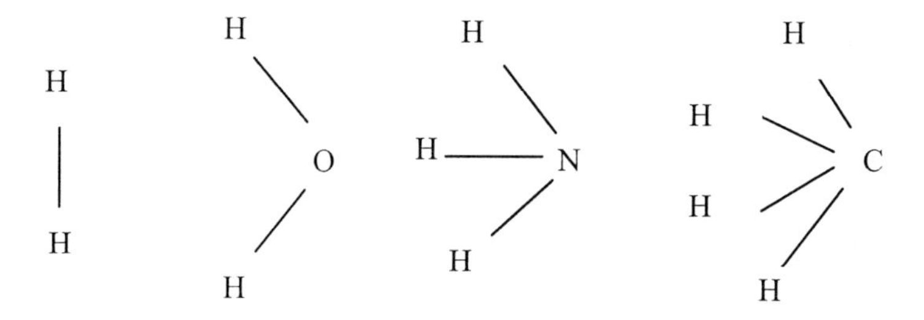

This was extremely useful in the development of chemistry, but it did not last for long. The types 'could be considered only as parts of a scaffolding, which was demolished when the construction of a system of organic chemistry had sufficiently prospered, so that it could do without it', wrote Lothar Meyer in 1863.[95] Gerhardt would have agreed and have said: I never meant it to be anything else! 'Chemical formulae', Gerhardt says, 'do not express and cannot express anything but relations, analogies; the best ones are those which give sense to the greatest number of relations and most analogies.'[96]

It is interesting that this verdict of Gerhardt's closely resembles Fresnel's similar statement about the *physical* truth of his undulatory theory. By striving after the most-embracing 'mathematical' theory, Gerhardt was on the way to a 'physical' theory. In the same way Haüy's 'mathematical' subtractive molecules almost ousted the integrant molecules because they were more 'general' and pushed analogy further than three or four types of integrant molecules could do. And almost inevitably their essential feature (homogeneous distribution of units in space) was turned from a 'mathematical' device into a 'physical reality' by Seeber and Delafosse and — with many refinements and additions — by modern crystal chemistry.

It is an irony of history that once Gerhardt's conception had demolished the old realistic system of chemistry, his own conventionalist system was so successful that in 1884 Lothar Meyer spoke of 'the system of organic chemistry which today is accepted and which in its foundations is Gerhardtian.' He stated that this became the basis of the new, *realistic* chemical theory when his followers set themselves the task of finding the laws of atomic linking.[97] Thus a successful 'mathematical' theory became almost inevitably a 'physical' theory, just as an artificial classification, the more properties it takes into account, the closer it comes to a natural system.

Finally, however much the hypotheses (fictions) and the metaphysical standpoint (faith) of the rival factions in optics, crystallography and chemistry differed, they did not disagree about their basis, the observed facts.

9. THE CONTROVERSY OVER RESONANCE THEORY

Being long accustomed to saying that water *is* H_2O and rock salt *is* NaCl we easily forget that chemical formulae are but symbols of reality and not reality itself. Furthermore an empirical formula such as C_2H_6O expresses symbolically only part of the reality of 'alcohol'; i.e. which elements entered into its composition and in what proportions. A structural formula indicates in addition the sequence of the atoms and symbolizes several properties of the compound — for those who know how to interpret them. In some cases the classical conventions of symbolic language are insufficient for expressing the main properties of a substance. The benzene molecule (C_6H_6) was represented by Kékulé (1865) by a ring of six carbon atoms with three double bonds alternating with three single bonds. This leads one to expect it to be a strongly unsaturated compound which will easily add chlorine and hydrogen atoms (to form respectively $C_6H_6Cl_6$ and C_6H_{12}, cyclohexane). Its properties, however, are intermediate between those of a saturated compound and those to be expected from an unsaturated compound. Moreover, as all the bonds in benzene turn out to be equivalent, Kékulé (1872) assumed that the double bonds oscillate:

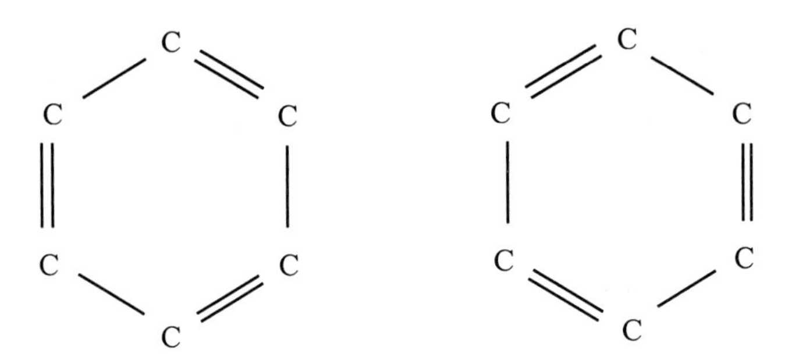

Fig. 39. Diagrams of benzene rings.

In the 20th century it was found that, whereas the distance between neighbouring C-atoms connected by a single bond is 1.54 Ångstrom units (1 Å = 10^{-8}cm) and between those connected by a double bond is 1.34 Å, the benzene-bonds have a C–C distance of 1.39 Å. As the true state of benzene molecules could not be represented by any classical single-bond structure, it was evident that in such a case a new type of bond was at stake, which had not been envisaged by the classical structure theory with which the names of Couper (1858), Kékulé (1858) and Butlerov (1861) are associated.

In the 1920s F. Arndt (1924) and C.K. Ingold and R. Robinson (1925) put forward the theory of mesomerism (intermediate forms), which underwent further development in the 1930s by L. Pauling and others into the resonance theory.

Unfortunately the protagonists of the theory of resonance introduced a confusing terminology. When they said that the substance 'resonates between the structures in question', this was easily mistaken for a claim that there is a 'dynamic equilibrium' of two substances with the same empirical formula ('tautomerism').[98] In fact it is not claimed that the mesomeric state lies intermediate between two actually existing structures nor that it is the average of a mixture of two substances with those structures. The several structures used in the description (such as those of benzene mentioned above) are, according to Pauling, 'idealizations, that do not have existence in reality.'[99] To say that the substance is in an intermediate or mesomeric state between certain valence-bond structures reduced the risk of misunderstanding.

In order to prevent any misunderstanding G.W. Wheland made the (inadequate) comparison of the chemical 'hybrid' to a mule which is 'a hybrid between a horse and a donkey'. 'This does not mean that a given mule is a horse part of the time and a donkey the rest of the time. Instead, it means that a mule is a new kind of animal, neither horse nor donkey, but between the two and partaking to some extent of the character of each.'[100] It is a nice analogy, but it may give rise to the further misunderstanding that the two formulae used to represent the 'hybrid' represent two substances which really exist, just as the horse and the donkey are as real as the mule. In a later edition (1955) of his book the author recognized that the analogy is misleading, for the Kékulé structures I and II are 'merely intellectual constructions which do not correspond to any molecules that actually exist'; only their hybrid, the benzene molecule, has physical reality. Therefore Wheland came to prefer a different analogy (suggested by J.D. Roberts): A medieval traveller when describing a rhinoceros (an animal unknown to his contemporaries) might find it convenient to convey an approximate idea of its appearance by saying that it is intermediate between a dragon and a unicorn, purely mythical animals with which they were familiar through the zoological works of the time. Just as the fictitious dragon and unicorn called forth clear pictures to medieval people, so the equally fictitious Kékulé structures call forth clear pictures to the modern chemist: 'In each case the unfamiliar reality is explained by reference to a familiar fiction.'[101] Since, however, not just two, but any number of structures may be included in the resonance, Wheland suggested a third analogy: A man's personality may be described by means of a comparison with familiar characters in fiction, e.g. when we say that a certain person is a cross between Sherlock Holmes and Don Quixote, but is more like the former, and that he also shows some resemblance to Sir Galahad: 'There is no restriction either on the total number of the individual fictional characters or on the relative weights that are assigned to them.'[102]

Some characteristics of the theory of mesomerism — *viz.* a 'mathematical' device invented to describe a physical reality, and the possibility of using alternative devices according to the chemical reactions under consideration — remind us of Ch. Gerhardt's several 'rational formulae' for one and the same substance. In the 19th century Gerhardt's conception had been gradually superseded by the structure theory of Couper, Kékulé and Butlerov. In this latter theory, to each substance corresponds a certain arrangement of the atoms symbolized in the structural formula and expressing in which sequence the atoms are linked together by 'valence bonds'. The insider then knows to a large extent which properties are inherent in such an arrangement.

Though the protagonists of the classical valence and structure theories in general considered the formulae as reflecting a physical reality, they were aware of the fact that this was not a perfect reflexion. No classical formula corresponds wholly with reality, no formula expresses the properties of a compound completely. Already before the discussion about the new theories became a hot topic, a chemist of great repute, Walther Hückel (1934), who considered the structural formulae as representing the real linking up of the atoms within the molecule, nevertheless spoke of 'the exhaustion of the formal means of expression.'[103] He wrote: 'The structural formulae thus are a visual (*anschaulich*) systematics with concrete images, as far as the situation of the atoms is concerned, and therefore more than a mere 'schema' (scheme, scaffolding); what, however, is the meaning of the chemical bond represented by a line (*Strich*) cannot be imagined and comes to the fore in a theoretically incomprehensible way in the manifold transformations.'[104] The exhaustion of our symbolic means is, however, more clearly revealed in the theory of resonance than in the chemical valence-bond formulae, because it uses for one and the same molecule two or more formulae at the same time and denies that any transformation between the two or more forms occurs, thus making these formulae into *ideal* limits. Ingold and Pauling never assigned several structures to one chemical species, but they held that for a 'hybrid' a single chemical representation is impossible and so — adapting themselves as much as possible to the current symbolic language of chemistry — they introduced more than one classical formula as 'boundary structures'. But Pauling emphatically declared that a substance showing 'resonance between two or more valence-bond structures does not contain molecules with the configurations and properties usually associated with these structures.'[105] In this way formulae drifted farther away from the claim to physical reality.

Pauling was criticized for stating that the several structures used in descriptions of the mesomeric state are 'idealizations' that do not exist in reality. In defence he reminded his critics of the fact that all carbon-carbon and carbon-hydrogen simple bonds, too, have no 'existence in reality', but are theoretical constructs which during the past hundred years have been of great value.

While recognizing that the current theories, images and models contained 'idealistic' elements, one should not, however, consider them as on the same level of 'idealization' with the formulae of the resonance theory.[106] The classical formulae may not give an *adequate* image of reality, but the vast majority of chemists since Couper, Kékulé and Butlerov certainly considered them to be more than 'idealizations' introduced for the sake of convenience. As Wheland recognized: 'Resonance is a man-made concept in a more fundamental sense than most other physical theories. It does not correspond to any intrinsic property of the molecule itself, but instead it is only a mathematical device, deliberately introduced by the physicist or chemist for his own convenience.'[107] At any rate, when the classical theory (in its electron formulations also) fell short, the chemists of Western and Eastern Europe were quite willing to use Ingold's and Pauling's theories as an 'intellectual scaffolding' (Ingold). In general the scientist is a pragmatist: he wants 'objective', 'realistic', 'physical' truth (whatever that may be), but if he cannot get it, he is content with an intellectual scaffolding, in particular if its images go together with a description and connection of the phenomena in the language of mathematics.

Some of the believers in Marxist-Leninist dialectical-materialism were less tolerant. They found the conceptions of Pauling and his like incompatible with verdicts of Friedrich Engels and V.I. Lenin. In June 1951 the USSR Academy of Sciences organized a conference in Moscow on the structure theory in organic chemistry. There were about 400 participants, mainly chemists but also some physicists and philosophers. Among the latter the main spokesman was B.M. Kedrov who expounded how idealistic, mechanistic, reactionary, obscurantist, sterile and pseudo-scientific were the theories of Ingold and Pauling, and how they served the bourgeois ideologies in propagating idealism and clericalism, as did Weismannism-Morganism in biology and Schroedinger, Dirac and other idealists in physics. The idealists Dirac, Bohr, Jordan and Compton, so he said, had furthered the doctrine of free will, and Schroedinger helps to 'demonstrate the existence of God and the immortality of the soul.' [108]

The attack was avowedly a follow-up of the campaign against the deviations of the 'Mendel-Weismann-Morgan bourgeois genetics' of Vavilov and others. These had been condemned in 1948, whereas the truly Darwinistic [in fact Lamarckist!] doctrine of T.D. Lysenko was declared to be in the official party line. A similar campaign against modern physics had followed and now (in 1951) the chemist G.V. Chelintzev and the philosopher Kedrov had chosen the 'Ingoldite-Paulingites' J.K. Syrkin and M.E. Dyatkina as their main targets.

Apart from invectivizing against their aberrations from the currently prevalent interpretation of dialectical-materialism and then falling back on Gerhardtian theory, there was the chauvinistic accusation that they had mentioned Kékulé and Couper as the fathers of structure theory but had neglected its 'true' originator, the Russian chemist Butlerov, who was a protagonist of the classical realistic conception of structure and held that to each chemical species there corresponds one structural formula. The preliminary report of a Commission of the Academy was severely criticized: its authors were accused of being ancient Ingold-Paulingites who in their critique of Syrkin, Dyatkina and their supporters paid only lip service to dialectical materialism. Kedrov told the meeting that the work of J.V. Stalin on Marxism and linguistics had given a powerful impetus to *all* sciences and that in Britain and the U.S.A. an 'idealistic, mechanistic, reactionary theory had spread, which had been a serious hindrance to the development of organic chemistry, by putting fictitious notions in place of the real object of chemistry, the molecules.'

Syrkin and Dyatkina then admitted their ideological errors, declaring that 'over 40 years ago V.I. Lenin in his work of genius *Materialism and Empirio-criticism* liberated the natural sciences from mysticism and Machism', and that 'Comrade Stalin's work *Marxism and the Problems of Linguistics* teaches all those who are active in Soviet culture how to apply veritable partisanship to science, this partisanship being diametrically opposed to bourgeois objectivism' ... 'really scientific, objective explanations of the laws governing the world of reality are always and in every case found to confirm the correctness of dialectical materialism.'[109] Moreover, they promised, 'We shall in our future work make every effort toward the rectification of our mistakes. Our future task will be to contribute to the development of the knowledge of the structure of molecules in line with the teachings of dialectical materialism and the ideas of Butlerov, Mendeléev, Markovnikov, and other coryphaei of Russian science.'

It should be mentioned that both Butlerov (a spiritualist) and Mendeléev fostered beliefs that were, in fact, rather 'idealistic'. Lenin had said that ideas and sensations are copies or images of objects and that 'thought divorced from matter' is 'philosophical idealism'.[110] It was therefore indeed ideologically incorrect to describe molecules in terms of ideal or physically impossible structures. Yet a certain 'idealism' is inevitable as soon as man's *science* of nature is involved. One of Mendeléev's keenest insights was the distinction between 'element' (material component of the compound, not perceptible to the senses) and 'simple body' (a certain homogeneous substance).[111] His famous periodic system was a system of things not observed, of 'pure ideas, defying any possible description' (Urbain 1925).[112]

Of course similar abstractions could not be banned even by Kedrov. No 'scientist' remains free from 'idealistic stains'. Kedrov considered 'perfect gas' as a 'truly scientific abstraction, which above all supposes conformity to reality', whereas (in his opinion) Pauling's limit abstract system, which does not really exist, is a 'fiction' that is not an image of reality.[113] There is a difference indeed between an 'ideal gas' (considered as a real gas of which the molecules are stripped of volume and mutual attraction) and the formulae of resonance. The former is in a quasi-Platonic sense a real thing; the latter is 'idealistic' in the pejoratively meant Marxist sense because there is no corresponding concrete thing. But they have in common that in neither case are they 'copies' of things that exist in concrete reality; the one an abstraction from concrete reality, the other a convenient artifice which has many traits in common with reality. The chemist, however, does not care about such niceties; to him both the perfect gas and the mesomeric limits are fictions of great value.

The conference adopted a resolution declaring that 'the decisions of the Central Committee of the Communist Party of the USSR about ideological questions ... have helped to discover the errors made in chemistry and to delineate the way for the further development of chemical science on the basis of the dialectical-materialist conception of the world, the only true one,'[114] and it accepted the main conclusions of the report made by the Commission of the Academy, though it was deemed to lack ideological precision,[115] e.g. by not pointing out that the deformation of chemical theory has close ties with the wrong theories in biology and physiology and 'are part of the united front of the reactionary bourgeois ideology against materialism'. The alternative theory of Chelintsev, however, was also rejected; so he did not become the Lysenko of chemistry. In fact one can read between the lines that the members of the Commission, for a part 'former Paulingites' themselves, were not wholly sincere in their condemnations and accusations.[116] A wind of change was blowing: ten years later, in November 1961, L. Pauling lectured in Moscow before a large and sympathetic audience of 1200 on the new theories.

10. Conclusion

The contrast of mathematical and physical theories is no longer primarily that between theories which are true in nature and artificial devices serving only practical ends. It is mainly that between the mathematical aspect of physical things and an explanation by means of analogies with other physical things.

The analogy between light and vibrations of an electric medium, according to Maxwell, is founded only on a resemblance in form between the laws of light and those of vibrations. 'By stripping it of its physical dress and reducing it to a theory of "transverse alternation" we might obtain a system of truth strictly founded on observation, but probably deficient both in the vividness of its conception and the fertility of its method.'[117]

Maxwell hoped to attain 'generality and precision' by 'avoiding the dangers arising from a premature theory professing to explain the cause of the phenomena' by referring everything to 'the purely geometrical idea of the motion of an imaginary fluid'. He also hoped that this would help experimental philosophers to arrange their results and form a mature theory 'in which physical facts will be physicallly explained' and thus to find 'the only true solution of the questions which the mathematical theory suggests.'[118]

Some scientists evade the whole problem by a more or less positivistic conception of a physical theory as a description and classification of phenomena in mathematical terms. The French physicist P. Duhem, who held that physical science should be free from metaphysics, considered physical theory as an 'economical' representation of experimental laws and a classification of these laws.[119] This 'économie intellectuelle' implied the beauty of the physical theory; in his opinion it is the best warrant that the classification is 'natural' indeed.[120] He held that the so-called 'explanation' of optical phenomena by reducing them to vibrations of a hypothetical aether does not bring us any closer to nature, but the mathematical forms (wavelength, frequency, wavefront, etc.) that are introduced to describe, to unify and to represent the phenomena correspond to something in the unknown 'reality'. In Duhem's opinion we are invincibly convinced that the physicist's classification of the physical laws in an orderly whole is not purely artificial.

> Without being able to account for our conviction, but also without being able to get rid of it, we see in the exact order of the physical system the mark by which a natural classification is recognized; without claiming to explain the reality hidden behind the phenomena whose laws we arrange, we feel that the groups established by our theory correspond to real affinities between the things themselves.[121]

However much Duhem told himself that theories cannot grasp reality and that they exist to give nothing but a concise and ordered representation of experimental laws, he yet could not *believe* that a system that simply and easily orders so many, at first sight divergent, laws could be an *artificial* system. 'An act of faith ... assures us that these theories are no purely artificial system, but a natural classification.'[122]

All this goes to show that an 'idealistic' conception of physical theories (which reduces them to mathematical relations between abstract entities) does not necessarily accompany a wholly 'mathematical' concept (in the medieval sense) of method. It is possible to assume, like Duhem, a positivist attitude in science and to have at the same time a firm belief in its results. A similar conception is met with in another French scholar, Henri Poincaré.[123] In his opinion, when Fresnel's theory of light was replaced by Maxwell's, this might imply a change of the physical images (vibrations of aether became 'alternating electrical and magnetic fields'), 'which had been substituted for the real objects which nature will eternally hide from us', but the relations expressed in the mathematical equations remain the same: 'The true relations between those real objects are the only reality we could ever reach, and the only condition is, that there be the same relations between those objects as between the images we are obliged to put in their place'. When these relations are known, it does not matter whether one image takes the place of another one: 'this is the truth which will always remain the same under all the vestments in which we could think it useful to dress it up.'[124] Similarly, Sir James Jeans stated that a 'mathematical formula ... and nothing else, expresses ultimate reality.'[125]

Physicists with such a frame of mind have a tendency to minimize the importance of 'images' and to emphasize the mathematical abstractions that parade in the formulae describing and connecting phenomena. It would be going too far to call them agnostics in physics just because they are sceptical about the possibility of finding an adequate image of the sub-phenomenal (or super-phenomenal) world behind the world of phenomena. They realize that the chasm between nature herself and the science of nature cannot be covered up by images and words, however 'scientific' they may be.

IX. Works of Nature, Works of Art

1. Introduction 265
1.1. Nature - Art (physis - techne) 266
1.2. Aristotle 267
2. The Liberal Arts 268
2.1. Pierre de la Ramée (1515 - 1572) 269
2.2. The Quadrivium 271
2.3. Olivier de Serres 272
2.4. Ramus on Astronomy 273
2.5. Ramists 274
2.6. The Renaissance 275
2.7. The Literary and the Visual Arts 277
3. The Mechanical Arts 279
3.1. Alchemy, Chemistry 281
3.2. Bacon's New Approach to Nature 284
3.3. Descartes 285
3.4. The Synthesis of Minerals 286
3.5. Leibniz 288
3.6. Artificial Mineral 'Genesis' 288
3.7. Experiments on Rock Formation 291
3.8. Synthesis of Organic Compounds 295
4. Models 297
4.1. Cosmographical Models 300
4.2. Geological Models 304
4.3. Mechanical Geological Experiments 305
4.4. Artificial Production of New Animal Species 308
4.5. Aether Models 309

1. Introduction

We have contrasted scientific theories that claim to represent the reality of nature with those that are just useful fictions establishing connections between phenomena, without any claim to physical truth (chapter VIII). We now tackle a related problem — the comparing of natural products with similar ones made by human art.[*] In the case of chemical compounds, minerals and rocks, for instance, we could try to find out their composition by means of chemical analysis and then confirm this analysis by a synthesis out of the components. Supposing we find that human art is indeed capable of making things produced also by nature, immediately the question arises: can we find a procedure to make a natural product (e.g. sugar), minerals or rocks in the same way as nature does?

*1.1. Nature - Art (*physis-techne*)*

The ancient idealistic philosophers and their medieval followers held that artificial production of things and processes of 'Nature outside Man' is impossible. Man may be able to *know* the 'logos', the 'nature' (*physis*) of natural things, but he is unable to *generate* them. The principle was: 'all that generates, generates things similar to itself.' The Form[†], the seed, is self-perpetuating; Man can only generate man and is unable to bestow any other 'natural' Form on matter.

On this view the mechanical arts (e.g. mechanics and chemistry) and the manual trades (like those of the sculptor, the carpenter and the smith) do not occupy themselves with 'generation' but with 'fabrication' of their products. Human hands transform things of the world of phenomena according to plans formed in the human mind. In these 'arts' the hand struggles with a world outside man, and the product is a non-natural thing, made by art (arti-ficial). They possess 'artificial Forms', generated not by Nature but by Man's mind that steers Man's hand, and can change them at will.

Before we tackle this problem we draw attention to the universality of the contrasting of natural to artificial products and procedures. Even where it has disappeared from the platform of official scholarship it lives on in daily life.

> 'Natural' foodstuffs, 'natural' medicines, 'natural' methods of healing often are recommended as more effective and less violent than their artificial counterparts. Some decades ago Belgian ecclesiastical authorities prescribed

[*] [*Editors' note:* Originally, the lecture that corresponded to this chapter was preceded by a lecture in which (in his own words) Hooykaas distinguished natural classification systems representing the objective order of natural things from artificial systems which are useful means for identification of things without any pretence of expressing natural affinities. Having thus compared the allegedly natural with artificial ways of classifying natural phenomena, Hooykaas continues in this chapter with the similar problem in imitating nature.]

[†] We shall write 'Form' with a capital 'F' when we mean the Aristotelian 'substantial Form' (*morphè*; *eidos*) and not the geometrical shape of a thing. This Form needs 'Matter' (*hylè*) in order to bring forth a real substance (*ousia*).

'natural' water (from fountains, rivers, etc.) for baptism, implying that synthetic water would not do. It seems as if the old argument used against the alchemists still lingers on: though a metal might be produced that had all the properties of gold, this artificial ('sophistical') gold would not be real gold because it lacked its natural Form, its 'soul'.

On the scholarly level we find that formerly the contrast nature-art was repeated *within the arts*, i.e. within the liberal arts (grammar, logic, rhetoric) and the mechanical arts (mechanics, agriculture) and even the fine arts (architecture, sculpture, painting). This may seem strange, for the Artificial is the opposite of the Natural; how could there then be something 'natural' in *arts*? Whereas Nature is conceived as guided by the Godhead himself, without human interference, in the arts Man is in a certain sense the creator. Yet, particularly in the 16th century, the contrast between artificial logic and natural logic, between building while keeping to the rules of Nature or deviating from them, gave rise to fiery debates. In the science of Nature the problem was whether an artificial Nature is possible which is really 'nature' — as the 'philosophy' of the alchemists claimed, but in the liberal and mechanical arts there was diversity of opinion about the possibility of a natural Art.

1.2. Aristotle

In order to find a connection between such divergent problems we go back to Aristotle, until the 16th century the common teacher of all 'educated' people, his works or those of his disciples studied by everyone, including those who, following the fashion of the day, had become ardent teachers of others in their post-school life. The notion of *generation* of things by things with a similar character was transferred in a metaphorical sense — by Aristotle himself — from natural entities to artificial products. Because Nature generates things that are as they are by necessity whereas Art produces contingent things according to the will and whims of man, Aristotle advocated a 'natural' Art, which is *not* contingent but bound to certain rules. He pointed out the similarity as well as the essential difference between generation by Nature outside the human mind, and that in which the human mind is involved: 'Things are generated either by art or by Nature, or by chance or potentiality. Art is a generation principle in something else; Nature is a generation principle in the subject itself (for man begets men).'[1] This implies that the essential contrast between Art and Science was weakened by a notion of the 'natural' which functioned in the humanities and in the mechanical arts as well.

In his *Metaphysics* Aristotle made a clear distinction between 'generation by Nature' (*genesis*: man from man, horse from horse, etc.) and 'production' (*poièsis*), i.e. artificial 'generation' of things whose Form (*eidos*) is contained in the (human) soul (*psyche*). This production consists of 'cogitation' (*noèsis*) and 'production in the proper sense'.[2] The production or fabrication of artificial things thus has two stages: the conception within the human mind and, next, the concretisation of the concept in matter outside man. The spoken oration in its relation to the planned oration fits to a certain extent also in this concept.

Artificial production thus is a kind of 'generation'. So, Aristotle could say that 'a house is generated from a house, for the Art (*technè*) is the Form (*eidos*).'[3] This seems to express in the first place that the 'art' is the formation of the concept of 'house' in the mind; which next leads to the concretisation by the hand in the phenomenon 'house'. The real generation takes place when the 'formal' nature is embodied in the world of phenomena: then the concrete thing is in the appropriate matter.[4] In contrast to Plato's conception this is not a step away from reality, but its full accomplishment; a substance cannot exist without Form and without Matter.

The essential contrast between Arts and Science was thus weakened by a notion of the 'natural' which functioned in the humanities and in the mechanical arts as well.

The analogy between art and nature is also stressed when it is pointed out that art as well as nature works towards a purpose: the earlier stages of their operations are for the purpose of leading to the later: 'Thus, if a house were a natural product, the process would pass through the same stages that it in fact passes through when it is produced by art; and if natural products could also be produced by art, they would move along the same line that the natural process actually takes.'[5] Failures in the arts made by writers and physicians, then, may be compared with monstrosities in Nature, which are also like failures of purpose.[6]

Later generations could find passages in the works of their predecessors that could easily be interpreted as supporting theories opposite to the general trend of the philosophy of the erstwhile great authorities like Plato and Aristotle. When mechanistic corpuscularian theories lessened the influence of Plato and Aristotle on the natural sciences, this did not imply that they lost their influence in the humanities. Moreover, one could combat Aristotle but venerate Cicero and Seneca who had inserted many of their predecessor's ideas in their own philosophy.

Passages like those quoted above helped to weaken (and sometimes almost to obliterate) the contrast between products of art and of nature; analogy became quasi-identity, natural art tended to turn into artificial nature.

2. THE LIBERAL ARTS

The account above concentrates on mechanical arts (building). The human mind has to realize its plan in a material outside itself, which limits its freedom and needs the help of instruments, first of all of human hands. In the liberal arts of the trivium (grammar, logic, rhetoric), and — to a large extent — in mathematics too the situation is different. In later Antiquity it was held by the idealistic philosophers, like Cicero and Seneca, that the laws of logic and the truth of mathematics exist in 'nature', i.e. in the universal Divine Mind, and thus, in principle, are also in the human mind which is part of the universal Mind. It was therefore assumed that the human mind, when it follows 'right reason' or 'natural light' (*lumen naturale*) will be able to think rightly (logic) and to speak rightly (rhetoric). As Cicero put it: 'True law, then, is right reason in agreement with nature; it is diffused in all people, unchanging, everlasting.'[7] The question remained, however, whether what is present in perfection in the mind, can also be expressed in perfection in what the ear can hear (e.g. rhetoric) or the eye can see (e.g. a painting).[8]

In Cicero's conception the perfect natural art is in the mind of the artist rather than in the product. This is the more so in neoplatonism, the last great Greek philosophical system. The neoplatonists held that the Ideas in the mind cannot be fully expressed in the material world, because matter is in itself imperfect.

The Renaissance philosophers were so strongly imbued with Platonic, Aristotelian and Stoic philosophy, that to bridge the gap between Athens and Jerusalem, they amalgamated the concept of the human mind as sharing in the divine logos, with the biblical doctrine of man created 'in God's image and likeness' (Genesis 1:26), while referring 'the image' to man's spiritual and rational character. By conceiving the Fall as having corrupted the will rather than the intellect, they maintained the notion of man's capacity 'to think rightly by nature'. The right words, in the right arrangement and with the greatest convincing power are spontaneously brought forth ('generated', so to say) by those people who are best in speaking and reasoning; 'natural' systems of grammar, logic and rhetoric may, in the Ramist opinion, be extracted from their writings.

Consequently, the humanists who were influenced by Plato and the Stoics accepted that liberal arts like logic, rhetoric and mathematics — the humanities in the strict sense — could entirely follow Nature (which in this case was their own logos). They blamed the scholastics for having constructed totally artificial grammar and logic founded on arbitrary rules, and they appointed themselves the task of finding for such arts rules that are wholly 'according to nature'.

2.1. Pierre de la Ramée (1515 - 1572)

In Peter Ramus we find a particularly explicit exponent of the view that Art *should* follow Nature. The growing philosophical tendency to weaken the contrast between Nature and Art was stimulated by Ramus' utilitarian conceptions.

The liberal arts, being in his opinion as much as the mechanical arts rooted in practice, should be defined by their application. Thus grammar is the art of speaking well (i.e. correctly), logic the art of reasoning well ('art de bien disputer et raisonner de quelque chose que se soit'), rhetoric the art of discoursing well, arithmetic the art of calculating (or numerating) well.[9]

The emphasis on the practical application of the liberal arts implied for Ramus that he hardly recognized their separation from the mechanical, 'applied' or illiberal arts. Geometry's aim of measuring well, so he says,

> appears with much greater splendour in the application (*usus*) ... than in the precepts and rules, as one sees astronomers, geographers, geodetians, navigators, mechanicians, architects, painters and sculptors using nothing but geometry in the description and measurement of the stars, the lands, the earth, machines, seas, buildings and paintings — just as grammar, rhetoric, logic are more fully and abundantly seen in their use by poets, orators and philosophers than in the *prescriptions* of grammarians, rhetoricians and logicians.[10]

Not art itself, but much more its application and practice make the artisan.[11] It would be better to have the use (application, exercise) without the art than the art without the application.[12] Consequently, Ramus does not like the division of the quadrivium into arithmetic, geometry, astronomy and music, for as mathematics is applied in astronomy, the latter cannot be separated from it, just as grammar cannot exist apart from its applications.[13]

There is a streak of rationalism in Ramus, however, when he speaks of the 'sovereign light of reason'.[14] His unplatonic emphasis on practice then turns out to go together with a rationalistic metaphysics. Man knows God, himself, and all things by Reason, and Reason is the Artificer (*artifex*) who exercises the innate gifts of speaking, numerating, measuring, singing, etc.[15] He believes that the 'natural dialectics' (*naturalis dialectica*) consists of 'reason, the mind, the image of God who brings forth all things ... which emulates the eternal light, is proper to man and is born with him.'[16] He even goes so far as to speak of the 'divinité de l'homme'.[17]

On the other hand Ramus declares that God is 'the only perfect Logician', the only One who always used reason with perfection.[18] For, in spite of his inflexible statements, he was keenly aware of the imperfection of his own thought, which needs never-ending self-criticism.

Dialectics, according to Ramus, consists of three things:

a) Nature, by which God wrote the principles of reasoning or discoursing in our souls;
b) Science or Art (*doctrina*, *ars*), by which the observer of nature makes these eternal principles visible and imitates them, and
c) Exercise, putting them into practice 'by hand and tongue'.[19]

Nature and the skill to put into practice are innate, but Art has to be taught by teachers or by self-contemplation.[20] Through art Nature sees her own pure essence, so that she may be purified from stains and faults, and thus 'Art which was at first the pupil of Nature becomes in a way her teacher.'

Where do we find this 'nature' at its purest? Ramus asks. Not with the scholastic philosophers, who established 'artificial' (in the sense of 'arbitrary') rules of logic, etc., but with those people whose mind was not deformed by such 'art'. Therefore, even illiterate people, who never heard of dialectics (such as wine-growers pursuing their trade), show the image of nature in their common sense deliberations better than those school philosophers. The practice of the art (its *usus*) will reveal its right rules.[21]

Yet, Ramus thinks that innate Nature is seen at work at its best in the greatest minds (Cicero), and therefore their works should be analyzed and reduced to a system of rules which enables us to transfer it more easily to other people. Such a system is 'nature made into an art'. Ramus, therefore, seeks for the principles as well as for the practice of this art in the writings of the best authors, rhetoricians, philosophers and poets.[22] They are mainly classical authors, but in his French *Dialectique* (1555) he includes many examples from the poets of the Pléiade, in particular Ronsard and Du Bellay. Thus Ramus' rationalism goes together with what one could call 'literary empiricism'.

2.2. *The Quadrivium*

In the quadrivium (arithmetic, geometry, astronomy, music) too, Ramus expected to find the 'art of nature' or 'natural art' in the application of natural reason. Accordingly, he set himself the task of observing how the merchants and bankers of Paris performed their calculations, how carpenters applied geometry, arithmetic, etc., and he prided himself that there was no mechanical workshop in the whole town of Paris which he had not 'meticulously searched' several times.[23] Yet, disappointingly, he did not obtain his own mathematical method from those practitioners, but from one of the greatest authorities in 'natural' mathematics, *viz.* Euclid.[24] Just as in dialectics, he followed here, too, the great classical writers and not the wine-growers and peasants.

The choice of the *problems to be treated* however, was indeed determined by that 'mathematical' practice he so highly praised: only what can be used in the practice of life is worth dealing with in 'art'. So, because of its lack of practical usefulness he criticized Euclid's tenth book. This aroused great indignation in Johannes Kepler, who called him a 'Euclid-killer' and who censured the Netherlands' Ramist mathematician Willebrord Snellius for sharing Ramus' view on the tenth book, because: 'he wants to make a mathematician into a merchant'.[25] This being engrossed in mercantile matters of the Netherlandish scholars did not please Kepler at all: in his opinion it impeded their formerly outstanding astronomical achievements. Small wonder that Ramus' numerous opponents nicknamed him an 'usuarius'.

Ramus advocated a kind of rationalistic realism with a utilitarian aim: one has to investigate concrete reality to see Reason at work, as sound practice is rational in itself.[26]

2.3. Olivier de Serres

Ramus' thoughts on nature and arts are found, rather unexpectedly, with the Huguenot country nobleman Olivier de Serres, seigneur du Pradel (1539 - 1613), who in his *Théâtre d'Agriculture et Mesnage des Champs* (1600) asked himself why he should seek in *books* for knowledge of agriculture, which he could find in himself by common sense and from his tenants' 'natural skill'; a question, so he adds, one could put with regard to all liberal sciences. Olivier de Serres, in common with many of his contemporaries, believed that the principles and germs of all things are innate in the human soul. In his opinion man can learn from scholarly books only what he already knows from his mother's womb onward, albeit in a confused way that needs to be put in order by some artificial method.[27] This method can be found in books on physics, ethics, husbandry and politics. Man is born with the principles necessary to know these disciplines, but because of the divergent opinions of the expert labourers — dependent on the local conditions in which they live — it is better to revert to learned books that join reason and practice: 'Art is a summary of Experience, and Experience is the judgment and use of Reason.'[28] 'The art of agriculture arises so ingenuously (*naivement*) from the Book of Nature and by such manifest effects, that in her Reason allows itself to be seen by the eye and touched by the hand.'[29] It does not exist without its practical application: 'as use (*usage*) is the aim of each laudable enterprise and science is the enabling to right use, the rule and the compass for doing well is the connection of science and experience.'[30] In Serres's preface the whole Ramist theory of knowledge thus is contained in a nutshell.

No tidy ordering of natural knowledge was to be found with the 'uneducated peasants' (*paisans sans lettres*) and therefore Serres had more confidence in his own wide experience and that of the great agriculturists of Antiquity, like Columella and Virgil, who, like himself, were educated landlords who could express 'what nature taught them' in a more articulate way than his simple farmers.[31]

Ramus likewise had recommended Virgil's *Georgica* as a source for knowledge of plants, animals and agriculture, just as he followed Euclid rather than the artisans and merchants of Paris in his exposés of mathematics. And similarly, he had recommended classical authors as sources for logic and rhetoric, while at the same time praising the dialectics of the simple viniculturists.

Being no absentee landlord, seigneur Olivier tried to be a good practitioner who realized that words are not enough, but assiduous work and reasoning are necessary: 'we want corn in the granary and not in a painting'. Yet, the humanist conviction that agriculture must have been at its summit and 'natural light' at its brightest among the Ancients made his prescriptions for practice sometimes rather bookish.

In this respect his work contrasts with that of another Huguenot, the potter and agriculturist Bernard Palissy (1510 - 1590) who, unhampered by classical learning, set out the results of a wide experience of different regions of France (not Italy!) in a more original, though less systematic, way, in a style less polished but charmingly 'naive'.

2.4. Ramus on Astronomy

The fact that there are different systems of hypotheses to account for the apparent motions of the planets, indicated to Ramus that they are 'artificial' — merely devices for computation of the stars' positions in the sky. There can be only one 'natural' astronomy, an astronomy conformable to nature. True astronomy, according to Ramus, must reflect the planets' true motions and therefore the various hypotheses of Greek Antiquity ought to be rejected.[32]

Nor did Ramus accept Copernicus' theory. He held it to contain — as in fact it did — many hypotheses. Moreover, its main novelty, the motion(s) of the earth, he deemed an absurdity. The fact that Copernicus had proposed a *structure* of the universe that more or less answered the Vitruvian rule of harmonious proportions does not seem to have impressed him.

Instead of all these astronomical systems based on the Greek systems, Ramus wanted an 'astronomy free from hypotheses', that is, without fictitious circles. But Ramus himself could produce nothing better when he urged first Tycho Brahe and afterward Rheticus to develop such a system.

Kepler, then, at the beginning of the 17th century could claim that by his reformation of the whole of astronomy he had fulfilled the task Ramus had proposed.[33]

Kepler was convinced that he had found the general plan God had in mind when creating the cosmos (1596). Before He created the world according to a geometrical plan, geometry was in the Divine Mind, and 'with the image of God it was transferred into Man, and received not by the eyes but inwardly.'[34]

He believed that he had found the 'natural' art of astronomy Ramus had not dreamt of as being attainable. He discarded the remnants of the 'artificial' circles which even Copernicus had needed to fulfil his task as a 'calculating' astronomer and introduced the elliptical orbits (Kepler's first and second laws, 1609).

And, finally, he believed he had accounted for the variation of the velocity of a planet in its orbit by the discovery of the 'harmony', the music inaudible to the human senses but enjoyed 'in intellectu'. Here his belief that the human mind is an image of the divine (which should not be confounded with the idea of divinity of man) found for him its confirmation.

Ramus' untimely death in the massacre of St Bartholomew (1572) prevented him from tackling the subject of the natural and the artificial in the last of the seven liberal arts, *viz.* 'musica'. There the judgment by the human ear, the physical experiments with strings and flutes, and the *a priori* conceptions of numerological speculation would lay bare the tensions within a field that on one side belongs to sensory perception of nature outside man and on the other side to the human intellect and the human 'heart', the sense of beauty and the theory about it.

2.5. *Ramists*

In the last decades of the 16th century a fierce battle raged at Cambridge university between Ramists and their opponents. According to the Aristotelian Everard Digby logical method and natural method should coexist, whereas the Ramist William Temple (1584) retorted that, if this were true, Digby would not have recognized *two* methods but a hundred methods: if 'artificial' implies that a certain arbitrariness is possible and admissible — provided the system is practical and useful — there must be *many* artificial methods; if, however, 'artificial' means the art of following nature exactly, there can be only one method, the true one, which is 'art according to nature'.[35]

This contrast is analogous to that between the one natural, physically true, system of astronomy, and the many artificial hypotheses, which are just mathematical devices, or that between the one natural system of classification of plants and the many artificial methods of arranging them in order to recognize them easily.

This doctrine of 'natural art' or 'art of nature' (*technologia*, as the theologian William Ames called it) had a considerable influence on English Puritans and, afterwards, on New England Congregationalists.[36] As late as 1672 Milton thought it worthwhile to rewrite Ramus' *Dialectica*.[37]

Ramism in Newton

From Milton's to Newton's time is but a step. It appears that Ramist influence may be traced in the preface of Newton's *Principia* (1687). There we find a verdict one would expect from a Ramist rather than from the man who finished and perfected what Kepler had begun. Newton announces that he will deal with mathematics in so far as it concerns the science of nature. He points out that the Ancients made a distinction between *rational* mechanics (which proceeds accurately with strict demonstrations) and *practical* mechanics, handwork (*artes manuales*), from which the name 'mechanics' has been taken. But as artificers do not work with perfect accuracy, mechanics has thus been distinguished from geometry, so that which is less accurate is called 'mechanical'. The errors are, however, not in the art, but in the artificers. He who works with less accuracy is an imperfect mechanic; if anybody could work with perfect accuracy, he would be the most perfect mechanic: 'for the description of straight lines and circles, upon which geometry is founded, belongs to mechanics.' Geometry does not teach us to draw these lines, but requires them to be drawn and then 'by geometry the use of them ... is shown.' 'Therefore, geometry is founded in mechanical practice, and is nothing but that part of universal mechanics which accurately proposes and demonstrates the art of measuring.'[38]

Geometry a part of mechanics! The sequence accepted by both the Ancients and the Moderns is completely turned over: to the Ancients mechanics was 'applied geometry'. Ramus, however, had defined geometry as 'the art of measuring well' (*ars bene metiendi*) and he founded the pure sciences on their practical applications, in which the intelligible principles are unconsciously used. These ideas may have been strengthened by the belief (which Newton also shared with Ramus) that the origins of sciences and arts do not lie with the Greeks but with the Babylonians and Egyptians whose geometry according to Herodotus arose from the need of measuring the bounds of their fields.

On the other hand, Newton borrowed from Ramus only what he could use. Ramus' *Scholae Physicae* was nothing but a critical commentary on Aristotle's *Physica* and not at all the result of personal research of nature, whereas Newton applied mathematics in astronomy, mechanics, optics and music ('the more physical of the mathematical arts', as Aristotle called them) to give a realistic description of the mathematical aspects of physical phenomena.

2.6. The Renaissance

In the figurative arts we enter the realm of aesthetics and at the same time we have to do with manual skill in imitation of 'outside-nature': the 'apeing' of nature without the possibility of pretending to bring forth really 'natural' things.

In the fine arts (painting, sculpting, architecture) the problem 'nature and art' was to a certain extent similar to that in the liberal arts (dialectic, rhetoric). Poetry is painting in words and painting is a depiction of scenes described in poetry and historiography. The Italian Renaissance brought forth a great multitude of studies on this topic. The problem was not new. The affinity of the liberal arts with the fine arts was dealt with by the ancients. Cicero in the fifth of his *Tusculan Disputations* (45 BC) said that, though Homer was blind, yet 'it is his painting, not his poetry that we see'; 'what place in Greece, what force of combat, what tugging at the oars, what movements of man and animals has he not so depicted that what he did not see himself, is made to us as if we see them?'[39] The Roman poet Horace (65 - 8 BC) succinctly put the same in the often quoted words 'Ut pictura poesis.'[40]

So, although cultivators of the fine arts (painting, sculpting, architecture) had to cope with problems similar to those faced by the cultivators of the liberal arts (dialectic, rhetoric), the matter was perhaps even more complicated because they belonged to the mechanical arts. They too worked with their hands. They yearned, though, to attain the status of free, 'liberal' arts and even of sciences, in spite of the fact that they could not deny that they needed their hands.

A special difficulty was that most of them were 'platonizing' more or less. In Plato's opinion the painter and the sculptor are 'the third generation with respect to Truth and Nature'.[41] For example, the 'real bed' is the Idea of the bed, which exists in Nature (i.e. the world of Ideas, invisible to bodily eyes); the beds made by hands are 'artificial beds' which enclose the Idea to some extent; the bed in a painting is but an 'imitation', by which the painter could give only the appearance, but could not depict the whole thing in the same picture.[42] But the new Platonist, who had undergone the influences of Aristotle and of the Stoics, modified this doctrine, so that, in particular, the painters and sculptors developed a doctrine which at the same time acknowledged the importance of observation of the world around us and that of the 'ideal world' implanted in our minds.

The artificers of the Renaissance aspired to *natural* art (true to Nature), so that the sculptors and architects introduced 'harmonious' proportions; the painters in addition to this remained 'true to nature as we see it', by strictly observing the laws of perspective. Consequently, the Renaissance sculptors and painters claimed to be no mere 'mechanicians', but rather cultivators of 'free' (liberal) arts and even 'philosophers' (i.e. scientists) ranking equally with other scholars who occupied themselves with 'natural' reality.

With L.B. Alberti and Albrecht Dürer they recognized that the perfection of a species is present in the phenomenal world though no individual is free from defects. Alberti and Dürer even held that painters and sculptors in a certain sense *surpass* Nature, for she rarely reaches her ends in perfection, whereas the artificer, like Zeuxis, will combine the most beautiful parts of the most perfect bodies into a new harmonic whole.[43]

The endlessly repeated ancient story of Zeuxis tells how he chose from five of the most beautiful maidens of Croton the most perfect parts (from one the limbs, from another the head, etc.) and then combined them according to the rules of 'symmetry' (the right proportion) in the picture of Helen. That is to say that the several parts have to be adjusted so that these mutual dimensions are conformable to the fixed canon, laid down by Vitruvius: the distance of the fingertops from the right hand to the left hand with stretched arms must be the same as the length of the body; the distance from the crown of the head to the chin, must be to the total length of the body as one to eight, etc. (see figure 24 on page 169). Vitruvius also gave rules for the relation of dimensions in temples.

A somewhat different canon was prescribed by Albrecht Dürer, though he allowed a certain freedom in the application of the rules. Moreover, we saw above how Copernicus applied the doctrine of the 'common measure' in the construction of his astronomical system (see chapter VI, page 173).

The artistic product, based on an Idea in the mind of the artist, is bound by objective rules of proportion; the data observed in nature around us are combined, in such a way that the perfect image in the soul — the real Nature — is finally extracted from nature. So, Art surpasses nature (the phenomenon) precisely by following 'Nature as it should be'.[‡]

If there are Platonic elements in this approach they are essentially changed by the influence of Aristotle (who seeks the Forms *within* the world around us) and the Stoics like Cicero (who stressed that the perfection of the Forms is within the artist's mind.) This can go so far that the emphasis is strongly shifted to the inner vision, and that — as Michelangelo put it — one paints with the brain, not with the hands. The Zeuxis story implies that the reality of nature as we see it, is depicted as much as possible, but that it is corrected by the reality of the Form present in the mind, that is by Nature's image in its perfection.

Some people stressed that though the latter may exist in the Design of the master, the master's hands cannot fully repeat its perfection. Even so, this Art according to Nature never shows the defects which are present in the individual we see in nature, and in this sense 'natural art' not only emulates 'nature' but even surpasses her. In the extreme conception, the human mind, as part of the universal Mind, creates its objects in a perfection never seen by the senses in nature around us. Here again the art arises because the creator compares his work with that of the Creator of the world, and conversely, God is said to be the greatest Painter, when he devised this world.

2.7. The Literary and the Visual Arts

Peter Ramus, in stressing that true doctrine should never deviate from nature, but always follow her religiously, said that, as Apelles (fl. 350 BC), when he was to paint Alexander the Great, looked at his head, arms, etc., so the dialectician has to

‡ This was the same as Ramus' ideal of the literary arts: observe the writings of the best authors and led by the light of reason, choose the best out of it, so that your art which at first had learned from nature, now surpasses it; from being its pupil it has become its teacher.

observe all the hints which are immediately collected in the mind and then laid down in the picture (*tabula*) and image (*imago*).[44]

Again and again Ramus says that, like 'our Apelles', we should have the 'living example' before our eyes, and give the closest possible description of natural dialectic, and try to 'imitate it by depicting it'.[45] Thus the dialectician has made an image in the mind in which there are no defects, so that in the artificial mirror nature may be seen in its full dignity. In this way 'art, which at first was the pupil of nature, in a way becomes her teacher.' This 'pupil-cum-teacher' (natural dialectic) has to include the instruction by Nature amongst its precepts.[46]

Ramus obtained the data for 'natural' dialectics by introspection of what was in his mind as well as by observation of its application in practice, but in the latter aspect retouching was needed to move the data towards closer conformity with the ideal of purity and perfection — an ideal which he recognized he had not reached. The phenomenal, non-purified nature was to be cleansed from its defect by 'natural arts' in order to restore it to the perfection of ideal Nature.

The humanist rhetoricians and dialecticians compared their discipline with painting and the painters from their side did the reverse. Francisco de Holanda (1518 - 1584) pointed out the great affinity between the literary and the visual arts.[47] Michel Angelo, he said, called a great man of letters a good painter, and the good painter a literate man.[48] The Ancients were of the same opinion: 'read the whole of Virgil, what else will you find than the work of a Michelangelo?'[49]

The art of drawing (which is common to the three fine arts) is praised by Holanda's Michel Angelo as perhaps even the *only* art, for it is the foundation not only of the visual arts but also of the mechanical arts of the pilots, the farmer, and so on.[50]

The real painting is already in the mind before it is made physical by the expert hand. According to Holanda 'Invention' or 'idea' is the first step, 'the noblest part of painting'; it exists in the 'thought' (*pensamento*).[51] It is seen with the internal eye, and shown externally by the work of the hands;[52] it is a 'science' and a 'second Nature'.[53] In a later work Holanda reached an even higher level of enthusiasm and stressed the divine character of the art of painting: 'It is a science not learned from a master only, but a natural and free gift of God, the highest Master, proceeded from His eternal science, which is called DESEGNO and not 'drawing' (*debuxo*) or painting (*pintura*)'. And this Design is the origin of all other works, arts and crafts of mankind.[54] And he then goes on: '... design is that Idea created in the understanding ... which imitates or wants to imitate the eternal and divine un-created sciences by which Almighty God created all things we see, and it embraces all the works that are based on Invention, form, or beauty and proportion, or hope to have them — the interior ones in the Idea, as well as the exterior ones in the work.' [55]

The unity of Renaissance patterns of thought, however, covers not only liberal arts and figurative arts, but also a new physics. There is an analogous thought running through Ramus, Alberti and Galileo: Ramus constructs a natural dialectic from elements found in various reasoning people; Alberti conceives of a figurative art which paints or sculpts an ideal man by putting together 'ideal' parts and 'ideal' proportions observed in living reality, though never all in one individual; Galileo conceives of a physical event as consisting of component parts which are perhaps never found in a state of purity, yet nevertheless are present in a concrete way in diverse physical phenomena.

All three efforts testify to the desire of bringing back the observed phenomena to an 'ideal reality', existing, perhaps 'under a veil', in the world of phenomena.

3. The Mechanical Arts

Can human art produce things that are *identical* to products of Nature? And if so, can it produce them in the same way as Nature does? In general, how far can human art make natural things or repeat natural processes at will?

The figurative arts of the Renaissance boasted of being 'natural'; their cultivators never got tired of repeating the story of the Greek painter who painted grapes so naturally that the birds came to pick at them, and horses so naturally that real horses started neighing upon seeing them. But these stories could not bridge the fundamental gap between actual physical objects and their images.

As long as the organistic conception prevailed one could at best hope to make things that ape Nature as did the automata made by the engineers of Antiquity. In that conception, an artificial mechanism cannot adequately re-produce, but only imitate the natural organism. As soon, however, as the natural world itself is conceived of as a mechanism, there is hope of reproducing it by a mechanistic procedure.

A step forward in the imitation of nature was made when engineers ('people of great ingenuity') made not only three-dimensional images of the outer form of their originals, but also managed to 'animate' their productions by providing them with the capacity of moving. The story of Archytas, who made a wooden dove that could fly and come back to the hand from which it set off — a commonplace in Renaissance literature — shows that artificers were haunted by the idea of constructing automata which could have at any rate the movements of natural living beings. Their activities comprised the fabrication of moving human figures, which blew trumpets and danced. Like the engineers of Greek Antiquity, the Arabs and the Byzantines, the engineers of the Renaissance often were the same people as the architects and the painters (Leonardo da Vinci, Michelangelo, Jan van Scorel, Albrecht Dürer).

These mechanicians in a certain sense tried to escape the restrictions put by nature on mankind and even to surpass nature by their clever devices. *Mèchanè* or *machina* often came close to 'trick' and 'machination'. If a genuine imitation was impossible one could at least create the illusion in order to amuse or to baffle the spectators. In the fourth century BC the author of the *Mechanica* of 'Aristotle' pointed out that technology works *against* Nature: 'Miraculous is ... all that comes to pass by Art, for the need of mankind ...; *mèchanè* is that part of the arts that helps the solution of such absurdities ... as lifting great burdens by a small weight.' Even when the engineers acknowledged that what they did was 'rational' (Heron, Philo), they yet underlined that mechanics is popular because it 'brings forth admiration' (Heron), and that therefore the mechanicians were formerly called 'miracle-workers', 'thaumatourgoi' (Pappus, 3rd century). But in spite of the rationalization by the engineers, mechanics remained something unnatural; after all, automata were not animate beings but only imitated (aped) them. This may be one of the reasons why experimentation played a subordinate role in ancient physics: an artificial and mechanical experiment cannot give reliable information about nature as long as nature is conceived of as a living being.

According to the organistic world view of Plato, Aristotle and the Stoics, nature is rational, but this rationality does not work in every respect in the same way as it works in human (mechanical) art. Whereas human art works on material alien to the artificer himself, nature realizes her plans in matter belonging to herself. She plans and *generates*, that is, she continues her own eternal Forms.

Plato personified the coming-to-be of things as the work of a divine Artificer (*Dèmiourgos*); Aristotle compared Nature (*physis*) with an artisan executing his plans. The comparison implied that Nature does not work by chance: if even the human artisan makes his products according to a rational plan, how much more the most perfect artisan — which is Nature — will do so.

The causal relation, however, is the other way round: the rationality of human art consists mainly in that man imitates the art of Nature and, consequently, works also in a rational way. Moreover, the analogy is far from perfect. The potter *fabricates* things not similar to himself; Nature, on the other hand, *generates* things. This generation means that each 'nature', each essence or Form, produces things similar to itself: a horse brings forth a horse and not an ass. Through reason man is able to recognize the rational aspect of natural things, but the generation of individuals that are not similar to himself is impossible to man. He cannot give other things their natural Form; he cannot animate things except beings similar to himself.

3.1. Alchemy, Chemistry

In chapter III on the Philosophers' Stone we have pointed out that Aristotle distinguished true compounds (which have their proper Form and are perfectly homogeneous) from 'compounds to the senses' — mixtures of finely powdered substances of which the parts do not posses a common Form. The true compound is something other than the sum of its parts, whereas in pseudo-compounds the parts become one 'by Chance', not 'by Nature'. The controversy about alchemy (i.e. whether it is a science or a pseudo-science) turned upon the question: can art give the natural Form of gold to something that is not gold? And the answer of the alchemists was in general: we do not pretend to do such a thing, but lead and gold belong to the same species in the way that caterpillar and butterfly belong to the same species, or in the way that a healthy man and a sick man belong to the same species. They held that they did not generate gold, but only perfected, or took away impediments to the development of, a thing that was on its way to becoming perfect gold. Their art, so they said, helps Nature to reach her end, as when artificial heat hatches eggs. This placed their theory (in spite of some un-orthodox conceptions) in the framework of Aristotle's philosophy: 'arts either (on the basis of nature) carry things further than nature could do, or they imitate nature'; 'if then artificial processes are purposeful, so too are natural processes, for the relation of antecedent to consequent is identical in art and in nature.'[56]

The alchemist, then, maintained that, like farmers who further the growth of corn or the hatching of hen's eggs, they only supported nature and so did not work against her. Their opponents accused them of doing the latter: alchemical 'gold' is spurious gold; 'art is the ape of nature'. But both the alchemists and their opponents were agreed that Nature has to do the real work in the generation, the giving of the Form, of gold.

When the mechanistic world view was taking the place of the organistic conception, the problem became different. If gold was a compound it was, at least in principle, possible to compose it out of its constituents, but if it was a simple body one could at best extract it from compound bodies whose constituent it was.

The problem was not, of course, confined to gold: on the organistic world view there was no possibility of making any natural substance artificially. In science, however, it is not only the *a priori* view of the world that plays a role. There is also plain fact, and it seems that the laboratory experience had a considerable share in the overthrowing of the old conceptions. In 1605 Jean Bodin stated that 'electron' (a natural gold-silver alloy) made in the laboratory is wholly identical with the electron found in nature. In 1617 Angelo Sala decomposed the mineral 'copper vitriol' [$CuSO_4 \cdot 5H_2O$] into copper (or 'copper ash', [CuO]), vitriolic acid [SO_3] and water [H_2O], and next synthesized it again from these components.[§] Having proved that artificial vitriol possesses all the properties of the natural one, and being no revolutionary who wanted to attack the prevalent world view, he had to conclude that copper vitriol, the artificial as well as the natural, did not have a Form of its own but was only a 'compound to the senses', a mechanical mixture. It was thus no homogeneous body but 'an apposition of particles' of the three components, a unity, not by Form or nature, but by chance. If this was so for artificial vitriol, it must also be true for natural vitriol, which has precisely the same properties.

There was even an ancient authority to back up such a conclusion. The Greek physician Galen had said that true compounds are homogeneous and are made only by Nature, whereas 'compounds to the senses' are heterogeneous and are made by Nature as well as by Art.

However this may be, it is evident that *chemical practice* led Sala in the direction of corpuscular theory. Remaining within the bounds of Aristotelian orthodoxy, this non-revolutionary man made a revolutionary step. For his practical sense made him realize that 'by chance' was not a wholly correct description of copper vitriol: artificial as well as natural vitriol had a fixed composition: Sala found that one third of its weight was water. He knew of course that a true mechanical mixture (let us say: sand and sulphur) is not bound to any fixed proportion of its components. Moreover, though Sala said that the three components came 'accidentally' together, he also said that this happened through 'divine providence'. On another occasion he said that the 'archeus' (a spiritual principle immanent to nature in Paracelsist philosophy) has the same function in bringing the three components together in nature, as the human design has in bringing them together in the synthesis of artificial vitriol. This indicates that he had an open eye for the special character of vitriol as being a compound of a character between a 'natural unity' and an arbitrary mixture.

On the other hand, Sala held that a (to him!) indecomposable mineral like rock salt must be a truly homogeneous compound. The distinction between true compounds and 'compounds to the senses' thus became dependent on the state of chemical analysis and lost its absolute character. More and more minerals could be decomposed and synthesized again from their component parts, so that more and more minerals could be considered as each an 'apposition of particles' in the sense of the corpuscular theory.[57]

With the triumph of the mechanistic world view the distinction was finally abandoned by Joachim Jungius and, after him, by Robert Boyle: *all* mineral substances, including those which as yet resisted decomposition by art, were considered by analogy to be appositions of particles of their constituents. Thus the

[§] See also page 99.

practice of the chemists furthered a mechanistic interpretation of phenomena and the destruction of the essential contra-distinction of products of art and of nature.

Whereas this destruction of the old School doctrine took place step by step, as practice advanced, there was a more sudden and radical destruction by a 'revolution' in philosophy. As long as it was believed that art can but ape nature or help nature, technology had no place in the philosophy of *nature*. Nor could it have much, even if it might be admitted that it could surpass nature in working against her by magic.

This changed, however, as soon as it was believed that it is possible to dominate nature (and in some respects even to surpass her) by obediently following her. That is: man should not only be a fabricator or imitator, but also God's fellow-worker in 'creating' new things, conducting events without violating the fundamental laws of nature.

In general the 'thaumatourgoi' and mechanicians of Antiquity did not claim to produce 'natural' Forms, let alone *new* 'natural' Forms. A true domination of nature by art could not appear unless art itself created Forms identical with those already existing, or wholly new Forms equivalent to the existing ones. We may think here of chemical products that do not occur in nature, but nevertheless fit in with a series of natural substances (e.g. the hydrocarbons).

> With hindsight we see that such a change became possible when the organistic world view was replaced by a mechanistic view. This change was prepared by the nominalists. They held that the only things that really exist are the individuals (which meant that the Forms lost much of their importance.) Heinrich von Langenstein (1325 - 1397) said that for transformations of metals one ought to know not only the right proportions of the components, but also the configurations of the heat applied (see chapter V), which determine whether a certain mixture will yield gold or some other metal. But as there is an infinite diversity of combinations of qualities and an infinite range for their intensities, this opens the door to possibilities one cannot foresee: 'Man cannot decide by the light of nature alone whether an astonishing fact does or does not come forth from the forces of nature.' Man cannot prescribe the limits of nature and say: 'thus far shalt thou go.'
>
> This philosopher opened new vistas: in his opinion possibilities as yet unrealized are slumbering in nature. So, new species, living as well as non-living, will arise by combinations of natural forces, for more species are possible in nature than are found in the present dispensation; with God more things are possible than He has called into being up till now.[58]
>
> Such an attitude takes away the prejudices against artificial production of new substances, as the Forms evidently do not hover as metaphysical entities over matter, but just are the results of certain proportions of components and configurations of qualities.

All this, however, was of little value for practice, for it is highly improbable that by trial and error one will hit upon the right proportions and configurations of heat necessary for producing a certain substance. Nevertheless there is no absolute impossibility for human art to imitate 'eternal' Forms, though in practice it remains well-nigh impossible to give a wanted Form to matter. In Langenstein's conception experience hits upon surprising results purely by chance. Alchemy remains an empiristic art without plan.

3.2. Bacon's New Approach to Nature

Perhaps no other early 17th century author has proclaimed so clearly that art may be able to emulate nature as Francis Bacon did. He accused the School philosophy of causing an 'artificial despair' by making people believe 'that nothing is to be hoped from human art by which man could subdue and dominate nature.'[59]

Bacon rejects as too narrow Aristotle's idea that art is at best but a servant of nature and that it can accomplish only what nature has begun, or that it only corrects nature when she commits an error, or that it sets her free when she is bound by outward circumstances (as when we weed the fields). He attributes to art a power of changing more radical than that. In his opinion the artificial event does differ essentially from the natural; art can only *direct* natural motions; it can cause combination and separation of natural things.

But this is no 'going against nature', for Bacon rejects the Aristotelian distinction of natural motion and forced motion: every artificial motion is but a directed natural motion. Nor is there an essential difference between the products of art and of nature, art is 'nature under constraint'. In his *New Atlantis* (1624) Bacon sums up what art will be able to do when free from bounds of ancient philosophy. In a famous passage, reminiscent of Heinrich von Langenstein, he predicts that art will make chemical compounds that up till then could be made only by nature; it will grow minerals in artificial mines which will imitate natural ones; transform vegetable and animal species; make new metals with desired properties; make rain and change the climate.[60] (Prophesies now fulfilled, alas not all for the good of mankind!).

And all this would not be *against* nature, but it would remain within the limits put to nature. This dominion would be realized not perchance — as were the great discoveries of the compass, printing and gunpowder – but according to a plan based on scientific knowledge. Knowledge and power would not be aquired by reasoning alone, but by submission to the fundamental laws discovered by experimental research. 'One cannot command nature except by obeying her.'[61] We have to stay within her bounds, so that in some cases we will be able to change things and in others to make better use of them — as the farmer uses the earth and the seasons in a 'right way'.

Bacon could wipe out the essential distinction between the natural and the artificial because to him everything 'mechanical' was also 'physical' ('natural'): 'all that man can do is to put together or to part asunder natural bodies. The rest is done by nature working within.'[62] Consequently there is in his opinion no difference between a 'history' (account) of 'nature wrought, or mechanical' and 'history of nature', nor between natural and artificial compounds, nor between violent and natural motion.[63]

Bacon's submission to the *facts* — whether reason can grasp them or not — had a religious background. And this was not so much inspired by some innate natural religion, as by the revealed religion he believed in. Just like Buridan and Oresme he chose for voluntarism and against intellectualism, that is, for unconditional surrender to the incomprehensible *will* of God. Man has been given dominion over the works of God's hands — 'Thou hast to put all things under his feet' (Ps 8:6) — and therefore there is nothing illegitimate in man's interference with and dominion over the world, but he can only effectuate this by obedience, and should not claim sovereignty of his own reason.

In Bacon's opinion the first Fall consisted in man's losing his innocence by the desire to be like God and to establish for himself what is good and what is evil, instead of humbly obeying the command of God. Yet God let him retain the domination of nature given to him in Eden. But he lost this, too, by a second Fall, when he wanted to decide how nature must be, instead of accepting her as it had pleased God to make her.[64] Current philosophy was, according to Bacon, hindered by two impediments: overestimation of the power of our reason in understanding things, underestimation of our power in making things.

Evidently, Bacon's religious optimism about human domination over nature did not take into account the possibility of a third Fall, *viz.* that human arrogance and greed, after having acquired knowledge in following the right way of submission to the laws that God had put into nature, might use this knowledge to dominate nature not as God's lieutenant, but as a sovereign over nature, so turning the boon of better technology and better medicine into the bane of destruction of fellow creatures (humans as well as animals and plants) and the violation of the earth's beauty and harmony. The third Fall, then, like the first, was to bear the character of moral evil, the *hybris* of Man, 'the great Magician' who thinks that he is 'a mighty God' and that 'all things between the quiet poles will be at his command.'

3.3. Descartes

Among the adherents of the mechanistic conceptions a new dogma now arose, which suppressed *a priori* any essential difference between nature and art. Descartes said that he did 'not recognize any difference between the machines made by artificers and the diverse bodies which nature alone compounds.'[65] In his opinion 'it is certain that all rules of Mechanics belong to Physics ... so that all things that are artificial are at the same time natural'; 'it is no less natural for a watch to indicate the hours than it is for a tree to produce fruits.'

In this way mechanics, an 'illiberal art', was incorporated into physics, the science of nature, for the 'nature' of things (their 'physis') *is* mechanical. If 'nature outside man' and human art follow the same rules, the phenomena of nature can be wholly explained by the phenomena of art (which themselves are products of 'nature'); there is no essential difference between the Art of Nature and the Art of Man. Both are 'arts', both are 'natural'. Up till now mechanics had been at best a branch of 'applied mathematics', whereas physics was a speculative science, which contemplated the essence of things. Henceforth mechanics was to be the most fundamental part of experimental (mechanical) physics, of which it formed the theoretical base and the experimental touchstone.

According to the Ancients mechanical work distracts the mind from superior philosophy of nature; according to Descartes it is precisely by occupying himself with mechanical things that man is made more able to cultivate physics. Thus mechanistic philosophy laid the theoretical foundation of the domination of nature by art in the modern, non-magical sense.

It is true that today scientists no longer consider all physical phenomena as wholly analogous to the mechanisms of the machines we make or the billiard balls we play with. 'Energy' rather than 'matter' has become the fundamental notion, and the Rutherford-Bohr theory does not reduce the phenomena of electricity to material devices but rather reduces units of matter (atoms) to units of electricity (electrons, protons). But in its experimentalism and its mathematical method and in its efforts to apply the same patterns of thought to all things, natural and artificial, it is of the same kind as so-called classical-modern mechanistic science.

3.4. The Synthesis of Minerals

Is it possible artificially to make substances similar to products of nature: non-living products of non-living nature (e.g. ruby); non-living products of living nature (e.g. sugar); and even living beings themselves? In a rigorously mechanistic system all three cases must be, in principle, realizable. Yet, even after the triumph of mechanistic philosophy the tenet that art can emulate 'nature left to herself' has not been held without reservations. For, though this possibility has to be admitted in principle, this need not be so in practice, because the fabrication of the thing or event might be beyond our technological capacity. Moreover, one may recognize the mechanical character of products of living nature, but at the same time maintain that 'life' is in some cases indispensable for their production.

A second problem is whether, if we can make a natural product by means of art, this happens in ways used by nature in its own production. In particular the effort to repeat large scale natural phenomena on a small scale in the laboratory may lead to wrong conclusions. As Aristotle remarked: if twenty men can move a ship at a certain speed, this does not imply that one man can do it at a twentieth of that speed.

The possibility of synthesizing some minerals (electron, cinnabar, vitriol) was, as we saw above, recognized in about 1600. In the 18th century a larger number of substances with the same chemical composition as their counterparts among the natural 'fossilia' was made in the laboratory. The mining engineer J.F. Henckel (1679 - 1783) in his *Pyritologia* (1725) claimed to have done so in the case of cinnabar [HgS] and antimonite [Sb_2S_3], but he acknowledged that he had no success in making artificial pyrite [FeS_2] and copper pyrite [$CuFeS_2$] directly from the metals and sulphur.[66] He found, however, that even in those cases where he was more successful, the artificial compounds in general had an appearance different from that of their natural counterparts, e.g. because they lacked the crystal form. Nevertheless the crystalline product obtained from silver powder and sulphur perfectly imitated the corresponding silver ore. On the other hand, the needles of artificially made antimonite were always shorter than those of the natural mineral; in this case Henckel supposes that if art could use the same time as nature had, the needles would be of the same length.[67]

Why then, he asks himself, do we not always succeed in making artificial substances of the same composition as metal ores, and why — when we are able to do so — have they in most cases an appearance different from their natural counterparts? His answer is that the procedures of the 'laboratory of Nature' are in most cases widely different from those of our laboratories. Following Stahl's theory, Henckel believed that sulphur and the metals are not elementary bodies but secondary or tertiary substances. In nature the sulphureous metal ores are formed directly from vapours of a more primitive character (a lower degree of composition) than metals and sulphur.[68] In general minerals have not been formed directly from the composita of which they now consist.

In some cases the product of the reaction between these mineral vapours will be the same as that of the reaction between metal and sulphur, in other cases it will be different. Moreover the fact that nature has at its disposal a much longer time may lead to difference in outward form. Only in the direct interaction of quicksilver and sulphur 'art most perfectly imitates nature'.[69]

Henckel found that in nature metal-sulphur ores give rise to vitriols (copper- and iron sulphate) that are perfectly identical with artificial vitriols. Evidently the same vitriol-forming 'spirit' is active on the surface as in the interior of the earth.[70] He combats the prejudice of those 'people who believe that mineral substances are always perfect and infinitely better than those made by human labour,' e.g. for medical use: 'there is no difference between the principles composing substances made by Art and those of substances formed by Nature.'[71]

Henckel was a soberminded and practical scholar; he rightly claimed: 'my ideas are based on observation, on probabilities and on facts;'[72] 'one simple, true experience — however "mechanical" — is preferable to all vain speculations and metaphysical subtleties.'[73] He could not make artificial pyrite, but he denied that this would never be possible or that artificial procedures would never correctly imitate the natural ones.

3.5. *Leibniz*

In the *Acta Eruditorum* of 1693 G.W. Leibniz, with his usual optimism, declared that 'the offspring of the laboratories' very often is similar to what nature brought forth subterraneously: 'there often appears a wonderful similarity between things born and things made.' Everyone knows that the mineral cinnabar may also be made from quicksilver and sulphur and that auripigmentum [As_2S_3] can also be made from sulphur and arsenic, and that artificial antimonite has the same composition as the natural product.[74] In blast furnaces at Goslar are found deposits which also occur as zinc ores in nature, where 'the mountains are the alembics and the volcanoes the ovens.'[75] In Leibniz's opinion 'nature is nothing else than large scale art' and artificial products are not always essentially different from natural ones; it does not matter whether the same thing has been made by some fireworking Daedalus in a furnace or is brought to the daylight out of the bowels of the earth.[76]

To Leibniz, then, in principle all minerals could be reproduced in the laboratory, though he did not elaborate upon the problem of whether the procedures were similar in both cases.

3.6. *Artificial Mineral 'Genesis'*

By the beginning of the 19th century it was generally recognized that in principle all mineral substances could be fabricated artificially, because in nature and in the laboratory the same chemical affinities are at work. Experimental mineralogy, however, reaches further than just the fabrication of chemical bodies of the same composition as natural ones. Minerals are also characterized by their crystal form and this may assume various outward appearances within the limits of their symmetry class, depending on difference in size, different combination of faces and degree of development of individual faces, and different occurrence of 'forms' (e.g. whether alum assumes the cubic or the octahedral form or a combination of both.) Whether minerals occur in needles, plates or any other special outward appearance depends upon such factors. Moreover, small additions of alien substances may give rise to varieties, e. g. when corundum [Al_2O_3], by containing traces of certain metals appears as ruby or as sapphire. Different minerals may have the same chemical composition: the crystal symmetry, hardness, specific weight and optical properties of calc spar and arragonite are not the same, though both are calcium carbonate [$CaCO_3$]. Graphite and diamond have little in common as to their properties; the former is black, soft and 'hexagonal', the latter colourless, extremely hard and 'regular': yet, chemically speaking both consist of pure carbon.

The beginning of systematic research on the growth of crystals (1786) was made with soluble salts, mainly sulphates (alum, the vitriols of copper, zinc, cobalt, and 'mixed crystals' of copper and zinc vitriol), when Nicolas Leblanc created the new discipline of 'cristallotechnie'.[77] The rising star in crystallography, R.J. Haüy, warmly recommended these experiments in his report to the *Académie des Sciences*, pointing out that they were not haphazard, but that 'this art of commanding, in a certain sense, the results of the operation' was to contribute to 'the theory of crystallization' (by which he meant his own theory of 1784 of crystal structure), and that the new art was very important for 'chemists and naturalists.'[78]

Neither Leblanc nor Haüy made any distinction between 'natural' and 'artificial' substances, for they considered all these processes of growth as natural, independent of human interference in the circumstances under which they took place. By variation of the circumstances different forms of the same species were formed.[79] Leblanc hoped to find some indication how (free) nature brings forth various forms; in this way his new 'art of cristallotechnie' would be useful to the mineralogist and the naturalist in general.[80] By giving a collection of saline crystals a place in their cabinet, the study of his 'cristallotechnie' — 'this important branch of natural history' — would be furthered.[81] If anything, this shows that for Leblanc there was no borderline between artificial and natural crystals. In his reports on Leblanc's continuing work, Haüy (1792) was of the opinion that Leblanc's collection should be displayed in public for the instruction of students of 'natural history'.[82]

The experiments of Leblanc were restricted to soluble salts and they concentrated on the outward circumstances during the crystallization. In order to synthesize minerals in the way followed by nature, geological observations and hypotheses played a role in planning artificial methods of their production. In particular the volcanoes provided material for the working plans, as their nature is seen clearly at work and, in general, at not too slow a rate.

In 1821 the physicist and chemist Gay-Lussac brought together vapours of iron chloride [$FeCl_3$] and water and obtained crystals of specular iron (*fer oligiste*, [Fe_2O_3] perfectly similar to the natural ones.

Conversely, chemical experiments could provide new geological information. Gay-Lussac and Thenard found that, when heated, neither the mixture of sea salt [$NaCl$] and dry sand [SiO_2], nor sea salt and water produced hydrochloric acid

[HCl], but that it was abundantly developed when water was led over a mixture of sand or clay and sea salt at a high temperature. This proved to them that water and salt must already have been present in the volcanic magma.[83]

In particular the industrial procedures of glass ovens, blast furnaces etc. gave rise to artificial 'minerals' and 'rocks' that strongly resembled counterparts in nature. In such cases there was no deliberate planning, no working hypothesis, behind this formation, so that they could be considered as 'natural', were it not that the circumstances which unintentionally led to their formation were 'artificial'.

In 1829 E. Mitscherlich found in a potter's furnace crystals of iron oxide which were undistinguishable from the natural ones, 'so that we are justified in considering its way of formation as similar to that of volcanic crystals'. In the oven the vessels — whose clay contained iron oxide and siliceous earth — had been glazed with sea salt, while the atmosphere in the oven contained water vapour. Mitscherlich concluded from this that 'the formation of crystallized iron oxide in now active volcanoes or in regions that took their origin from or were altered by volcanoes ... is caused by sea salt and water vapour ... that acted upon siliceous earth or other compounds, and yielded hydrochloric acid which, with or without water, meets iron oxide or iron compounds; this yields iron chloride which again is decomposed by water vapour and — when it proceeds slowly — leaves behind iron oxide in large crystals.'[84] In this way the unintentional formation of iron oxide crystals as a byproduct of a human industrial process pointed back to the 'natural' formation of the mineral as the product of a geological 'fire'.

As the chemist F. Wöhler put it: 'When talking about slags and other products from smelting — what, in a wholly inadequate way, is usually called *art* and *artificial* — is always the same ancient Nature. In these so-called artificial processes the same natural forces and natural laws are at work as were active since the beginning of things (*seit dem Uranfang der Dinge*).'[85]

The number of synthetic minerals increased rapidly, but in most cases there was no complete certainty that nature had followed the same procedure as the laboratory in forming them. In 1872 C.W.C. Fuchs wrote a comprehensive survey of the artificial minerals in which he pointed out that many of them could be synthesized in various ways.[86] The problem was to prove that the conditions of the artificial formation existed when the natural formation took place. The production of artificial minerals may be a physical and chemical problem, but for the imitation of their natural genesis geological knowledge is indispensable.[87] He hoped that by combining laboratory results with geological study of the diverse deposits, we may perhaps find the way in which the phenomena took place in the earth's interior.[88]

3.7. Experiments on Rock Formation

James Hutton's theory, known as 'plutonism', stated that crystallized rocks are not deposited from aqueous solutions (as the 'neptunists' maintained), but are formed from an 'igneous solution', i.e. from a molten mixture of minerals. The cooling of molten rocks in the laboratory, however, always yielded vitreous masses, and Hutton was inclined to attribute this to the fact that in nature much higher pressure prevailed during the process of congelation. His friend James Hall (1762 - 1831), on the other hand, supposed that the slow tempo of geological processes also played a role. This became the more probable to him when in 1790 in a glass-works in Leith a mass of molten bottle-glass that had been cooled very slowly, changed into a stony body which had lost all the specific properties of glass, being opaque and crystalline. When it was melted again, however, and then cooled quickly, it became common glass as before.[89] Hall now concluded by analogy, that even molten granite, if cooled down slowly enough, would not form a 'glass' but a crystallized mass identical with natural granite, and he attributed the crystalline character of lava, basalt, granite, etc. to very slow crystallization from the molten state.[90] Hutton, however, disagreed with him: the fact that his violent opponent Richard Kirwan had disproved by experiments the possibility of congealing molten rocks into a crystallized mass, probably contributed to Hutton's aversion to testing his theory by experiment. He spoke deprecatingly about the superficiality of men 'who judge of the great operations of the mineral kingdom from having kindled a fire and looked into the bottom of a little crucible.'[91] He was of the opinion that the heat to which the minerals had been exposed in nature was 'of such an intensity as to lie far beyond the reach of our imitation', and that the 'operations of nature were performed on so great a scale, compared to that of our experiments, that no inference could properly be drawn from the one to the other.'[92]

It is remarkable that not only the plutonist Hutton but also the neptunist d'Aubuisson de Voisins held that the huge masses and the long time nature has at her disposal make the effects and products of the same causes totally different; moreover, we do not know all the means nature uses for her formative processes.[93]

Hutton evidently was *a priori* afraid that failure of the experiments would damage his theory. Kirwan concluded *a posteriori* from the failure of the experiment that it disproved the theory, while the French geologist held *a priori* that success of the experiment would not be enough to prove the theory.

In spite of Hutton's discouraging attitude Hall started experiments in 1790 on basalt and lava, hoping thereby to be able to confirm the truth of Hutton's theory of the igneous origin of these rocks. In 1798 he succeeded in letting molten basalt congeal into a mass with the same texture as the original rock, whereas rapid cooling yielded a vitreous product.[94]

Already in 1790 Hall had pointed out that lava and basalt showed great resemblance, though they differed in that calc spar (limestone, [$CaCO_3$]) did occur in basalt but not in lava's. Now another friend of Hutton, the chemist and physician Joseph Black, had found that, when heated, limestone is decomposed into carbonic acid gas and quicklime. How then could undecomposed limestone crystallize together with the other ingredients of basalt? This difficulty was met by the crucial tenet of Hutton's theory that neither water nor fire were causes that by themselves would sufficiently explain the changes in the earth's crust, but that an enormous *pressure* was needed besides: the modifications of the effects of heat caused by pressure — which, according to Hall, no geologist except Hutton had taken into account — distinguished Hutton's from all other fire theories.[95] Accordingly, thanks to the high pressure, the limestone contained in the molten magma could remain undecomposed.[96] From this being-undecomposed the opponents of plutonism, however, concluded that the limestone had not been produced from a fused mass. Hall therefore wanted to demonstrate by experiment that limestone, exposed to heat under pressure, might be fused without decomposition and that on cooling it could re-crystallize. But Hutton rejected this proposal because he believed that the immense force of the natural agents could not be imitated in a laboratory.[97] After Hutton's death, however, Hall made his famous 'Experiments on the Effects of Compression in Modifying the Action of Heat'.[98] He showed that calcium carbonate, when strongly heated in a closed gun-barrel, was not decomposed and even re-crystallized, so that it got the consistency of marble. Moreover, he found that in similar circumstances coal partially decomposed into 'blind coal' which burned without flame.

He was of the opinion that he had succeeded in reproducing 'the perfect representation of the natural fact': the production of sandstone from sand, of limestone from shells, and of coal from animal and vegetable substances.[99]

Hall's experiments did not amount to an artificial synthesis of minerals and rocks; he had demonstrated only that molten basalt could crystallize directly and need not to take its origin from an aqueous solution, and that limestone could have an igneous origin, because of its not-being-decomposed under high compression. It was the plutonist theory which he supported, and by doing so he showed more confidence in the capability of art's emulating nature than the creator of this theory possessed.

Hall's publications had a large influence. In Britain Gregory Watt (1804) made his experiments on a larger scale than Hall. He described the successive stages during 8 days' cooling and saw that even when the mass had become apparently solid there was still an inner change of structure going on.[100]

In France Dartigues (1804) and Fleuriau de Bellevue (1805) were stimulated by Hall's example. It is significant that both of them had close connections with glass factories. Dartigues pointed out that devitrified glass had much in common with lava. He expected that this 'might provide the means for explaining many geological phenomena about which there was difference of opinion up till now, because nothing could make it believable that rocks had been glass.'[101]

Fleuriau de Bellevue remarked that if the procedures and products of the (industrial) arts, like those of iron foundries, do not give sufficient material for comparison with the volcanic products and the circumstances of their formation, one should question nature by *deliberate* experiments.[102] He attributed the dissensions among geologists to the fact that 'very few naturalists of those who have seen active volcanoes were within the reach of foundries, where they would have been able to make the observations and experiments necessary for solving these difficulties.'[103] He demonstrated the identity of the events occurring in the fires of the furnaces and in those of the volcanoes and he pointed out that not only the chemical and physical properties of the products of both categories should be compared but also their structures, i.e. the relations of the diverse crystals of which the artificial and the natural rocks consist.[104] He recognized that the artificial products of the furnaces were 'essentially analogous' and only slightly different from volcanic lava's, the difference arising because there had been no compression in the crucible, and, consequently, no resistance to the escape of gases and volatile salts.[105] He concluded that our furnaces produce not only simple or apparently homogeneous rocks [i.e. minerals], but also rocks analogous to composite natural rocks, and that 'the mode of action of subterraneous fire cannot differ from that of furnaces.'[106] In his opinion one of the reasons that these facts had been overlooked was that the immense flames and torrents of fire emanating from volcanoes made such an impression that it was difficult to recognize that they were but laboratories greater than ours: the mind opposes itself against identifications that 'destroy the marvellous'.[107]

In general, those who held with Fleuriau de Bellevue that mineralogy, chemistry and geology should cooperate for solving the problems of the origin of rocks, liked to compare industrial foundries with volcanoes. The famous geologist J.B.L. Hausmann believed that the

> investigation of the history of our terrestrial globe is possible only by inference from changes that occur before our eyes to those which happened to the earth in the time when no human observation saw them ... The large metallurgical operations may not lie in the realm of phenomena of nature, but in them the same natural forces are working which — without being used by human art for certain purposes — earlier or later worked on our earth constructively or destructively.[108]

Shortly afterwards (1858) K.C. von Leonhard wrote his classic on 'Furnace Products and other Artificially Produced Minerals as Supports of Geological

Hypotheses.'[109] The title expresses what the great founder of experimental geology, James Hall, had in mind when (against Hutton's will) he supported the 'plutonic' (igneous) theory by his experiments. The motto of von Leonhard's work was a long quotation from Fleuriau de Bellevue which stressed the necessity of comparing the products of volcanoes with those of the fires by which man changes minerals, and, if there is no congruity between them, of making planned experiments. The book described a multitude of artificial minerals that were similar to natural ones and also many without natural counterparts. The author had no doubt, however, that sooner or later these 'artificial minerals' would be found in nature.[110]

Von Leonhard was encouraged by his 'ingenious friend', the chemist and geologist Th. Scheerer, who wrote to him that in general 'our smelting furnaces are little volcanoes and, conversely, our volcanoes are big smelting furnaces.' Scheerer added, however, that for the 'more ancient' crystalline rocks (like granite and gneiss) 'volcanism' was insufficient. The analogy to furnace products would be much weaker in the case of such rocks, for they were not products of the 'fire' alone, for water and high pressure had also been active in their formation in nature.[111]

After a period of ultra-Huttonism during the first three decades of the 19th century, the tides were turning and it became widely accepted that not only heat, but also high pressure and the action of water had played an important role in the formation of many rocks, and in the secondary changes (metamorphism) they underwent. Scheerer had developed this theory (1846) in what has been called an 'epochemachende Abhandlung', and indeed a new epoch now began, as A. Breithaupt in Germany (1849) and Élie de Beaumont in France (1847) accepted this 'hydrato-pyrogenous' theory, which was neither purely 'plutonic' nor purely 'neptunic'.

This theory needed experimental confirmation, a task which G.A. Daubrée (1814 - 1896) set himself in his *Studies and Synthetic Experiments on Metamorphosis and on the Formation of Crystallized Rocks*.[112] In this prize-essay he not only gave a survey of earlier experiments 'on mineral and rock synthesis, but also what the *Académie des Sciences* had asked 'above all', a description of experiments 'for verification and extension of the theory of metamorphic phenomena.'[113] Heat, water and high pressure are said to be the cause of most metamorphic phenomena. Daubrée heated minerals strongly in sealed tubes, so that the enclosed water which was developed reached an exceedingly high pressure.[114] He claimed to have made the component minerals of granite, but not granite itself. He was, however, optimistic enough to predict that this would happen soon. As granite, in contrast to e.g. the minerals formed in thermal springs, is not generated before our eyes in nature, the only way to find out how it arises in nature is in his opinion the experimental one.[115]

Daubrée believed that a new period in geological science had begun, a change like that inaugurated by Galileo in physics, thanks to the chemical, physical and mechanical experiments illuminating the earth sciences.

Experimental geology developed rapidly, in particular through F. Fouqué and Michel-Lévy (1882). In France since the time of Haüy there had been a close relation between geometrically-descriptive mineralogy and chemistry, and this cooperation was now continued in the genetic approach by the 'triple alliance' of crystallography (mineralogy), chemistry and geology. [116] The synthesis of rocks was considerably refined when it was demanded that artificially produced rocks should contain not only the same minerals in the same proportion as their natural counterparts, but that their crystals should also have the same size, habitus (i.e. external appearance) and mutual arrangement, and that even alien inclusions should not be neglected. For checking these aspects micrographic mineralogy rendered great services (H.C. Sorby 1863; F. Zirkel; W. Vogelsang 1867). In 1877 Fouqué and Michel-Lévy, by heating in a platinum crucible a glass that had the stoechiometric composition of basalt, managed to obtain a mass containing the component mineral crystals that agreed with a natural basalt of Auvergne; 'thus our experiment definitively resolves the question of the origin of basalts: they are rocks of a purely igneous origin.'[117] The satisfaction is understandable; yet, if there ever was a petrological problem that had found, already much earlier, its definitive solution by purely geological means, it was the discovery of the volcanic origin of basalt by Desmarest on a journey in Auvergne in 1763.[118]

3.8. Synthesis of Organic Compounds

In 1800 the French chemist Fourcroy stated that chemistry was *as yet* unable to synthesize vegetable matter, though one might hope that some day this would happen.[119] Such a hope perhaps came to him because he considered plants as 'machines organiques'. He did not speak of 'vital forces': the fundamental processes in animals and plants depend wholly on the chemical forces between the molecules. L.J. Thenard was less hopeful that organic substances would ever be synthesized. With him, too, this was not a matter of principle: the fact that carbon is a solid body and hydrogen and oxygen are always gases would make it necessary to work at high temperatures, which would immediately destroy the organic compound again.[120]

The vast majority of scientists at that time, however, believed that the laws of organic nature differed essentially from those of inorganic nature because 'vital forces' are indispensable for producing substances that do not occur in inorganic nature.[121] Even Friedrich Wöhler, when in 1828 he had synthesized urea, hardly dared claim to have performed a purely artificial synthesis, for it could be said against him that the vital force had been hidden in the animal coal he had used to make the cyanic acid that with ammonia was needed for the formation of the urea.[122] Ten years later, however, Wöhler and Liebig believed that the future 'production of all organic substances ... in our laboratories ... should be considered not only probable, but perfectly certain.'

The famous chemist Charles Gerhardt underwent the same change of mind. In 1844 he held that a 'vital force' builds up organic compounds, whereas 'chemical forces' destroy them (a doctrine also propounded by Lamarck.)[123] In 1853, however, he declared that organic chemistry is the 'knowledge of the means appropriate for the construction of organic substances apart from the functioning of living things.' The only restriction he now admitted was that one cannot synthesize organic compounds *in the same way* as living organisms do.[124] Organic and mineral substances 'are subject to the same laws of attraction and decomposition, which are wholly independent of the phenomena of life.'[125] Organic chemistry teaches us 'the means to imitate or rather to reproduce in its retorts and crucibles the compounds made by living nature, instead of extracting them from animal or vegetable parts in which they are completely formed.'[126] 'Natural compounds and artificial products of our laboratories are links of the same chain, which the same laws keep riveted together.'[127]

'Organic chemistry' thus had become 'carbon chemistry', a science which synthesized not only many organic compounds, but also a multitude of carbon compounds that are analogous to them (and belong to the same 'series') but do not necessarily occur at all in nature.

It should be emphasized that the disappearance of the barrier between *products* of inorganic nature and those of living beings and between imitable and non-imitable chemical substances in general, in no way implied any necessity to deny that there is an essential difference between living beings and non-living things.

The rapid development of chemistry in the early decades of the 19th century had considerably heightened the self-confidence of its adepts.[128] A great number of natural compounds and an even greater number of new compounds never met in nature, had been produced by the art of chemistry. As J.F. Daniell (1843) put it: 'the forces of nature, indeed, are the powerful, but submissive, servants of Man; and through their agency he has been endowed with a subordinate power of creation, but their respective actions have been limited by laws which cannot be transgressed, and in a knowledge of these limits consists the secret of their application.'[129] In these words we hear an echo of Bacon's verdict that Man, God's fellow creator, may command nature by obeying the laws laid down in her.

The imitation of nature was not, of course, the main object of chemists. The synthesis of the natural dyes of madder (alizarine) and woad (indigo), were great commercial successes. Of fundamental importance, however, was the fact that the theory of the structure of carbon compounds became the basis of the synthetic methods by which compounds were built as one builds a house, that is, according to a premeditated plan, restricted only by the properties of the building materials.

The great successes of chemical synthesis led to very bold assertions. In 1860 Marcellin Berthelot highly praised the 'creative faculty' of chemistry: it creates phenomena by making artificial beings that are the images of abstract laws brought into reality.[130] 'We may lay claim without transgressing the limits of legitimate hope ... to make anew all substances that have developed since the beginning of things, to make them under the same conditions, according to the same laws, by the same forces as nature causes to work together for their formation.'[131] Chemistry is, in his opinion, an 'art' creating its objects, in contrast to the 'natural' and 'historical' sciences which are bound to already existing things.[132]

There was some truth in Berthelot's bold statements. The modern chemist who makes dyes of higher quality than the natural ones, metal alloys with properties desired beforehand, compounds with structures planned on paper, sometimes has the feeling of power such as experienced by some of the ancient alchemists and adepts of natural magic. With them Berthelot shared a belief in the omnipotence of human art.

4. MODELS

We have dealt with several cases of art imitating, or rather emulating Nature by bringing forth products not distinguishable from their natural counterparts, and we have even met with cases for which it is claimed that Art does the bringing forth in the same way as Nature, as in the artificial production of minerals in the 'natural' way.

The problem may be given a wider scope if we let it embrace not only the artificial production of natural substances but also the artificial imitation by means of models. These models may imitate both natural structure (planetary systems, atoms and molecules in a crystal; atoms in a molecule) and the motions within such structures (e.g. a planetarium). The ideal would be an artificial construction made of the same materials arranged in the same way, performing similar movements as the original, albeit on a different (enlarged or diminished) scale.

In fact this ideal cannot be realized: differences in size, proportions and materials are inevitable, so that the imitation is imperfect. Even if the models were made of the same materials as the original and the geometrical proportions were the same, this would be no guarantee that all other properties would be faithfully reproduced. One has to be satisfied with a model imitating or possessing *some* of the properties of the original but not all of them.

The strength of smaller models of mechanical contrivances (such as a crane) will not be the same as that of the original, though the cohesive forces are the same. Galileo pointed this out in his *Discorsi* (1638):

> Revoke the opinion that Machines, and Forms composed of the same Matter, with punctual observation of the self-same proportions between their parts, ought to be equally, or to say better, proportionately disposed to Resist ... For it may be geometrically demonstrated, that the greater are always in proportion less able to resist than the less, so that *in fine* there is not only in all Machines and Fabricks Artificial, but Natural also, a term ... beyond which neither Art nor Nature may pass ... always observing the same proportions and the identity of the matter.[133]

A model of the solar system inevitably is on a greatly reduced scale: it may illustrate the possible positions and motions of the planets, but the natural motive forces have to be replaced by mechanical contrivance and the forces that hold the natural system together have to be replaced by rigid connecting rods.

Similary, the model of a crystal lattice presents, on an enormously enlarged scale, the arrangements of atoms, ions or molecules, but it does not purport to be an imitation of the materials and of the forces that hold the structure together.

Such models give a geometrical or kinematic picture, but do not claim even in an analogical way to represent dynamical causes, and therefore it is irrelevant whether the 'building stones' are made of metal, wood or some plastic material. Such models serve a didactic purpose, while explaining or illustrating some of the properties of the orginals without claiming identity. In general there is only some *analogy* between the model and the original.

An analogy in the original sense of the word is an equality of two ratios, i.e. a proportion: 2 : 3 = 4 : 6. In this case the four terms are of the same kind and there is indeed equality between the two relations. The only thing compared is the numerical value of each term. In the establishing of physical laws analogical reasoning plays an important role. Now if even in mechanics the analogy (between a larger machine and its smaller imitation) in the strict sense breaks down, this will be the more so in the case of models of more complicated (e.g. chemical and geological) phenomena, when only a few aspects of the originals are mirrored by the model. For which aspects may resemblance be claimed and how far does this allow us to expect resemblance of other aspects of the things compared? Aristotle was of the opinion that the lungs of a mammal have the same relation to the air as the gills of a fish have to the water (lungs : air = gills : water).

This analogy seems correct to us because we read it so that the organs represented by the first and the third term respectively, absorb the free oxygen that is contained respectively in the bodies indicated by the second and the fourth terms.

To Aristotle himself, however, the water had the function of cooling the gills, so that he concluded that the air served to cool the lungs.

The truth of an analogical reasoning depends in the first place on the truth of the relation from which we start and the properties chosen for comparison. In Aristotle's case this was the seemingly self-evident 'truth' that water, whose main quality is coldness, cools the gills. In modern chemistry the fundamental truth is the absorption of the oxygen — respectively present in the atmosphere and in watery solution — by the blood respectively in the lungs and in the gills.

The analogy between a human arm and a lever does not go farther than their mechanical functions; it cannot be extended to their inner structure. In natural science analogy is founded upon resemblance, not upon identity. The resemblance may be rather superficial and the choice of the things compared an unfortunate one. The risk of making serious mistakes in 'demonstration' by means of analogical reasoning is great. When the alchemists observed the loss of lustre and ductility of metals heated in the air, this struck them as resembling the death of organic beings. They concluded that 'calx of iron' : iron = 'ashes of a plant' : the living plant. Next the analogy was reversed: as it is possible to 'reduce to life' the dead metal (*reductio ad vitam*), it must be possible to revive plants from their ashes. In this way 'palingenesia', one of the pseudo-sciences of the 16th and 17th centuries, took rise.

Lavoisier made use of analogical reasoning when he inferred that the same relation as that existing between sulphur, carbon, phosphorus and nitrogen and the acids corresponding with them (sulphuric, carbonic, phosphoric and nitric acid) must exist between an as yet unknown element, 'murium', and 'muriatic acid' [HCl; hydrochloric acid]. As the acids of sulphur, etc. were the oxides of these elements, the 'muriatic acid', too, must be an oxide and thus contain oxygen.

The conclusion was wrong, but by a similar reasoning Lavoisier rightly concluded that, as the action of acids on metallic oxides yields 'salts', and the action of acids on lime [CaO], too, yields salts, lime must be the oxide of an as yet unknown metal.[134]

An analogical reasoning may lead to absurdities, to errors, and to important truths; it is left to the intuition of the scientist and, finally, to the test of observation and experiment whether its conclusions are right or wrong.

Analogies become the more risky the more different the phenomena that are compared: arithemical analogy is perfect; the analogy between phenomena (e.g. a current of water and the electrical current) may be fruitful to a certain limit; an analogy between physical and biological or between biological and sociological phenoma should be treated with utmost circumspection.

4.1. Cosmographical Models

Cosmographical models serve in general to demonstrate the positions and motions of the original.

In Antiquity Archimedes was credited with the invention of a planetarium, a mechanical model of the system of homocentric planetary spheres. The later system of Ptolemy, in which the spheres of the epicycles penetrated those of the deferents was difficult to imitate in a three-dimensional model. This did not disconcert Ptolemy, for there was a more profound reason for this impossibility than merely technical difficulties: the heavens consist of a heavenly, incorruptible matter that cannot be imitated by a model made of terrestrial, corruptible, elements. 'It is not proper to apply human things to divine things, nor to get beliefs concerning such great things from such dissimilar examples. For what differs more from the things that always are the same than those that never are the same, and what differs more from those which are hindered by all kinds of things than those which are not impeded, not even by themselves?'[135]

The only way out was to depict only the mathemical relations of the heavenly motions, with no greater claim to physical reality than a painting has to a natural thing that it represents.

Efforts to make three-dimensional models of the Copernican universe were quite frequent in the 17th century (e.g. those of Jost Bürgi and Kepler). In Leiden University Willebrord Snel demonstrated the motion of the earth round the sun (a notion he himself rejected) by a model moved by hand. In Groningen University Nicholas Mulerius (also not a believer in the Copernican system, though he was the editor of the third edition of Copernicus' work) had a model made of the Copernican system for demonstration to his students. Evidently he had the general public also in mind, for he advised hiding the driving clockwork in the base of the apparatus, so that the astonishment of the onlookers would be greater.

The famous cartographer Willem Jansz. Blaeu fabricated (and offered for sale) planetaria according to the Copernican system (see figure 40). In a book written in the vernacular he described its use and meaning. The humanist Caspar Barlaeus wrote a laudatory poem for the Latin edition in which he praised Ptolemy for drawing the world as God *made* it, and Blaeu for imitating it as Copernicus had *excogitated* it. It seems that Blaeu (who must have known the impish character of his auditor) had no objection to asking him to provide this preface to a system the latter did not like.[136]

Models like those of Blaeu were indispensable for didactic purposes, though they imitated only the geometrical aspects of the original. The same is the case with static models of atomic arrangements by Dalton of circa 1807 (see figure 51 on page 325), and the stereo-chemical models of Van 't Hoff (1875, see figure 41).

Fig. 40. Tellurium constructed by Willem Jansz. Blaeu. (Courtesy Netherlands Maritime Museum, Amsterdam).

These models claim to elucidate at least the relative positions of the invisible atoms. As to the material of the original, this bears no analogy to that of the model, for the matter of the atom is the theoretical basis of the materials we use in making the models, and we cannot make something similar to 'atomic matter' in materials which we can see and touch. It must be recognized, however, that to Dalton (and perhaps to later chemists as well) a sulphur-atom was just a minute piece of brimstone.

A closer analogy between the model and the original is found in the works of Petrus Peregrinus (1269). He considered the outer sphere of heaven as a huge magnet that turned round in 24 hours. He held that an artificial spherical loadstone, directed with its pole towards the celestial north pole, would share the heaven's rotation around its axis. There must be a sympathetic influence of the macrocosm (the universe), itself a large magnet, on the spherical loadstone, a small magnet. Mechanistic philosophy may be more favourable to experimentation, but evidently an organistic conception is not wholly incompatible with it.

Fig. 41. Paper stereochemical models of molecules with carbon atoms made by J.H. van 't Hoff (1875). (Courtesy Museum Boerhaave Leiden).

A similar model was devised by William Gilbert (1600). He made a spherical lodestone, this being in his opinion the most perfect form, and moreover, most like a spherical earth in shape.[137] Such a magnet was to him in the strictest sense a model of the earth: not an analogical imitation possessing *some* of the earth's properties, but a small-scale earth, which he called a 'terrella' or 'micro-earth' (*mikro-gè seu terrella*)[138] (see figure 42).

In the first place it has the earth's magnetism, for the Earth is, Gilbert recognized, itself a huge magnet that turns round on its axis with magnetic poles coinciding (in his opinion) with the geographical poles.[139] Gilbert rejected Peregrinus' opinion that the outer heaven is a rotating magnet,[140] and, in opposition to Aristotle, he maintained that the earth must be animated, as it could not be inferior to the animate beings, like worms and ants, which it brings forth out of its womb.

In the second place Gilbert's terrella is said to consist of the same materials as the magnet Earth. This again is an essential difference from Peregrinus' conception, in which the *terrestrial* matter of the spherical magnet is totally different from the *celestial* matter of the rotating heaven.

Yet Gilbert was also familiar with models in which the analogue showed a greater difference from the original. He conceived the character of (static) electricity as totally different from that of magnetism: the former has a material efficient cause (*effluvia)*, whereas the latter is caused by the 'Form'.[141] Electrical

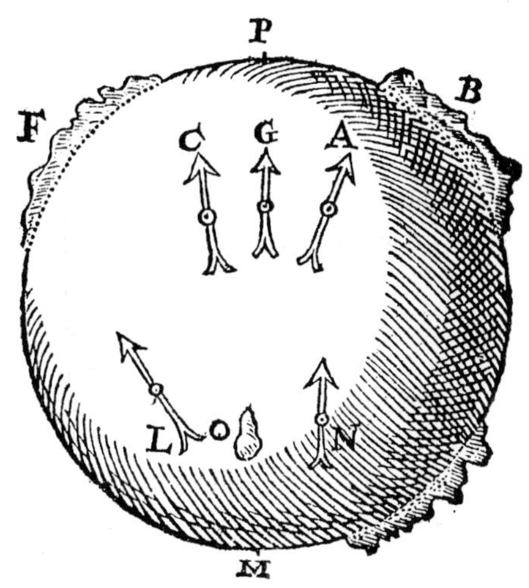

Fig. 42. Gilbert's *terrella*. (Gilbert, *De Magnete*, lib.IV, p.57)

attraction he considered to be analogous to that between a wet and a dry rod or between water bubbles (see figure 43). In electrical attraction a specific fluidum is involved, while in the analogous models of it (wet and dry rods; water bubbles) water vapour is held to play a similar role.

The perceived similarity of the terrella and the earth's globe prompted parallel experiments on them. The 'variation' (i.e. declination) of the magnetic needle is explained by the fact that the earth is not a perfect globe: large land masses (i.e. an excess of magnetic matter) cause the deviation from the 'true north'. In order to demonstrate this by a model Gilbert made a large terrella with elevated and concave parts and he found that small needles placed on the terrella did not point to the pole (i.e. did not follow the direction of the meridian), but deviated from it in the direction of the projection on the surface. In the same way the compass needle on the Earth deviated from the meridian in the direction of a large continent. Similar phenomena thus were expected to be observable on the terrella and on the Earth if the magnetic force is not evenly spread over its surface.[142]

Peregrinus had claimed that his globular magnet would turn round in 24 hours, (though he did not say that he had seen this happen). Gilbert rejected such a claim precisely because he had not *observed* it to be true.[143] Why then this lack of similarity between the terrella and the Earth? Not because the properties were not the same, but because the *circumstances* were different. They impede the rotation in the case of the terrella: the loadstone is heavy with regard to the main body of the Earth in which there is an electrical attraction and, moreover, the Earth is moved by itself and every part of it already shares in this movement. Similarly the

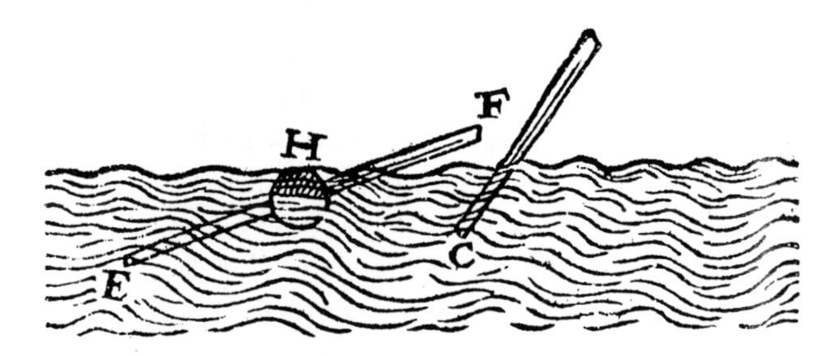

Fig. 43. Gilbert's analogy between (static) electricity and the attraction between a wet and a dry rod (Gilbert, *De Magnete*, lib.II, p.57).

daily rotation of the whole earth is the cause of the maintenance of life on its surface, which is not the case with any part of it, like the terrella.[144]

Whether right or wrong in fact, it seems that Gilbert's use of analogical reasoning and of models for support of his theory bears a truly scientific character.

Whereas Gilbert's model claimed to be an almost identical artificial simulation of the original, the model of the tidal movements planned by Isaac Beeckman had a weaker analogy. Galileo had considered tidal movement as one of the three main proofs of the daily rotation of the earth. In his opinion the periodic alternation of ebb and high tide must ensue from a combination of the daily and yearly motions of the earth. He said that he had made a model to demonstrate this, but he failed to give its description (see page 210).

Dr. Beeckman, to whom a manuscript of Galileo's *De Fluxu et Refluxu Maris* had been shown in 1631,[145] proposed to test Galileo's theory by a model. A glass sphere, in the centre of which is placed a strong spherical magnet, is covered on its surface by filings ('vijlsel') of iron or magnetite which represented the ocean. If the glass globe is given movements similar to those of the earth, the filings will 'doubtlessly' represent the motions of the sea (and those of the winds), if Galileo's theory is true.[146] The plan was not realized, but the idea of it shows again that models were devised not only for didactic purposes but also to test a theory about natural phenomena.

4.2. Geological Models

One of the oldest examples of model-making in geology is presented by the 13th century philosopher Albertus Magnus, a famous commentator on Aristotle's works. The Greek philosopher had explained earthquakes and volcanic eruptions by the pressure of air accumulated in subterraneous caves and finally bursting forth. It might be that heat was developed and an eruption of air and fire ensued.[147]

Albertus (circa 1193 - 1280) demonstrated such an eruption by a spherical copper vessel provided with legs; on the top and at the bottom are holes closed by wooden stoppers. The vessel is full of water and placed in the fire. The explosion then generated will push out the upper stopper and drive hot water out through the upper opening. If the lower stopper is pushed out, water falls into the fire and the impetus of the vapour will throw coal and ashes round the fire.[148]

In this case the whole arrangement is more 'artificial' or 'unnatural' than in the experiment performed almost five centuries later by the French apothecary Nicolas Lémery. His model of volcanic action was based on a *chemical* theory; it was supposed to be the consequence of the combustion of brimstone in the earth's interior. He mixed iron filings with sulphur powder, moistened the mixture with water, and put it in a long and narrow pot, which was set in the ground and covered with earth. A spontaneous reaction [*viz.* the formation of iron sulphide] then yields much heat: 'after 8 or 9 hours I saw that the earth expanded, grew hot, and burst open; next, hot sulphureous vapours emerged and then flames which enlarged the opening and spread around that place a yellow and black powder.'[149] Lémery claimed to have demonstrated by his experiment the causes of the eruptions of Mt Vesuvius and Mt Etna.

4.3. *Mechanical Geological Experiments*

In 1781 the *Observations sur la Physique*, edited by Rozier and Mongez, published extracts from letters of Dr Paccard on the various directions of geological layers and 'the way of artificially imitating ore deposits.'[150] The article of this 'neptunist' is vague and it contains no data of lasting value. Yet, Paccard is worth mentioning as being one of those who performed geological experiments before James Hall (1790). Paccard deliberately connected analogical reasoning and model-making: 'walking along the path of analogy, let us try to make Nature produce synthetically on a small scale what she produces on a large scale in the vast bottom of the seas ...'[151] He let suspensions in water of five sorts of earth (taken from different strata) subside and he found that they were deposed in layers according to their specific weight. By adding copper sulphate to the water he obtained black mineral veins resembling natural lodes with their ramifications.[152] He concluded that 'nature has unveiled to me quite a few beautiful things about the formation of ore deposits, because of the great analogy between them and the products of my experiments.'[153]

Mechanical experiments in geology, like the imitation of the folding and compression of strata, do not yield a fully reliable picture of the phenomena they are said to reproduce. In general the fact that nature works on a much larger scale causes greater discrepancies between the original and the model than in the case of chemical experiments. Moreover, for mechanical experiments one has to use in many cases materials other than the natural sedimentary rocks. In experiments on the folding of strata, layers of clay or even metal sheets have been used. James Hall, in his experiments on folding, used layers of cloth and clay which were under pressure vertically by a heavy weight and at the same time submitted to lateral pressure. In such experiments a small model does not give a reliable picture of the original, because the relation of the pressure applied by art to the resistance against bending and breaking of the materials used by art, will not be the same as that of the natural counterparts.[154] Linen and wool may to a certain extent demonstrate the formation of folding layers in the earth's crust, but they cannot break. The use of materials different from the natural ones, implies already a deviation from the natural process.

Moreover, apart from exogenous phenomena (erosion, sedimentation), the geological events have to be 'illustrated' by hypothetical reconstruction of events that took place, mainly in the past (mountain-building) or in the earth's interior (formation of magma). This means that in general the geological experiment is a process with causes and effects analogous to *supposed* natural processes. Hall's folding experiments demonstrated Hutton's theory that pressure on the various layers caused their folding. Alphonse Favre (1878), however, followed Elie de Beaumont's contraction hypothesis (a shrivelling of the earth's crust was said to be caused by the cooling down of its kernel). Consequently, he used layers of clay on stretched sheets of rubber and did not use any vertical pressure. Decrease of tension led to folding and inflection of the strata which showed some resemblance to that of the Jura.[155]

A. Daubrée, one of the foremost geological experimenters, held that the imitation of mechanical geological phenomena (like the folding and breaking of the earth's crust) had less value than that of geological phenomena of a chemical and physical nature.[156] Following up Hall's experiments on folding, he did not claim to have arrived at rigorous demonstrations. Instead of Hall's layers of clay or felt, he used metals, and also wax mixed with gypsum, resin or turpentine, on which he exerted horizontal or vertical pressure and also unevenly distributed vertical pressure, which led to asymmetrical foldings (see figure 44).[157]

However, the difference between the compacted strata, their huge dimensions and the long time of compression in nature on the one hand, and the sheets of lead, etc. in the laboratory experiments, on the other hand, was so great that many geologists attributed little value to such experiments. In his comprehensive survey of the geological sciences, their methods, results and problems (1913), L. de Launay spoke of 'the coarse, sometimes almost caricatural, representation of the phenomena', which, in his opinion, does not really prove the hypotheses in the manner of a good physical experiment. Only the chemical synthesis of minerals and rocks he considered to be a true imitation of natural events, but in general the demonstrations by experimental geology should be treated with great scepticism.[158]

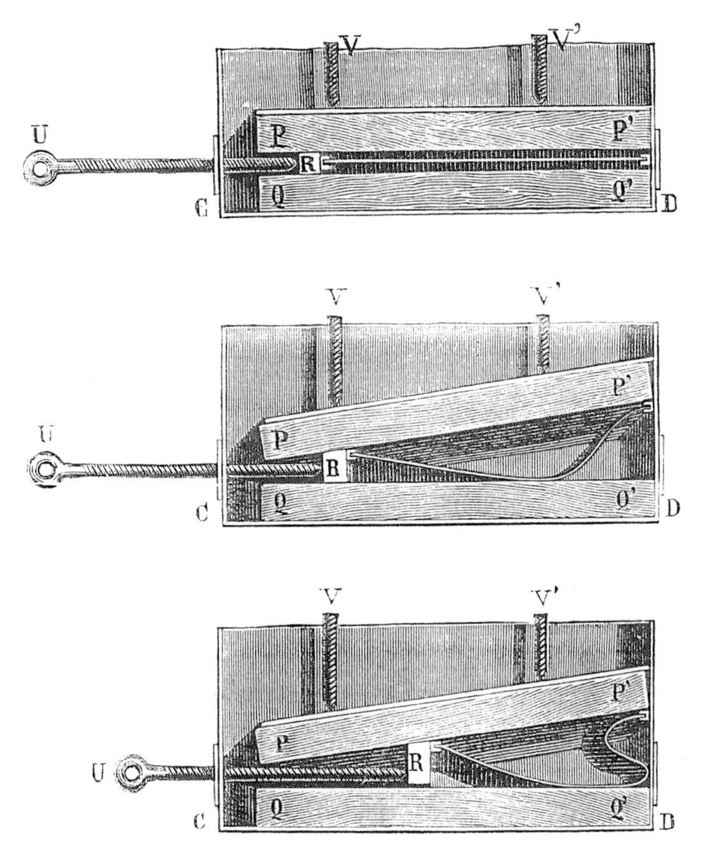

Fig. 44. Daubrée's models of mechanical geological phenomena (A. Daubrée, *Études Synthétiques de Géologie Expérimentale.* Paris 1879, ch.IV, fig.78-80 (p.294)).

The Scottish physicist James David Forbes,[159] on his numerous geological travels investigated glaciers in the Alps and Norway, and explained the observed phenomena by his 'viscous theory of glaciers'. 'A glacier is an imperfect fluid, or a viscous body, which is urged down slopes of a certain inclination by the mutual pressure of its parts.'[160] In support of this theory he tried to show that 'the obscure relations of the parts of a semifluid or viscous mass in motion (such as I have attempted to prove that the glaciers may be compared to) may be illustrated by experiment.'[161] A mixture of plaster of Paris with glue glided by its own weight down a narrow channel: 'the relative velocities of the top and bottom, the sides and centre of such a pasty mass were displayed by the alternating layers of two coloured pastes, which were successively poured in at the head of the model valleys.'[162] (See figure 45).

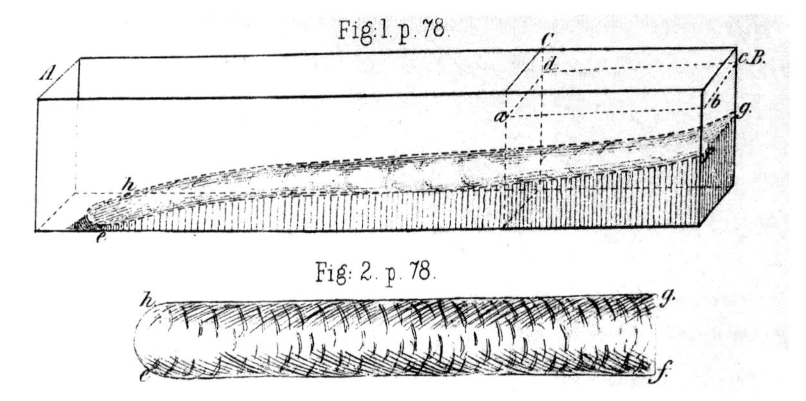

Fig. 45. J.D. Forbes, 'Experiments on the Flow of Plastic Bodies and Observations on the Phenomena of Lava Streams', in: *Philosophical Transactions* 1846; Figures from the reprint in: J.D. Forbes, *Occasional Papers*, Edinburgh 1859, Plate I, figs.1, 2.

Forbes was fully aware of the analogous character of artificial models as 'illustrating' natural phenomena, as well as of the fact that a certain natural phenomenon may serve as a model for another. In the paper quoted above the first paragraph bore the title 'Plastic Models', and the second 'Analogy of Glaciers to Lava Streams'.[163]

4.4. Artificial Production of New Animal Species

Leaving aside the rather speculative topic of transformation of inanimate into living matter, we now turn our attention to the subject of the transformation of organic species, which Bacon expected to be realized in the near future. The breeders of plants, fancy pigeons and cattle had already made small steps in this direction, before more ambitious schemes arose under the influence of theories that broke away from the dogma of the fixity of species.

The French zoologist Étienne Geoffroy St Hilaire (1772 - 1844) combined a mechanistic conception of life and functions of animals with 'idealistic morphology' (1825; 1828; 1831). All animal forms are considered to be variations of one simple archetype. Within this conception a transformation of species was considered possible in the way that transmutation of metals was deemed possible if one considered all metals as variations on one perfect metal. Moreover Geoffroy held that in the past there had been natural animal transmutations by leaps (saltatory evolution). The essential change then was regarded not so much as a change of structure of the animal body as a change in functions of its organs. So he held that lower oviparous vertebrates (reptiles) could become birds at one leap by a sudden transfomation (in the embryonic state) of the lungs under influence of a sudden change in the environment (in particular the composition of the atmosphere and the temperature.)[164] As a *natural* model (analogue) for this process of transformation he adduced the change — even into a higher class — of a 'fish', named tadpole, into a 'reptile', named frog.[165] As the saltation in nature was attributed not to an inner cause, but to a 'mechanical' outward cause and as it was thought to be a sudden event, it was natural to seek for an experimental model. He tried indeed to breed 'viable monstrosities' with the help of various 'external modifications' (of temperature, humidity, composition of the 'atmosphere') during the artificial incubation of eggs.[166]

In Geoffroy's opinion the monsters one could breed in this manner were as 'natural' and as much in the design of Providence as the existing 'normal' animals; the one difference being that they cannot live in so-called normal circumstances.[167] Small wonder, then, that Geoffroy St Hilaire, the great protagonist of idealistic morphology, was at the same time the founder of teratology (science of monstrosities).

One might ask which attitude Charles Darwin assumed on this issue. His theory of the origin of species by variation and natural selection admitted only very small variations in nature, so a long time was needed for species formation and only very small leaps were deemed possible.

His experiment on artificial selection (which served as an analogue for natural selection), starts, like those of the fancy pigeon breeders, from *natural* variation. The difference was in the manner of selection: in one case by human desire, in the other by the circumstances in the natural enviroment. But in neither case was the result the *observation* of the rise of a new species. Moreover, apart from the time problem, Darwin did not expect that artificial selection would ever emulate its natural counterpart: 'Natural Selection ... is as immeasurably superior to man's feeble efforts, as the works of Nature are to those of Art.'[168]

4.5. Aether Models

When the British (mainly Scottish) physicists of the 19th century set themselves the task of elucidating their theories of electricity by means of models, they met with the difficulty that the hypothetical aethereal matter that is the bearer of the sensible phenomena was not found discoverable in the world of these phenomena and moreover has to be endowed with properties not met with in this concrete world.

The problem was not new; it had bothered some philosophically-minded physicists who devised theories in which they introduced 'matter behind the scenes' to explain the phenomena of sensory experience.

In the early 17th century Isaac Beeckman, a staunch pioneer of the new 'mechanical philosophy', stated that the atomic theory implied a difficulty which neither he nor anybody else could solve. The atoms had to be absolutely hard (having no parts) and at the same time perfecly elastic (which implied that they had parts that temporarily changed their position during a collision). Yet he declared that he preferred these mechanistic explanations involving atoms with incompatible properties to the scholastic explanations by 'Matter and Form' which he deemed empty words that did not say anything to the imagination.

In the early 19th century Augustin Fresnel met with a similar problem when he had to conceive of light as transverse vibrations of the aether; the latter then had to be an infinitely penetrable thin fluid and at the same time it must have the properties of a perfectly solid body. J.D. Forbes, who deemed Fresnel's hypothesis 'at least very ingenious' and leading to satisfactory computations and important discoveries, nevertheless remarked that an ether that is 'not only highly elastic, but absolutely solid' is 'perfectly appalling'.[169]

As soon as we try to explain phenomena by means of more fundamental entities, supposedly behind the scenes of the phenomenal world, we have to resort to abstractions from the world of sensory experience. While aiming at a scientific world picture of things more 'objective' (and supposedly more 'real') than those of the world of daily life, we create a world of abstractions which, whatever it may have won in 'objectivity', has lost immediate reality. This was clearly seen by scholars so different as Pascal and George Berkeley.

The makers of mechanical models of electrical phenomena inevitably had to cope with this problem. Their efforts to make mechanical models of electrical phenomena show how much they wanted to reduce electricity to mechanics. These efforts became obsolete when at about the beginning of the 20th century the electrical particles were conceived of as the primary entities and the world of the physicist was no longer that of classical mechanics.

James Clerk Maxwell (1831-1879) would have liked to reduce all physical phenomena, those of electromagnetism included, to mechanical ones.[170] He considered explanations by means of mechanisms as the only truly *physical* hypotheses. Yet, in his own opinion the pictures he gave of electricity and magnetism did not fully represent the reality of nature: one had to be satisfied with a 'physical analogy' or a 'physical illustration'. 'By a physical analogy I mean that partial similarity between the laws in one science and those in another which makes each of them illustrate the other' (1855 - 56).[171] So, in order to establish analogy between hydrodynamical and electrical phenomena, he developed a 'theory of the motion of an incompressible Fluid'. He stressed that it is a purely imaginary substance, which does not possess any of the properties of ordinary fluids except those of freedom of motion and resistance to compression: 'it is not even a hypothetical fluid which is introduced to explain actual phenomena. It is merely a collection of imaginary properties which may be employed for establishing certain theorems in pure mathematics in a way more intelligible to many minds and more

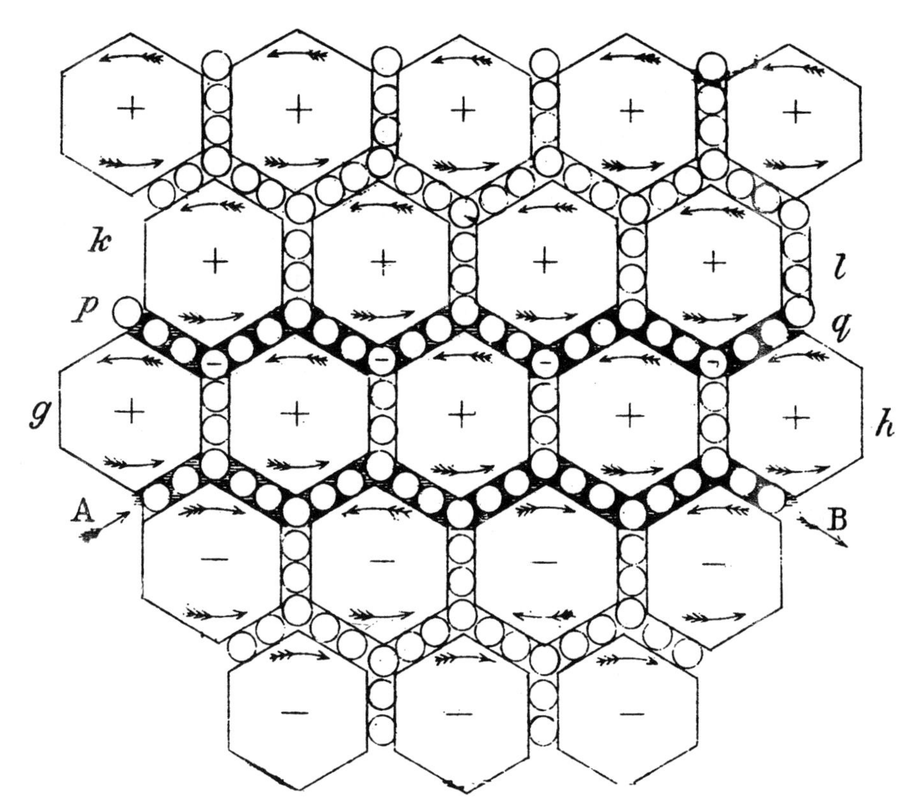

Fig. 46. Maxwell's model of electromagnetic force. The little balls represent particles of electricity, flowing from A to B; the ether vortices rotate around axes parallel to the lines of force. James Clerk Maxwell, 'On Physical Lines of Force' 1861 - 2. From: *The Scientific Papers of James Clerk Maxwell.* Ed. W.D. Niven. Cambridge 1890. Vol.I, Plate VIII.

applicable to physical problems than that in which algebraic symbols alone are used.'[172]

In 1861 - 62 Maxwell applied the theory of vortices to electric currents.[173] He represented the electromagnetic field as consisting of vortices of ether, much smaller than ponderable molecules (see figure 46). Between the vortices are interspersed small moveable 'idle wheels'[174] whose rotation transmits the motion of vortices from one part of the field to another. In a conductor their motion of translation constitutes an electric current. The model represents a mode of connection not 'existing in nature', but it is 'mechanically conceivable'. It has only a 'provisional and temporary character'. Indeed it will 'rather help than hinder' in the 'search after the true interpretations of the phenomena.'[175] The model was not

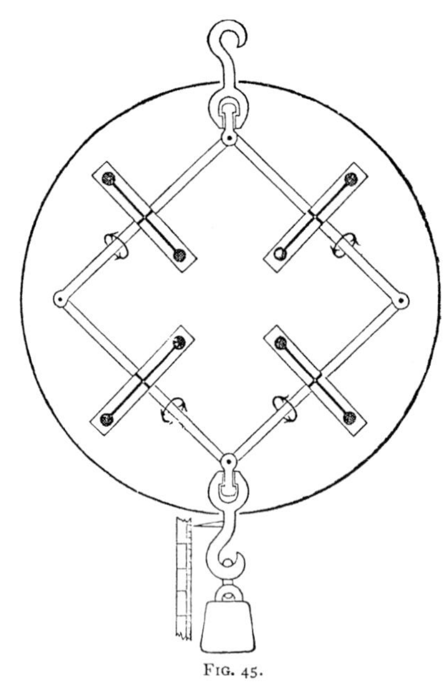

Fig. 47. Thomson's gyrostatic system imitating the properties of a spring. From: William Thomson, *Popular Lectures and Addresses* vol.I, London 1889, p.239, fig. 45.

intended to be executed in any existing material substance; it was wholly fictitious, but it was the starting point for Maxwell's celebrated equations.

After excogitating imaginary models, the next step is often the attempt to execute them in a concrete apparatus. Its material, however, is not the imaginary stuff of the thought model; it is metal, wood, plastic, etc. which cannot be endowed with the properties of this ethereal matter. In order to imitate the latter we have to find some concrete 'machine' that delivers similar effects to the hypothetical ether. Such a complicated concrete mechanism could at best imitate (or 'ape') some of the properties of the abstract ether model, but not its structure. It was expected, however, that for reasons of analogy, some consequences from the concrete model would be similar to those of the theoretical one.

It was not necessary to attribute too great a realistic value to such machines: one could quite easily exchange one model for another, provided the same properties were effectuated. William Thomson (1884) showed 'that any ideal system of material particles, acting on one another mutually, through massless connecting springs, may be perfectly imitated in a model consisting of rigid links joined together, and having rapidly rotating fly-wheels pivoted on some or on all of the links'[176] (see figure 47). 'Here, then, out of matter possessing rigidity, but

Fig. 48. Cogwheels representing electricity. From O.J. Lodge, *Modern Views of Electricity* (1892), p. 203, fig. 34.

absolutely devoid of elasticity, we have made a perfect model of a spring in the form of a spring-balance.'[177] It was not claimed that the 'incompressible fluid' had indeed a similar construction, but only that this construction showed the strange combination of properties no known substance possesses.

In 1889, in an address to electrical engineers, he showed a model of 'a medium which has the properties of an incompressible fluid, and no rigidity except what is given to it gyrostatically,' and he said that this fulfilled 'the almost inconceivable conditions for a dynamical model of electromagnetic induction in iron.'[178]

Thomson did not naively consider his models as wholly realistic. Nevertheless the general idea that some mechanism or other lay behind the phenomena was realistically meant: 'It seems to me that the test of "Do we or do we not understand a particular point in physics" is "can we make a mechanical model of it?"'[179]

Thomson stated that the demand for something like a mechanical explanation of electrical phenomena was 'growing in intensity every year.'[180] This is confirmed by the work of Oliver Lodge (1892) who asked the readers to 'think of electricity in the molecules of insulating mediums as something connected like so many cog-wheels gearing into one another, and also gearing into those of the metal conductor.'[181] The cogwheels represent positive and negative electricity, as they rotate alternately in opposite directions (see figure 48.) According to Lodge such models are much more practicable than the original model of Maxwell of 1861, which was, according to Lodge, 'the basis of all these modes of representing the equations of an electromagnetic field.'[182]

Lodge also made a model that was a hydrostatic analogue of the Leyden jar,[183] and invited his readers to think of electrical phenomena as produced by an all-permeating liquid embedded in a jelly, and to think of conductors as holes and pipes in this jelly and of attraction as due to strain and of discharge as bursting.[184] This, so he said, would give you more insight into the actual processes occurring in

Nature, 'unknown though these may still strictly be, than the old idea of action at a distance, or contenting yourselves with no theory at all on which to link the facts. You will have made a step in the direction of the truth.' It would be 'unwise to drift along among a host of complicated phenomena without guide other than that afforded by hard and rigid mathematical equations.'[185] It is evident which side Lodge had chosen in the controversy between the rival protagonists of physical explanation by imaginable models and of mathematical description by abstract formulae.

In Germany and France 'action at a distance' remained much longer *en vogue* than in Britain, where Faraday had successfully developed his field theory. This changed, however, in particular when Ludwig Boltzmann (1844 - 1906) became Maxwell's enthusiastic adherent. He applied to him Goethe's words: 'Was it a god who wrote these signs, which, with a mysteriously hidden urge, unveil the forces of nature around me and fill my heart with silent joy.' Utterance of aesthetic delight by scientists has become scarcer since the 18th century, yet, though the style of scientific prose may have become dryer, the 'joie de connaître' (as the French geologist P. Termier called it) has not abated. The unqualified admiration of the renowned physicist Boltzmann for Maxwell and his work, is a testimony of his own greatness. It shows also that not only nature is a subject of admiration but also the way in which knowledge of nature is acquired.

In a truly Maxwellian spirit Boltzmann emphasized that the hypotheses of Maxwell's theory need not represent the *real* elements and forces of nature: they are mechanisms that have a certain analogy to them.[186] But, Boltzmann says, though the constructions may not be wholly conformable to reality, they have a great heuristic value; they may be improved in the future, and what they have in common with reality will remain. Boltzmann did not see any essential difference between the Maxwellian model of ether and electricity and the representation of a gas by elastic spheres, or the hexagonal model of a benzene molecule: these too are, in his opinion — though perhaps less conspicuously — mechanical analogues, dynamical illustrations.[187] The model and the electrical current which it imitates, have the same equations, and this must be 'founded in the essence', as they are supposed to belong to precisely the same 'class of mechanisms'. They are analogous because they are of 'the same mechanical fundamental type'.[188] Yet, 'the mechanism of the electric current is not only totally different from such simple mechanism, but it is also completely unknown.' Perhaps it deviates in a for us unimaginable way from all mechanisms we construct from solid materials, elastic fluids, etc. The only certainty is that all that follows from the basic equations of mechanics will be valid not only for our models but also for the electrical mechanism.[189]

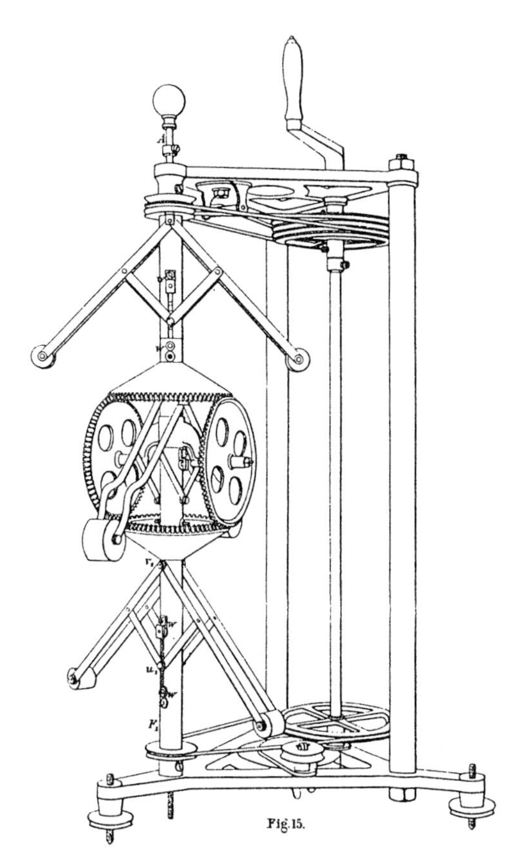

Fig. 49. Boltzmann's machine to demonstrate self-induction of electrical currents. (From: Ludwig Boltzmann, *Vorlesungen über Maxwells Theorie der Elektrizität und des Lichtes* (1892) fig. 15, Taf. II).

These words show that the 19th century models of the engineering type, however naive and coarse they may appear at first sight, were not based on a coarse theory. At the end only the 'mathematical equations' link the model with its theoretical ethereal counterpart. In Boltzmann's opinion the 'ideal' models (i.e. the imaginary models like Maxwell's) are sufficient for our purpose, which is 'representation' (*Veranschaulichung*) and *not* experimental checking.[190]

Nevertheless, Boltzmann could not resist the temptation to make a *real* model; 'representation' was followed by 'concretization': 'here, too, there is a powerful urge to real existence,'[191] and therefore he constructed a complicated machine of steel, pipes, etc., to demonstrate self-induction of electrical currents.[192] (See figure 49). In this way he imitated some of the effects, but could not pretend that there was similarity between the structure of the hypothetical ether and that of his machine. He recognized, however, that it is impossible to fill up space and make a brass model similar to Maxwell's ether, consisting of hexagonal cells and to make concrete their interaction, and that it is no less impossible to have a clear picture of it in our imagination.[193]

The opposition between the physical models and the purely mathematical descriptions was inherited from the 18th century. The 'mathematicians' deemed the models too speculative and unfit for bringing the physicist closer to the truth; the 'physicists', while recognizing that their models were fictitious and inadequate, held that they represented at least an analogue to the truth of nature: 'a step in the direction of the truth' as Lodge put it. They had *a priori* a strong faith in the mechanical 'nature' of electricity, magnetism, etc. They stresed the heuristic value of their fictions for finding the mathematical laws by which they described the facts of sensory experience. With Maxwell they believed that similarity of mathematical description of different phenomena pointed to an essential affinity.

In the second place they claimed that their models had a psychological value. As Maxwell wrote: 'the human mind is seldom satified ... when it is doing the work of a calculating machine.'[194] And according to Lodge the mathematical theory of potential had enabled mathematicians

> to dispense for the time being with theories of electricity and with mental imagery. Few, however, are the minds strong enough thus to dispense with all but the most formal and severe of mental aids; and none, I believe, to whom some mental picture of the actual processes would not be a help if it were safely available.[195]

The contrast between the two currents did not go so deeply as it might seem at first sight. Maxwell recognized that many physicists prefer a purely mathematical description of phenomena without any image. In this way, however, in his opinion 'we lose the phenomena to be explained completely out of sight'. He also pointed out (1855 - 56) that the analogy between light and vibrations of an elastic medium 'is founded only on a resemblance in form between the laws of light and those of the vibrations.' 'By stripping it of its physical dress and reducing it to a theory of "transverse alternations", we might obtain a system of truth strictly founded on observations, but probably deficient both in the vividness of its conception and the fertility of its method.'[196] These last words demonstrate convincingly that the emphasis was laid upon the psychological and the heuristic advantages of the models. According to Maxwell different types of personality present scientific truth in different forms, which should be regarded as equally scientific, whether appearing 'in the robust form and the vivid colouring of a physical illustration, or in the tenuity and paleness of a symbolic expression.'[197]

Maxwell would not be involved in a controversy about methodological principles as long as their results were conformable to the facts of nature.

The tide seems to have turned against model-making in physics. Whereas to Oliver Lodge (1892) a model was a step in the direction of the truth, James Jeans (1930) was of the opinion that 'the making of models or pictures to explain mathematical formulae and the phenomena they describe, is not a step towards, but a step away from reality.'[198] He pointed out that the objects of mathematical physics are unlikely to coincide *in toto* with the properties of any single macroscopic object of our daily life: we can think of light as either particles or waves 'according to the convenience of the moment'; 'it exists in a mathematical formula; this, and nothing else, expresses the ultimate reality.'[199] In science, the simplest explanation (i.e. the mathematical) has the highest probability of being the nearest to truth.[200] It is almost unanimously agreed that the stream of knowledge is heading towards a non-mechanical reality; 'the universe begins to look more like a great thought than like a great machine.'[201]

Nevertheless, after these seemingly bold assertions, follows the modest warning that 'every conclusion that has been tentatively put forward, is quite frankly speculative and uncertain.'[202] Jeans even says that most scientists would agree that also the 'mathematical pictures' of nature are but fictions, 'if by fiction you mean that science is not yet in contact with ultimate reality.' We can only study the shadows on the wall of Plato's cave, classify them and explain them in the simplest possible way.[203]

This may be true, but shouldn't we rather say 'never' instead of 'not yet'? We understand nature in an analogical way and cannot get rid of the feeling that the full reality is beyond our ken. To put it otherwise, all science is human science about non-human nature. The more we refine our notions about nature and the more we turn to abstractions, the more nature seems to evaporate and to slip out of our grasp. Four centuries ago, one of the greatest founders of modern science, Johannes Kepler, felt that the closer we come to nature, the more she plays with us, and 'tries to escape from him who wants to catch her — yes, almost holds her — and yet she never ceases to invite us to understand her.'

However much ether models in the fashion of Thomson, Lodge and Larmor have become obsolete, in other departments many scientists still like to have a 'konstruirbare Vorstellung'. The freedom Maxwell granted himself and others is still a mark of modern science, where models are still used when they 'rather help than hinder the search for truth.' When they become a real hindrance to progress, they are as easily abandoned.

So, the scientist will go on to enjoy the colourless and soundless and eventually shapeless world of mathematical and physical 'fictions' and he will also enjoy the world of phenomena from which he starts and to which he comes back in the end:

the world of colours, scents and forms, of plants, animals and men. Which of those two is the most real, which the more shadowy, is left to his personal choice.

X. CLEOPATRA'S NOSE

1. NEWTON'S STATUS 320
2. NEWTON'S HYPOTHESIS CONCERNING 'ELASTIC FLUIDS' 321
3. DALTON AND ATOMS 322
4. CHEMICAL COMPOUNDS 324
5. MOLECULAR VOLUMES 326
6. AVOGADRO'S HYPOTHESIS 329
7. CHEMICAL FORMULAE IN CONFUSION 331
8. TAKING STOCK 334
9. HISTORIOGRAPHY 337

'The nose of Cleopatra: had it been shorter, the face of the entire world would have been changed.'[1] This famous aphorism of Pascal's — all the more memorable for the pun it contains — raises a much-debated question which has its relevance for the history of science: does history inexorably run its course, determined mainly by social and economic forces describable in terms of fixed socio-historical laws or is it rather capriciously determined by contingencies like the sudden death of a prince without legitimate offspring, the murder of a prospective heir to the throne by the hand of a madman, or a natural disaster that devastates a country? A shorter nose would not only have defaced the face of Cleopatra, it could have changed the political face of the world. For Mark Antony might not have fallen in love with this last Queen of Egypt; his conflict with Caesar Octavian would then have taken a different form; the history of the Roman empire and consequently that of Western Europe might have followed a quite different course.

Traditional historiography would have accepted this latter conception. It laid emphasis on influential personalities and dealt largely with wars and peace treaties, dynasties and revolutions. Emphasis nowadays has often shifted towards the role of society as a whole and of economic circumstance. The individuals involved are often considered as causing relatively small and incidental ripples on the surface of an inevitable flow of events — the resulting picture, it must be said, sometimes depending as much on the beliefs of the historian who selects and arranges the facts as on the events themselves. To illustrate this theme of 'Cleopatra's Nose' I propose to consider a striking case from the history of science. We shall then return to the question of historiography. The example concerns Newton's hypothesis about the constitution of gases and its overwhelming influence on Dalton and on Dalton's attitude to Avogadro's hypothesis.

1. Newton's Status

That the works of Newton afford an example of small causes having great effects is by no means surprising. Anything Newton wrote enjoyed great authority even in his own lifetime. Indeed, once general hesitation in the face of an unfamiliar doctrine had been overcome and opposition by the Cartesians had broken down, enthusiasm for Newton became almost hysterical. 'God said: "Let Newton be," and all was light' wrote Alexander Pope. Voltaire, addressing the cherubim before the throne of the Almighty, to whom He entrusts his secret counsels, asks whether they do not feel a little jealous of Newton.[2] Boerhaave spoke of 'the miracle of our time'.[3] Newtonianism became for the the next century a literally 'catholic' faith shared by all people of all countries, from the cynical 'free-thinking' Voltaire to the pious Quaker John Dalton. The highest revelation ever given to mortal in natural science, nay, in natural theology, had been vouchsafed to Newton, who had thus become immortal. Those who followed him did not follow a man, but Nature herself: 'You wrongly speak of Newtonianism,' said Voltaire, for 'Truth has no party name.'[4]

For the scientists too the greater part of Newton's work became 'gospel truth', and rightly so, because it turned out to be conformable to nature, or — if that sounds too metaphysical — to be a trustworthy basis for a self-consistent system of classical mechanics and hence for a unification of terrestrial and heavenly physics. Even the conjectures and the 'queries', those parts of his work where Newton had cautioned his readers, acquired the same status of incontrovertible truth. Newton himself knew well that he could not give solid proofs of these, and therefore (however strongly he may have believed in them himself) did not put them forward assertively as scientifically demonstrated. His canonization, however, by the Church scientific, made every word he had written sacrosanct. Henry Brougham, a contemporary of Dalton's, used a review of Dr Black's chemical lectures as an occasion for eulogising Newton and his freedom from defect: 'The most astonishing intellect that has ever been permitted to enlighten mankind ... It is in vain that we search every corner of the Newtonian writings for some trifling proof that their author was, like ourselves, liable to the common intellectual failings of the species ... The chief characteristic of Newton is the degree of superiority in which he towers above every other natural philosopher, so as to form a class by himself.'[5] The distinction which Newton himself clearly drew between *theses* warranted by facts and those more or less probable *conjectures* needing further confirmation, came to be ignored by many people.

It goes without saying that in the long run this Newton-worship, instead of furthering the free development of scientific theory, became a burden that immobilized it. The fact, for example, that Newton held that rays of light consist of streams of corpuscles emitted from the light-source made it well-nigh impossible in the eighteenth century to secure a hearing for an undulatory theory of light. This was indeed one of Newton's main tenets so that reluctance to abandon it is understandable. And we may admire all the more the courage of the Quaker physician, Thomas Young, who eventually questioned it at about the end of the eighteenth century, for at that time Newton's authority was not yet on the wane. Even in France, Fresnel's wave theory of light (1816) met with considerable resistance for similar reasons. Haüy, in his physics textbook prescribed for French lycées, stated in 1821 a preference for Newton's theory over Fresnel's, and expressed himself proud that his compatriot Biot had adapted it to the latest discoveries in optics.[6]

2. Newton's Hypothesis Concerning 'Elastic Fluids'

If even the queries in the non-canonical part of Newton's *Opticks* had gained such authority, it is not strange to find that a 'hypothesis' inserted almost surreptitiously into the main text of the *Principia* could also acquire great weight. It was this, his hypothesis as to the constitution of 'elastic fluids' (i.e. gases) which became the foundation stone of the *New System of Chemical Philosophy* (1808 - 1827) by the Quaker physicist John Dalton (1766 - 1844).[7] The hypothesis is set out in the second book of the *Principia*:

> If a fluid be composed of particles fleeing from each other, and *the density be as the compression*, the centrifugal forces of the particles will be inversely proportional to the distances of their centres. (...) Conversely, particles fleeing from each other with forces that are inversely proportional to the distances of their centres, compose an elastic fluid, whose density is as the compression (italics RH).[8]

Both propositions start from the assumption that the 'elastic fluid' consists of near stationary particles seemingly *repelling* each another. In the first proposition, Newton puts forward a second condition, that 'the density be as the compression'. Newton says that he has put it 'ex hypothesi' (incidentally one of several indications that his famous 'hypotheses non fingo' needs some qualification). But was it for him a mere supposition, or was he in fact thinking not of a *fictitious* elastic fluid but of a very *concrete* one? Only a few pages earlier we find him saying that experiments show that 'our air' has a density that is 'almost or wholly' proportional to the compressing force. Thus Newton felt that he had 'explained' an empirical law (Boyle's law) by a corpuscular image. This was the real situation in spite of his presenting the whole problem as a purely theoretical game in terms of a wholly fictitious, anonymous 'elastic fluid'.

Having given his demonstrations, Newton reveals that he had also introduced the additional condition that the 'centrifugal' forces between the particles do not reach farther than 'those particles that are next to them, or are diffused not much farther'. This is a strange supposition to come from the creator of a theory of a gravitational force whose action was supposed to extend to an infinite distance. Significantly, he compares the centrifugal law of his fictitious particles not with the centripetal law of gravitation, but with 'the attractive force of magnetic bodies which is terminated nearly in bodies of their own kind that are next to them'.

The conclusion is now formulated anew with a third 'if' — intended presumably to remind his readers of the tentativeness of his words: 'If in this manner particles repel each other *of their own kind* that lie next to them, but do not exert force on the more remote ... they will compose such fluids as are treated of in this Proposition' (italics RH). But is there any certainty that this supposition is conformable to nature? Other possibilities might also explain the observed facts and are not to be excluded *a priori*. Newton evidently realized this, for he closed the paragraph by saying that 'whether elastic fluids do really consist of particles so repelling each other, is a philosophical (i.e. physical) question'. In other words, the model, together with the mathematically formulated law included in it, is still for him problematical. He continues: 'We have here demonstrated *mathematically* the property of fluids consisting of particles of this kind, that hence philosophers (i.e. physicists) may take occasion to discuss the question' (italics RH). Clearly the physical question is not yet answered.

More than one 'mathematical' hypothesis could be — and later was — advanced for the same physical problem. Leonhard Euler (1727), for example, deduced Boyle's law from the assumption of an elastic fluid whose particles are vortices of a more subtle matter. There was also Daniel Bernoulli's *kinetic* theory (1738) in which the particles of a gas are supposed to be in translatory movement and cause the pressure by their perfectly elastic collisions with the walls of the vessel — a theory afterwards reinvented several times in course of the nineteenth century.[9]

3. DALTON AND ATOMS

But it was Newton's hypothesis alone, even though it had avowedly not yet reached the status of a well-grounded *physical* theory, that was to influence John Dalton (1766 - 1844). He founded his atomic theory on Newton's famous query 31 which asks: 'is it not probable that God in the beginning formed Matter in solid, massy hard, impenetrable, moveable Particles... of such Sizes and Figures ... as conduced to the End for which He formed them ...' Though earlier efforts had been made (e.g. by William Higgins[10]), it is Dalton who deserves the credit for making this *a priori* faith in atoms become one of the most fruitful *fictions* in chemistry: a fiction, that is, which was able to coordinate numerous experimental *facts*. Dalton forged a direct connection between the atomic theory and Lavoisier's analytical elements — those substances which are not (as yet) capable of further chemical division. Whereas most of his predecessors had left the number and kinds of atoms undefined and undefinable, Dalton held that there are just as many kinds of atoms as there are 'simple bodies' (final products of chemical analysis). Here we have an odd combination: the notion of absolutely indivisible atoms coupled with that of the end-products of a technique of chemical analysis which was still in course of development. Philosophical weakness, however, turned out on this occasion to be Dalton's greatest strength. (At the other extreme, one thinks of the scholastics whose philosophical rigour and strength proved to be their greatest weakness — as far as Science was concerned).

Dalton used the atomic theory to derive the quantitative laws of chemical combination. To this end he characterized each kind of atom by a specific (relative) 'atomic weight': that is, the weight of the atom of an element divided by the weight of the atom of the lightest element, hydrogen. Now in the meteorological experiments of his earlier days Dalton had not found any difference in composition between the atmosphere at the base and at the top of a mountain.[11] His conception, however, of the air as a mechanical mixture of particles of oxygen, nitrogen, carbon dioxide and water vapour had led him to expect that at a greater height the heavier gases would be relatively less abundant than the lighter ones. Finding no significant differences in the composition of the air, he began an investigation of gas diffusion in the hope of discovering an explanation for this strange phenomenon (1801). His conclusion was that in a mixture of different gases the pressure of each component is independent of the pressure of the others (see figure 50): 'One gas is as a vacuum to another.'[12]

To explain this he assumed that the particles of one gas are not repelled by those of a different gas but only by those of their own kind, and it is at this point that he supports his assumption by referring to Newton's *Principia*. Quite misunderstanding what Newton had claimed, Dalton represented him as having 'demonstrated from the phenomena of condensation [i.e. compression] and rarefaction that elastic fluids are constituted of particles, which repel one another by forces which ... are reciprocally as their distances.'[13] As we saw above, Newton had done — and claimed — nothing of the kind; he had merely set out a hypothesis as to a possible constitution of gases, leaving the truth of it an open question. Nevertheless Dalton incautiously goes on: 'This deduction will stand as long as the Laws of elastic fluids continue to be what they are. What a pity it is that all who attempt to reason, or to theorize respecting the constitution of elastic fluids, should not make themselves thoroughly acquainted with this immutable Law (that the

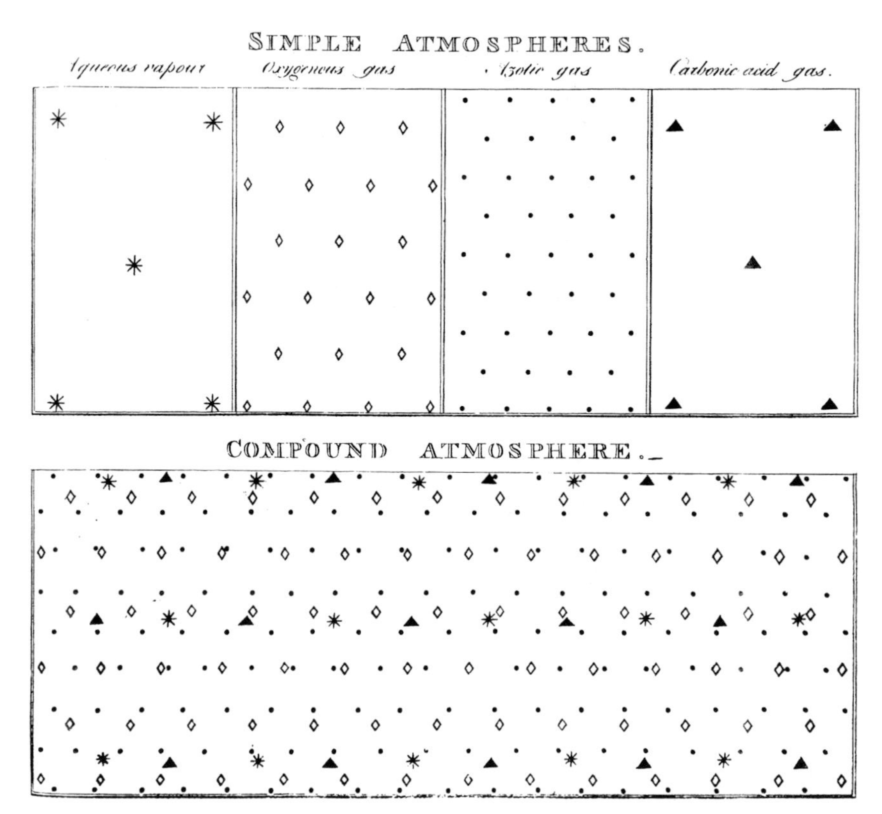

Fig. 50. The atmosphere consisting of the four separate gases of water vapour, oxygen, nitrogen and carbon dioxide (law of partial pressures of mixed gases). From: Dalton, *Manchester Memoirs* 52, p.602.

particles of a gas repel each other with a force inversely proportional to their distance), and constantly hold it in their view whenever they start any new project!'[14]

4. Chemical Compounds

It was, then, on this uncertain Newtonian basis that Dalton built his theory of chemical combination. He laid great emphasis on the tenet that all molecules of a given substance are equal, i.e. identical. His theory implies that when atoms A and B form more than one compound, there must be *simple* numerical relations between the quantities of the same element in these several compounds (e.g. AB, A_2B, AB_2, A_2B_3, etc — the law of multiple proportions, see figure 51). Dalton then assumes that when two elements form (as far as we know) only one combination, 'it must be

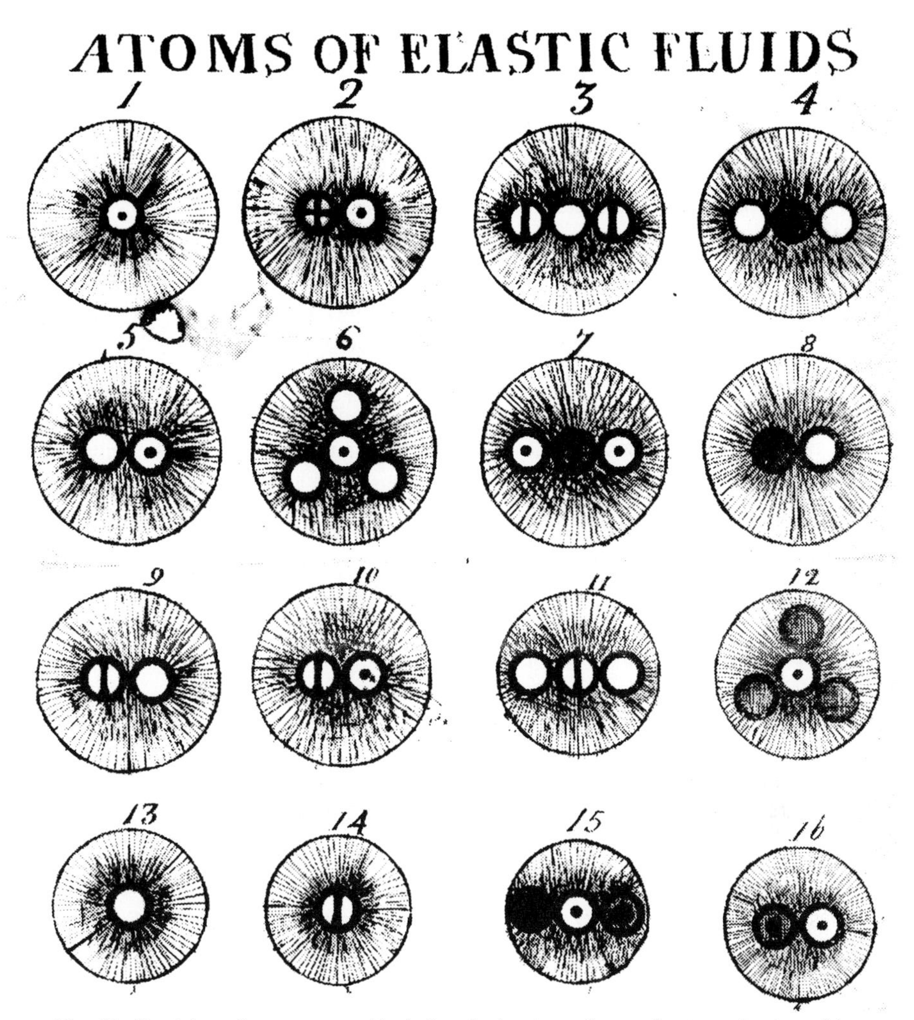

Fig. 51. Particles of gas atoms with their caloric atmospheres. Lecture sheet used by Dalton circa 1807. (From H.F. Coward and A. Harden, *Manchester Memoirs* 59, nr.12 (1915) pl.5, p.49). 1. hydrogen (H), [H_2]; 2. sulphuretted hydrogen (HS), [H_2S]; 3. nitrous oxide (N_2O), [N_2O]; 4. carbonic acid gas (CO_2), [CO_2]; 5. water vapour (HO), [H_2O]; 6. muriatic acid (HO_3), [HCl]; 7. carburetted hydrogen (CH_2), [CH_4]; 8. carbon oxide (CO), [CO]; 9. nitrous gas (NO), [NO]; 10. ammonia (NH), [NH_3]; 11. 'nitric acid' (NO_2), [NO_2]; 12. alcohol (C_3H), [C_2H_6O]; 13. oxygen (O), [O_2]; 14. azote (N), [N_2]; 15. ether (C_2H), [$C_4H_{10}O$]; 16. olefiant gas (CH), [C_2H_4]. (In brackets the (anachronistic) notation by letters of Dalton's formulae; in square brackets the correct formulae).

presumed to be a *binary* one, unless some cause appears to the contrary.'[15] In a ternary compound such as A_2B or AB_2 there is, besides this attraction of each of the two equal atoms by the unlike third atom, the counter effect of Newton's supposed mutually repulsive force between the two equal (i.e. identical) atoms.

In this theory the law of multiple proportions was implied *a priori*. As the law was not immediately borne out by the experimental data at his disposal, Dalton adapted these to his preconceived ideas. His notebook records that he found (21 March 1803) that under certain circumstances one volume of oxygen can combine with 1.7 volumes of nitrous gas (NO, nitrogen oxide), whereas under other circumstances 2.7 volumes were bound.[16] The proportions were thus 1.7 : 2.7. Some months later these proportions were changed in his notebook into 1.7 : 3.4 i.e. 1 : 2.[17] Evidently he simply decided not to *admit* a deviation from simple multiple proportions. One certainly had to have a belief in multiple proportions in order to find tidy integer ratios, especially in the case of the oxides of nitrogen, where in practice chemical actions almost always generate mixtures of the oxides in variable proportions! Fortunately, Dalton went on to more thoroughgoing investigations into multiple proportions, which led to more reliable results.

Dalton's theoretical tenets, however, led to conclusions that were to cause great troubles. His rules for chemical combination forced him to assume that — water being the only then known compound of hydrogen and oxygen — the water molecule consists of one hydrogen and one oxygen atom (HO rather than H_2O), and ammonia had to be NH (instead of NH_3), while the two carbon oxides were correctly represented as CO and CO_2.[18] The belief that atoms of the same kind repel each other convinced him that the molecules of simple bodies had to be monoatomic. Thus the free oxygen molecule cannot be O_2 but must consist of a single atom O.

On these theoretical assumptions and following rather crude chemical analyses, Dalton drew up, on 6 September 1803, his first table of relative atomic weights.[19] In this the weights of hydrogen and oxygen which combine together are given as 1 : 5.66; later on, in his *New System* this became 1 : 7. If the formula of water is held to be HO, the latter statement implies that the atomic weight of oxygen is 7. (With the correct formula H_2O it would have been 14. A somewhat more accurate analysis yields 1 : 8, which with the formula H_2O leads to the currently accepted atomic weight 16 for oxygen).

5. Molecular Volumes

Most of the atomic weights in Dalton's table were in fact wrong, both because his theory was mistaken and because his analytical work was inexact. This had repercussions at the next stage of his researches. For from his atomic weights, together with the specific weights (or densities) of the gases, he calculated (1804) the relative volumes (i.e. relative to hydrogen) of molecules of several gases. He found these volumes to be different for different gaseous molecules. These volumes are not the volumes of the molecules in the strictest sense but of 'the nucleus, together with that of its surrounding repulsive atmosphere of heat.'[20] Equal 'molecules' (i.e. molecules of the same kind) repel each other because the 'rays of heat' which they emit touch one another (figure 52). Consequently it now became of great importance to him that particles of different gases should *differ in size*.

From the above it is evident why Dalton could not accept the 'volume law' announced by Gay-Lussac in 1808: 'the volumes of gases reacting together are in simple ratios'. As a result of exact experiments Gay-Lussac stated:

1 volume nitrogen gas + 1 volume oxygen gas → 2 volumes nitrous gas (NO).

Dalton held *a priori* :

1 atom N + 1 atom O → 1 'compound atom' NO.

If Gay-Lussac's data were correct, one atom of N-gas and one atom of O-gas would be of equal size. This clashed with Dalton's conviction that particles of different gases have different sizes. Moreover, assuming that 1 atom N + 1 atom O form 1 'atom' NO, he expected the NO-particle to have a much greater volume than the N-particle or the O-particle alone.

When confronted with the choice between the simplicity of Gay-Lussac's Law of volumes, and the simplicity of his own rules of combination, Dalton voted for the latter. He thus felt he had to deny the correctness of Gay-Lussac's measurements. The ratio found experimentally by Gay-Lussac's best measurements — those for the reaction of hydrogen with oxygen — was practically 2 : 1, but Dalton was quick to point out that the precise figures he had found himself had been 1.97 : 1 which is not a ratio of simple integers. Yet this same Dalton was, as we have seen, willing to change a proportion of 1.7 : 2.7 to 1.7 : 3.4 in the interests of his own theory, and had no difficulty in claiming that the weights of oxygen in the two oxides of sulphur are as 1 : 2 (thus conforming with the Daltonian formulae SO and SO_2) whereas they are in fact 2 : 3 (fitting the correct formulae SO_2 and SO_3).[21] Similarly for the weights of hydrogen and nitrogen in ammonia he found the ratio 1 : 4¾, 'but as an integer is more easily remembered we shall prefer the ratio 1 to 5, till a more accurate one can be ascertained.'[22]

Thus when his own cherished theory was threatened he became meticulous about experimental data in order to ward off change; but when it was to his advantage, he was prepared to treat the results of Gay-Lussac's very exact

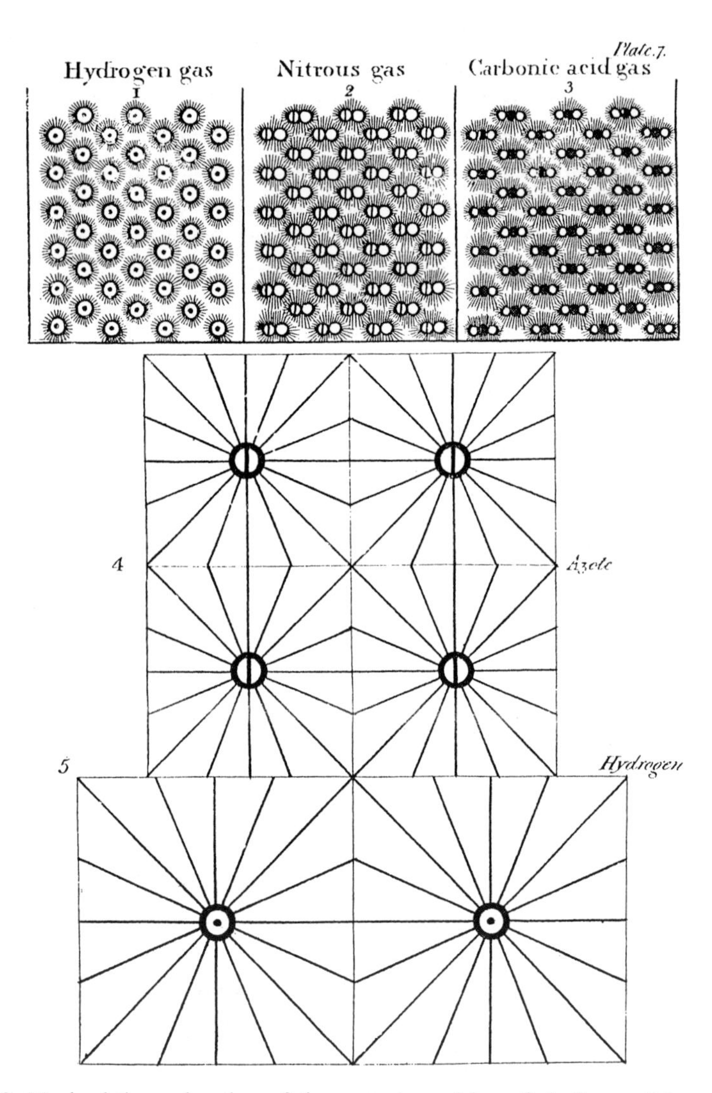

Fig. 52. Mechanistic explanation of the mutual repulsion of similar particles and of the mutual indifference of particles of different species. 4. 'Particles of azote with their elastic atmospheres, marked by rays emanating from the solid central atom; these rays being exactly alike in all the four particles, can meet each other and maintain an equilibrium'. 5. 'two atoms of hydrogen drawn in proportion to those of azote, and coming in contact with them; it is obvious that the atoms of hydrogen ... can not apply to those of azote, by reason of the rays not meeting each other in like circumstances; hence, the cause of the intestine motion which takes place on the mixture of elastic fluids, till the exterior particles come to press on something solid' (Dalton, *New System* I, 2 (1813), pl.7, p.548).

measurements in a highly cavalier fashion. It is only too obvious that his law of multiple proportions was essentially an *a prioristic* theory; it is a remarkable if regrettable fact that scientists often fight harder for the fictional products of their own brains than for the facts, the results of measurements and the experimental laws founded upon them.

Dalton's own chemical measurements were much less exact than those of Gay-Lussac. Yet he maintained that the volumes of nitrogen and hydrogen which form ammonia are as 1 : 2½ although Gay-Lussac (correctly) found 1 : 3.[23] And in his *New System of Chemical Philosophy* we find Dalton condescendingly excusing Gay-Lussac:

> When the mind is ardently engaged in prosecuting experimental enquiries, of a new and extraordinary kind, it is not to be expected that new theoric views can be ... formed so as to be consistent with all the well-known and established facts of chemistry; nor that the facts themselves can be ascertained with that precision which long experience ... and a comparison of like observations made by different persons, are calculated to produce.[24]

6. Avogadro's Hypothesis

Even less acceptable to Dalton was another important gas law, a law which at that time was just a hypothesis since (unlike Gay-Lussac's law) it did not state an experimental *fact*. Avogadro (1776-1856), in order to explain the simple ratios of volumes of gases partaking in reactions, realized that it was *simplest* to assume that *equal volumes of different gases [at the same temperature and pressure] contain the same number of molecules* (1811) i.e. that the particles of elastic fluids are all of the same size.[25] This hypothesis implies, as Avogadro says, that the 'constituent molecules of any elementary gas ... are not formed of a solitary elementary molecule [i.e. atom], but are made up of a certain number of these molecules united by attraction.'[26] His reasoning is as follows. The experimental finding

2 volumes H gas + 1 volume O gas → 2 volumes water vapour,

implies, according to Avogadro's hypothesis, that

2n molecules H + n molecules O → 2n molecules of water,

and so

2 molecules H + 1 molecule O → 2 molecules of water.

If so, one molecule of water contains one molecule of hydrogen and *half* a molecule of oxygen. The oxygen molecule must thus contain an *even* number of

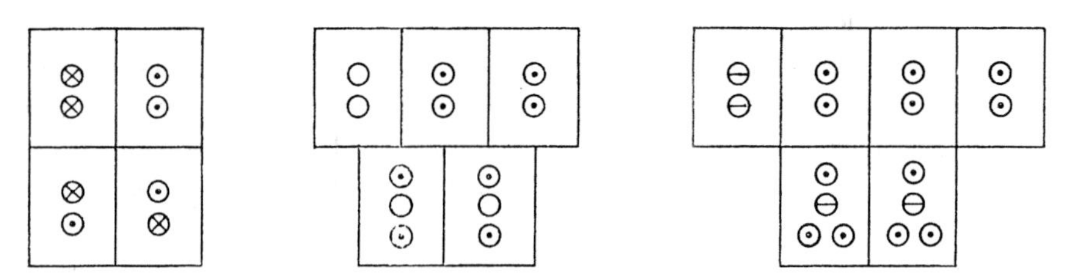

Fig. 53. Schematic model of reaction (from left to right) of $Cl_2 + H_2 \rightarrow 2HCl$; $O_2 + 2H_2 \rightarrow 2H_2O$; $N_2 + 3H_2 \rightarrow 2NH_3$ (from: M.A. Gaudin, *Ann.Ch.Phys.*[2] 52 (1833), p.113).

atoms. As no cases are known in which oxygen splits into *more* than two parts, and moreover the simplest is the most probable, we may conclude that the oxygen molecule is bi-atomic [O_2]. The same conclusion was reached with regard to the H_2 and N_2 molecules.[27] (The formula for water thus became H_2O, and the atomic weight of oxygen not 8 but 16.)

In Avogadro's hypothesis, then, two things were unacceptable to Dalton: equal volumes of different gases had to contain the same number of molecules (or, molecules of different gases occupy the same space) and (implicitly) particles of elementary gases may be bi-atomic. The first, basic, hypothesis was less repellent to him than the second: he had once accepted it himself. But the second ran contrary to the great Newtonian thesis that similar particles repel each other; and this remained for him the main obstacle.

In the ensuing years Avogadro's hypothesis met with considerable difficulties. Dumas found for sulphur vapour (1832) three times the density he expected (because the molecules in this case are not S_2 but S_6), and for mercury vapour (1826) half the expected density (because the mercury molecules are not Hg_2 but Hg).[28] Avogadro's assumption, grounded upon arguments from analogy and simplicity, that molecules of *all* elementary gases or vapours are bi-atomic was here a great impediment.[29] For the next two decades nobody cared about his hypothesis, with the exception of M.A. Gaudin (1804 - 1880), who in 1832 expounded the law and its consequences in a more consistent way than Avogadro himself had done (see figure 53).[30]

Little attention was paid to Gaudin, however, and it was not until 1843 that Gerhardt assumed the law for all *compounds*, though he held that all elementary bodies had to be mono-atomic: again a rejection of the bi-atomicity of the molecules of simple bodies.[31] As molecular weights and atomic weights based upon

Avogadro's law lead to correct and therefore self-consistent values and correct formulae, while those based on Dalton's assumptions (and those of Berzelius) very often led to wrong and divergent results, theoretical chemistry became during these years a chaos of opinions. There was no consensus about atomic weights: mercury, for example, could find itself credited with the weight of 100, or 200 (correct) or 400.[32] Moreover, new difficulties arose because some gaseous bodies showed a density of half that expected. This was disturbing until it was eventually realized that their molecules 'dissociate' (split up) into two parts so that an 'abnormal' volume of the gaseous mixture must ensue.[33]

7. CHEMICAL FORMULAE IN CONFUSION

Back in the 1790's, Lavoisier had declared with great satisfaction that the revolution in chemistry was accomplished; but since that time order had turned into disorder. The historian who tries to relive in imagination the situation in the 1840's wonders how anybody at that time could be attracted to the study of chemistry. Already in 1835 one of the outstanding chemists of the day, Friedrich Wöhler, almost abandoned hope: 'Organic chemistry can make one completely mad. It appears to me a jungle full of the strangest things; an enormous brushwood, without exit, without end, that one had better not enter'. By 1860 this Babylonian confusion had not yet been cleared up: Kékulé could point to eighteen so-called 'rational' formulae for so simple a compound as acetic acid (figure 54),[34] and Berthelot mentions *four* 'empirical' formulae for this same compound made possible by the different atomic weights used:

$C_4H_4O_4$ if H=1, C=6, O=8;

$C_2H_4O_4$ if H=1, C=12, O=8;

$C_2H_4O_2$ if H=1, C=12, O=16;

$C_8H_8O_4$ if H=1, C=6, O=16;

Looking back upon this chaotic situation Kékulé wrote in 1877: 'At that particular time, the chemists who mainly set the tune were caught by a general discouragement...; it was believed that all speculation should be banned from chemistry and in particular that all atomistic considerations should be abandoned.'[35]

Towards the end of the 1850's, however, it became evident that change was imminent. More cases of abnormally low vapour density — that is, of an abnormally large volume of a gaseous substance — proved explicable by

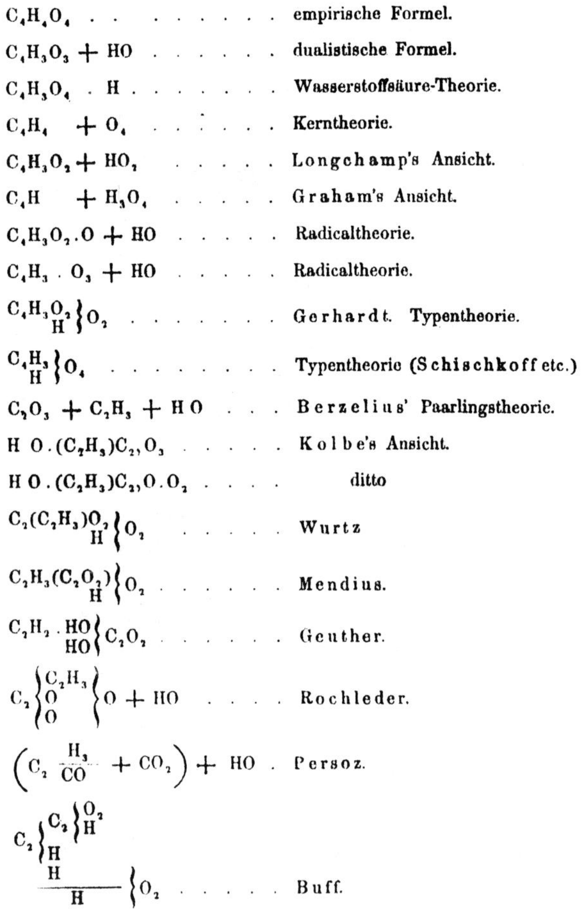

$C_4H_4O_4$ . . . . empirische Formel.

$C_4H_3O_3 + HO$ . . . . . . dualistische Formel.

$C_4H_3O_4 \,.\, H$ . . . . Wasserstoffsäure-Theorie.

$C_4H_4 \quad + O_4$ . . . . Kerntheorie.

$C_4H_3O_2 + HO_2$ . . . . . Longchamp's Ansicht.

$C_4H \quad + H_3O_4$ . . . . . Graham's Ansicht.

$C_4H_3O_2.O + HO$ . . . . . Radicaltheorie.

$C_4H_3 \,.\, O_3 + HO$ . . . . . Radicaltheorie.

$\left.\begin{matrix}C_4H_3O_2\\H\end{matrix}\right\}O_2$ . . . . Gerhardt. Typentheorie.

$\left.\begin{matrix}C_4H_3\\H\end{matrix}\right\}O_4$ . . . . Typentheorie (Schischkoff etc.)

$C_2O_3 + C_2H_3 + HO$ . . . Berzelius' Paarlingstheorie.

$HO.(C_2H_3)C_2,O_3$ . . . . . Kolbe's Ansicht.

$HO.(C_2H_3)C_2,O.O_2$ . . . . ditto

$\left.\begin{matrix}C_2(C_2H_3)O_2\\H\end{matrix}\right\}O_2$ . . . . . Wurtz

$\left.\begin{matrix}C_2H_3(C_2O_2)\\H\end{matrix}\right\}O_2$ . . . . . . Mendius.

$C_2H_2 \,.\, \left.\begin{matrix}HO\\HO\end{matrix}\right\}C_2O_2$ . . . . . . Geuther.

$C_2\left\{\begin{matrix}C_2H_3\\O\\O\end{matrix}\right\}O + HO$ . . . . Rochleder.

$\left(C_2\,\frac{H_3}{CO} + CO_2\right) + HO$. Persoz.

$C_2\left\{\begin{matrix}C_2\left\{\begin{matrix}O_2\\H\end{matrix}\right.\\H\end{matrix}\right.$

$\left.\frac{H}{H}\right\}O_2$ . . . . . Buff.

Fig. 54. Rational formulae for acetic acid ($C_2H_4O_2$) current about 1860. From: A. Kékulé, *Lehrbuch der Organischen Chemie oder der Chemie der Kohlenstoff-verbindungen* I, Erlangen: Enke 1861, p.58.

'dissociation' (AB → A + B). Then in 1858 the Italian chemist Stanislao Cannizzaro (1826-1910) stated that all disagreement about atomic weights of chemical elements and about the molecular weights of volatile substances could be solved by accepting Avogadro's law.[36]

At about the same time some physicists too arrived at the conclusion that equal volumes of different gases contain the same number of molecules. Unlike Newton (and also Avogadro) they did not consider gases and vapours as consisting of particles at rest, but rather of freely moving, perfectly elastic molecules whose motion caused the gas pressure ('expansive force') of the gas. From this hypothesis they mathematically deduced Boyle's law. Now there had been several earlier kinetic theories by which Boyle's law had been mathematically deduced (1727; 1738; 1821; 1848)[37], but the new thing in the theories of the 1850's was that, in addition, a law similar to Avogadro's was deduced from them — though the authors were unaware of the fact that it had already been propounded by Avogadro (1811), Ampère (1814), Gaudin (1832) and — with some waverings — by some French chemists.

It was in 1856 that A. Krönig derived from his kinetic hypothesis not only Boyle's law but also the thesis that equal volumes of different gases contain an equal number of molecules.[38] Quite independently in the following year (1857) Rudolph Clausius published a much more sophisticated kinetic theory of gases.[39] Clausius concluded that molecules of all gases at a given temperature have the same kinetic energy of their translatory motion and that in the gaseous form equal volumes not only of simple bodies (hydrogen, oxygen, etc) but also of compounds contain the same number of molecules. He went further than Krönig by explicitly stating as an hypothesis that molecules of simple (gaseous) bodies are as a rule bi-atomic.[40] Shortly afterwards (1858) Clausius admitted to French critics of his article that at the time he wrote it, he was ignorant of the fact that Dumas, Laurent and Gerhardt had — on chemical grounds — put forward the thesis that molecules of gaseous simple bodies may consist of more than one atom, so that he had to recognize that his conception was not so new as he had thought it.[41]

Both Clausius and his contemporaries remained ignorant, however, of the fact that on similar grounds the Scottish physicist J.J. Waterston (1811-1883) had made the same statement on poly-atomic molecules in 1845. Waterston acknowledged that the objection against half molecules of oxygen, etc. is 'plausible from the natural repugnance to the idea of dividing what has been considered as an ultimate element into parts, and of supposing that an element should have a strong affinity for itself.'[42] He suggested nevertheless that simple bodies can form bi-atomic molecules and in some cases, referring to earlier experiments by French chemists, can form molecules of four and of six atoms (e.g. phosphorus and sulphur). Moreover he left open the possibility that atoms of a given element might assemble in *various* numbers.[43] In opposition to the growing scepticism about the reality of atoms and molecules, Waterston maintained that the theory of gases was 'philosophical' (i.e. physical) and not merely 'mathematical', giving Newton as his reference for this distinction.[44] Clausius, who like Waterston combined the theory of gases with the theory of heat, was less confident, saying cautiously that the subject touched so many fields that the same result might well ensue from chemical and physical approaches, but that being so hypothetical each new support of that result must be welcomed.[45]

At the first international chemical congress (Karlsruhe 1860) Cannizzaro demonstrated that Avogadro's law, when consistently applied (without the superfluous analogies introduced by Avogadro himself), could solve at one stroke all the difficulties in chemistry about atomic weights and chemical formulae. One of the participants, Lothar Meyer, wrote: 'it was as if scales fell from my eyes: the doubts vanished and a feeling of calm certainty came in their place.'[46] It was indeed as if the magic formula of Avogadro's law had awakened the truth that had been sleeping in the facts amassed between the first (Lavoisier's) and this second chemical 'revolution'.

8. Taking Stock

It seems then that if Newton had not propounded an incidental hypothesis on the physical constitution of gases, and if Dalton's adoration of Newton had not induced him to turn this hypothesis into a physical *fact* ('an immutable law that must stand for ever'), Avogadro's law could have been accepted half a century earlier; the correct atomic and molecular weights and molecular formulae would have been introduced, the enormous confusion in chemistry would not have arisen, and the learned world would not have had to wait until 1860 before the chaos was shaped into a harmonious cosmos. But was Dalton's dogmatic attitude the only reason for the neglect of Avogadro's hypothesis? It must be said that Avogadro did not present his ideas in a particularly clear way. He spoke of 'molécules intégrantes' when referring to the free particles of compound bodies, and of 'molécules constituantes' when referring to the free particles of simple bodies (the elements); whereas he called the *atoms* of the elements 'molécules élémentaires'. To make things worse he sometimes omitted the adjective and simply spoke of 'molécules', leaving the reader to decide which kind of 'molécule' he meant. One can imagine how confusing this must have been for readers accustomed to the notion that free particles of simple bodies are mono-atomic. Moreover, as remarked above, Avogadro extended his theory farther than could be warranted by the facts. By analogical reasoning he concluded that not only oxygen, hydrogen and other gaseous simple bodies had bi-atomic molecules, but that this was the case for *all* simple bodies: C_2, Cu_2, Hg_2, P_2, S_2. In consequence his theory, too, would have led in several cases to wrong calculations of atomic weight. Finally, he took the atomic weight of the hydrogen atom (H) as the unit for what we now call relative *atomic* weights, and the weight of the hydrogen molecule (H_2) as the unit of relative *molecular* weights which means that the atomic weight of hydrogen was 1, but the molecular weight of hydrogen was also 1. Nevertheless, since Gaudin's version (1832), which was clear and unambiguous, was also practically ignored, we must seek the main reasons for its neglect elsewhere, and then would be inclined to blame Dalton alone.

This, however, would be going too far, for Dalton's conception that similar atoms repel each other and only different ones may chemically attract each other, is not the only reason for the delay. A great stumbling block in the way of acceptance of Avogadro's hypothesis must have been the *dualistic* or binary character of chemistry in the first decades of the nineteenth century. For this frame of thought, what gave offence was not so much the main thesis of the equality of size of all gas molecules, but rather its consequence, that atoms of the same element may combine with *one another*. Lavoisier had introduced into chemistry a binary nomenclature (*sulphate de cuivre*, *oxide de plomb*, etc., see figure 55) and in so doing had followed his own tenet that nomenclature should be a faithful mirror of the state of scientific knowledge.[47]

The dualistic principle was extended by him to organic compounds consisting of C, H and O by assuming that so-called compound radicals or 'organic elements' (C_xH_y) formed partnerships with oxygen in compounds such as sugars [$(C_xH_y)O_z$]. Berzelius (1818) gave this dualistic conception a theoretical basis in his electro-dualistic theory: each compound consists of a more positive in combination with a more negative component. But into such a scheme molecules consisting of two similar atoms (e.g. H_2) do not fit. Eventually the theoretical basis of the dualistic theory was undermined when Dumas discovered that in organic compounds the electro-positive H-atoms are liable to be replaced by the electro-negative Cl-atoms [$CH_4 + Cl_2 \rightarrow CH_3Cl + HCl$].[48] It is understandable that it was precisely the 'unitarian' chemists (Dumas, Gerhardt, Laurent) who made efforts to introduce Avogadro's hypothesis, though for some time their ideas met with little approval: the influence of the dualistic Lavoisier-Berzelius way of thought was too strong.

Clearly, then, even if Dalton's theory had been adapted to Avogadro's law at an earlier stage, there would have been other obstacles to the latter's acceptance. As the authority of Newton weighed too heavily with Dalton, so that of Lavoisier weighed with Berzelius, and that of Lavoisier and Berzelius together weighed with the school of Berzelius. As well as activating the growth of science, great scientists have often had a stifling influence upon it. However much modern science can still profit from their work, the Newtons, Lyells and Darwins become a dead weight when they are taken as unquestionable authorities. A further moral of this tale is that scientific heretics should not always be rejected out of hand.

This brings us back to the question of Cleopatra's nose. If Dalton had not been so prejudiced in favour of Newton's hypothesis and had not stopped his ears against what might be said in favour of Gay-Lussac and Avogadro, the face of chemistry would have changed half a century earlier than in the event it did; for by the 1860's Daltonian atomism found itself revised in just the way Avogadro had proposed. In the development of science truth *has* to come out sooner or later. In politics it is not clear to me that people grow much wiser (though the slogans may become more lofty or more hypocritical); in the development of art one could not say that Van Gogh is better than (but only that he is different from) Rembrandt or the artists of ancient Greece or Egypt.

238 COMBINAISONS DE

TABLEAU des combinaisons de l'Acide sulfurique ou de leur affinité avec cet acide,

	NOMENCLATURE NOUVELLE.	
Nos.	*Noms des bases.*	*Sels neutres qui en résultent.*
1	La baryte...........	Sulfate de baryte.......
2	La potasse...........	Sulfate de potasse......
3	La soude...........	Sulfate de soude.......
4	La chaux...........	Sulfate de chaux.......
5	La magnésie..........	Sulfate de magnésie.....
6	L'ammoniaque........	Sulfate d'ammoniaque...
7	L'alumine...........	Sulfate d'alumine ou alun.
8	L'oxide de zinc.......	Sulfate de zinc........
9	L'oxide de fer........	Sulfate de fer.........
10	L'oxide de manganèse..	Sulfate de manganèse....
11	L'oxide de cobalt......	Sulfate de cobalt.......
12	L'oxide de nickel......	Sulfate de nickel.......
13	L'oxide de plomb......	Sulfate de plomb.......
14	L'oxide d'étain........	Sulfate d'étain.........
15	L'oxide de cuivre.....	Sulfate de cuivre.......
16	L'oxide de bismuth....	Sulfate de bismuth.....
17	L'oxide d'antimoine...	Sulfate d'antimoine.....
18	L'oxide d'arsenic......	Sulfate d'arsenic.......
19	L'oxide de mercure....	Sulfate de mercure.....
20	L'oxide d'argent.......	Sulfate d'argent.......
21	L'oxide d'or..........	Sulfate d'or..........
22	L'oxide de platine.....	Sulfate de platine......

Combinaisons de l'acide sulfurique avec :

Fig. 55. Dualistic nomenclature of salts according to Lavoisier (from: A.L. Lavoisier, *Traité Elémentaire de Chimie*, 2. éd. T.I, Paris 1793 (1789), p.238.)

The history of science, however, certainly shows progress: it is accumulative — which is not, of course, to say that we are *better scientists* than our ancestors. This progress of science even Newton's great authority could not stop for long. His corpuscular theory of light may have *delayed* the acceptance of the wave theory; it could not prevent it from coming in the long run. Moreover, even in the original context the state of 'Cleopatra's nose' was not the only 'cause' of the course of events. If Mark Antony had been less easily impressed by the charms of the Egyptian Jezebel, and if Octavian had not been ... In any historical event *many* 'ifs' are at stake. In our case, however, one thing is certain. If Dalton and Berzelius had accepted Avogadro's law, the chaos would have been reduced to order much earlier, and many theoretical efforts by other scientists would have been unnecessary.

9. HISTORIOGRAPHY

Our question touches also on the significance of outstanding individuals in the history of science. Is it they who determine the main course or is there a general development of which they are only the exponents? Romanticizing historiographers, or rather hagiographers, tend to make each of their heroes the pivot around whom history turns. The works of the several fathers of geology, chemistry, crystallography and physics seem to be so many miraculous births. In our day the pendulum has swung back and hero worship, though not extinct, is unfashionable. The 'forerunners' of a development are not now seen as subservient to the great, but rather as proofs that the time was ripe, that the great discoveries *had* to come and that they did not originate in a 'catastrophic' way as by a flash of lightning. It must indeed be recognized that the great finds of science did not come out of the blue. The law of gravitation, for example, was 'in the air' when Newton put it forward, and the law of falling bodies was found not only by Galileo but also by Beeckman. It would nevertheless be wrong to put Hooke on the same level with Newton, or Beeckman with Galileo. By their powers of synthesis, by their mathematical and systematic talents, the great ones far surpassed the rest. Yet they were explorers on the same beach. Beeckman, for example, expressed the same urge to unite mathematical and physical methods as Galileo.

The history of science shows indeed that certain discoveries come about because 'their time has come'. A good example is the case of the periodic system of the elements (1869). Numerous efforts had been made to classify the elements, but they came to nothing, mainly because of the uncertainty about atomic weights. Then suddenly we find Mendeléev and Lothar Meyer thinking on similar lines quite independently of each other. It was as if a period of incubation had come to an end, and the new system *had* to come. The same is true of the invention of calculus by Newton and Leibniz in the seventeenth century; the rise of stereochemistry on the basis of the tetrahedral model of Van 't Hoff and on different lines in the work of Le Bel in 1874; the publication of kinetic theories of gases by Krönig in 1856 and Clausius in 1857 (with earlier efforts by Waterston in 1845 and Joule in 1848); the theory of natural selection independently advanced by Darwin and by Wallace; the statement of the law of conservation of energy by Robert Mayer, Joule, Colding and Helmholtz; the wave theory of light by Young and by Fresnel.

Those who minimize the individual will be inclined to conclude that all this proves their contention that the great scientists are but a mouthpiece of their time. It must indeed be recognized that the people with creative minds are children of their age and to a large extent products of their time and environment. If they were not, how could anyone have understood and followed them? In reaction against the overestimation of individuals, however, we should miss the point of the examples given above if we were to underestimate the uniqueness of their achievements. For those who have learned from the first pages of a textbook the principles of mechanics laid down by Newton, it is difficult to realize the extraordinary power of thought behind their formulation, the outstanding capacity for observation and experimentation involved in following up their consequences. Even if we succeed in piecing together the course of development of their work from what has been done and said by their predecessors, it should be realized that the Galileos, Newtons and Darwins alone had that touch of genius which could lead to the synthesis that constituted their achievements in the development of science. And if they themselves stood, as Newton said, on the shoulders of giants (think of Plato, Aristotle, Archimedes) and if they also rested on the work of numerous scholars of lesser status, it would be quite wrong to explain them away on these grounds as mere exponents of the inevitable course of development.

It is unfortunate that when the late R.G. Collingwood (1946) rightly criticized writers who lack any conception of the historical origins or processes leading to the modern scientific approach, he appears to include Pascal among their number. For he chose as a 'typical' example 'the remark of Pascal that if Cleopatra's nose had been longer (sic) the whole history of the world would have been different — typical, that is, of a bankruptcy of historical method which in despair of genuine explanation acquiesces in the most trivial causes for the vastest effects.'[49] But there is causality and causality. When an avalanche descends a snow slope the general cause is gravitational force. It may be, however, that what set it going was a small disturbance in the air. This is what the German scientist Robert Mayer has called 'Auslösungskausalität' (trigger-causality).[50] If some triggering event had not happened, perhaps the snow would have melted away and the face of the locality would not have been changed. Pascal had a keen eye for such small causes (Cromwell's sudden end[51]), but he did not belong to those writers for whom, as Collingwood put it, 'The central point of history is the sunrise of the modern scientific spirit.'[52] Nor did Pascal believe that 'Before that, everything was superstition and darkness, error and imposture', the detailing of which is 'a tale told by an idiot signifying nothing.'[53] On the contrary, he was one of the rare scientists

— and that in the seventeenth century — who had a positive appreciation even for some of the notorious errors of the Ancients; only he could not stand those same errors in his contemporaries. He pointed out that when the Ancients said that nature does not admit a vacuum, they were right, for as far as their experience went, a vacuum could not be realized and they could speak about nature only insofar as they knew it. Pascal says that we can have other opinions than our predecessors 'without despisal and without ingratitude, for the first knowledge which they have given to us has served as steps to ours'[54] and we are indebted to them even for the advantages we have over them, as they brought us to such a level that the smallest effort was enough to make us climb higher. We have to admire them for the conclusions they drew from the few data they had and to excuse them because they lacked the luck of experience *rather than the strength of reasoning*. 'Without contradicting them, we can affirm the opposite of what they said.'[55] This is hardly the language of one who thinks ancient science 'a tale told by an idiot'!

This truly scientific attitude is to be found also in the Scottish geologist and physicist James David Forbes (1848). The historian Macaulay, speaking in Edinburgh, had declared that a girl from boarding school nowadays knows more of Geography than Strabo, and implied that the superficial knowledge of the child was superior to the so-called 'profound' knowledge of the ancients and medievals. Forbes retorted that the parrotry of the child had nothing to do with 'knowledge' in the strict sense and that creative thought is something more than knowing what is in the text book. Though certainly 'ideas which ... tasked the highest powers of mankind in one age, became familiar even to the minds of the young in another ...; in reality the first steps are the most difficult ... If anyone feels rising in his mind even a casual and almost involuntary sentiment of superiority to his great predecessors (to whom in fact he owes everything)' ... let him remember that 'but for the gigantic perseverence of those hard-working men who preceded us ... we had been toiling in the quarry instead of carving the pinnacles of our temple'. 'The truly great are those who originate ... it would be bold to affirm that the highest types of the human intellect improve as the world grows older.'[56] So Macaulay is wrong when he says that 'the intellectual giants of one age become the intellectual pygmies of another'.

Finally, each historical event is the coincidence of several independent causal series; and the more complicated this coincidence, the more difficult it becomes to disentangle essential from accidental components. In particular, human history, where the course of events is not only undergone but is also influenced by the subjects themselves, disentanglement is especially difficult. We should not forget that the history of science is not the history of nature, but part of the history of mankind. Pascal's saying is just a pithy remark; it was never intended to express the

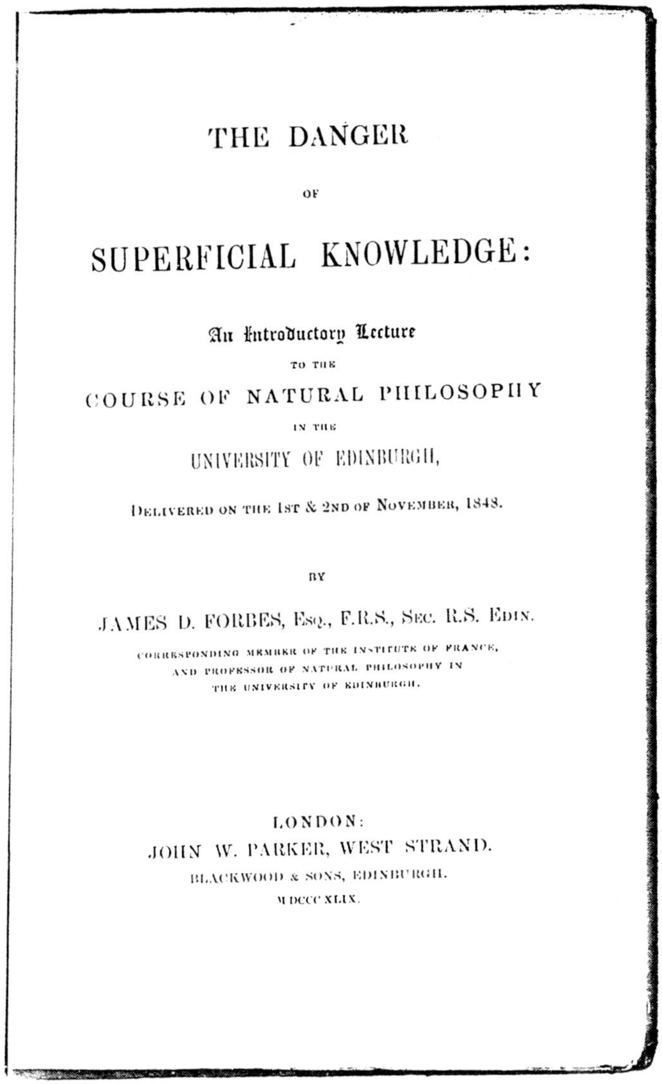

THE DANGER

OF

SUPERFICIAL KNOWLEDGE:

An Introductory Lecture

TO THE

COURSE OF NATURAL PHILOSOPHY

IN THE

UNIVERSITY OF EDINBURGH,

Delivered on the 1st & 2nd of November, 1848.

BY

JAMES D. FORBES, Esq., F.R.S., Sec. R.S. Edin.

CORRESPONDING MEMBER OF THE INSTITUTE OF FRANCE,
AND PROFESSOR OF NATURAL PHILOSOPHY IN
THE UNIVERSITY OF EDINBURGH.

LONDON:
JOHN W. PARKER, WEST STRAND.
BLACKWOOD & SONS, EDINBURGH.
MDCCCXLIX.

Fig. 56. Title page of J.D. Forbes's refutation of Macaulay.

full complexity of the question, and he, with his open eye for the several sides to any problem, would have been the first to acknowledge this.

What then, after all these 'ins' and 'outs', is the final answer to the problem of Cleopatra's nose? I must confess that I have none to give; nor am I ashamed of this. Like old Boyle[57] I am content to be an 'underbuilder' and to 'dig in the quarries', providing the materials out of which the great masters of historiography, who suppose that they can see the grand pattern in history, may construct their monumental edifices. Not that I expect them to succeed. For me, to be a historian means to yield all my powers and imagination to the enterprise of reliving the past, and entering into the minds of those who went before us. I see it as neither possible nor pertinent to draw from history the kinds of lesson that can be expressed in laws and systems. To have lived with these people, to have come to recognize ourselves to a certain extent in their human strengths and weaknesses — for me this is the reward of the study of history.

XI. THE 'THINKING REED'

1. PASCAL, HIS SCIENCE AND HIS RELIGION 343
2. PHYSICS 346
3. FICTIONS 346
4. REALISM 347
5. AUTHORITY 348
6. NO SYSTEMS 349
7. LIMITS OF REASON 349
8. THE HEART 350
9. THE SENSE OF BEAUTY 351
10. NATURAL THEOLOGY 351
11. DIRECT REVELATIONS IN NATURE 353
12. SPECIAL REVELATION 355
13. THE NECESSITY OF REVELATION 356
14. ALLEGORY AND HISTORY 357
15. THE HIDDEN CHRIST 358
16. THE LITERARY STYLE OF THE GOSPELS 358
17. CHRISTOCENTRIC REVELATION 359
18. THE 'CHIQUENAUDE' 360
19. METHODS OF PHYSICS AND OF HISTORY COMPARED 361
20. 'DOGMATISM' AND 'PYRRHONISM' 364
21. FROM SCIENCE TO RELIGION 367
22. PASCAL'S CHARACTER 369
23. PASCAL'S STYLE 369

1. PASCAL, HIS SCIENCE AND HIS RELIGION

For the title of this, my last chapter, I turn to Pascal (1623 - 1662): 'Man is but a reed, the weakest thing in nature: but he is a thinking reed. There is no need for the whole universe to take up arms to crush him: a vapour, a drop of water is enough to kill him. But even if the universe were to crush him, man would still be nobler than his slayer, because he knows that he is dying and the advantage the universe has over him; the universe knows none of this. Thus all our dignity consists in thought.'[1] 'By means of space the universe encloses me and swallows me up as if I were a point; yet by thought I enclose (*je comprends*) the universe.'[2]

This theme of man's greatness and his lowliness comes back again and again in Pascal's *Pensées* (1659). It is a theme faintly echoed by Darwin at the end of his *Descent of Man* (1871): 'Man with all his god-like intellect which has penetrated into the movements and constitution of the solar system — with all these exalted powers — Man still bears in his bodily form the indelible stamp of his lowly origin.'[3]

Yet human pride should beware of too hastily applauding Pascal's praise of reason! For he goes on to point out that there is not only a contrast between the insignificance of the body and the greatness of the mind but also that, within that mind itself, there is a similar contrast of greatness and insignificance. Even Reason, in which consists 'our whole dignity', shares in human frailty. It shows its shortcomings in all fields, science and theology included. Therefore, says Pascal, Man should not be too proud of himself; 'if he exalts himself, I humble him; if he humbles himself, I exalt him, and I go on contradicting him until he understands that he is an incomprehensible monster.'[4]

There is no Pascalian philosophical system — only a Pascalian manner of philosophising. He has been claimed by several parties, though fundamentally he belonged to none of them.[5] His admirers generally choose what appeals to them, ignoring the likelihood that Pascal himself would immediately have advocated the other side! Being imbued with biblical thoughts he cannot be caught in one formula: his positivism can be characterized neither as rationalism nor as irrationalism, neither as dogmatism nor as scepticism but only as an uncompromising realism.

Pascal has remained more 'modern' than most writers of the 17th century. He is known as a great mathematician and physicist. He is perhaps even more widely appreciated as one of the best French authors of his century, mainly because of the eighteen satirical *Lettres Provinciales* (1656-57) in which he defended the Augustinian orthodoxy of his friends at Port-Royal against the casuistic morality of their Jesuit enemies.

But above all, his fame rests on his *Pensées*, the notes and aphorisms jotted down for his planned apologetics of the Christian religion. Even those people on whom his plea for Christianity has little impact, admire his analysis of human nature, of human society and of human knowledge.

Fig. 57. Portrait of Pascal as a young man. Artist unknown.

2. PHYSICS

Pascal's earliest publications were on mathematics (1640) and science (1647). The main problem which occupied his short career as a physicist was that of the 'horror vacui'.

The philosophers of the ruling Aristotelian school held that a vacuum could not exist because 'nature abhors a void'. Descartes likewise — because of his *a priori* tenet that space and matter are identical — would not admit the possibility of a space devoid of matter. He decreed that the 'vacuum of Torricelli' (above the quicksilver in a barometer tube) must be full of subtle matter (*matière subtile*). Although atomists like Gassendi held the opposite view — that empty space *does* exist — they too founded their opinions on argument alone. There was no tangible proof that a very subtle fluid, imperceptible by the senses, filled the space of Torricelli; but on the other hand nothing could prove that it did not.

Pascal kept aloof from the dispute, because he held that 'in physics, experiments are more convincing than arguments.'[6] He refused to make any apodictical statement as to whether an absolute vacuum does or does not exist, because it was impossible to decide this by experiments. The only thing he claimed to know was that above the mercury of the barometer there is a space devoid of any matter capable of affecting the senses.

In 1648, however, he performed the experiment which provided the solution to the problem and demonstrated that no 'horror vacui' — that fiction of theoretical reason — was required to explain the ascent of quicksilver in a barometer tube. The Puy de Dome in Auvergne was climbed carrying a barometer. The level of the mercury in the tube (as measured from the surface of the mercury in the vessel in which the tube was standing) was found to be lower when at the top of the mountain than it had been at the foot. Pascal deduced that the height of the quicksilver column is proportional to its cause, the weight of the air of the atmosphere pressing on the quicksilver in the vessel.[7] This insight enabled him to write in 1654 a pair of elegant treatises on 'the equilibrium of liquids' and 'on the weight of the mass of the air' (published posthumously in 1663) dealing respectively with hydrostatics and aerostatics. In these he consistently uses analogical reasoning, the leading thought being that air behaves like a liquid. The atmosphere when in equilibrium with a column of liquid may be compared with a liquid in equilibrium with another liquid.

3. FICTIONS

To Pascal the issue here at stake was not only one of *facts*, but also one of scientific *fictions*. He teasingly remarked of the thesis 'that there is an imperceptible substance unknown to the senses' in the space of Torricelli, that 'it is not very difficult to explain how an effect can be produced if one invents the matter, the nature and the qualities of its cause, especially if they are so well adjusted that from the invented imaginings truths are cogently concluded that were already evident beforehand.' The most puzzling phenomena, such as high tide and magnetic attraction, would seem easily explained if we were allowed to create substances and qualities expressly for the purpose. When Pascal's opponents say: first show us that this subtle matter does *not* exist, his answer is that, on the contrary, it is *their* duty to prove that it *does* exist: for all these 'substances' of magnetism, gravitation, light, etc. which have been invented for the occasion 'cannot be accepted simultaneously without making nature a monstrosity.'[8]

4. Realism

Pascal is thus on the side of positive facts and distrusts hypotheses based on fictions derived from Aristotelian or corpuscular philosophies. Being, as we have seen, extremely critical of human reason, it is no wonder that he deems it practically impossible to find physical hypotheses that are absolutely certain. In astronomy he does not go so far as to take the apparent motions of the planets to be the real ones, but he refuses to make a choice between the rival hypotheses of Ptolemy, Tycho and Copernicus. The retrogradations of the planets may be deduced from all three, though not all three can simultaneously be correct: 'But who will dare to make such an important decision and who will be able to support one of them to the exclusion of the others without the possibility of error?'[9]

Pascal's rather positivistic approach is also evident if we compare him with Robert Boyle. In *The Skeptical Chymist* (1661) Boyle puts the question as to whether a vacuum exists in an *absolute* sense, and consequently finds himself unable to make a choice for or against its existence. Pascal on the other hand, says that the space of Torricelli is empty, not in an absolute, but in an empirical sense: 'By this word "void" I always mean a space void of all bodies that fall under the senses.'[10] He emphasizes that he does not declare 'in decisive terms' that a vacuum does exist, but only that he will regard that space as empty as long as no matter has been detected in it.[11]

This quasi-positivistic attitude implies a rejection of dogmatism in science, but it also protects him against scepticism — the view that real knowledge of a kind is unobtainable.

5. AUTHORITY

In Pascal's physics experience occupies first place, reasoning comes second, but to authority no place at all is granted.[12] It was not, therefore, from fear of Rome that he avoided making a choice for or against the motion of the earth.

After Galileo's condemnation in 1616 and 1633 the theological argument had indeed become the decisive one in Roman Catholic circles: Rome had spoken and the case seemed to be closed. Catholic scholars complied with the verdict of the Inquisition, some half-heartedly like Gassendi, some with duplicity like Descartes who threw a thin veil of verbal conformity over his essentially Copernican convictions. The Jesuits, who were considered by Galileo and Pascal to be the main instigators of the decree against the motion of the earth, yet tried to be as modern as possible without admitting the motion of the earth and therefore in general opted for the Tychonian system.

The reason Pascal did not decide for the earth's mobility was that, in his eyes, the empirical data did not yet justify any definitive choice. It was not because the authority of the Church he belonged to was against it. The *facts* of nature are, in his opinion, something upon which authority is least able to lay down the law, for they are what they are, whatever a Pope might decree.[13] He tells the Jesuits:

> In vain have you obtained that decree from Rome against Galileo, which condemned his opinion touching the motion of the earth, for that will not prove that the earth stands still. If one had solid observations proving that she does revolve, then all people together would be unable to prevent her from revolving or to prevent themselves from revolving with her.[14]

Like some Copernicans before him, he reminded his opponents of Pope Zachary's condemnation of the idea that there are antipodes: had not the hard fact of their discovery made such a condemnation null and void?

Much of Pascal's quarrel with the Jesuits was about the respect due to *facts*: facts in physics and also facts in history. Just as one should not add anything that is not in nature to the facts of nature, so one should not add anything to the writings of a certain author as if it were contained therein. But the Jesuits had put together five theses, allegedly taken from the *Augustinus* of Bishop Jansenius, though they are not to be found in that book; and then obtained a decree from Rome condemning them. Pascal, pursuing the parallel, points out that it was foolish not only to establish by decree that the earth does not revolve, but also to behave in a similar way in the persecution of the 'Jansenists': 'Thus, you see, Father, the nature of matters of fact; from which it follows that it is not possible to extract the five propositions from Jansenius, if they are not therein.'[15]

Pascal's especial wrath was roused by those who, in the Name of Him, who is worshipped as *the* Truth, made up 'facts' with an appeal to authority in science and theology: 'the Inquisition and the Society [of Jesus], the two scourges of the truth.'[16]

6. No Systems

One of the things that have disappointed many scholars, is that Pascal has no closed philosophical system claiming to explain all things, nor a theology explaining away mysteries by words, nor a complete system of natural science. On the contrary, he stands in downright opposition to them. Reason, which he alternately exalts and humiliates, can, in his opinion, be used correctly only when it has recognized its own limitations. In this respect, for Pascal (as afterwards for Newton) the great offender was Descartes, who treated physics as if it were, in its main outlines, a deductive science like mathematics, because he believed that if one uses reason rightly, true physics must be the definitive outcome. In Pascal's eyes, however, the Cartesian system is a 'roman de la nature' and when he announced his intention to write against those who want to get too much out of science, this was aimed mainly at Descartes.[17]

Pascal pointed out that the knowledge of nature is not deductive but inductive in character. In all matters in which proof rests in the first place on observations and experiments and not on logical arguments, the incompleteness of the available data makes definitive statements impossible. To Pascal physics will always remain an open discipline, for Nature, in its infinite wealth, grants us no rest. Sooner will the human mind 'become tired of conceiving than Nature of furnishing [new data and problems, RH]' (fr.72). And as Nature is a whole and to Pascal any knowledge is based on only a part of it, revision in the light of further discoveries may become necessary and a final system can never be constructed: 'Do not draw from your apprenticeship the conclusion that nothing remains for you to know, but rather that infinitely much remains for you to know' (fr.231).

7. Limits of Reason

Pascal pointed out how little knowledge we possess of the *nature* of things.[18] Like Galileo, he recommended that for the moment we should describe them exactly and find out how they behave, rather than saying *why* they behave thus. He waged war against the false claims of 'science so-called' and spoke of the 'vanity of sciences' (fr.67), and the 'folly of man's science and philosophy' (fr.74). Philosophers who fabricate theses and antitheses resemble those 'who make blind windows for the sake of symmetry' (fr.27). After relating, in his treatise on 'the equilibrium of fluids', how the makers of pumps had long known that water cannot be pumped up from more than a certain depth, he scornfully exclaimed: 'simple workmen were able to convict of error all the great men that are called philosophers.'[19]

Reason cannot encompass reality: 'We may blow up beyond imaginable spaces our conceptions — we only bring forth atoms, at the expense of the reality of things' (fr.72). What do we know about the propagation of light through a vacuum, about magnetism and about the fundamentals (e.g. what matter really is, or the soul or the interaction of the two)? Reality in its depth and extension is incomprehensible to Reason, yet we have to recognize the existence of things beyond Reason: 'It is not by our capacity for understanding things that we should judge about their truth.'[20] It is not only *physical* facts that are incomprehensible and have to be accepted because they press themselves upon us with irresistible force; the same may be said of the principles of mathematics. Because people cannot imagine infinite divisibility, they think it cannot *exist*, but 'it is a natural disease of man to ... deny everything he does not understand.'[21]

8. THE HEART

What then legitimately gives us the certainty of the existence of incomprehensible things in spite of our inability to 'demonstrate' or to grasp them by reason? The fundamental facts in natural science or mathematics are undeniable; we do simply accept them without logical demonstration. Reason has no authority here; what then gives evidence of their truth?

The answer is: 'We know the truth not only through reason, but also through the heart. It is through the latter that we know first principles, and it is in vain for reason, which takes no part in this, to attempt to combat them' (fr.282). Natural science comes to a halt before facts which, however incomprehensible, yet are evident. Mathematics comes to a halt before axioms and undefinable objects, which are inexplicable and yet evident.

The 'heart' (*le coeur*), with Pascal, stands for immediate intuitive knowledge, which (self-deceit apart) is not less certain and is even more fundamental than rational knowledge. The term covers all kinds of immediate inner experience and direct insight, which play a role not only in science and mathematics, but also in aesthetics, religion and morals. Unfortunately, however, sentiments of the moment are often mixed with correct intuition: 'Men often take their imagination for their heart' (fr.275).

In mathematics, the most certain of sciences and one about which all people agree, Pascal considers the heart hardly errs. It is most liable to go astray in religion and morals, where the interest in self-justification and self-vindication is greatest and a special divine grace is indispensible for finding the truth.[22]

9. THE SENSE OF BEAUTY

Quite special is the functioning of the heart with regard to the experiencing of the beautiful. In Pascal's view there must be 'a certain model of agreement' between our heart and the thing we consider beautiful.[23] (An opinion which we should by no means confuse with the platonic belief in absolute Beauty, which has no analogy with Pascal's own belief in absolute Truth!) In his opinion — though it is not the custom to do so — we should speak not only of 'poetical beauty' but also of 'medical' and 'geometrical' beauty. His answer to the question 'why don't we do so?' is that we know the object of geometry (*viz.* demonstration) and of medicine (*viz.* healing), but we do not know what is the 'agreement' ('the charm') that is the object of poetry, nor do we know the natural model that should be imitated (fr.33). Consequently, there cannot be a general consensus about what is beautiful: 'man finds in his own heart the model of that beauty which he seeks outside himself.'[24]

Pascal's skill in deciding how far analogical reasoning should be applied is shown here again. In physics he used the analogy of hydrostatics and aerostatics, up to a certain point: compressibility he wisely left out of it. With even more reservations he compares the infinite in mathematics, in nature and in theology. The heart functions in a different way in these diverse realms: in mathematics it recognizes intuitively the fundamentals, in aesthetics it is the personal agreement between the individual and the object, in religion the heart is being overwhelmed by divine grace.

In 17th century physics the situation was rather complicated: there were fundamental differences between the principles of the atomists, the Cartesians and the Aristotelians. Pascal pointed out that it is nevertheless possible to build reliable *partial* systems in physics, e.g. in hydrostatics. In metaphysics and ethics, however, the differences between the various philosophical systems are so great that the heart here is no longer reliable: what is good on one side of the frontier is not good on the other side. The very diversity of philosophical systems shows how little the heart can here be trusted, corrupted as it is by our *will*. Therefore, although one may philosophize in a critical way one cannot construct a reliable all-embracing philosophical system. 'Se moquer de la philosophie, c'est vraiment philosopher' (fr.4).

10. NATURAL THEOLOGY

The above brings us to the topic of natural theology, or perhaps natural religion, which (as Henry Brougham pointed out) is not precisely the same thing. Can Reason produce such a thing as natural theology, or, to keep to our theme of 'nature outside us': do the senses provide us with relevant data for it? There is in Pascal's works on physics not the slightest mention of natural religion, nor any appeal to revealed religion. Evidently in physics he wanted to keep to the physical facts. At first sight it may seem strange that neither in his scientific works nor in his *Pensées* does he ever proclaim nature or natural science as unmistakeably pointing to a Creator of nature. To Pascal, although God is always and everywhere active, this is activity in a manner above and beyond the ken of philosophers and physicists as such. Pascal's physical works deal only with small parts of reality, and then only from the limited viewpoint of mathematical-physical description. The claims he made for the truth of physical theories and descriptions are so modest; he could hardly found a 'demonstration' of God's existence and activity on what he deemed so narrow and uncertain a basis.

Yet with Pascal science and mathematics are not wholly useless for 'natural theology'. They may not demonstrate God's existence and presence in a direct and apodictic way, but they point to Him in an indirect way by undermining any suggestion that He does *not* exist.

Man's rationalistic certitude — his cocksureness — is undermined by the incomprehensibility of the infinite in mathematics and the vastness and contingency of nature. We are, says Pascal, between the abysses of infinity and next-to-nothingness.[25] He who realizes this 'will tremble ... and as his curiosity changes into admiration, he will be disposed rather to contemplate [the things of nature] in silence than examine them with presumption.' Men try to 'comprehend the principles of things and to arrive from there at the knowledge of everything, inspired by a presumption as infinite as their object.' Every foundation that has been laid, requires its own foundation: every point to which 'we intend to attach ourselves ... eludes our grasp, slips away and flees eternally before us ... This is the state that is natural to us, and yet it is the most contrary to our inclinations — we burn from desire to find a firm footing, an ultimate, lasting base on which to build a tower rising up to infinity, but our whole foundation cracks and the earth opens up into the depth of the abyss' (fr.72).

Here we get a glimpse of man's yearning for the essence of things, for an understanding that is at the same time an identification with the object of knowledge as well as a rational comprehension of it. It is the vain pursuit of the horizon, the search for the philosophers' stone, which is always 'almost' found, but which vanishes when the hand stretched out to grasp it has almost touched it.

But at the same time there is an indication of the Numinous: the boundless space which besets us on all sides, and terrifies us by its eternal silence.[26] Clearly the 17th century consciousness of the vastness of the universe did not diminish religious sensibility: it may have undermined a certain comfortable feeling of encompassing the whole world 'God had made for man', but it increased the feeling of what Boyle called 'awe and veneration'. It was a 'revelation' of a rather negative character, a testimony of the Great Unknown. It was still a far cry from the God Pascal so fervently sought: the God who unites Himself with man.

This, then, is Pascal's indirect 'natural theology'. As God is (by definition) incomprehensible, the incomprehensibility and infinity of Nature (physical and intellectual) may be the imprint of the character of its Creator. And just as God's work contains something of His image (infinity), so man's understanding of it (by science) is tinged with the character of man's mind (finiteness; insufficiency). Science, then, the contemplation and investigation of nature, by its very limitations and impotence could be a way to the 'hidden God'.[27]

11. Direct Revelations in Nature

But does not nature point in a more direct and positive way to God? Pascal has no time for those apologists of Christian religion who, when addressing non-believers, write 'their first chapter to prove the Divinity from the works of nature' (fr.242). He marvels at the boldness with which these people dare to speak about God: 'I would not be surprised at their enterprise if they were addressing their discourse to the believers, for those with living faith in their hearts can certainly see at once that everything which exists is entirely the work of the God they worship' (fr.242). But he thinks that it is useless to say to those in whom the light of faith is extinguished and who find only obscurity and darkness in nature 'that they have only to look at the least thing around them and they will see in it God plainly revealed; to give them no other proof of this great and weighty matter than the course of the moon and the planets ... is giving them cause to think that the proofs of our religion are indeed feeble' (fr.242).

To the question 'But do you not say yourself that the sky and the birds prove God? ... And does your religion not say so?', his answer is: 'No. For though it is true in a sense for certain souls to whom God gives his light, yet it is untrue for the majority' (fr.244).

When Pascal thus says that in order to recognize God in nature one's eyes must be opened, he assumes a standpoint which for centuries has remained quite exceptional among Christian apologists. Nearly all of them started by demonstrating God from nature, and then continued by making it acceptable that this God of nature is the same as the God of the Bible. A typical case is Hugo de Groot (Grotius) in his *On the Truth of the Christian religion*, a book widely read throughout two centuries. Apparently with reference to this work Pascal says: 'No canonical author has ever used nature to prove God. David, Solomon, etc., never said "there is no void, therefore there is a God"' (fr.243).

Pascal seems to be right indeed: David's psalms of spontaneous praise for God's power, wisdom and goodness in His creation have nothing in the nature of a logical demonstration. Pascal was certainly correct to distinguish the psalmist's laudatory utterances from Grotius' appeal to the 'horror vacui', demonstrated from Aristotelian philosophy by sophisticated proofs. It is interesting and understandable that from among all Grotius' natural theological arguments Pascal picked out this topic. For had not he himself submitted the 'horror vacui' to unrelenting criticism? 'Water', so Grotius had written, 'contrary to its nature, is raised upwards, lest through a vacuum intervening, the fabric of the universe should part asunder... Now this universal end could not be intended and a power be implanted in things to that end, except by an Intelligence, to which this universe was in subjection.'[28]

Pascal, who had contributed so much to the overthrow of the 'horror vacui', must have felt this a conspicuous example of the weakness and danger of arguments for God's existence borrowed not even from nature but from natural philosophy, man's fallible interpretation of nature. Of course the critique had little effect. Jean Leclerc, who edited Grotius' work in 1709, added to the passage under discussion a reassuring note. He unabashedly wrote that the ascent of the water in the pump is nowadays explained by the gravity of the air and this shows no less the order of the universe and the wisdom of the Creator![29] Pascal did not believe that nature (let alone the science of nature which is a product of man) had been made for the purpose of religious instruction: 'If the world existed in order to instruct man about God, His divinity would shine forth on all parts in an incontestable manner ... What does appear marks neither total exclusion nor a manifest presence of divinity, but the presence of a God who hides Himself.'[30] God, as we know Him from nature, is a hidden God, *deus absconditus.*[31]

But even if by natural reason proof could be given of God's existence, or of the Trinity, or the life hereafter this knowledge would be 'useless and sterile': 'If a man were persuaded that the proportions of numbers are immaterial, eternal truths, depending upon a primary truth in which they subsist and which is called God, I should not consider him advanced very far towards his salvation' (fr.556). And only this Pascal thinks worthwhile: he wants a God with whom he has a personal relation; therefore he feels not the slightest urge to discover 'harmonic relations' or particularly meaningful numbers, like 3, 7 or 10 in nature. He rejects the natural theology of Raymond de Sébonde, which finds all Christian tenets mirrored in the Book of Nature, as well as the more soberminded natural theology that considered Scripture as a specifically Christian addition to the conclusions human reason could draw from metaphysics and the contemplation of nature. The God whom he seeks is not 'a God who is simply the author of geometrical truths and of the order of the elements', but the God 'of love and consolation'. This God is known in the first place not by reason, but by the Heart: 'it is the Heart (and not Reason) that experiences God'. He is a God who enters into personal relationship with men, and 'who unites Himself with them in the depths of their soul' (fr.556).

Of course, this 'God of Abraham, Isaac and Jacob, not of the philosophers and scholars', was also recognized by Christians who had more respect for natural theology than had Pascal. Newton held that the task of science was to follow up the chain of causes, until it reaches the First Cause, by 'the Light of nature'[32] and made much of the demonstration of God's wisdom through final causes. But finally his emphasis was on the same personal relationship as Pascal's when he wrote: 'for we say, my God, your God, the God of Israel ... but we do not say, my Infinite or my Perfect.'[33]

On the issue Pascal certainly did not expect as much from the sciences as did Newton. At best he regarded them as in a certain sense a help to religion: the sciences make us acknowledge that things beyond human comprehension yet do exist and that there is some analogy of the impersonal infinite in mathematics and in the world of nature with the personal Infinite hidden behind them. But this is a rather poor result; only just enough to prevent dogmatic rejection of God. It is even impossible to elicit from Pascal any wholehearted affirmation that in nature we meet with a Great Mathematician as was held by many scientists before and after him, from Kepler to Jeans. In Pascal's physics the role of mathematics is descriptive rather than ontological.

12. SPECIAL REVELATION

Critical reason had shown by analysis of scientific procedures that the facts of nature are contingent and that fundamental notions and axioms are not obtained by discursive reasoning but that they are intuitively apprehended. If then the fundamentals of the science of nature (notions and objective 'facts') are not demonstrable, how can metaphysics and natural theology demonstrate the foundation of religion, even God Himself? Here Reason is powerless, as it is also when confronted with Nature which cannot be excogitated, but has to be accepted as it is given.

Like Francis Bacon before him, Pascal advocated the submission of reason to facts even if they are beyond reason, as being a most reasonable act: 'Nothing is so much in conformity with reason as this disavowal of reason' and its submission when it must submit.[34] But this recognition of facts by the 'heart' meets sometimes with a resistance inspired by the fear of giving up one's cherished scholarly constructions; a fear that leads to self-delusion. The 'heart' is then easily confounded with fancy or imagination, which resembles the heart and nevertheless is contrary to it.[35] In religion these obstacles are strongest, so that God's special grace is needed to convert the heart and lead the *will* in a new direction.[36]

Though Pascal is fully aware that nobody can plant religion in another person's heart, he thinks that there are two 'objective' tests for the truth of a religion: 'any religion that does not say that God is hidden is not true,' and 'for a religion to be true it must have known our nature.'[37] The Christian religion satisfies both conditions. It emphasizes the ambiguous character of human nature, for which it is equally impossible to know with certainty and absolutely not to know, a nature which is always wavering between scepticism (which is confounded by the heart in which some light of truth has remained) and dogmatism (which is confounded by our incapacity in 'demonstrating').[38] The origin of all these contradictions and dissatisfactions lies, Pascal repeatedly says, in the Fall of Man; the only way out of this situation is redemption given by Jesus Christ (fr.242, 556).

But this is incomprehensible! Pascal fully recognizes this: the so-called 'proofs' of Christian religion are not 'absolutely convincing'; divine matters are not subject to the 'art of persuasion'; they are far above nature and only God Himself can lay them in the heart in the way that pleases Him.[39] They enter from the *heart* into the mind and not from the reason into the heart: 'in order to humiliate this proud power of reason.'[40] For 'the heart has its reasons which are not known to reason.'[41] 'It is the heart which senses God, and not Reason ... Faith is God sensed by the heart, not by Reason.'[42]

Pascal's intellectual honesty did not permit him to go further than claiming that it is not unreasonable to recognize the truth of Christianity and to accept it by the heart. At the same time he emphasized that faith is 'a gift of God' and he vigorously denied that he (Pascal) could make the mind *willing* to accept it or that he could plant it in another man's heart.[43]

What he means is perhaps best understood by comparison (given due reservations) with aesthetic experience. We can all agree, on the basis of physical measurements, about the frequencies of the vibrations that are present in a piece of music; we can also agree about the technical achievement of the composer, but we cannot prove to somebody else the beauty of the music as it is appreciated by our 'heart' — the beauty which is, though undemonstrable, the most essential part of its reality. Those who are deaf to it will only hear sounds without a deeper meaning to them. However different aesthetic experience and religious experience and belief may be, they have in common that they are not transferable. We must add, however, that Pascal would have stressed where the analogy stops: religious experience must be put to the outward test of Holy Scripture, whereas for aesthetic experience no checking is possible.

13. The Necessity of Revelation

Pascal's way of thinking corresponds closely with biblical conceptions. Christian religion acknowledges that God is hidden; there is no way from man to God; our most lofty ideas about Him are as far from His perfection as the most primitive ones. Consequently, if God exists and if He can be known, it must be by His coming to us. Otherwise we are left to agnosticism, for, as far as man can go by his own force, 'scepticism is right'.[44] If we are ever to meet Him, it is because He stoops down to us, speaks to us in human words, using human concepts, and this in whatever way pleases Him, a way that may be 'foolishness' to the 'learned' and degradation of divine dignity to the 'religious'. Now the Scriptures tell us that He has revealed Himself in Jesus of Nazareth, a carpenter's son, and that in Him God has communion with us and has become one with us.[45]

The above paragraph is, in paraphrase, Pascal's faith, and if people with more lofty and enlightened religious ideas object that it is unbelievable that God should unite with man, Pascal's retort would be: 'What entitles this animal, so well aware of his own weakness ... to keep God's mercy within limits suggested by his own fancy?'[46] And to the objection that all this is incomprehensible, his answer would be that 'everything that is incomprehensible does not cease to exist.'[47]

Christian faith, then, finds its *foundation* not in a moral code, not in a mystical experience, not in an ideology, but in the reality of historical events: the revelation of Christ, the birth of the incarnate God, his life, his death on the cross and his resurrection. As in science, so in religion, the basis lies in facts not excogitated by the human mind. Those who have not witnessed these events have to resort to the reports written by his disciples. And though their inner evidence is, in Pascal's opinion, great enough for Reason to accept them as truthful, the 'perversity (*malice*) of the heart' will resist. Therefore, in those who accept these reports, 'it is grace and not reason' which makes them do so (fr.564), for historical proofs are 'not absolutely convincing' in the way mathematical proofs are.

Nevertheless Pascal does not feel himself exempted from the duty to remove 'rational' obstacles by expounding why acceptance of the historical reliability of the gospel narratives is quite 'reasonable', even if they are 'incomprehensible'.

14. Allegory and History

The interpretation of a biblical passage has to establish first whether it is meant allegorically ('spirituellement') or historically ('littéralement') (fr.648). That is, whether it should be taken as a 'figure' or as a 'reality' (fr.687).

The next step is to establish whether that which is in Scripture 'literally false' is 'spiritually true', e.g. that it is speaking about God in an anthropomorphic way (which is inevitable because 'divine things are inexpressible and cannot be said in another way', fr.687).

As to the historical truth of the gospel narratives, we hardly find support from ancient profane writers, who are only interested in political matters (fr.786-787) and who leave Jesus Christ 'in obscurity' ('according to what the world calls obscurity', fr.786). The gospel stories have been written by men who claim to have witnessed the events related or to have obtained first hand information from the witnesses. So, if they bear the character of historiography and claim to be historic truth, they are in Pascal's opinion either truth indeed or they are phantasies. Even an allegorical interpretation would not make them less false if the authors clearly intended that they should be taken as historical truth.

15. The Hidden Christ

We have seen that Pascal considered one of the marks of the truth of a religion to be that it says God is a 'hidden God'. But if it is true that the Infinite stoops down and *reveals* Himself in Jesus Christ, would this unexpected fact not be easily recognizable? Pascal, however, is not slow in pointing out that even in this 'revelation' (disclosing, unveiling) there is hiddenness. God, so he says, was hidden under the veil of nature until the Incarnation, and when He appeared He was hidden in that He covered Himself by being-human.[48] If it had not been otherwise, everybody would have accepted Him, but Christ's divine character was not recognized even by his own disciples. And when, after his resurrection, He fully revealed himself, His character and mission were (as had been prophesied) revealed only to his disciples and not to the 'world'. What do the prophets say about Jesus Christ? That he will *manifestly* be God? No, but that He is a really hidden God, that He will be misjudged, that people will not at all think that He is it, that He will be a stumbling stone (fr.751).

Pascal deems the greatness (*grandeur*) of Jesus Christ to consist in his lowliness (*bassesse*) 'in his life, in his passion, in his obscurity, in his death, in his choice of friends, in his being abandoned by them in his *secret* resurrection' (italics RH) (fr.793).

Of course, Pascal was not the first to make these points. Before him, also in the wake of Augustine, the early 17th century poet John Donne, for example, had pointed out that it was not easy to believe that man should look for salvation from him 'who could not save himself from ignominy, from the torment, from the death of the cross'. And Donne, too, pointed out that a man-made religion would have given a more glorious picture: 'if any wise *man* had been to make a religion, a gospel; would he not have proposed a more profitable, a gospel more credible to man's reason than this?'[49]

16. The Literary Style of the Gospels

The style of the gospels is, so Pascal says, 'admirable' (fr.797): the soberness of the writers (*les historiens évangéliques*) is a proof of the veracity of their story. The evangelists do not invectivize against the enemies of Jesus; they do not depict the apostles as heroes. Pascal thinks that it has been overlooked with how great a coolness (*froideur*) the facts have been reported (fr.798). If the evangelists had invented their story, quite a different tale would, he thinks, have emerged. He asks himself who could have taught them to relate 'the character of a heroic soul' so perfectly as when they depict Jesus as so strong a man and yet 'make him so weak in his agony'. Were they not able to depict a more constant martyrdom? The answer is: yes, they could, for the same St Luke describes Stephen's martyrdom (Acts 6:9ff., 7:54ff.) as 'stronger' than that of Jesus of whom he reports the weakness when alone in Gethsemane (Lk. 22:39, Matt. 26:39, Mark 14:36) as well as the great strength shown in public (fr.800). In particular Jesus' absolute loneliness in the night of Gethsemane made a deep impression on Pascal: Jesus sought some consolation from his three best friends, but they slept.[50]

If the story of the resurrection was a criminal conspiracy by the apostles, Pascal goes on, what a risk they would have taken; for if one of them had given in to the pressure of the authorities, all of them would have been lost (fr.801). They were not deceivers nor were they deceived. As long as Christ was with them he could keep them going, but after his death [when to the world his cause seemed lost], how could they have gone on if he had not indeed appeared to them? (fr.802) It is evident that in Pascal's view the apostles either spoke the truth or concocted a story that in several respects was detrimental to its own recruiting power from the standpoint of the 'world'.

17. Christocentric Revelation

Pascal's vindication of Christian religion is wholly Christocentric: the Scriptures, the Old as well as the New Testament, the prophets as well as the evangelists and the apostles wholly concentrate on Jesus Christ. We know God only through Christ; without that Mediator all communication with God is taken away (fr.547) and, indissolubly connected with this, 'without Scripture, which has but Jesus Christ as object, we know nothing and see only obscurity and confusion in God's nature and in Nature herself' (fr.548).

Pascal points out that the Scriptures speak of Mary's virginity only up to the birth of Christ.[51] Evidently the question as to whether she remained a virgin in her subsequent life played no role in the drama of salvation, and therefore does not enter into its story. This may explain the fact that in Pascal's apologetics neither Mary nor other saints are mentioned as mediating between God and man.

The Fall of Man from his original perfect state explains, so Pascal points out, his present ambiguous moral nature. Pascal does not enter into the historical details of the Fall in the way he does about the historical details of the redemption through Christ. Why not? Here again, as in physics, his self-imposed restriction to facts operates. We can conceive, so he says, neither the glorious state of Adam nor the nature of his sin, nor the transmission thereof to his descendants; these are things which happened in a situation of nature totally different from ours, and one which wholly surpasses our present capacity of understanding. All this is useless for knowing how to get out of our present situation; what matters is to know that we are miserable, corrupted, separated from God, but bought free by Christ (fr.560).

With a sober conception — what matters most is not to know how man fell into the abyss, but how he can get out of it — such questions as 'how could God allow the Fall to happen, and what was the situation before it happened' were avoided. Pascal was as averse to theological speculation as he was to natural-philosophical speculation. He declined to enter into the question as to how the Fall could take place, in the same way that he refused to answer the question as to how light can pass through a vacuum when there is no matter to transmit it. What counts is reality: light goes through the space of Torricelli and, as far as we know, there is no matter in this space. Similarly, there is sufficient proof in real life that we are in a fallen state, and we see also that there are traces left of a better state (fr.425).

18. THE 'CHIQUENAUDE'

It goes without saying that an empiricist in physics like Pascal neither tries to give a 'scientific' interpretation of the biblical story of creation, nor to imagine an alternative scientific story of it as Descartes had done. In Pascal's Christocentric conception the problem of the interpretation of the Genesis story is marginal and is never touched.

While admitting with Descartes that 'these things happen through their shapes and motion,' he refused to specify *which* shapes and motions, because that is 'useless and uncertain and awkward' (fr.79). And this is the more so when the events took place without witnesses and are not repeatable. Yet it would be in his spirit that *if* such an imaginative physical story were put forward, it should bear, like experimental physics, a purely physical character and remain free from metaphysics.

Why then did Pascal write: 'I cannot forgive Descartes: in his whole philosophy he would have liked to do without God; but he could not help allowing him a snap of the fingers (*chiquenaude*) to set the world in motion; after that he had no more use for God' (fr.77)?

I think it is because Pascal held that, if one insists upon bringing God into a physical account, one should not do this only at the beginning and then in the later stages let things follow automatically. If God has indeed set the world into motion, He must also uphold this motion.[52] Pascal implicitly reproaches Descartes with having produced a semi-deistic system, and with having mixed up two levels of 'explanation'. On the higher, 'metaphysical', level God is always and everywhere behind all physical events; on the physical level one ought to restrict oneself, as a matter of method, to the rules and the mechanisms that are found by observation.

One might ask: why the emphasis on the *historical* character of the revelation in Jesus Christ? Isn't Christian religion 'God experienced by the heart' ('Dieu sensible au coeur') and doesn't the following of Christ manifest itself in charitable action: the two, in Pascal's own view, 'inseparable from each other'? (fr.442) And doesn't he, the passionate fighter for Truth, warn against making Truth an idol?[53]

Indeed yes, but the source and the guarantee of the genuineness of Christian inner experience as well as of Christian action and morals lies according to the apostles in the concrete acts of Christ himself, in his redemptive sacrifice and his resurrection. To the realist Pascal the historical basis is indispensable to make faith more than an ideology and a myth. Religious experience could be self-delusion, for we often take our 'imagination for our heart' (fr.275) and what we consider sound morality is often determined by what we deem good for ourselves. What is called 'good' on one side of the border, is considered evil on the other (fr.375). As a touchstone for Christian experience and Christian action Pascal resorts to Scripture. The 'historical' faith, then, is to him the basis of the other apects of Christianity; his religion is a belief, though inseparably connected with an experience and a way of life.

19. Methods of Physics and of History Compared

In about 1930, when for the first time I was perusing Pascal's works, I formed the impression that though the critical change in his life in 1654 may have abated his interest in physics, his frame of mind remained that of a physicist even in his religious writings: 'there is only a change in the "matter" but not in the "Form" ... the statements from his younger years about natural science, mathematics and religion are not in contrast with those of the *Pensées*.'[54] Thus we see him drawing analogies from his earlier fields of interest — the mathematical concept of infinity serves to illuminate the infiniteness of time and space and of 'The Infinite'.

Both, science and theology, are based on *facts*; in physics facts of nature, recognized by observations and experiments; in theology there ought to be submission to facts. Facts of history and facts of nature are not 'demonstrable'; they are accepted by the 'heart' and reason has to recognize that this is rightly done.

All analogies, however, break down at a certain point. The facts of physics are repeatable, those of history are not. Only a few pages of the Book of Nature have as yet been read by mankind whereas those of the Book of Scripture are known from the first to the last page. The Book of Nature has not been written in human language, in contrast to the Book of Scripture.

The difference mentioned above has been expounded in Pascal's fragment of a preface to a treatise planned (but never written) on the Vacuum. He deals there also with the meaning of 'truth' in science and in religion, in particular with relation to the authority of the Ancients in science and theology. We should make a distinction, Pascal says, between sciences based on memory — which are purely 'historical', as they relate what authors have written — and 'dogmatical' sciences, which discover hidden truths. The former deal with simple facts (e.g. historiography), or human and divine institutions (law, theology) of which a complete knowledge is possible. In them one has to refer to the authority of the authors, in particular in theology. But in the disciplines that depend only on observation and reasoning (physics, medicine, mathematics) authority is useless and their inventions are growing with time and efforts. Even if we are less inventive than our predecessors, our knowledge, consisting of *their* knowledge together with our additions to them, is greater than theirs. Therefore we should deplore the blindness of those who refer to authority alone in physical matters instead of reasoning and experiments and should abhor the deceptions (*malice*) of others who use only reasoning in theology instead of Scripture and the Church Fathers. 'We should strengthen the courage of those timid people who don't dare to invent new things in physics, and confound the insolence of those temerarious people who introduce novelties into theology.'[55]

We should not forget, urges Pascal, that if the Ancients had not dared to add to the science inherited from *their* predecessors, we would not have enjoyed today the fruit of their inventions. Therefore, let us follow their example and make their achievements the auxiliaries and not the end of our studies, and 'thus try to surpass them by imitating them'. This implies that we should not hold an inviolable respect for our predecessors which *they* did not have for *their* predecessors ...

Pascal then remarks how the works of nature are hidden, so that we do not always discover their effects: time reveals them from age to age, and 'though always equal to herself, she is not always equally known.'[56] The experience and experiments which are the only principles of physics are continually increasing and the conclusions drawn therefrom increase proportionally. In this way we can have new opinions without being ungrateful and without looking down upon them, for the first knowledge they delivered to us, serves as a stepping-stone towards our knowledge and, as they have lifted us to a certain degree, the slight effort makes us climb higher: with less exertion and less glory, we are standing above them. We have a wider outlook, and though they knew as well as we do all they could perceive in nature, nevertheless they did not know so much about her, and we see more than they saw.[57]

As a true historian of science Pascal thus states that physical knowledge has increased but that our inner process of acquiring it is not qualitatively superior to that of our forebears. We may find when standing on the tenth step of a ladder that we see more than when standing on the ninth, but the act of proceeding from the ninth to the tenth is of the same kind as that of mounting from the first to the second step. In other words, science is increasing in quantity, but scientific *thinking* is not of a higher quality than that of our predecessors.

If one wants to follow the general custom of speaking of the 'evolution' of science, this would be admissible provided one realizes the danger of pushing too far the analogy with the biological theory. But some devotees of the belief in the increasing perfection of scientific man who speak of the 'evolution of scientific *thinking*', go too far. The *results* of scientific thinking (and acting) do increase, but this is no proof that in the short period that *homo sapiens* and his 'science' have existed there has been perceptible 'evolution' in his thinking.

The urge to improvement is inherent in the human character. We want to climb the ladder precisely because the human mind (which remains essentially the same), having made the first steps in the knowledge of nature, wants to mount further. This, in Pascal's opinion, is the great difference from the instincts of the animal. Bees, he points out, make the same well-measured hexagonal cells nowadays as they did a thousand years ago. They stay within the limits nature prescribed them in order to maintain their existence. Man, however, has been made for infinity; he learns in all stages of life, not only from his own experience, but also from that of his predecessors, for their achievements stay in the memory by means of books. Not only man as an individual daily increases his knowledge, but all men together do the same, so that the whole of mankind increases knowledge throughout the ages. This implies that, as a whole, mankind gets older with time and that what we call 'antiquity' is in fact the youth of mankind, and that 'it is in us that this antiquity which we revere in the Ancients can be found.'

This leads Pascal to a view that should be taken to heart by all 'modern' writers on the history of science. He advises us to admire the Ancients for the conclusions they have drawn from the few data at their disposition, and to excuse them for those matters in which they lacked the advantage of experience rather than the strength of reasoning. [Here the result, rather than the quality of thinking, is sometimes said to have 'evolved'.] For, was it not excusable that they, who had not the artificial instruments to support their eyesight, believed that the Milky Way was a more solid part of the heavens, whereas *we* know, thanks to the telescope, that it consists of an infinite number of stars? Similarly, they had every reason to say that all corruptible things are enclosed within the lunar sphere, as they had seen no corruption or generation beyond that space. But we know now that new stars do arise and disappear.[58]

And then Pascal comes to his own subject, the vacuum. The Ancients were right, he says, to hold that nature does not admit it, for all their experience taught them so. In claiming that nature does not allow empty space, they spoke about nature as far as they knew it. In all matters in which the proof exists in experiments and not in 'demonstration', one cannot make a universal assertion but by an enumeration of all cases. So, if we say that diamond is the hardest and gold the heaviest substance, this means respectively the hardest and the heaviest of the substances we know. Similarly, when the Ancients said that nature does not admit a vacuum, this meant that it did not admit it in the experiments they had seen.[59] Pascal concludes that whatever force Antiquity has, the truth is always stronger, even if it has been discovered only recently, for it is always older than all opinions that have been held about it; 'it would be ignoring its nature to imagine that it has begun to exist at the time when it has begun to be known.'[60]

This is a simple truth about truth. It is in line with his later remark that whatever we think about the motion of the earth, *if* it is moving it will keep on moving notwithstanding our opinion (or that of the Church catholic in 1633) to the contrary. Or, to take a modern example: if the continents are drifting apart, they will continue doing so in spite of the opinion of the church scientific (1926) to the contrary.

Pascal's discussion implies a simple truth. While relativizing the finality of truth in experimental science and maintaining the finality of the contents of a written text and of a historical event, he implicitly says that our natural science is not the same thing as nature, and our theology is not the same as Scripture.

20. 'DOGMATISM' AND 'PYRRHONISM'

To whom was Pascal's account of the Christian faith directed? Not to unsophisticated fellow-believers, who simply put their trust in Christ without asking any questions. Nor to those who only by custom, without feeling or thinking further, follow tradition. The main target was his own circle of intellectuals — philosophers and those who pretended to be philosophers — who follow their own devices mixing their scholastic or mechanistic philosophy with Christian doctrine, or, perhaps even more, to the 'esprits forts' who reject any special revelation, calling themselves deists or atheists. He divided them into 'dogmatists' and 'pyrrhonists' (sceptics, agnostics).

With the dogmatists he had already clashed when he attacked the gratuitous assumptions of scholastics and Cartesians in science. He was as much against the 'laws' and the imaginary substances put forward by physicists as against the moral principles and metaphysical theses produced by moralists according to their own taste, which was mainly, in his eyes, 'to satisfy their vanity by the ruin of verity'. The quasi-deism of Descartes' cosmogony was almost as reprehensible in his eyes as the eternal world of Greek philosophers.

Pascal held that 'everybody' feels urged to choose between dogmatism and pyrrhonism. He who thinks that he can remain 'neutral' will be a 'pyrrhonien par excellence', for neutrality is the essence of the dispute — most 'pyrrhonists' are not choosing for themselves but they are neutral, indifferent, even towards their own standpoint (fr.434).

The dogmatist, who believes that he possesses the truth, can easily be refuted by Reason. The pyrrhonist, who doubts whether we are awake or asleep, is refuted by 'nature', which refuses to abandon all certainties (fr.434). But the fact that good and evil, truth and untruth, are indistinguishable and are inextricably mixed is the ground of our uncertainties.[61]

It is clear that Pascal felt more affinity with agnosticism than with any of the man-made 'dogmatisms' in religion and ethics. He tells us that during a large part of his life he had believed that 'justice' (moral good) exists and that he rightly believed this, for there *is* justice — that is to say, insofar as God has willed to reveal it. 'But I did not conceive it in this way, and therein I was wrong, for I believed that *our* justice was essentially right and that I was capable of knowing it and able to judge about it' (fr.375). But then he so often found his judgement wrong that he started to doubt about himself and about others: in all countries and in all people he found diversity and change in their opinions about true justice, and this led him to the conclusion that our 'nature' is not a fixed thing, but is continually changing. 'And since then I have not changed any more, though if I changed this would confirm this opinion of mine' (fr.375). This implies that he originally was a 'dogmatist' (who held that he knew the 'good' by his own nature) but finally had discovered that 'pyrrhonism is true' (fr.432), however paradoxical this may sound.

From his own experience Pascal knew the utter loneliness of one who feels himself thrown in a recess of a mute universe: 'When I see a man without light, left to himself, without knowing who put him there, what he has to do there, what will become of him after death, incapable of any knowledge: I get frightened.' He compares this nightmarish situation with that of a man who in his sleep has been brought to a desert island, and awakens without knowing where he is and how he could get out of it (fr.693).

He asked other people whether they knew the answers to these questions, but they said 'no' and then they suppressed these feelings (fr.693). After some short theological discussions and a couple of hours reading in Scripture they say that they have tried to find an answer, but without success and then they passionately attach themselves to worldly diversions. They spend their life without thinking of the end and the purpose (fr.194). Pascal is irritated by this attitude: the acquiescence in this ignorance is 'a monstrous thing'. He cannot understand how, while knowing that their life and its pleasures are ephemeral, and expecting that they will either be eternally destroyed or have to appear before God, they indulge in their pleasures 'as if they could annihilate eternity by diverting their thought from it' (fr.195).

Pascal sees an enormous difference between these 'libertines' and people who, though equally uncertain, are tormented by these doubts and do not acquiesce in their situation, but are concerned about their state and yearn for truth. He has been one of them, so, when he says that he has 'com-passion' with them, this should be taken in the literal sense of suffering with them. The attitude of the others, however, 'rather irritates than endears' (fr.194).

To Pascal, then, the final choice is not between dogmatism and agnosticism, but between a perspectiveless agnosticism and the ignorance of those who recognize that they cannot find rest unless, 'exhausted and tired by the futile search for the true good, they hold out their arms to the Liberator.'[62]

What then is the function of apologetics? Just the removal of obstacles, so that we may recognize that a solution cannot be given by ourselves but through divine grace alone. By critical reason he combats the false claims of theoretical reason in science, philosophy and theology. He stresses that it is reason that recognizes the facts and decides to submit to them: 'submission is a use of reason, that is what makes true Christianity' (fr.269), and it is reason which recognizes when and where submission is due: 'We must know to doubt where it is necessary, to be certain where it is necessary, while submitting where necessary' (fr.268).

To dub Pascal's religious attitude as 'fideism' quite misses the point, and overlooks the dialectical character of his approach: *reason* decides when reason has to submit. Pascal advocates neither rationalism nor irrationalism, but a religion of 'reasonable' surrender to data and facta, things that have been done and given. Of course, he realized that reason is not always 'reasonable' and that the will (which he considered depraved by the Fall, much more than reason) makes it refuse divine grace.

Pascal's never-ending dialectic gives the impression that there is no 'order' in his 'Vindication of the Christian religion'. This apparent disorder has often been attributed to the fragmentary character of the notes which his untimely death at the age of 39 prevented him from arranging. But should we deplore this lack of systematization? Perhaps not, for it prevented him also from making a closed system of the kind he disliked so much: a system in which there have been made blind windows for the sake of symmetry. It would have marred the realistic character of his psychological approach; it would have impaired his method of entering into the mind of other people, whose thinking — as he perfectly knew — does not work in so orderly a fashion as the Cartesians would like it to do.[63] He himself declared: 'I will write down my thoughts (*pensées*) here as they come and in a perhaps not aimless confusion: this is the true order and it will always show my aim by its very disorder. I should be honouring my subject too much if I treated it in order, since I am trying to show that it is incapable of it' (fr.373).

Pascal, then, wanted to write down his 'thoughts' not in a well-arranged battle array, but in a dialogue with his readers. And as he often identified himself with them — as also with the opponent within himself — he does so with great honesty, realism and rationality, though sometimes one might have liked perhaps that the surgeon's knife of his critical analysis had been handled with more 'compassion' for the frailty and folly of people with whom he had friendly relations in society.

It is understandable that Pascal's method of apologetics met with less response than that of current Natural Theology. This, in a quiet and reasonable way, guided the reader through page after page of the Book of Nature which so clearly displayed the 'power, wisdom and goodness' of its Creator, and claimed to provide him with a solid foundation on which Christianity could be built. With Pascal, however, there is no easy exposition of the wonders of nature outside us — no coaxing appeal to our good intentions — but an unsparing disclosure of what is wrong with mankind: 'we are but lie and duplicity' and 'we hide and disguise ourselves from ourselves': most people who speak about humility glory in it, and who speak about their scepticism affirm it without the least doubt (fr.377). People would be pleased as they heard him call man the 'glory of the universe', but they would recoil when immediately afterwards man is dubbed 'the scum of the universe', a 'chaos' full of contradiction, 'glory and refuse of the universe'.[64]

The apologetics of people like Boyle, Wilkins and Ray was entirely orthodox, but it was less disturbing and fitted better with the latitudinarianism prevailing since the Restoration and the Glorious Revolution. Even further away from Pascal was the easy-going, compromising semi-Pelagianism of the French Jesuits. Similarly, the 18th century Enlightenment, latently or openly deistic, with its self-congratulation and its belief in the fundamental goodness of man, turned a deaf ear to Pascal's stern message. It did not like his method, a method not of calm persuasion by gradually developed arguments but a disquieting dialogue, with endless ins and outs, with paradoxes and antinomies, not aiming at assent but at surrender. 'You are embarked' (engaged in a hazardous undertaking) Pascal calls to his reader.[65] And in the sails of the ship on which he takes them blows not a soft breeze but a tumultous whirlwind. It was only with early 19th century soul-searching romanticism — which however tended to identify the 'heart' and 'sentiment' with sentimentality, thus overlooking the rational element in his 'thoughts' — that 'the world' began to take his *Pensées* seriously.

21. From Science to Religion

Until November 1654 Pascal's main interests lay in mathematics and experimental physics. He regularly attended the meetings of an 'Académie' of scientists in Paris, invented a calculating machine and performed experiments on the pressure of the atmosphere. After the religious crisis that is known as his 'second conversion' he seems to have abandoned physical research and turned to the study of man as an individual and as a social being. He himself confirms this shift of interest when he relates that during many years he had devoted himself to 'abstract sciences', but it annoyed him that they provided so little 'communication'. Consequently, he turned to the 'study of man', hoping that he would find there more companions. But the reverse was the case (fr.144). Mankind, so he says, though made for thinking, nevertheless is engrossed in diversions (dancing, music, poetry, fighting, etc.) without thinking about what it means to be human (fr.146).

Under the influence of Pascal's biography (or rather hagiography) by his sister, Marguérite Périer, it has often been said that his conversion meant a total loss of interest in scientific and worldly affairs. This seems, however, a gross exaggeration.

He had a sense for business which showed itself when in 1644, having invented a calculating machine, he immediately tried to produce more of them and offered them for sale in a brochure (1645).[66] In 1649 he obtained a royal privilege, according to him the monopoly for its fabrication and sale.[67] After his conversion he retained this desire to protect his rights (or those of his heirs). Shortly before his death he obtained a royal patent (1660) for the public omnibus service ('carriages at five sols per ride') which he had introduced in Paris.

In spite of the growing intensity of his religious life, his love for mathematics was not lost. Now and again bouts of assiduous study in this field occurred. In one of his most interesting treatises, *De l'Esprit Géométrique* (1658-59), in which the problem of the infinite was discussed, philosophical interest and apologetic aims went together. Some of his most important publications on pure mathematics (calculus of probability, calculus of indivisibles, etc.) stem from this late period. In 1658 he had solved infinitesimal problems relating to the cycloid (*roulette*), and he challenged other mathematicians to do the same. This led to the publication (under the pseudonym Amos Dettonville[68]) of some tracts and to rather sharp controversies with his competitors.

After 1659 his mathematical activity rather abruptly abated. He fell seriously ill and passed his last years mainly in meditation, prayer and charitable work. Even then, however, his mathematical interest did not vanish completely. In 1660 he still corresponded with other mathematicians. To Fermat he wrote in August of that year that he found mathematics the highest occupation of the mind, but at the same time so useless that he would make little difference between a man who is only a mathematician and an able artisan. 'It is the most beautiful trade (*métier*)'; it is good for trying our force but not for employing it.[69] He then declared that he was engaged in studies far away from it and did not expect ever to return to it, his health being, moreover, very weak. Yet he was 'human' enough to be agreeably surprised by the deference Huygens showed him, which gave him 'une joie extrême'.[70]

It is interesting that when discussing in his *Pensées* the differences between the realms of the flesh, the spirit and the heart, he chose as representatives of the second and third categories Archimedes and Jesus Christ. Both without 'éclat' in the world, the one having his glory in the kingdom of learning, the other in that of sainthood; both were princes, but without outward glory. But however great the distance between the order of the flesh and that of the spirit, maybe infinitely greater is the distance between the order of the spirit and that of charity and wisdom.[71]

22. Pascal's Character

His conversion neither effaced his scientific interest, nor did it kill 'the old Adam' in him. His frequent warnings in the *Pensées* against desire for worldly things (*concupiscence*) and pride (*orgueil*) and his insistence on 'humility' are directed to all people, himself included. It is true that he rarely shows this — on one occasion when pointing out the vanity of an author who knows that he has written some good lines; on another occasion in his passionate dialogue with God, when he confesses his 'abyss of pride'.[72] But he knew too well that much talk about one's own unworthiness is also a form of pride. His criticism of the human character was based not only on study of the words and acts of his fellow-men but also on self-contemplation: his behaviour with relation to the competition about the cycloid shows that the desire to be the first lingered on, if only in 'disguise'.[73]

It has been said (in my opinion rightly) that 'Pascal was a complex person, whose pride constantly contended with a profound desire to submit to a rigorous Augustinian insistence on self-denial.'[74]

In his apologetics, however, the 'désir de vaincre' was certainly not the driving force; he was too deeply convinced that faith is a 'gift of God' (*don de Dieu*) and not of Pascal. Shortly before his death he wrote to a correspondent (1661):

> We behave as if it were our mission to make truth triumph, whereas our mission is only to combat for it. The desire to win is so natural that, when it covers itself with the desire to make the truth triumphant, we often take the one for the other and think that we seek the glory of God when seeking in fact our own glory.

23. Pascal's Style

Pascal's succinct literary style did not change after his 'second conversion'. His irony and his irritation at the arrogance of the opponents of 'empty space' were similar to his irony and indignation at the duplicity and laxity of the moral instructions of the Jesuit opponents of Pont-Royal.

However much his life centred on religion, he still found joy not only in mathematics but also in language. It is with obvious pleasure that he weighed his words, and it is not only because of what he had to say but also because of how he said it that his work lives on to the present day.

It is precisely the 'simplicity', the avoiding of rhetorical phrases, which is the lasting charm of Pascal's writings.[75] He realized that true eloquence, which does not bore in the long run,[76] is based on a correspondence of the 'esprit' and the 'coeur' of those who are addressed with the thoughts and expressions of the speaker. We need to have studied the hearts of men to enter into the minds of those who, we hope, will understand us, and must try this out on ourselves (fr.16). Pascal aimed at what he called 'natural simplicity' — a natural simplicity which, it may be said, is the highest art.[77] He realized this ideal in his own work. Unwittingly his advice depicts the effect he was to have on a diverse multitude of future readers, who were 'delighted' by his 'natural style' and who while perhaps expecting to see only an *author*, 'found a man'.[78]

There are many images of Pascal: the master of language, the great mathematician, the rationalist, the sceptic, the fideist, the ascetic, the universal genius. Yet it is 'the man' who is most fascinating. He said that we picture Plato and Aristotle only in their long 'robes de pédant' though they were just decent persons, 'who laughed with their friends' like the rest.[79] The greatness of scholars is not diminished but enhanced when we recognize that, however much endowed with special talents, they were just human beings.

EPILOGUE

[*Editors' note:* The epilogue that follows stems from three sources: (1) a loose sheet we found in another file, of which we are not therefore certain whether the author would have agreed with its placement here (or indeed anywhere), even though the content strongly suggests it; (2) sentences noted down (verbatim) by VMcK during a conversation with RH in 1988 about the message of this book; (3) in translation, the concluding passage of Hooykaas's booklet, *L'Histoire des Sciences, ses Problèmes, sa Méthode, son But.* Coimbra 1963.]

These chapters have for a large part dealt with the relations between Reason and Experience; and we have indeed seen these two 'officially' recognized elements of science. Yet, there is a third element, or rather a complicated entanglement of elements whose existence Pascal has recognized perhaps better than any other scientist, and called 'the heart'. It cannot be sharply defined, it comprises religious experience, aesthetic feelings, scientific intuition and also moral convictions. It lies beyond the power of control by reason and experience, though they have every right to examine its products critically to see where it degenerates into wild fantasy. It works not only consciously but also subconsciously, so that it is not easily recognizable. Nevertheless the 'heart' has played an important role in the history of science and in the greatest scientists has been a powerful stimulus to their discoveries. Not every scientist possesses all three elements in the same degree: some lack respect for experience and trust too much in their own reason, others are mainly collectors and classifiers of facts and lack scientific imagination. This may be deplorable but it has to be accepted: 'one cannot pluck feathers from a frog'. In spite of their shortcomings such people made valuable contributions to science. It is a striking fact, however, that the greatest, people like Kepler, Newton or Maxwell are highly endowed with all three elements.

In all science there is a human factor. We are always ready to deceive ourselves — but we are always corrected by what, to us, is fundamentally a revelation: they are given to us — the facts. Just as in seeking God we cannot say 'God has to be thus and so,' but must listen to Him; so also in science: we have to sit down before the facts as little children,' — as T.H. Huxley reminds us.

We have to keep to the facts whether they are comprehensible or not. Rationalism is apt to lead us astray when it loses an emphasis on experience. The facts are sacrosanct. Nevertheless the scientist is walking a tight rope between his duty to adapt his theory to the facts and his inclination to adapt a fact to his theory. No recipe can be given against too readily abandoning a theory because some facts do not seem to fit in with it and too boldly proclaiming 'the facts'.

Our resulting science is a human construct: theory tested by facts. Science is *not* nature, it is our rational system, our interpretation of nature. In the last resort (as Pascal observed when discussing 'learned ignorance') we know nothing (fr.327).

Therefore, we should remember that there is more in life than facts. Some would see for example the poetry; some, like Boyle, approach nature with 'awe and veneration' for Him who made it. Real science will not say that facts are all that can be said about ourselves and about the world we live in.

The history of science may 'humanize' the education of men of science, and offer to those with a background in the humanities an understanding of their own times. It provides material for a critical examination of science and for the philosophy of science; it demonstrates the inanity of scientism and of every *metaphysics* which claims to be erected upon an exclusively scientific foundation, and also of every *science* built upon metaphysics rather than upon the investigation of nature by means of observation. It reveals the connections between the sciences and the humanities, and shows how the sciences of nature are human sciences, sharing in the human values of our times. It creates in the mind of the physicist a wholesome scepticism towards his own theories.

Every scholar runs the risk of exaggerating the importance of the branch of learning dear to him. If therefore it might be thought that these claims are not exactly modest, there is in any case the fact that there are scholars who take pleasure in investigating not only the contents of theories but also their *genesis*, and who study the history of science for the intellectual and aesthetic joy it provides.

An intellectual *and* aesthetic joy indeed, for in scientific works there is prose which merits a place in literature, e.g., Thomas Digges's description of the Copernican world, or Pascal's sarcastic 'fragment d'un Traité du Vide', or Kepler's lyrical outburst when he thinks he has discovered the plan of the world and sings a hymn that without exaggeration might be called psalm CLI, or Lavoisier's 'Préface' to his *Traité Élémentaire de Chimie*, written in the clear and elegant prose of the 18th century.

But also an intellectual joy, for demonstrations and modes of reasoning sometimes are of an incomparable beauty, as in Pascal's just and elegant use of analogical reasoning in his two treatises on the equilibrium of fluids and on the weight of the air, or the vivid dialogues in which Galileo develops his ideas, or the magnificent reasoning, a tissue of induction interwoven with deduction, by which Haüy's magic evokes from the chaos of the phenomenological world of crystals, the cosmos of the ideal world of crystallography.

Finally, the history of science makes us recognize in our ancestors men like ourselves. It reveals man in his littleness and in his greatness, and thus yields us a pleasure which has some affinity with that from reading a play by Shakespeare. In that sense there is indeed a 'humanization' of the sciences of nature. History of science has a peculiar charm because of its inner tension. On the one hand it is the history of disciplines which are progressing as experience of nature increases and as fruitful methods of investigation emerge from the crowd of possible methods; but on the other hand it is the history of sciences produced by the human mind, which in the entire course of written history has displayed a structure equal to our own. In those sciences which are focused upon the outer world we see farther than our predecessors, not because we are larger than they were but because we stand upon their shoulders.

NOTES

[*Editors' note*: References have been checked in particular by using Hooykaas's own work.]

PREFACE

1 *Gifford Lectureship*; Extracts from the Trust Disposition and Settlement of the late Lord Gifford, dated 21st August 1885, p.4.

2 *Edinburgh Review*, Oct.1802 - June 1809, Vol.I, n.3, 7th ed. (1810), p.289.

3 Henry Brougham, *Natural Theology*. London-Glasgow 1856, p.21.

4 Karl Barth, *The Knowledge of God and the Service of God According to the Teaching of the Reformation Recalling the Scottish Confession of 1560*. London 1938, pp.3-6.

5 *Ibidem*, p.9.

6 A.E.Taylor, *The Faith of a Moralist*. The Gifford lectures delivered at the University of St Andrews 1926 - 1928. London 1930.

INTRODUCTION

1 D. Brewster, 'Review of the Vestiges of Creation', in: *North Brit. Rev.* III (1845), p.480.

2 J.D. Forbes, *The Danger of Superficial Knowledge* (An Introductory Lecture to the Course on Natural Philosophy in the University of Edinburgh, delivered on the 1st and 2nd November 1848). London 1849.

3 *Ibidem.*

4 J.C. Sharp, P.G. Tait, A. Adams-Reilly, *Life and Letters of James David Forbes*. London 1873, p.193.

5 It was generally believed that the barnacle (a cirripede: *Lepas anatifera*) developed into a goose (either Branta bernicla or Branta leucopsis: species that were not clearly distinguished). The emperor Frederick II, who did not share this belief, sent an expedition to the north, which did not find eggs of these birds. He concluded that they must breed even farther to the north. Albertus Magnus (1193 - 1274) had seen the birds copulate and lay eggs in captivity. The Hollanders in 1696 found the 'rotganzen' in the far north, sitting on their eggs. The Utrecht professor A. Senguerd (*Physicae Exercitationes*, Amsterdam 1658), when dealing with the 'scottish geese' declared the story a myth. Yet the myth lived on till the beginning of the 18th century among scholars of good reputation. The Netherlanders call the barnacle: 'eendemossel' (duck mussel). See E. Heron-Allen, *Barnacles in Nature and Myth*. London 1928.

6 C.E. Raven, *English Naturalists from Neckam to Ray*. Cambridge 1947, pp.130-131.

7 R.G. Cant, *The University of St Andrews*. Edinburgh-London 1970, p.81.

8 W. Chambers, *Memoir of William and Robert Chambers*, Edinburgh-London: W.& R.Chambers 12th ed. 1883, p.309.

9 Hebrews 11:1.

[10] D. João de Castro, *Roteiro de Lisboa a Goa* (circa 1538). Reprint in Vol.I of *Obras Completas de D.João de Castro.* A.Cortesão and L.de Albuquerque ed., Coimbra 1968 - 1980; —, *Roteiro de Goa a Diu (1540).* Reprint in: *Obras,* Vol.II, iii. Cf. R. Hooykaas, *Science in Manueline Style*, Coimbra 1980, p.36.

[11] Ptolemy, *Almagest* Bk.III, p.4: 'It would be more reasonable to stick to the hypothesis of eccentricity which is simpler and completely effected by one and not two movements.'

CHAPTER I

[1] Cf. R. Hooykaas, *Religion and the Rise of Modern Science*. Edinburgh and London: Scottish Academic Press 1972. The first two chapters of these Gunning lectures given in 1969 at the University of Edinburgh touch problems of Natural Theology.

[2] R. Hooykaas, *ibidem*, pp.1-6.

[3] Cicero, *De natura deorum*, Bk.II, XLVI, 117-118.

[4] *Ibidem*, XXXVII, 93.

[5] *Ibidem*, 94.

[6] *Ibidem*, II, 4.

[7] *Ibidem*, 5.

[8] *Ibidem,* VII, 20.

[9] *Ibidem*, XI, 29.

[10] *Ibidem*, XI, 32.

[11] *Ibidem*, 29-30.

[12] *Ibidem*, XV, 42.

[13] *Ibidem*, XXXIII, 84.

[14] *Ibidem*, XLVI, 117-118.

[15] *Ibidem*, XXX, 76.

[16] *Ibidem*, 77.

[17] *Ibidem*, XLVII, 120ff.

[18] *Ibidem*, LII, 130 - LIII.

[19] *Ibidem*, LIII, 133; LXII, 154.

[20] Cf. Hooykaas, *Religion and the Rise of Modern Science*, pp.31-35.

[21] J.D. Bernal, *Science in History*. London 1954, B.I, pp.30, 139, 149; B. Farrington, *Greek Science: Its Meaning for Us*. London 1944, pp.15, 55, 80. Both authors assumed a Marxist standpoint, but whereas Bernal remained a Stalinist diehard, Farrington had a much less dogmatic attitude, in particular in his later publications.

[22] Hooykaas, *Religion and the Rise of Modern Science*, pp.17-19.

[23] Henricus Monantholius, *Aristotelis Mechanica*, Parisis 1599, Epistola dedicatoria fol.aIIIr. Cf. R. Hooykaas, *Das Verhältnis von Physik und Mechanik in historischer Hinsicht*, Wiesbaden 1963, p.11. (reprint in: R. Hooykaas, *Selected Studies in History of Science*, Coimbra 1983, pp.167-189).

[24] Descartes's position is more complicated. See Hooykaas, *Religion and the Rise of Modern Science*, pp.42-44, and from the same author: '*Experientia ac Ratione*', *Huygens tussen Descartes en Newton*, Leiden 1979.

[25] Isaac Newton, *Philosophiae Naturalis Principia Mathematica*, London [2]1713. Scholium Generale.

[26] Henry Brougham, 'Observations, Demonstrations and Experiments upon the Structure of the Cells of Bees' (Appendix to the 'Dialogues on Instinct'). In: Henry Brougham, *Natural Theology*, London - Glasgow 1856, pp.312-364; see also p.191.

[27] The Duke of Argyll, *The Reign of Law*, London 1867. Ch.III: 'Contrivance a Necessity arising out of the Reign of Law' Example in the machinery of flight, pp.128-180. Also chapter IV: 'Apparent Exceptions to the Supremacy of Purpose', pp.181-216. 'Here again the Laws of Nature are seen to be nothing but combinations of Force with a view to Purpose: combinations which indicate complete knowledge not only of what is, but of what is to be, and which foresees the End from the Beginning' (p.216).

[28] Charles Kingsley wrote on November 18th 1859 a letter of support to Charles Darwin, reprinted in: F. Darwin ed., *The Life and Letters of Charles Darwin*, Vol.II, London 1887, *q.v.* p.287. See also: *Charles Kingsley, His Letters and Memoirs of his Life* (ed. by his wife) 7th abridged ed., Vol.II, London 1888, pp.198, 254, 258. 'I am very busy working out points of natural theology, by the strange light of Huxley, Darwin and Lyell' (Ch. Kingsley to F.D. Maurice, 1863 p.155). See of Asa Gray: *Darwiniana, Essays and Reviews pertaining to Darwinism*. New York 1876 (contains art. III: 'Natural Selection not inconsistent with Natural Theology' (1860).) Much information on this subject can be found in: J.R. Moore, *The Post-Darwinian Controversies. A study of the Protestant Struggle to Come to Terms with Darwin in Great Britain and America 1870 - 1900*, Cambridge 1979.

[29] Johannes Kepler, *Mysterium Cosmographicum*, Tübingen 1596, chapter II. Reprinted in *Werke (ed.* Caspar) *Vol.*I, p.23ff.

[30] Ramundus Sabundus, *Theologia Naturalis seu Liber Creaturarum* (before 1436). (Facsimile of edition Sulzbach 1852: Stuttgart-Bad Canstadt 1966).

[31] *Ibidem*, Prologus, pp.27-28.

[32] 'Unde duo sint libri nobis dati a Deo, scilicet liber universitatis/ creaturarum sive/ liber Naturae; et alius est liber Scripturae sacrae. ... quaelibet creatura non est nisi quaedam littera, digito Dei scripta; et ex pluribus creaturis, sicut ex pluribus litteris, est compositus liber unus, qui vocatur liber creaturarum' (*Ibidem*, p.35). This passage shows a striking resemblance to article 2 of the Confessio Belgica (the 37 articles of the faith of the Reformed Churches in the Netherlands); which shows that after 200 years the book was still current.

[33] *Ibidem*, p.36.

[34] *Ibidem*, p.37.

[35] *Ibidem*, p.38: '... nullus potest videre nec legere per se in dicto libro semper aperto, nisi/ sit a Deo illuminatus et a peccato originali mundatus'.

[36] Romans 1:20.

[37] Hebrews 11:3.

[38] Sabundus, *Theologia Naturalis*, titulus XVII (p.25): 'Quod ipsum esse Deus produxit totum esse mundum de non esse et de nihilo voluntarie et per modum artis et non naturaliter, nec ex necessitate' (Not by necessity but by free will, not by natural generation but by artificial fabrication the world was created). These views go back to the biblical view of creation (Hooykaas, *Religion and the Rise of Modern Science*, pp.7-9) and are also found with Tempier (p.32), Oresme (p.33) and Newton (pp.19,49).

[39] Sabundus, *Theologia Naturalis*, tit. XVII, p.26.

[40] Hooykaas, *Religion and the Rise of Modern Science*, pp.10-12.

[41] Sabundus, *Theologia Naturalis*, tit.XVII, p.27: 'Nec mundus potest per se stare et conservari absque praesentia incessabili Dei per unum momentum, immo statim rediret in non esse et in nihil totus mundus, quia de nihilo venit.'

[42] *Ibidem*, pp.27-28: '... ipsum sustinet sine labore, quia volendo ... ipsum continue creat, sicut sol suos radios continue creat in aëre. Aliter si non produceret continue, radii deficerent: ita mundus esse desineret, si non a Deo continue crearetur et conservaretur.'

[43] *Ibidem*, tit.CLXXXVI, p.257.

[44] *Ibidem*, tit.XCVIII, p.124; XCIX, p.128.

[45] *Ibidem*, tit.XCVII, pp.122, 123.

[46] *Ibidem*, tit.CXCIV, p.275.

[47] *Ibidem*, tit.CCXI, p.311.

[48] *Ibidem*, tit.CCXIII, p.315.

[49] *Ibidem*, tit.CCXIII, p.316.

[50] *Ibidem*.

[51] *Ibidem*, pp.317-318.

[52] *Ibidem*, p.319. Sabundus repeatedly stresses that all things in the universe have been made for man ('propter hominem', pp.38, 12, 126, 129). But man has been made for God's honour (p.129). And God can be known by man, in whom He impresses his image as a seal is impressed in wax: 'sicut sigillum imprimit totam suam imaginem in cera; ita Deus impressit totam suam imaginem in homine' (tit.CXXI, pp.163 and 292).

CHAPTER II

[1] '... omnia in mensura, et numero, et pondere disposuisse' (*Liber Sapientiae*, ch.11, v.21)

[2] Thomas Norton, *Ordinall of Alchimy*, ch.5, p.57. In: Th. Ashmole, *Theatrum Chemicum Britannicum*, 1652, pp.1-106; also in J.J. Manget, *Bibliotheca Chemica Curiosa*, Vol.II, 1702.

[3] The Pythagorean tenet of the sphericity of the earth and the other cosmic bodies always met with an almost general approval by the learned. There were strong arguments for it, e.g. the roundness of the shadow cast by the earth during eclipses of the moon (Anaxagoras, 5th cent. BC). The sphericity of the whole universe was based mainly on aesthetical and metaphysical arguments (T.L. Heath, *Aristarchus of Samos*, Oxford 1913, p.48), which were adduced up till the 17th century.

[4] Aristotle, *Metaphysica* Bk.I, ch.5, 986a2-12.

[5] *Idem*, *De Coelo* Bk.II, ch.13, 293a25-29.

[6] W.B. Fowler and N.P. Samios, 'The Omega-Minus Experiment'. In: *The Scientific American*, Oct. 1964, p.36-45. See also C.A. Littlefield and N. Thorley, *Atomic and Nuclear Physics: An Introduction*, Van Nostrand 2nd ed. 1968.

[7] M. Gell-Mann pointed out in 1961 that in a ten-particle triangle, one particle was as yet missing.

[8] P.T. Mathews, 'Order out of Subnuclear Chaos', in: *New Scientist* 379 (20 Feb. 1964), p.460. Matthews enthusiastically exclaimed: 'This is the type of situation about which scientists dream ... a startling new theory at an absolutely fundamental level, which coordinates and clarifies a previously confused experimental situation; which is in agreement with the known facts; and which makes an absolutely precise prediction, specific to the new theory and appearing quite weird on the basis of previous ideas.' At that moment an announcement of the discovery was expected every day, and it was said that 'the confirmation of uniting Symmetry in strong interaction will be a dramatic advance in our understanding of the laws of nature.'

[9] Plato, *Oeuvres complètes de Platon.* L. Robin ed. Paris: Pléiade 1950, Vol.I, p.1362 n.57.

[10] Boethius (480 - 525), Macrobius (circa 400), John Dee (1527 - 1598).

[11] Plato, *Republic* Bk.VII, 531b.

[12] Aristoxenus, *The Harmonics.* H.S. Macran ed. Oxford 1902, Bk.II, 32-33, pp.188-9.

[13] Plato, *Republic* Bk.VII, 531a. Cf. *Oeuvres* (Robin ed.), p.1362 n.62.

[14] Plato, *Republic* Bk.VII, 531c.

[15] Cf. C.V. Palisca, in: H.H. Rhys ed., *Seventeenth Century Science and the Arts*, Princeton 1961, p.118.

[16] Vincenzo Galilei, *Discorso* 1589; Palisca, *ibidem*, p.128.

[17] Palisca, *ibidem*, p.122.

[18] *Ibidem*, pp.132-5.

[19] Galileo Galilei, *Discorsi* (1638) (first day) in: *Opere* VIII, 142.

[20] Plato, *Republic* Bk.VII, 530d; cf. *Phaedrus*, 247c-250bd.

[21] In later works Plato was not such an extreme idealist (Heath, *Aristarchus of Samos*, p.139). In *Timaios* (his work which has had most influence on 'science'), he recognized that no word would have been said about the universe, if we had never seen the sun and other heavenly bodies (*Timaios,* 47a). Mathematical Ideas are the fundamentals of the visible world. The four elements consist of particles that have the shapes of four regular bodies: tetrahedron (fire), icosahedron (air), octahedron (water) and hexahedron or cube (earth). Plato borrowed from the Pythagoreans the tenet that the universe is essentially mathematical. It has a spherical shape, for the Most Perfect could not but make the most perfect world. Plato shared with the Eleatics the belief that true being is not liable to any change. This means that it is not the 'dumb' elements that are true nature (*physis*), but the Ideas, the transcendent world of eternal, immutable law (*nomos*), of Order and Beauty, which is above and beyond the visible world. The world of phenomena is but a faint shadow, for phenomena are liable to change.

[22] Plato, *Laws* Bk.VII, 821b.

[23] Anaxagoras was persecuted in Athens because he had said that the moon is a large stone of the size of the Peloponnes.

[24] Plato, *Laws* Bk.VII, 822a-c.

[25] Plutarch, 'Life of Nicias', in: I*dem*, *Les vies des hommes illustres Grecs et Romains*; transl. Jaques Amyot in 1559; edition Paris 1603, Vol.I, p.1066.

[26] Simplicius, *De Coelo* Bk.II, ch.12, 292b10; Heath, *Aristarchus of Samos*, pp.195, 221.

[27] See below, chapter VI on Copernicus and chapter VIII on Physical and mathematical theories.

[28] P. Kraus, *Jabir ibn Hayyan, Contribution à l'Histoire des Idées scientifiques dans l'Islam.* Vol.I; Le Corpus des Écrits Jabiriens, *Mém. Institut d'Egypt* 44, Le Caire 1943; Vol.II, Jabir et la Science Grecque. *Mém. Institut d'Egypt* 45, Le Caire 1945 According to Kraus Jabir's writings bear the stamp of the shi'ite sect of the Ismaelites (Vol.I, p.LXV).

[29] *Ibidem* Vol.I, p.XXXV.

[30] *Ibidem* Vol.II, p.45.

[31] *Ibidem*, p.95.

[32] *Ibidem*, p.203.

[33] In each compound the intensities of the four natures are as 1 : 3 : 5 : 8; their sum 1 + 3 + 5 + 8 = 17, is Jabir's 'tetraktys', in which the harmony between all things on earth consists; it is the basis of the whole System of Balance (*ibidem*, pp.195; 227-8).

[34] *Ibidem*, p.224. The analysis of the word establishes the qualitative and quantitative structure of the thing represented by it.

[35] *Ibidem*, pp.305; 228.

[36] *Ibidem* Vol.I, p.239; Vol.II, p.228.

[37] *Ibidem* Vol.II, p.30.

[38] Johannes Kepler, *Mysterium Cosmographicum*, Praefatio (Tubingae 1596); In: —, *Gesammelte Werke*, M. Caspar ed., Vol.I, München 1938, p.9.

[39] *Ibidem*, p.10. Cf. *ibidem*, ch.4, p.30: 'ut ludam aliquantisper in re seria'.

[40] *Ibidem*. Cf. *ibidem*, p.14. See also below, page 43 (Law of Titius-Bode)

[41] *Ibidem*, p.13.

[42] *Ibidem*, p.11. Cf. *ibidem*, Nachbericht, p.405.

[43] Ibidem.

[44] Ibidem, 2nd edition (Francofurti 1621), Epistola dedicatoria. In: Werke VIII, p.9

[45] 'Instar illius qui clamavit Εὕρηκα' (Kepler to Maestlin, 2 Aug.1595. In: Werke XIII, p.28-9). 'Eureka', Archimedes' exclamation when he had solved the problem of determining the relative weights for solid bodies. The same was written down in his notebook by Christiaan Huygens ('our Archimedes', as his father called him), after he had discovered the explanation for double refraction by calc spar.

[46] Ibidem, p.35: 'Nihil enim sine ratione optima facit.'

[47] Kepler to Maestlin, 9 April 1597. In: Werke XIII, p.113.

[48] Kepler, Mysterium Cosmographicum, 2nd ed. (1621), ch.11 (Notae Auctoris), in: Werke VIII, p.62.

[49] Ibidem, In Praefationem Notae Auctoris; in: Werke VIII, p.30.

[50] Kepler to Tanckius, 12 May 1608.In: Werke XVI, p.161: 'In hoc solo offendit pietas, quod bonitatis leges et necessitatem videt collocari non in Dei creatoris arbitrio, sed extra in Idea geometrica.' Cf. Plato, Timaios, 30.

[51] Kepler, Mysterium Cosmographicum, ch.11 (Werke I, pp.37-38).

[52] Ibidem, 2nd. ed., ch.11 (Werke.VIII, p.62).

[53] Ibidem (1st ed), Praef. (Werke.I, p.13). Kepler had to allow a certain thickness to the spheres in order to take account of the eccentricity of the planetary orbits.

[54] Tycho to Kepler, 1 April 1598; Tycho to Maestlin, 21 April 1598; Tycho to Kepler, 9 Dec. 1599. In: Kepler, Gesammelte Werke I; 'Nachbericht' von M. Caspar, p.412.

[55] Kepler, Astronomia Nova (1609), P.III, ch.40; In: Werke III, p.263.

[56] Kepler, *Harmonice Mundi*, Lincii Austriae 1619; in: *Werke*, Vol.VI.

[57] Kepler, *Mysterium Cosmographicum*, ch.23 (*Werke* I, p.79). The last line is from Romans 11:36

[58] Kepler, *Harmonice Mundi*, Bk.V, ch.10 fin (*Werke* VI, p.368).

[59] Kepler, *Mysterium Cosmographicum*, 2nd ed., ch.1, In Caput Primum Notae Auctoris (*Werke* VIII, p.39).

[60] Kepler, *Harmonice Mundi*, Bk.IV, ch.1 (*Werke* VI, p.223).

[61] Kepler to Maestlin, 3 Oct.1595 (*Werke* XIII, p.40).

[62] Kepler, *Mysterium Cosmographicum*, Epistola Dedicatoria (*Werke* I, p.5).

[63] *Ibidem*.

[64] *Ibidem (Werke* I, p.6).

[65] *Ibidem, ch.4 (Werke I, p.30).*

[66] *Ibidem*. The same will be found with Cicero, Sabundus, Copernicus, Donne, Hutton, i.e. with pagan, Roman Catholic, orthodox Protestant and deist.

[67] Bonnet, Charles, *Contemplation de la Nature*. Amsterdam 1764 - 1765, Translated in 1766 in German by Johann Daniel Titius.

[68] In general the cleavage of crystals will directly reveal the dihedral angles of the primitive form but not the respective dimensions. A cube of sea salt can be split into rectangular parallelepipeds of all kinds of respective dimensions. Here the principle of symmetry helps out: the essentially cubical form of the primitive form (the equivalence of the six sides) is demonstrated by the formation of secondary forms. If one of its sides develops into a four sided pyramid, the other five will be changed into pyramids with the same dihedral angles and a solid with 24 faces will ensue, unless the adjacent faces of the neighbouring pyramids are in the same plane, so that the number of faces is halved and a rhombic dodecahedron is formed. Abbé R.J. Haüy, *Essai d'une Théorie sur la Structure des Crystaux*. Paris 1784, pp.57-58; 75 (plate I, fig.3); pp.16; 18-21.

[69] Ibidem, p.2.

[70] R.J. Haüy, *Traité de Cristallographie*, Paris 1822, Vol.I, p.IX.

[71] Haüy, *Essai*, p.23.

[72] 'On reconnaît ici ce qui caractérise en général les lois émanées de la puissance et de la sagesse du Dieu qui l'a créé et qui la dirige. Economie et simplicité dans les moyens, richesse et fécondité inépuisables dans les résultats' (Haüy, *Traité de Cristallographie* I, p.IX).

[73] '... lorsqu'un des angles saillans ou des angles plans d'un crystal est sensiblement droit, je le suppose tel en toute rigueur'; '... la Nature s'arrête dans le cours de ses opérations et de ses mouvemens: telle est la direction suivant le perpendiculaire; telles sont les égalités entre certaines quantités du même ordre' (Haüy, *Essai*, p.26).

[74] R.J. Haüy, 'Observations sur la Mesure des Angles des Cristaux'. In: *Journal de Physique, de Chimie et d'Histoire naturelle* 87 (1818), p.247.

[75] 'Mais il faut à la théorie des points fixes d'oú elle puisse partir pour arriver à son but, et qui ne soient pas susceptibles de plus ou de moins' (Haüy, *Traité de Cristallographie* II, p.343).

[76] 'La méthode que j'ai adoptée pour obtenir ces rapports sous la forme la plus avantageuse, consiste à représenter par des quantités radicales, les deux termes qui les composent' (Haüy, 'Observations', p.249).

[77] R.J. Haüy, *Traité de Minéralogie*, Vol.II, Paris 1801, p.492; 408.

[78] Haüy, *Traité de Cristallographie*, Vol.II, pp.387-388.

[79] R.J. Haüy, *Tableau comparatif des Résultats de la Cristallographie et de l'Analyse chimique*, Paris 1809, p.122.

[80] *Ibidem*, p.124.

[81] R.J. Haüy, 'Observations', p.243.

[82] H.J. Brooke, *Annals of Philosophy* 14 (1819), p.453.

[83] '... le cristallographe qui a calculé un de ces angles liés étroitement à ce résultat fondamental, ne le mesure ensuite que comme pour se satisfaire. Il ne doutait pas d'avance que l'observation, si elle était exacte, ne dût parler comme la théorie' (R.J.Haüy, 'Observations', p.243).

[84] Haüy, *Traité de Cristallographie,* Vol.II, p.498.

[85] C.S. Weiss, 'Über eine verbesserte Methode für die Bezeichnung der verschiedenen Flächen eines Krystallisationssystemes', in: *Abh. Preuss. Akad., Phys. Kl.* 1816/17, pp.286-336. (*q.v.* 'Zweiter Abschnitt: Über eine Bezeichnung der Flächen eines Krystallisationssystemes, welche von der Annahme der Primärform völlig unabhängig ist', pp.305-336).

[86] 'Wer sich mit dem geometrischen Studium der Krystalle beschäftiget, der wird gleichsam *a posteriori*, d.i. durch den Erfolg überführt, dass die Verhältnisse in den Dimensionen der Körper schwerlich anders, als in Quadratwurzelgrössen ausdrückbar, anzunehmen seyn dürften, und er wird es Haüy Dank wissen, dass er für diese Art von Annahmen die Bahn gebrochen hat' (C.S. Weiss, 'Krystallographische Fundamentalbestimmung des Feldspathes', in: *Abh. Preuss. Akad.* 1816/17, p.253).

[87] *Ibidem*, p.254.

[88] *Ibidem*, p.253.

[89] This tenet of Weiss (and Haüy) was short-lived. In 1825 E. Mitscherlich found that the crystal angles change with temperature, which implies the impossibility of expressing the parameters by constant values (rational numbers or their radicals). Cf. P. Groth, *Entwicklungsgeschichte der mineralogischen Wissenschaften*. Berlin 1926, pp.70-71, n.1.

[90] C.S. Weiss, 'Betrachtung der Dimensionsverhältnisse in den Hauptkörpern des sphäroedrischen Systems und ihren Gegenkörpern, im Vergleich mit den harmonischen Verhältnissen der Töne', in: *Abh. Akad. d. Wiss. Berlin* 1818/19, Berlin1820, pp.227-241; *q.v.* p.228.

[91] *Ibidem*, p.233.

[92] E.g. Victor Goldschmidt (1853 - 1933), *Über Harmonie und Complication*. Berlin 1901.

[93] T. Bergman, *Opuscula Physica et Chemica*, Vol.II (Commentatio XII. De formis crystallorum, praesertim de spatho ortis) Upsaliae 1780; p.9 L.

[94] Laue 1912, W.H. and W.L. Bragg 1913.

[95] J.B. Richter, *Anfangsgründe der Stöchyometrie oder Messkunst chymischer Elemente* (3 Teile 1792 - 1794) Tl.II, Breslau-Hirschberg 1793, pp.32, 35 (par.24), 39, 42, 281 (par.22). 'Die Massengrössen womit die bisher bekannten alkalischen Erden mit der Salzsäure die Neutralität behaupten sind demnach Glieder einer wirklichen arithmetischen Progression deren Glieder entstehen, wenn dem ersten Gliede ein Produkt aus einer gewissen Grösse in eine ungrade Zahl zugesetzt wird, nur dass dazwischen viele ungrade Zahlen z.B. 5, 7, 9, 11, 13 ausgelassen sind' (*ibidem*, p.31, par.22).

[96] J.B. Richter, *Über die neuern Gegenstände der Chemie* (11 Theile, Breslau-Hirschberg-Lissa 1791-1802); Tl.VI (1796), pp.181-182. Richter is not the only one to use both arithmetical and geometrical progression. John Dalton put forward the following law: 'The force of steam from pure liquids, as water, ether, etc, constitutes a geometrical progression to increments of temperature in arithmetical progression' (John Dalton, *A New System of Chemical Philosophy*, Vol.I, Pt.I, Manchester 1808, p.13).

[97] Richter, *Anfangsgründe der Stöchyometrie*, Tl.II, p.41 (par.26).

[98] Richter, *Über die neuern Gegenstände der Chemie*, Tl.VI, p.181.

[99] Richter, *Anfangsgründe der Stöchyometrie*, Tl.II, pp.44, 49.

[100] *Ibidem*, p.49; p.50 (par. 26): '... man wäre bey so wenigen bekannten Gliedern nicht im Stande gewesen das Gesetz derselben ausfindig zu machen.'

[101] *Ibidem*, p.50.

[102] J. Schiel, *Annalen der Chemie und Pharmazie* 43 (1842), p.107.

[103] 'Nous appelons *substances homologues* celles qui jouissent des mêmes propriétés chimiques et dont la composition offre certaines analogies dans les proportions relatives des éléments' (Ch. Gerhardt, *Précis de Chimie organique*, Vol.I, Paris 1844, p.25).

[104] A.L. Lavoisier, *Traité élémentaire de Chimie*, Vol.I, 2nd ed., Paris 1793, pp.125, 197; 'Tableau des radicaux ou bases oxidables et acidifiables composées, qui entrent dans les combinaisons à la manière de substances simples' (p.196). 'Tableau des combinaisons d'oxygène aves les radicaux composés' (p.208).

[105] J.R. Partington, *A History of Chemistry*, Vol.III, London 1962, p.686.

[106] E. Cassirer, *Substanzbegriff und Funktionsbegriff.* Berlin 1923, p.274.

[107] J.W. Döbereiner, 'Versuch zu einer Gruppierung der Elementaren Stoffe nach ihrer Analyse' in: *Pogg. Ann. Phys.* 15 (1829), p.301.

[108] It should be noticed that Döbereiner often grouped together elements that are not analogous (beryllium and aluminium; boron and silicon). One of the main obstacles was the uncertainty of determination of the atomic weights (see chapter X, 'Cleopatra's Nose').

[109] J.A.R. Newlands, *Chem. News* 10 (1864), pp.59, 94; 'On the Law of Octaves', in: *Chem. News* 12 (1865), p.83.

[110] F.P. Venable, *The Development of the Periodic Law*, Easton, PA, 1896, p.10.

[111] *Ibidem*, p.8.

[112] 'Extract from Report of Meeting of the Chem.Society, March 1866' in: *Chem. News* 13 (9 March 1866), p.113: 'Prof. G.F. Foster humorously inquired of Mr Newlands whether he had ever examined the elements according to the order of their initial letters? For he believed that any arrangement would present occasional coincidences, but he condemned one which placed manganese so far distant from chromium, and iron from nickel and cobalt. ... Mr Newlands said that he had tried several other schemes before arriving at that now proposed ... and no relation could be worked out of the atomic weight under any other system than that of Cannizzaro.'

[113] D. Mendeléev, *Grundlagen der Chemie*. St Petersburg 1891, p.684; *Idem,* 'Faraday Lecture', in: *J. Chem. Soc.* 55 (1889), pp.634-656.

[114] Mendeléev, *Grundlagen der Chemie*, pp.681-2.

[115] D. Mendeléev, 'Die Beziehungen zwischen den Eigenschaften der Elemente und ihren Atomgewichten', in: *J. Russ. Chem. Soc.* 1 (1869); German transl. by Fehrmann in: Meyer and Mendeléev, *Das natürliche System der Chemischen Elemente*. 2 Aufl. Leipzig 1913 (Ostwald's Klassiker der exakten Wissenschaften nr 68, p.38).

[116] L. Meyer, 'Die Natur der chemischen Elemente als Function ihrer Atomgewichte', in: *Liebigs Ann.* (= *Ann. Chem. u. Pharm).*, Suppl.VII (1870). We quote, if not otherwise stated, from the reprint in Ostwald's Klassiker der exakten Wissenschaften nr 68, p.13.

[117] Mendeléev, *Grundlagen der Chemie*, pp.682-683.

[118] L. Meyer, 'Natur der Atome, Gründe gegen ihre Einfachheit' (= pp.134-9 in: —, *Die modernen Theorien der Chemie und ihre Bedeutung für die chemische Statistik*, Breslau 1864). Reprint in: Ostwald's Klassiker der exakten Wissenschaften nr 68, p.8.

[119] Mendeléev, 'Faraday Lecture'.

[120] Mendeléev, *Grundlagen der Chemie.*, p 693 n.13.

[121] *Ibidem*, p.693 n.14.

[122] *Ibidem,* p.684 n.8; 692-693 n.13; p.694.

[123] *Ibidem,* p 693 n.13; p.694.

[124] Meyer, 'Die Natur der chemischen Elemente', p.17.

[125] *Ibidem,* p.16

[126] D. Mendeléev, 'Über die Beziehungen der Eigenschaften zu den Atomgewichten der Elemente', in: *Z. f. Ch.* 12 (N.F.5) (1864). Reprint in: Ostwald's Klassiker der exakten Wissenschaften nr 68, p.19.

[127] *Ibidem,* pp.29, 30, 33.

[128] Mendeléev, 'Die Beziehungen zwischen den Eigenschaften', p.39.

[129] D. Mendeléev, 'Die periodische Gesetzmässigkeit der chemischen Elemente', in: *Ann. Chem. u. Pharm.* (1871), Suppl. VIII, pp.133-229; Reprint in: Ostwald's Klassiker der exakten Wissenschaften, pp.50, 54, 55.

[130] *Ibidem*, p.103; —, *Grundlagen der Chemie*, p.692 n.13.

[131] In many versions of the Periodic System nickel preceded cobalt: not only their atomic weights but also other quantitative properties show little difference.

[132] Mendeléev, *Grundlagen der Chemie*, p.953 n.80; p.694 n.15; p.683 n.8.

[133] In 1904 G. Rudorf still maintained Mendeléev's standpoint: 'Es ist sehr wahrscheinlich, dass Tellur in der Tat ein Gemisch ist, und das wirkliche Atomgewicht des dem Selen folgenden Elements ungefähr 125 beträgt ... es ist wohl ganz unwahrscheinlich, dass das Periodische System hier nicht stimmen sollte' (G. Rudorf, *Das periodische System*, Hamburg-Leipzig 1904, pp.73-4).

[134] The difficulty was that a group of elements (the 'rare earths' metals) having sequentially increasing atomic weights nevertheless showed the same valency (3) and closely resembled each other in chemical properties. Already in 1871 Mendeléev wrote: 'Therefore, here, more than in any other place in the system of elements, new investigations are desirable to which the periodic law gives a guide line' (Mendeléev, 'Die periodische Gesetzmässigkeit', p.91).

[135] Mendeléev, *Grundlagen der Chemie*, p.695 n.15.

[136] Mendeléev, 'Über die Beziehungen', p.19.

[137] Quoted by F.B. Venable, *The Development of the periodic Law*, p.109 (from article by Mendeléev in Chem. News 1879).

[138] Cf. Th. Young, *Philosophical Transactions* 1800; A. Fresnel, *Oeuvres,* I, pp.629, 248. Fresnel points out that Huygens, Descartes and Euler had all made this comparison.

[139] J.J. Balmer, 'Notiz über die Spektrallinien des Wasserstoffs', in: *Verhandlungen der Naturforschenden Gesellschaft in Basel*, VII (1885), pp.548-560, *q.v.* p.551.

[140] *Ibidem*, pp.551-2.

[141] *Ibidem*, pp.552-3.

[142] A. Sommerfeld, *Atombau und Spektrallinien*, 4. Aufl., Braunschweig 1924, p.86.

[143] Balmer, 'Notiz über die Spektrallinien des Wasserstoffs', pp.558-9.

[144] J.J. Balmer, 'Eine neue Formel für Spectrallinien', *Verh. naturf. Gesellsch. in Basel* 11 (1897), pp.454, 460. Quote p.454: 'Für die geschlossene neue Formel spricht ihre grosse Einfachheit, welche nur die Einfachheit der erwähnten Wasserstoffformel an die Seite gestellt werden kann.'

[145] A. Hagenbach, 'J.J. Balmer und W. Ritz', in: *Die Naturwissenschaften* 9 (1921), pp.451-452.

[146] J.J. Balmer, *Des Propheten Ezechiel Gesicht vom Tempel. Für Verehrer und Forscher des Wortes Gottes und für Freunde religiöser Kunst übersichtlich dargestellt und architecktonisch erläutert.* Ludwigsburg 1858, p.2.

[147] *Ibidem*, p.5.

[148] Hagenbach, 'J.J. Balmer und W. Ritz', p.452.

[149] 'For my thoughts are not your thoughts, neither are your ways my ways, saith the Lord. For as the heavens are higher than the earth, so are my ways higher than your ways and my thoughts than your thoughts' (Isaiah 55: 8-9 (Authorized Version)).

[150] Balmer, *Des Propheten Ezechiel Gesicht vom Tempel*, p.49.

[151] J.J. Balmer, *Die Naturforschung und die moderne Weltanschauung*, Basel 1868, pp.5; 10, 12-15, 19, 26. The inner contradictions of the undulatory theory are expressed on p.13: 'Dieser Aether ... soll aüsserst elastisch sein und doch keinen Widerstand leisten; er soll eine Flüssigkeit sein und doch in den doppelbrechenden Krystallen nach verschiedenen Axenrichtungen verschiedene

Elastizität besitzen! Wer vermag solche Widersprüche wirklich zu vereinen?' Similar statements occur in: —, *Gedanken über Stoff, Geist und Gott*. Basel 1891, p.13.

[152] O.D. Chwolson, *Die Evolution des Geistes des Physik 1873 - 1923*. Braunschweig 1926 , p.143.

[153] 'Certe contra experimentorum tenorem somnia temere confingenda non sunt, nec a naturae analogia recedendum est, cum ea simplex esse soleat et sibi semper consonea' (Newton, *Principia Mathematica Philosophiae Naturalis* [3]1726, lib.III, Regula III, p.387). 'Causas rerum naturalium non plures admitti debere, quam quae et verae sint et earum phaenomenis explicandis sufficiant'. 'Dicunt utique philosophi: Natura nihil agit frustra, et frustra fit per plura quod fieri potest per pauciora. Natura enim simplex est et rerum causis superfluis non luxuriat' (*ibidem*, Regula I). In the first edition the sentence 'Causas rerum ... phaenomenis explicandis sufficiant' is presented as 'Hypothesis I' (*Principia Mathematica Philosophiae Naturalis*, Londoni 1687, p.402).

[154] N. Bohr, 'Über das Wasserstoffspektrum' (orig. lect. 20-12-1913, publ. in): *Fysisk Tidsskrift* 12 (1914), p.97; German transl. in: N. Bohr, *Drei Aufsätze über Spektren und Atombau* 2. Aufl., Braunschweig 1924, p.10: 'Indem wir die Plancksche Theorie annahmen, haben wir nämlich offen die Unzulänglichkeit der gewöhnlichen Electrodynamik erkannt und mit dem zusammenhangenden Kreis von Annahmen, die diese Theorie tragen, entschieden gebrochen ... der sicherste Weg aber ist natürlich, so wenige Annahmen wie überhaupt nur möglich zu machen.'

[155] A. Sommerfeld, *Atombau und Spectrallinien*, 'Aus dem Vorwort zur ersten Auflage (Sept.1919)', p.III.

[156] *Ibidem*, 'Vorwort zur vierten Auflage', p.VI; note on p.VII.

[157] M. Planck in: *Die Naturwissenschaften* 11 (1923), pp.535-6.

[158] M. Born in: *Die Naturwissenschaften* 11 (1923), p.538.

[159] *Ibidem*, pp.538, 542.

[160] N. Bohr, in: *Die Naturwissenschaften* 11 (1923), pp.606, 608-9, 612.

[161] *Ibidem*, pp.624.

[162] P.T. Matthews, 'Order out of Sub-nuclear Chaos, *New Scientist* 379 (20-2-1964), p.458.

[163] *Ibidem*, p.460.

[164] See above, note 6 of this chapter.

[165] J. Jeans, *The Mysterious Universe*, Cambridge 1930, p.134; cf p.149.

CHAPTER III

[1] Aristotle, Physica Bk.IV, ch.8, 215a25-216.

[2] Aristotle, *De Generatione et Corruptione* Bk.I. ch.10, 328a25-30.

[3] Geber, *Summa Perfectionis Magisterii*, ch.11. In: Gratarolus ed., *Verae Alchymiae, Artisque Metallicae Doctrina*. Basileae 1561, p.125. From internal evidence J. Ruska concluded that the Summa was written in Spain in the 13th century. There are alchemical works written in Arabic attributed to Jabir-ibn-Hayyan (ca 900) (J. Ruska, *Arabische Alchemisten*, Vol.II, Gafar-al-Sadiq, der sechste Imam. Heidelberg 1924. Also: J. Ruska and P. Kraus, 'Der Zusammenbruch der Dschâbir-legende', in: *Dritter Jahresber. d. Forschungs-Instituts f. Gesch. d. Naturwissensch. in Berlin*, Berlin 1930). On Ruska's important works about Arabic alchemy, see: R. Hooykaas, 'Ruska's Werk over de Arabische Alchemie. *Chemisch Weekblad* 35 (1938), pp.149-153.

[4] Bernardus Trevirensis (also: Trevisanus), *Peri Chimeias, Opus Historicum et Dogmaticum ... ex Gallico in Latinum Versum*, Argentorati 1567, fol.6vs, 8r, 8vs. He says that he succeeded in achieving the 'Great Work' when he was 74 years old, though he had begun at 17 (—, fol.38r.).

[5] 'The arts either, on the basis of nature, carry things further than nature can, or they imitate nature' (Aristotle, *Physica* Bk.II, ch.8, 199a).

[6] Geber, *Summa*, ch.51.

[7] Jâbir, quoted by E.J. Holmyard, *Chemistry to the Time of Dalton*, London 1925, p.10.

[8] Bartholomaeus Anglicus, *De Proprietatibus Rerum* (ca 1240), Bk.XVI, cap.7.

[9] Geber, *Summa*, ch.25. Cf. ch.10; ch.53.

[10] R. Hooykaas, *Het Begrip Element in zijn historisch-wijsgeerige Ontwikkeling*, Utrecht 1933, pp.41-49.

[11] Trevirensis, *Peri Chimeias*, fol.28vs. Cf. fol.25r, 26vs.

[12] Thomas Norton, *The Ordinall of Alchemy* ch.III (ed. Ashmole 1652, p.41).

[13] Goethe's Faust, 1042-1048:

Da ward ein roter Leu, ein kühner Freier,
Im lauen Bad der Lilie vermählt,
Und beide dann mit offnem Flammenfeuer
Aus einem Brautgemach ins andere gequält
Erschien darauf mit bunten Farben
Die junge Königin im Glas, ...

[14] Trevirensis, *Peri Chimeias,* fol.34vs.

[15] Norton, *The Ordinall of Alchemy* (ed. Ashmole 1652), ch.V, p.57, 60.

[16] Geber, *Summa,* ch.61: cf Hooykaas, *Het Begrip Element,* p.47; 45-49.

[17] E. von Meyer, *Geschichte der Chemie von den ältesten Zeiten bis zur Gegenwart*, 4. Aufl., Leipzig 1914, p.40.

[18] 'Ex praecedentibus igitur patet, quod multa quantitas argenti vivi est perfectionis, multa vero sulphuris, causa corruptionis' (Geber, *Summa,* ch.61, p.157).

[19] '... nam videmus argentum vivum argento vivo magis adhaerere et eidem magis amicari, post illud vero aurum et post haec argentum' (Geber, *Summa,* ch.59, p.155).

[20] Robertus Tauladanus, *In Braceschum Gebri Interpretem Animadversio*. In: G. Gratarolus, *Verae Alchemiae Artisque Metallicae ... doctrina.* Basileae 1561, p.53, 54: '(res) quibus cum aliquo affinitatis vinculo conjunctae sunt, tanta quaedam benevolentia complectantur, et eas sibi copulari gaudent, et laetantur ...'

[21] Norton, *The Ordinall of Alchemy,* ch.I, p.20; 18. See also page 90.

[22] Isaac Newton to Francis Aston, 18 May 1669, In: H.W. Turnbull ed., *The Correspondence of Isaac Newton*, Vol.I, Cambridge 1959, p.11.

[23] Hooykaas, *Het Begrip Element*, deals with the ambiguous character of 'element' from Antiquity to Mendeléev. R. Hooykaas, 'The Law of the Conservation of Elements' (originally in Dutch, 1947, as 'De Wet van Elementenbehoud', in: *Chem. Weekblad* 43 (1947), pp.526-531; translation in: *Selected Studies*, Coimbra 1983, pp.121-144).

[24] In the terminology of the atomic theory neither S_2 nor S_8, but 'S' is the element. All substances denoted as 'sulphur' (brimstone) are considered to be conglomerates of particles with the nuclear charge 16 and to all of them the same place in the periodic table is allotted, though there are 'isotopes' (occupants of the same place) of various atomic weights (32, 33, 34). There are different molecules of sulphur (S_8, S_6, S_4, S_2) and, in the solid state, sulphur molecules may be differently arranged in the crystals (rhombic and monoclinic sulphur). Cf. Hooykaas, 'The Law of the Conservation of Elements', p.139 ff.

[25] R. Hooykaas, 'Die Elementenlehre des Paracelsus', in: *Janus* 39 (1935), pp.175-187 (Reprinted in: *Selected Studies*, pp.43-57).

[26] R. Hooykaas, 'Die chemische Verbindung bei Paracelsus', in: *Sudhoffs Archiv f. Gesch. d. Medizin u. d. Naturwiss.* 32 (1939), pp.166-175. (Reprint in: *Selected studies*, pp.93-104).

[27] '... nun ist das wort auch dreifach gewesen, dan die trinitet hats gesprochen ... Und fürhin seind alle ding in drei gesetzt' (Paracelsus, *De Meteoris*, ch.2. In: Theophrast von Hohenhein, genannt Paracelsus, *Sämtliche Werke*, K. Sudhoff ed., Vol.XIII, München-Berlin 1931, p.135.

[28] '...ein ietliche kunst ... nicht mer dan dreierlei species suchen sol, als wenig minder oder mer die Zal in der gotheit ist. ... Und ein ietliche kunst die da mer suchet, die ist falsch, und irret in der Natur, sucht in ir das in ir nicht ist' (Paracelsus, *ibidem*, *Werke* XIII, p.136).

CHAPTER IV

[1] D. Sennert, *Hypomnemata physica* (1636). In: *Opera Sennerti*, Lugduni 1666. Vol.I, p.120, col.2.

[2] Cf R. Hooykaas, *Het Begrip Element*, Utrecht 1933; ch IX, §3, pp.145-159.

[3] G.E. Stahl, *Fundamenta Chymiae*, Norimbergae 1746; Vol.I, p.5; G.E. Stahl, *Specimen Beccherianum*, Lipsiae 1703, pp.7-37. This work, though called by its author just a commentary on J.J. Beccheri's *Physica Subterranea* (1667) is in fact quite original: 'Beccheriana sunt quae profero' (Praeloquium, 4vs.).

[4] 'Ut omnis scientia humana ... Analytica esse debet, nempe *a posteriori* ... ut Experientia precedat.' Stahl, *Specimen Beccherianum*, p.233; cf p.84.

[5] 'Ita enim v.g. quis a longissimis usque temporibus ignoravit, quod in ignobilibus metallis aliquid insit, quod *ignitione* absumi potest? Quis non in officinis omnibus metallariis ... videre potuit, quod violenta fusio et *immediatus contactus* carbonum, metalla ita *exusta* in consistentiam suam metallicam deducant ...' (Stahl, *ibidem*, p.84).

[6] '... quis saltem serio agnovit, quod illa ignitione incineratoria, revera hisce metallis *materialiter* aliquid *decedat* quod adeo simili *materia* resarciendum sit?' (Stahl, *ibidem*, p.85).

[7] Stahl, *ibidem*, p.298; cf p.140.

[8] 'Ego *Phlogiston* appellare coepi ...' (Stahl, *ibidem*, p.39).

[9] Kühnrusz. Stahl *ibidem*, p.148. '*materialis* principium ignescendi' (the material principle of inflammability) (p.150).

[10] Stahl, *ibidem*, p.148.

[11] '... *Phlogiston* seu Sulphureum Principium ...' (Stahl, *ibidem*, p.94).

[12] Stahl, *ibidem*, p.100.

[13] Stahl, *ibidem*, pp.50, 108, 301.

[14] 'Similia similibus quadrant' (Stahl, *ibidem*, p.27). Oily substances are dissolved by other oils (p.28).

[15] Stahl, *ibidem*, p.121, 204. '*cognatio materiae* solventium ad solvenda intercedat, dum v.g. ... *Aqua* tales materias, quibus intimiore nexu *insita* atque innexa est aqua, solvit' (p.18).

[16] '... haec duo, *sal* acidum, et nostrum *ignescendi* principium, nullo alio quam *terreo* intuitu, ingenio, et tam proprio quam mutuo, respectu, ut *terra cum terra*, connexionem seu mixtionem subeant' (Stahl, *ibidem*, p.150).

[17] 'Cum utique nemo sibi facile persuaserit, quod congressus *phlogistae* nostrae materiae, cum *acido*, fiat ex latere, aut qualicunque effectu, *aquositatis*: (Cum potius nostra phlogista materia, in quibuslibet mixtionibus, nescio quod *divortium* et *aversionem* ab aquositate introducere observetur) sed sane magis, imo vero unice, e latere *terrestris* substantiae, quae mixtum salis constituit' (Stahl, *ibidem*, p.151).

[18] A.L. Lavoisier, *Traité élémentaire de Chimie*, Vol.I, Paris ^{2}1793, p.69.

[19] A.L. Lavoisier, 'Sur le Phlogistique', in: *Mém. Acad. des Sci.* 1783; reprint in: *Oeuvres de Lavoisier*, Vol.II, Paris 1862, p.652. Cf. —, *Traité de Chimie*, Vol.I, p.69.

[20] A.L. Lavoisier, 'Sur l'Existence de l'Air dans l'Acide nitreux', in: *Mém. Acad. des Sci.* 1776; reprint in: *Oeuvres* II, p.130.

[21] Lavoisier, *Traité de Chimie*, p.XIII

[22] *Ibidem*, p.204. See: 'Tableau des combinaisons binaires de l'oxygène avec les substances'.

[23] A.L. Lavoisier, 'Décomposition et Recomposition de l'Eau', in: *Oeuvres* II, p.337, 339.

[24] Lavoisier, *Traité de Chimie*, Vol.I, 'Discours préliminaire', pp.IX-XI.

[25] Lavoisier, 'Expériences sur la Respiration des Animaux', *Mém. Acad. des Sci.* 1777; reprint in: *Oeuvres* II, p.181; —, *Traité de Chimie*, Vol.I, p.85.

[26] Lavoisier, *Traité de Chimie*, Vol.I, p.116.

[27] *Ibidem*, Vol.I, ch.XVII, p.176.

[28] *Ibidem*, Vol.I, p.116.

[29] J. Black, *Dissertatio Medica Inauguralis de Humore Acido a Cibis Orto, et Magnesia Alba*, Edinburgh 1754; —, *Experiments upon Magnesia Alba, Quicklime, and some other Alcaline Substances*, Edinburgh 1756.

[30] A.L. Lavoisier, Registre de Laboratoire 23 févr. 1773 (the beginning of Lavoisier's doubts). Cf. M. Berthelot, *La Révolution chimique, Lavoisier*. Paris 1890, pp.227-228; 233.

[31] A.L. Lavoisier, *Opuscules physiques et chimiques* (1774); reprint in: *Oeuvres* I, pp.612-613.

[32] *Ibidem*, Pt.II, ch.5: 'Du fluide élastique fixé dans les chaux métalliques' (*Oeuvres* I, pp.612-613). The 'gas' (probably mainly carbon monoxide) had not yet been well investigated at that time. Lavoisier says that the coal's main function is to give back to the 'fluide élastique fixé' (in the metal calx) its 'phlogistique, matière du feu, et lui restituer en même temps l'élasticité qui en dépend'. He warns, however: 'ce n'est qu'avec la plus grande circonspection qu'on peut hasarder un sentiment sur une matière si délicate et si difficile' (p.613).

[33] A.L. Lavoisier, 'Sur la Respiration des Animaux', in: *Mém. Acad. des Sci.* 1777; reprint in: *Oeuvres* II, pp.174-177.

[34] Lavoisier, 'Sur la Combustion en Général', in: *Mém. Acad. des Sci.* 1777 (*Oeuvres* II, pp.225-233).

[35] Lavoisier, *Opuscules physiques et chimiques (Oeuvres* I, pp.279-280). See also: —, 'Sur la Combustion en Général' (*Oeuvres* II, p.226)

[36] Lavoisier, 'Sur la Combustion en général' (*Oeuvres* II, p.226).

[37] 'L'existence de la matière du feu, du phlogistique, dans les métaux, dans le soufre, etc. n'est donc réellement qu'une hypothèse, une supposition ... (*Ibidem*, p.228).

[38] *Ibidem*, p.229.

[39] *Ibidem*.

[40] *Ibidem*, p.231: 'L'air pur, l'air déphlogistiqué de M. Priestley, est donc, dans cette opinion, le véritable corps combustible, et peut-être le seul de la nature'.

[41] *Ibidem*, pp.231-232.

[42] *Ibidem*, p.232.

[43] '... en attaquant ici la doctrine de Stahl, je n'ai pas pour objet d'y substituer une théorie rigoureusement démontrée, mais seulement une hypothèse qui me semble plus probable, plus conforme aux lois de la nature, qui me paraît renfermer des explications moins forcées et moins de contradictions' (*ibidem*, p.233).

[44] Lavoisier, 'Réflexions sur le Phlogistique, pour servir de suite à la Théorie de la Combustion et de la Calcination, publiée en 1777', in: *Mém. Acad. des Sci.* 1783; reprint in: *Oeuvres* II, p.652.

[45] *Crell's Chem. Ann.* 1793, pp.383ff.: Deiman, Paets van Troostwijk, 'Über die Entzündung des Schwefels mit Metallen ohne Gegenwart von Lebensluft'. Cf. M.P.M. van der Horn van den Bos, *De Nederlandsche Scheikundigen van het laatst der vorige Eeuw*, Utrecht 1881, pp.66-74.

[46] J.J. Berzelius, *Jahres-Bericht* 6 (1827), pp.189-198; 8 (1829), p.137. Cf. R. Hooykaas, 'Die Chemie in der ersten Hälfte des 19. Jahrhunderts', in: *Technikgeschichte* 33 nr.1 (1966), pp.1-24. Also in: W. Treue and K. Mauel (eds.), *Naturwissenschaft, Technik und Wirtschaft im 19. Jahrhundert* Vol.II, Göttingen 1976, pp.587ff.

[47] Lavoisier, 'Réflexions sur le Phlogistique' (*Oeuvres* II, pp.623-655)

[48] 'J'ai déduit toutes les explications d'un principe simple, c'est que l'air pur, l'air vital, est composé d'un principe particulier qui lui est propre, qui en forme la base, et que j'ai nommé *principe oxygine*, combiné avec la matière du feu et de la chaleur. Ce principe une fois admise ... tous les phénomènes se sont expliqués avec une étonnante simplicité. Mais si tout s'explique en chimie d'une manière satisfaisante sans le secours du phlogistique, il est par cela seul infiniment probable que ce principe n'existe pas; que c'est un être hypothétique, une supposition gratuite; ... une erreur funeste à la chimie, et qui me paraît en avoir retardé considérablement les progrès par la mauvaise manière de philosopher qu'elle y a introduite' (A.L. Lavoisier, 'Réflexions sur le Phlogistique' (*Oeuvres* II, p.623).

[49] Lavoisier, 'Réflexions sur le Phlogistique' (*Oeuvres* II, p.640).

[50] Lavoisier and Laplace, *Mém. Acad. des Sci.* 1780, pp.355ff; reprint in: *Oeuvres* II, p.283ff.

[51] Lavoisier, 'Réflexions sur le Phlogistique' (*Oeuvres* II, pp.651-2).

[52] Letter to Chaptal, quoted by É. Grimaux, *Lavoisier*, Paris 1888, p.126.

[53] 'La chimie est une science française; elle fut constituée par Lavoisier' (A. Wurtz, *Histoire des Doctrines chimiques*, Paris 1868). É. Grimaux wrote: 'Toute la science moderne n'est que le développement de l'oeuvre de Lavoisier' (—, *Lavoisier*, p.128).

[54] R. Marcard, *Petite Histoire de la Chimie et de l'Alchimie*, Bordeaux 1938, p.223: 'La chimie jaillit, comme jadis Minerve, toute équipée du cerveau d'un savant bien éminemment français, nommé Lavoisier.'

[55] I have dealt with this problem in *Het Begrip Element in zijn historisch-wijsgeerige Ontwikkeling*, Utrecht 1933, and in: 'De Wet van Elementenbehoud', *Chem. Weekblad* 43 (1947) pp.526-531 (English translation in: R. Hooykaas, *Selected Studies in History of Science*, Coimbra 1983, pp.121-143).

[56] Lavoisier, *Traité élémentaire de Chemie*, Vol.I, pp.176, 179.

[57] Cf R. Hooykaas, 'Die Chemie in der ersten Hälfte des 19. Jahrhunderts', in: *Technikgeschichte* 33, nr.1 (1966), pp.1-24. Also in: W. Treue and K. Mauel eds., *Naturwissenschaft, Technik und Wirtschaft im 19. Jahrhundert.* Göttingen 1976; Vol.II, pp.587-613; *q.v.* pp.594-595.

[58] K.A. von Zittel, *Über wissenschaftliche Wahrheit*, München 1902, p.3.

[59] R. Hooykaas, 'Wissenschaftsgeschichte: eine Brücke zwischen Natur- und Geisteswissenschaften', in: *Berichte zur Wissenschaftsgeschichte*, 1982, pp.164-166.

[60] R. Mayer, 'Bemerkungen über die Kräfte der unbelebten Natur', in: *Annalen der Chemie und Pharmazie*, 1842; reprint in: R. Mayer, *Die Mechanik der Wärme*, J.J. Weyrauch ed., 3e Aufl., Stuttgart 1843, p.31.

[61] Schroeder van der Kolk, 'Mechanical Energy of Chemical Reactions', in: *Pogg. Ann. Physik* 22 (1864), pp.434, 658. Quoted by G. Helm, *Die Energetik und ihre geschichtliche Entwicklung*, Leipzig 1898, p.139.

[62] G.N. Lewis, *The Anatomy of Science*, New Haven 1926, pp.164-165.

[63] W.D. Harkins, F.E. Brown, E.C.H. Davies, *J. Amer. Chem. Soc.*, 39, 1917, pp.354, 541. Quoted by H.R. Kruyt, *Colloids*, New York - London 1927, pp.45-49.

CHAPTER V

[1] J. Maitland, 'The Beginnings of St Andrews University 1410 - 1418', in: *Scottish Historical Review* 8 no. 31 (1911). At Lawrence's instigation in 1406 at Perth the first martyr fire was kindled in Scotland (p.239).

[2] James Halderstone, *Copiale Prioratus Sancti Andree* (The letter book of James Halderstone, Prior of St Andrews, 1418 - 1443) J.H. Baxter ed. Oxford 1930, pp.3, 382-384.

[3] *Acta Facultatis Artium Universitatis Sancti Andree 1413 - 1586.* A.J. Dunlop ed. Edinburgh-London 1964, p.12.

[4] Albert of Saxony brought this doctrine to Cologne; Heinrich of Langenstein to Vienna.

[5] K. Michalski, 'Les Courants philosophiques à Oxford et à Paris pendant le XIVe Siècle', in: *Bull. Intern. des Sciences et des Lettres de l'Académie Polonaise, Année 1916*, p.88.

[6] *Acta Facultatis*, pp.48-49.

[7] See e.g. W.H. Wallace, O.P., *Causality and Scientific Explanation.* Ann Arbor 1972, p.82, who speaks of Thomas's knowledge of 'projective geometry' as revealed by his proving the rotundity of the earth by its casting a round shadow on the moon. In fact this just repeats Aristotle's arguments (Aristotle, *De Coelo* Bk.II, ch.14; 297b24ff.).

[8] Petrus Apianus, *Cosmographia*, Antwerpen 1539: 'Coelum empireum habitaculum dei et omnium electorum.' Rembertus Dodonaeus, *Cosmographia in Astronomiam et Geographiam. Isagoge*, Antwerpen 1548: 'Coelum Empyreum, Beatorum sedes et habitaculum.' Thomas Digges, who let the heaven of the fixed stars extend to infinity, had no difficulty in identifying this starry heaven with the dwelling-place of the elect. Thomas Digges, *A Perfit Description of the Caelestiall Orbes.* London 1576: 'This orbe of starres fixed infinitely up extendeth itself in altitude sphericallye ...., the habitacle for the elect.'

[9] Aristotle's Prime Mover is a passive 'final cause' and not a working 'efficient cause'.

[10] It need hardly be said that at the same time the omnipotence of God was maintained. Medieval theologians were keenly aware of the fact that all their speaking about God was but a 'stammering' in an often inconsistent way.

[11] Thomas Aquinas, *Summa contra Gentiles*. Bk.II, qu.98, art.1. In: *Opera Omnia* XII, Romae 1901, suppl. p.213.

[12] E.g. in Francisco Maurolyco, *Cosmographia* (1535), dial.I, Introd.; Daniel Schwenter, *Deliciae Physico-mathematicae*, Nuremberg 1636, Tl.III, Auffgab XVI, p.186. In Dante's *Inferno* (14th cent.) Lucifer is seated on a throne at the centre of the earth. Dante's opinion that extreme cold reigns there (so that Lucifer is up to his waist in ice) is, however, quite exceptional (Dante, *Inferno*, XXXIV ll.70-111).

[13] Thomas Aquinas, *Summa contra Gentiles*. Bk.II, ch.1; ch.2.

[14] Aristotle, *De Coelo*, Bk.I, ch.2.

[15] Thomas Aquinas, Summa Theologica, Bk.III, suppl. qu.97, art.7, n.2. (Opera Omnia XII, p.243).

[16] *Ibidem*, qu.98, art. 1 (*Opera Omnia* XII, p.243).

[17] J. Clarisse ed., *Sterre- en natuurkundig onderwijs, gemeenlijk genoemd: Natuurkunde van het geheel-al, en gehouden voor het Werk van zekeren Broeder Gheraert*, Leiden 1847.

[18] 'Nu wil ic u doen ghewach / Waer die helle wesen mach / Bi scrifturen proef men wel / Dat si niewel el / Dan in midden van ertrike / Dats in centro sekerlike' (Clarisse ed., *Natuurkunde van het Geheel-al*, p.177, ll.1701-1706).

[19] '... et la plus basse chose et la plus parfonde qui soit au monde est li poins de la terre, ce est li milieu dedans, qui est apelez abismes, là où enfers est assis' (see: P. Duhem, *Le Système du Monde* IX, p.127).

[20] '... quanto espaço ha ... daqui ao centro do mundo e ao meio do inferno dos condenados, que he a medida do cemidiametro' (D. João de Castro, *Tratado da Spaera*. In: *Obras Completas de D. João de Castro*, A. Cortesão and L. de Albuquerque ed., Vol.I, Coimbra 1968, p.63).

[21] *Ibidem* (*Obras* I, pp.58-59).

[22] '*Alphius*: Plumbum nunquam perveniret ad centrum nisi liquefactum' (Erasmus, *Colloquia*, Basel 1533; In: *Opera Desiderii Erasmi Roterodami .... Ordinis Primi Tomus Tertius*, Amsterdam 1972, pp.714-715.)

[23] Aristotle, *De Coelo*, Bk.II, ch.7, 289a24-25.

[24] Johannes Buridanus, *Quaestiones super Libros Quattuor de Coelo et Mundo*, Bk.II, qu.16. Quoted from edition E.A. Moody, Cambridge, Mass. 1942, p.204.

[25] *Ibidem*, l.II, qu.16 (Moody p.200)

[26] Plutarch (circa 47 - circa 120), *Moralia*; quoted by L. Thorndike, *A History of Magic and Experimental Sciences*, New York 1929, Vol.I, p.219.

[27] Adelard of Bath, *De Eodem et Diverso*, ch.48; *Quaestiones Naturales*, ch.13-14 (cf Thorndike, *ibidem*, Vol.II, p.35).

Vincentius Bellovacensis, *Speculum Naturale*, VII, 7 (cf C.S. Lewis, *The Discarded Image, an Introduction to Medieval and Renaissance Literature*, Cambridge 1964, p.141).

For al-Khwazimi, see Clagett, *The Science of Mechanics in the Middle Ages*, Madison 1961, pp.58, 60.

[28] '... toda a coisa pesada em sumo grau, deseja o centro, e ali folga e cessa de se mover' (*Guia de Munique*, ch.I. In: L. Mendonça de Alberquerque, *Os Guias Nauticos de Munique e Evora*, Lisboa 1965, p.161). Cf Castro, *Obras* I, p.120.

[29] Cf R. Hooykaas, *Das Verhältnis von Physik und Mechanik in historischer Hinsicht*, Wiesbaden 1963, p.10; R. Hooykaas, *Religion and the Rise of Modern Science*, Edinburgh 1972, ch.III, pp.56-59.

[30] P. Duhem, *Études sur Léonard de Vinci*, Vol.III, Paris 1955, p.24.

[31] Aristotle, *Physica*, Bk.VII, ch.5.

[32] M.A. Hoskin and A.G. Molland, 'Swineshead on Falling Bodies: an Example of Fourteenth Century Physics', in: *Brit. Journ. Hist. Sci* 3 (1966), pp.150-182. The tract 'De loco elementi' is part of 'Liber Calculationum', written after 1328.

[33] Thomas Bradwardine, *His Tractatus de Proportionibus* (1328). H.L. Crosby transl. and ed., Madison 1954, pp.110-116. Cf M. Clagett, *The Science of Mechanics in the Middle Ages*, Madison 1961, pp.437ff.. In cap.III: 'The proportion of the velocities of motions follows the proportion of the power of the mover to the power of the moved thing' (Clagett, *ibidem*, p.438).

[34] Hoskin and Molland, 'Swineshead on Falling Bodies', p.438.

[35] A. Maier, *An der Grenze von Scholastik und Naturwissenschaft*, Essen 1943, ch.2, pp.288-348: 'Oresme's Methode der graphischen Darstellung'. Anneliese Maier introduced the phrase 'the mechanization of the world picture' (A. Maier, *Die Mechanisierung des Weltbildes im 17. Jahrhundert*, Leipzig 1938). See A. Maier, *Zwei Grundprobleme der scholastischen Naturphilosophie*, 2. Aufl. Roma 1951, p.88 n.1.

[36] Michalski, 'La Physique nouvelle et les différents Courants philosophiques au XIVe Siècle', in: *Bull. Intern. Acad. Polon. Sci.*, Cl. Lettres; (Année 1927), p.157.

[37] Nicole Oresme, *Tractatus de Configurationibus Qualitatum,* P.I, ch.22. Cf Duhem, *Le Système du Monde* Vol.VII, pp.582, 585. Cf Maier, *Zwei Grundprobleme*, p.105.

[38] Duhem, *Le Système du Monde,* Vol.VII, pp.583, 585; Maier, *Zwei Grundprobleme*, p.108.

[39] On the other hand it shows some affinity with the ancient theory of 'signatures'; the sympathy between things of the same shape, which led to such ideas as that liverwort (agrimony; hepatica) is a cure for liver disease. For now the 'configurations' of certain 'occult qualities' of precious stones etc., are effective on similar 'configurations' of bodies or parts of bodies.

[40] Henricus de Hassia, *Tractatus de Reductione Effectuum Specialium in Virtutes Communes et Causas Generales*, ch.1, 23, 25 (quoted from Duhem, *Le Système du Monde*, VII, pp.589; 594, 598). See also Duhem, *Le Système du Monde,* VII, p.585. For Oresme see: Maier, *Zwei Grundprobleme*, p.108.

[41] Henricus de Hassia, *ibidem*, ch.1 (quoted from Duhem, *Le Système du Monde*, VII, pp.587-588).

[42] William of Ockham (circa 1320), however, considered motion as just a sequence of places occupied by the moving body; once a body possesses motion it will keep it; the projectile is its own motor and not some implanted force of the air. If the air were the moving cause, what then would happen when two arrows moving along the same path in opposite directions are meeting? (Ockham, *Sentent.*, II, qu.18;26. Cf Maier, *Zwei Grundprobleme*, pp.155, 157.

[43] See for similar conceptions in the Arab and Latin Middle Ages: Clagett, *The Science of Mechanics*, pp.510ff; 520, 523.

[44] K. Michalski, 'La Physique nouvelle', p.157. Lawrence, however, wanted to save Aristotle as well as the phenomena, for at the same time he tried to twist the Philosopher into conformity with the impetus theory. One of his followers at Erfurt ('Magister de Stadiis') wrote: 'Others say, with Londorius, that when the Philosopher [Aristotle, RH] says that the projectile is moved by the air, this does not refer to the efficient cause properly so called, but to the necessarily accompanying cause (*causa sine qua non*)' (*Physica* Bk.VIII, qu.11, quoted by Michalski, *ibidem*, p.158.). That is, the presence of air is an indispensible condition for the motion, but the motion's *cause* is the impetus. Evidently Londorius thought that Buridan went too far in his abstractions.

[45] Buridanus, *Quaestiones super Libros Quattuor de Coelo et Mundo*, Bk.II, qu.12 (Moody p.180).

[46] Johannes Buridanus, *Quaestiones super Octo Phisicorum Libros Aristotelis*, Parisiis 1509, Bk.VIII, qu.12, fol.CXXI, col.1.

[47] Cf E.J. Dijksterhuis, *De Mechanisering van het Wereldbeeld*, Amsterdam 1950, Pt.II, §113 (pp.201-202), criticised Maier's arguments against this comparison (cf Maier, *Die Vorläufer Galileis im 14.Jahrhundert*, Roma 1949, pp.141ff).

[48] Buridan, *Quaestiones de Coelo et Mundo*, Bk. II, qu.12 (Moody p.180); —, *Questiones Phisicorum,* Bk.VIII, qu.12. fol.CXXr, col.2.

[49] Oresme, *Le Livre du Ciel et du Monde*, Bk.II, ch.2. (A.D. Menut and J. Denomy ed. London: Madison 1968, p.285).

[50] Buridan, *Quaestiones Phisicorum,* Bk.VIII, qu.12, fol.CXXIr, col.2.

[51] Gravity continually impresses impetus on falling bodies, and as their downward motion increases this impetus (*impetus accidentalis*) they will move swifter, and the swifter motion gives a greater impetus, etc., so that they fall with an accelerating velocity (Buridan, *Questiones Phisicorum*, Bk.VIII, qu.12, fol.CXXvs, col.2; Buridan, *Quaestiones de Coelo et Mundo*, Bk.II, qu.12 (Moody pp.180-181).

[52] Buridan, *Quaestiones de Coelo et Mundo*, Bk.IV, ch.7. (Moody p.267).

[53] Buridan, *Quaestiones Phisicorum*, Bk.VIII, ch.12.

[54] Oresme, *Le Livre du Ciel et du Monde*, Bk.II, ch.31 (Menut and Denomy, p.172)

[55] Oresme, *ibidem*, Bk.I, ch.18, (Menut and Denomy, p.144.)

[56] Oresme, *Quaestiones de Spera*. Quoted by Clagett, *The Science of Mechanics*, p.553.

[57] Albert of Saxony, *De Coelo*, Bk.II, qu.14; cf P. Duhem, *Le Système du Monde,* VIII, 291; Clagett, *The Science of Mechanics*, pp.566; 569.

[58] Cf Bradwardine, *De Proportionibus*, ch.3 (Crosby pp.110ff.); Crosby says of Bradwardine's 'law' that 'it is an axiom rather than a theorem' (p.38).

[59] The first suggestion that the motion of a falling body is uniformly accelerated was made at a very late date (1555); see Clagett, *The Science of Mechanics*, pp.555-556.

[60] Buridan, *Quaestiones Phisicorum*, l.VII, qu.8, ed. Paris 1509, fol.CVIIIvs, col.1.

[61] Oresme, *Le Livre du Ciel et du Monde*, Bk.I, ch.29 (Menut and Denomy p.196).

[62] L. Vives, *De Causis Corruptarum Artium*, l.V. Quoted from: Vives, *Opera*, Basileae 1555, Vol.I, p.410.

[63] G. Sarton, *Introduction to the History of Science*, Vol.III, Baltimore 1947, p.737.

[64] '... qui pene modum excessit ingenii humani' (J.C. Scaliger, *Exotericarum Exercitationum, Libri XV ad Hieronymum Cardanum* (1557). exerc.324. (ed. Francofurti 1612, p. 1028). Cf exerc.340 (p.1068); exerc.22 (p.104); exerc.28 (p.131).

[65] Scaliger, *Exotericarum Exercitationum*, exerc.16, 4 (p.83).

[66] Erasmus, *Colloquia* (*Opera*, pp.714-715). (Cf above, p.124n.22).

[67] Francisco Maurolyco, *Cosmographia* (1535). Quoted from ed. 1543, Dialogus I. See P. Duhem, *Études sur Léonard de Vinci*, Vol.III, pp.195-196. He says that only those without a sound knowledge of the problem will say that the stone immediately stops at the centre. The impetus of its weight, however, makes it pass the centre and perform oscillations which become smaller as the impetus gradually diminishes. In the same way a weight hanging from a cord, when displaced from the vertical position will perform diminishing oscillations finally stopping at the vertical position. The partner in the dialogue recognizes that this speculation is supported by a well-chosen illustration and reminds the other speaker of Erasmus' *Colloquia*.

[68] N. Tartaglia, *Nova Scientia*, Venice 1537, Bk.I. prop.I (transl. by S. Drake in: S. Drake and J.E. Drabkin ed., *Mechanics in Sixteenth Century Italy*, Madison 1969, p.76).

[69] J.B. Benedetti, *Diversarum Speculationum Mathematicarum et Physicarum Liber*, Torino 1585, p.369 (tranls. by S. Drake, in: S. Drake and J.E. Drabkin ed., *Mechanics in Sixteenth Century Italy*, Madison 1969, p.235).

[70] Galileo Galilei, *Dialogo sopra i Due Massimi Sistemi del Mondo, Tolomaico e Copernicano*, Argentorati 1632. Reprint in: *Opere* VII, pp.27-546. Quotations from —, *The System of the World in Four Dialogues. Wherein the two Grand Systems of Ptolemy and Copernicus are Largely Discoursed* ... in: Thos Salisbury, *Mathematical Collections and Translations* Vol.I, London 1661, p.117.

[71] *Ibidem* (Salisbury, p.207).

[72] The navigator D. João de Castro held (1538) that a leaden ball, dropped into a hole passing through the earth's centre, 'when arrived at the centre, would halt as if it were hanging there, quiet and at ease, without going any farther. The reason for this is that all heavy things descend to the middle, which is the centre, and if thence they would go farther this would be not descension but ascension' (D. João de Castro, *Tratado da Esfera*. In: *Obras* I, pp.58-59). Cf R. Hooykaas, *Science in Manueline Style*, Coimbra 1980, p.31.

Towards the end of the century (1592) the Netherlandish pilot and nautical writer Lucas Jansz Waghenaer (1553 - 1606) held that a stone would fall to the centre of the earth (were it not impeded by the solidity of the earth) and would remain there hanging and at rest (L.J. Waghenaer, *Thresoor der Zeevaert*, Leyden 1592, Bk.III, p.196).

The Altdorf professor of physics Daniel Schwenter, who was one of the pioneers in the introduction of experiments in the university teaching of physics, was in general a faithful follower of Aristotle, 'this miracle man' (D. Schwenter, *Deliciae Physico-mathematicae* (1636), p.390), e.g. when attributing some influence of the air in propelling a projected body (p.188) and when holding that the heavier a body is, the faster it will fall (p.391). On the topic of the stone falling into a hole that pierces the Earth, however, he took the side of Maurolyco and Walther Ryff (who followed Maurolyco) and held that the stone would oscillate until it came to rest at the centre of the universe. He also made the comparison with the pendulum (Vol.III, Auffgab.XVIII, p.187).

These differing examples clearly show that ancient ideas linger on alongside the new ones which — thanks to the selection we make when depicting the *progress* of science — are often believed to have been completely ousted long since.

73 William Gilbert, *De Mundo Nostro Sublunari Philosophia Nova*. (Opus posthumum) Amstelodami 1651, Bk.II, ch.10, pp.96, 154.

74 Gilbert, *ibidem*, Bk.II, ch.10, pp.32, 142, 147, 154, 164.

75 Gilbert, *ibidem*, pp.47, 48, 61. See also Gilbert, *De Magnete, Magneticisque Corporibus, et de Magnete Tellure, Physiologia Nova*, Londini 1600, pp.29, 65, 216, 219.

76 Francis Bacon, *Sylva Sylvarum or a Natural History* (1627), cent.I, 33.Quoted from Bacon, *Works*, Spedding, Ellis and Heath ed., London 1854, Vol.II, p.354.

77 Bacon, *Sylva Sylvarum*.

78 Bacon, *Novum Organum* (1620), Pt.II, XXXV (*Works* I, p.292.)

79 Mersenne, *Harmonie Universelle*, Paris 1636, Bk.II, prop.XII, pp.128-129.

80 Mersenne, *ibidem*, Bk.III, prop.XX, coroll.1, p.208.

81 Mersenne, *ibidem*, Bk.III, prop.XX, coroll.3, p.209.

82 Mersenne, *ibidem*, p.209.

83 Mersenne, *ibidem*.

84 Isaac Beeckman, *Journal*, 23 Nov - 26 Dec 1618, C.de Waard ed. La Haye 1939, Vol.I, p.264.

85 Beeckman, *Journal* (1614), (De Waard Vol.I, p.44).

86 Cf E.J. Dijksterhuis, *Val en Worp*, Groningen 1924, pp.311-313.

87 Cf Dijksterhuis, *ibidem*, p.375.

88 Isaac Newton, *Philosophiae Naturalis Principia Mathematica*, Londini 1687, Bk.I, sect.11. p.162. Cf: 'I shall therefore ... treat of the motion of bodies attracting each other, considering the centripetal forces as attractions; though perhaps physically speaking, they may more truly be called impulses' (*ibidem)*.

89 *Ibidem*, 3rd edn. London 1726, Bk.I, sect.11. Scholium. Quoted from: —, *Sir Isaac Newton's Mathematical Principles of Natural Philosophy and his System of the World*, F. Cajori ed., Berkeley 1947, p.192.

90 *Ibidem*, 3rd edn. Bk.I, prop..LXXIV, theorem XXXIV (Cajori, p.197).

91 *Ibidem*, 3rd edn. Bk.I, prop.LXXIII, theorem XXXIII (Cajori, p.196). Cf. props.LXX and LXXII (Cajori, p.193).

CHAPTER VI

1 Cf the replies to the enquiry by the magazine *Poland* ('Copernicus and the Contemporary World', *Poland* 1, 209 (Jan.1972), pp.35-47). In particular the contribution of H.J. Treder (Dir. Centr. Inst. f. Astrophysics of the (East) German Academy of Sciences): '... the reception of the Copernican system ... constituted part of the struggle between the materialistic and the idealistic outlook. The sharp edge of this struggle was above all directed against theology ... and finally also against the theistic positions of idealistic philosophers of later ages' (p.40). Prof. Jean Fabre (Sorbonne, Paris): '... by depriving Man of his central privileged place, Copernicus disturbed the traditional humanism, but simultaneously renewed it on a cosmic level...' (p.44).

2 Nicolaus Copernicus, *De Revolutionibus Orbium Caelestium*, Norimbergae 1543, fol.Ivvs. Paulus of Middelburg (1445 - 1533) wrote on calendar reform (1516). Born in Middelburg (Zeeland), he became bishop of Fossombrone (Italy) in 1494.

3 Copernicus, *De Revolutionibus*, praefatio fol.IIIr. Reference is to the advice of the Roman poet Horace: 'if ever you write anything ... then put your parchment in the closet and keep it back till the ninth year' (Horatius, *De Arte Poetica*, lines 386-389).

4 Copernicus, *De Revolutionibus*, praef. IIIr; IVr.

5 *Ibidem*, IIIr.

6 *Ibidem*, IVvs.

7 [Following the Horatian advice published more than nine years later:] R. Hooykaas, *G.J. Rheticus' Treatise on Holy Scripture and the Motion of the Earth; with Translation, Commentary and Additional Chapters on Ramus-Rheticus and the Development of the Problem before 1650*, Amsterdam: North-Holland 1984; cf R. Hooykaas, *J. Hist. Astron.* 15 (1984), pp.77-80, dealing with the discovery and identification of this anonymous treatise.

8 As, when borrowing from the superstition of the time, it is said that some people are 'like the deaf adder that stoppeth her ear, which will not hearken to the voice of charmers' (Psalm 58:4-5). Cf R. Hooykaas, *Religion and the Rise of Modern Science*, Edinburgh-London 1972, pp.118, 154n.27.

9 R. Hooykaas, *ibidem*, p.154n.27; R. Hooykaas, 'Calvin and Copernicus', in: *Organon (*1974), pp.139-148.

10 Thomas Ceva S.J., *Philosophia Nova-antiqua*, Venetiis 1732. The answer given to the question, 'cur heterodoxis praecipue systema terrae motae probatur?', was that the prevailing heresy does not recognize Holy Authority (pp.65, 68).

11 Cf Hooykaas, *Rheticus' Treatise*, pp.173-178.

12 *Ibidem*.

13 'To place the Sun in the centre of the universe not only seemed to contradict observation ... it was also opposed to all the accepted ideas and the image of the world made sacred by the Bible, but above all the concept of the special position occupied by the Earth' (A. Kauffeldt (Polytechn. Inst. Magdeburg, (East-)Germany), in: 'Copernicus and the Contemporary World' (*Poland* 1 (1972), p.46). G. Klaus, in his edition of Copernicus' 'first book', says in his 'Einleitung': 'Die Lehren der Bibel über Schöpfung, Weltall und Stellung des Menschen im All verwandelten sich in das, was sie tatsächlich waren, nämlich in naive Vorstellungen aus der Zeit der Auflösung der Urgemeinschaft ... Eine Gefährdung des totalen Herrschaftsanspruchs der feudalen Ideologie musste aber zwangsläufig, die Gefahr eines Angriffs gegen die ökonomische und politische Struktur des Feudalsystems mit sich bringen' ... 'Die Erde war durch Copernicus zu einem winzigen Stäubchen im unermeszlichen Weltall geworden ...' (Nikolaus Copernicus, *Über die Kreisbewegungen der Weltkörper, Erstes Buch*, G. Klaus ed. (East-)Berlin 1959, p.XLIX). [In fact the Earth was considered a tiny speck in the universe by Ptolemy, and also by medieval astronomers (Sacrobosco

etc.)!] A great influence was exerted by Haeckel, whose anti-religious bigotry inspired the following passage: 'Indem er das herrschende geozentrische Weltsystem des Ptolemäus stürzte, entzog er zugleich der herrschenden christlichen Weltanschauung den Boden, welche die Erde als Mittelpunkt der Welt und den Menschen als gottgleichen [sic!] Beherrscher der Erde betrachtete' (Ernst Haeckel, *Die Welträtsel* (1899), ch.20 (Reprint Leipzig: Kröners Taschenausgabe 1908, p.230)).

14 John Donne, *Ignatius his Conclave*. In: Donne, *Complete Poetry and Selected Prose*. J. Hayward ed.. Bloomsbury 1929, p.365.

15 John Donne to Sir Henry Goodere, April 1615 (*Complete Poetry*, p.468). Evidently an allusion to the fact that Copernicus attributed the lack of parallax of the fixed stars to their great distance from the earth.

16 Galileo, *Dialogo* (1632) In: *Opere* VII, p.62 (Salusbury I, p.25).

17 Copernicus, *De Revolutionibus* Bk.I, ch.6 ch.7fin., ch.10, p.71.

18 Thomas Digges, *A Perfit Description of the Caelestiall Orbes, According to the Most Ancient Doctrine of the Pythagoreans Lately Revived by Copernicus* (1576). Quoted by F.R. Johnson, *Astronomical Thought in Renaissance England* (1937); reprint New York 1968, p.165.

19 J. Kepler, *Epitome Astronomiae Copernicanae*, Lentiis ad Danubium 1618, Bk.I, p.II, De figura coeli (*Werke* VII, pp.42-47). Kepler often comes back to this problem, rejecting Cusanus' and Bruno's thesis about the infinity of the universe. See e.g. J. Kepler, *Dissertatio cum Nuncio Sidereo*, Pragae 1610 (*Werke* IV, p.308).

20 Cf R. Hooykaas, *Science in Manueline Style; The Scientific Work of D. João de Castro in its Historical Context*, Coimbra 1980, (separate edition of pp.231-426 of *Obras Completas de D. João de Castro*, A Cortesão and L. de Albuquerque ed., Vol.IV, Lisboa 1982). See also R. Hooykaas, 'The Reception of Copernicanism in England and the Netherlands', in: *The Anglo-Dutch Contribution to the Civilisation of Early Modern Society*, London 1976, pp.33-44 (reprinted in: R. Hooykaas, *Selected Studies in History of Science*, Coimbra 1983; pp.635-663).

21 R. Hooykaas, 'Thomas Digges's Puritanism', in: *Arch. Intern. Hist. des Sci.* 8 (1955), pp.145-149.

22 Cicero, *De Natura Deorum* II, 61 (154): 'Restat ut doceam atque aliquando perorem, 'omnia quae sint in hoc mundo quibus utantur homines, hominum causa facta esse et parata'.

23 Kepler, *Mysterium Cosmographicum*, ch.23 (*Werke* I, p.79). The Earth is placed in the middle of the planets, because all things have been made for man (*omnia propter hominem*).

24 Copernicus, *De Revolutionibus*, Praefatio Authoris, fol.IIIvs.

25 *Ibidem*, fol.IIIr.

26 Albertus Leoninus, *Theoria Motuum Caelestium*, Coloniae Agrippinae 1588; 1583.

27 Oresme, *Le Livre du Ciel et du Monde*, Bk.II, ch.25 (Menut and Denomy, p.539.)

28 G. Buchanan, *Sphaera* II, pp.146-148.

29 Copernicus, *De Revolutionibus*, Bk.I, ch.5, fol.3r-v.

30 *Ibidem*, Bk.I, ch.8, fol.6r.

31 *Ibidem*, Bk.I, ch.8, fol.7r.

32 *Ibidem*, Bk.I, ch.5, fol.3vs; ch.8, fol.7r.

33 *Ibidem* Bk.I, ch.8, fol.6r.

34 K. Michalski, in: *Bull. Intern. Acad. Polon. des Sciences et des Lettres, Cl. Hist., Philos., Philol., (1919 - 1920)*, Cracovie (1922), pp.86-88.

35 Copernicus, *De Revolutionibus* Bk.I, ch.8, fol.6v.

[36] *Ibidem* Bk.I, ch.8, fol.6vs.

[37] *Ibidem* Bk.I, ch.9, fol.7r.

[38] *Ibidem* Bk.I, ch.8, fol.6v.

[39] *Ibidem* Bk.I, ch.8, fol.6r.

[40] Copernicus, *De Hypothesibus Motuum Coelestium a se Constitutis Commentariolus* (1514), petitiones 2, 3, 5 and 6. German transl. by F. Rossmann: N. Copernicus, *Erster Entwurf eines Weltsystems,* München 1948.

[41] Even Rheticus' *Narratio Prima* (1540) does not mention them, though it is evident from his recently discovered work on 'Holy Scripture and the Motion of the Earth' that he was then acquainted with them. Cf R. Hooykaas, *Rheticus' Treatise,* pp.18; 91-92.

[42] Copernicus, *De Revolutionibus,* Bk.I, ch.8. The argument that a rotating earth would disintegrate is not adduced by Ptolemy in his *Almagest,* Bk.I, §7.

[43] Aristotle, *De Coelo,* Bk.IV, ch.3, 310b.

[44] Copernicus, *De Revolutionibus* Bk.I, ch.8; fol.6vs.

[45] Aristotle, *De Coelo* Bk.IV, ch.3; 310b; Plato, *Timaios,* 57c, 63e.

[46] Copernicus, *De Revolutionibus* Bk.I, ch.9; fol.7r. Similarly, Galileo writes that all parts of the earth 'move to the centre of the Earth, they move to their Whole, and to their universal Mother', and this is 'their natural instinct'. The same is true for parts of the Sun wanting to unite with the whole of the Sun, etc. (Galileo, *Dialogo* I (*Opere* VII, p.61; Salusbury I, p.25)).

[47] Copernicus, *De Revolutionibus* Bk.I, ch.8; fol.6vs. Galileo wrote: 'I conclude, therefore that the circular motion can only naturally consist with natural bodies, parts of the universe ...; and that the right is assigned by nature to its bodies, and that their parts, at such time as they shall be out of their proper places, constituted in a depraved disposition, and for that cause needing to be reduced by the shortest way to their natural state ...: there being nothing but rest and circular motion apt to the conservation of order' (Galileo, *Dialogo* I (Salusbury I, p.20)).

[48] Copernicus, *De Revolutionibus,* Bk.I, ch.8; fol.6vs.

[49] On the meaning of Copernicus' pun: R. Hooykaas, 'The Aristotelian Background to Copernicus' Cosmology', in: *J. Hist. Astronomy* 18 (1987), pp.111-116.

[50] Copernicus, *De Revolutionibus*, Bk.I, ch.7; fol.5r.

[51] *Ibidem,* Bk.I, ch.8; fol.5v.

[52] *Ibidem* Bk.I, ch.8; fol.6v.

[53] *Ibidem* Bk.I, ch.8; fol.6v.

[54] Aristotle, *De Coelo* Bk.IV, ch.3, 310b15, 311a5-10. In his *Physica* Aristotle made the same point.

[55] Buridanus, *Quaestiones de Coelo et Mundo*, Bk.IV, qu.2 (Moody, p.248).

[56] Copernicus, *De Revolutionibus,* Bk.I, fol.IIIv.

[57] 'neque rationi satis concinna' (Copernicus, *Commentariolus*, F. Rossmann ed., München 1948, p.10).

[58] Copernicus, *De Revolutionibus*, Praefatio Authoris, fol.IIIvs.

[59] Vitruvius Pollio (Marcus), *De Architectura*, Bk.III, ch.I, 1.

[60] Vitruvius *De Architectura,* Bk.III, ch.I, 2.

[61] Copernicus, *De Revolutionibus* Bk.I, ch.10; fol.9v., 10r. Cf Seneca, *Quaestiones Naturales*, Praef. Bk.I, 14: 'He [the Godhead, RH] is entirely Reason .... this [universe, RH] than which nothing is more beautiful or better ordered or more constant in plan ...'

[62] Copernicus, *De Revolutionibus*, Praef. Authoris, fol.IIIv.

[63] *Ibidem,* fol.IVr.

[64] *Ibidem,* Bk.I, ch.9, fol.7v.

[65] *Ibidem,* Bk.I, ch.10, fol.10r.

[66] *Ibidem,* Bk.I, ch.9, fol.7v.

[67] *Ibidem,* Bk.I, ch.10, fol.8v.

[68] *Ibidem,* Bk.I, ch.10, fol.9v.

[69] *Ibidem,* Bk.I, ch.10, fol.9r: 'in infinitam pene orbium multitudinem'.

[70] *Ibidem,* Bk.I, ch.10, fol.9v.

[71] *Ibidem,* Bk.I, ch.11, fol.12r.

[72] *Ibidem* Bk.I, ch.11: 'De triplici motu telluris demonstratio'.

[73] *Ibidem* Bk.I, ch.10; fol.9r.

[74] *Ibidem* Bk.I, ch.10; fol.9v.

[75] Aristotle, *De Coelo,* Bk.II, ch.13, 293b5ff.

[76] Aristotle, *De Partibus Animalium*, Bk.III, ch.4, 665b21-666a25.

[77] Vitruvius, *De Architectura*, Bk.III, ch.1, §3.

[78] Oresme, *Le Livre du Ciel et du Monde,* Bk.II, ch.24 (Menut and Denomy, p.516).

[79] Cicero, *De Re Publica*, Bk.VI, ch.17: 'Somnium Scipionis' ch.4, 17: 'Subter medium fere regionem sol obtinet.'

[80] *Ibidem*: 'dux et princeps et moderator luminum reliquorum, mens mundi ...'.

[81] Macrobius, *Commentary on the Dream of Scipio*, Bk.I, ch.XIX, 1 and 14 (Transl. W.H. Stahl, New York 1952, pp.162, 165).

[82] *Ibidem,* Bk.I, ch.XX, 1.

[83] *Ibidem,* Bk.I, ch.XX, 6.

[84] The early Pythagoreans placed the sun immediately above the moon, whereas Philolaus placed the sun between the spheres of Venus and Mars. Plato 'followed the Egyptians' (Macrobius, *Somnium Scipionis*, Bk.I, ch.XIX, 1) and took to the former opinion (*Timaios* 38d), whereas Aristotle held the latter, which finally prevailed. Plato, *Timaeus a Calcidio Translatus Commentarioque Instructus,* J.H. Waszink ed. Leiden 1962, Comment. LXXII, p.119.

[85] Kepler, *Mysterium Cosmographicum*, Tubingae 1596 (*Werke* I, p.70): 'cor mundi'; —, *Harmonice Mundi*, Lincii Austriae 1619, Bk.V, ch.12 (*Werke* VI, p.364).

[86] William Harvey, *De Motu Cordis et Sanguinis in Animalibus*, London 1628, ch.8.

[87] Copernicus, *De Revolutionibus*, fol.IIr. The ambiguity of the terms 'ancient' and 'Ancients' is evident when some years later (1539) Copernicus' only direct pupil, G.J. Rheticus, in his *Encomium Borussiae* wrote that Bishop Tiedemann Giese had said that Copernicus' principles and hypotheses were wholly contrary to those of the Ancients.

[88] Copernicus, *De Revolutionibus*, Praefatio Authoris fol.IIIv.

[89] Copernicus, *De Revolutionibus*, Praefatio Authoris fol.IVr.; Plutarch, *De Placitis Philosophorum*, Bk.III, ch.13.

[90] Copernicus, *De Revolutionibus*, Bk.I, ch.5; fol.3v.

[91] *Ibidem,* Bk.I, ch.10; fol.8v.

[92] Cf the facsimile edition of the manuscript: Copernicus, *Opus de Revolutionibus Caelestibus Manu Propria*, München-Berlin 1944, p.22. Also the text critical edition by F. and C. Zeller: —, *De Revolutionibus Orbium Caelestium Libri Sex*, München 1940, pp.29-30.

[93] Aristotle, *De Coelo* Bk.II, ch.13. Copernicus pointed out that in Philolaus' time few people understood astronomy and that the Pythagoreans were extremely secretive about their doctrine and did not put anything in writing, but delivered their ideas orally and only to a few trusted people. To prove this Copernicus inserted into his manuscript his Latin translation of a letter of the Pythagorean Lysis, in which he reminded a friend of the rule of silence. It goes without saying that in this way Copernicus and other admirers of the 'most ancient' philosophers (*prisci philosophi*) could attribute opinions to them without demonstrating their authenticity. See the facsimile edition of the ms., München-Berlin 1944, pp.22-24 and the text ciritical edition by F. and C. Zeller, pp.30-31.

[94] Copernicus, *De Revolutionibus*, Praefatio Authoris, fol.IIv.

[95] This is to be compared with the Reformers who claimed to go back to the most ancient Christian testimony (the Bible), whereas the papal party maintained what their fathers and grandfathers had accepted — or, in our own time, to those within the Roman Catholic Church who gladly reintroduce the ancient 'liturgical' position of the altar table, and the ultra-conservatives who cling to the later custom of placing the altar against the Eastern wall of the church.

[96] Copernicus, *De Revolutionibus*, Bk.I, 10; fol.9r.

[97] Copernicus, *De Revolutionibus*, Bk.I, 10; fol.9v. gives a simplified diagram of the world system.

[98] O. Neugebauer, who took the trouble to compare the number of circles used by Copernicus and by Ptolemy, concluded that 'the popular belief that Copernicus' heliocentric system constitutes a significant simplification of the Ptolemaic system is obviously wrong' (O. Neugebauer, *The Exact Sciences in Antiquity* (1962, reprint of 2nd edition of 1957), pp.204-205).

[99] The Copernican construction had been found already by the Persian scientist Nasir al Din al-Tusi (1201 - 1274) and by Proclus (5th cent. A.D.); cf I.N. Veselovsky, *J. H. Astr.* 4 (1973), pp.128-130.

[100] Cf R. Hooykaas, 'The Reception of Copernicanism', pp.33-44.

[101] See *Ibidem*, p.35; also —, 'The Aristotelian Background to Copernicus' Cosmology', p.111.

[102] Some people stuck to the central position of the earth and, like Tycho Brahe, turned the Sun around it with the other planets as its satellites, but admitted the Earth's daily rotation (David Origanus 1558 - 1628). And, finally, there were astronomers who kept to the main outlines of the Aristotelian structure of the universe but were 'Copernicans' in accepting Copernicus' interpretation of the librations as motions of the earth (Leoninus, Mulerius).

[103] To Galileo himself the strongest argument in favour of the earth's rotation was provided by (his interpretation of) the tides. Unfortunately on this topic his theoretical conclusions were wrong and, moreover, the facts upon which he based them were also wrong: he took the change in the tides of the ocean to be every 12 hours.

[104] Nicolaus Mulerius, *Tabulae Frisicae Lunae-Solares Quadruplices ...*, Amstelodami 1611, p.318.

[105] Nicolaus Copernicus, *Astronomia Instaurata, Libris Sex Comprehensa, qui de Revolutionibus Orbium Caelestium Inscribuntur*, N. Mulerius ed., Amstelodami, 1617, *q.v.*. Mulerius' notes to Bk.I, ch.10; Bk.I, ch.6; Bk.II, ch.1; Bk.III, ch.15.

[106] *Ibidem*, note to Bk.I, ch.1: 'paradoxon'.

[107] *Ibidem*, Bk.I, ch.1; Bk.III, ch.15.

[108] *Ibidem*, Bk.III, ch.15, p.203.

[109] H. Gellibrand, *A Discourse Mathematical on the Variation of the Magneticall Needle*, London 1635. It may be that Gellibrand borrowed the phrase 'absurdity' from Mulerius, whose edition of Copernicus' work was the most current one.

[110] For Fresnel see chapter VIII on 'Physical and Mathematical Theories'.

[111] See: R. Hooykaas, 'Science and Religion in the Seventeenth Century, Isaac Beeckman 1588 - 1637', in: *Free Univ. Quarterly* I, 3 (1951), pp.169-183, *q.v.* p.180.

[112] '[Diese Theorie, RH] welche zugleich mit dem neuen Jahrhundert das Licht der Welt erblickt und ihren trüben Stempel fast allen Abschnitten der neuen Physik dieses neuen Jahrhunderts aufgeprägt hat' (O.D. Chwolson, *Die Evolution des Geistes der Physik 1873 - 1923*, Braunschweig 1925, ch.VI 'Neue Theorien in neuem Geiste', p.143).

[113] *Ibidem*, ch.VIII, 'Der Rückschritt', p.186: 'Ein Nutzen ist wohl da, wenn mit Hilfe einer unverständlichen Hypothese eine Erklärung einer sehr grossen Zahl unverständlicher Tatsachen erreicht wird. Die vielen Unverständlichkeiten werden sozusagen durch die eine ersetzt.' The difference between this methodological simplification and that of Haüy's 'mathematical' theory of subtractive molecules (see chapter VIII on 'Physical and Mathematical Theories') is that Chwolson reduces the number of 'irrationalities' in science, whereas Haüy replaces a more complicated (but 'rational') physical theory by a less complicated (but also rational) fiction.

[114] Chr. Huygens, *Cosmotheoros*, Hagae-Comitum 1698, p.14 (*Oeuvres Complètes* XXI, Den Haag 1944, p.695). According to Huygens, Galileo, Gassendi, Kepler a.o. have overthrown by their reasoning the arguments that remained against Copernicus: 'ut omnes nunc Astronomi, nisi tardiore sint ingenio, aut hominum imperio obnoxiam credulitatem habeant, motum Telluri, locumque inter Planetas, absque dubitatione decernant.'

[115] A. Holmes in: *Nature* 171 (1953), pp.669-671.

CHAPTER VII

[1] J. Bostock, *J. Nat. Phil.* 11 (1805) p.75; 28 (1811), p.290; J. Dalton, *J. Nat. Phil.* 29 (1811), p.150; see also R. Hooykaas, 'The Historical and Philosophical Background of Haüy's Theory of Crystal Structure', in: *Academiae Analecta, Mededelingen van de Koninklijke Academie voor Wetenschappen, Letteren en Schone Kunsten van België*, 56, nr.2 (1994), pp.30-31.

[2] J. Dalton, *New System of Chemical Philosophy* Vol.I, pt.2. Manchester 1810, p.559.

[3] W.C. Henry, *Memoirs of the Life and Scientific Researches of John Dalton*, London 1854, .pp 38-39, 142-146.

[4] R.J. Haüy, 'Observations sur la Mesure des Angles des Cristaux', in: *Journal de Physique, de Chimie et d'Histoire Naturelle* 87 (1818), p.248; —, *Traité de Cristallographie* II, Paris 1822, pp.383-4.

[5] Stevin's biographer E.J. Dijksterhuis pointed out that there is a flaw in Stevin's reasoning. It is based on ideal mechanics in which friction and air resistance do not exist. In such a case the perpetuum mobile would not be impossible if an initial velocity was given to the system (—, *Simon Stevin, Science in the Netherlands around 1600*. The Hague: Nijhoff 1970, pp.52-54).

[6] Isaac Beeckman, *Journal tenu par Isaac Beeckman de 1604 à 1634*. C. de Waard ed. 4vols. La Haye 1939, 1942, 1945, 1953. *q.v.* Vol.II, p. 375.

[7] Ockham, *Sententiae*, Bk.II, qu. 18.

[8] Aristotle, *De Generatione Animalium*. Bk.III, ch.10, 760b3-4.

[9] J. Buridan, *Questiones super Octo Phisicorum Libris Aristotelis*. (Parisiis 1509) Bk.IV, qu.7, fol.LXXIII^{vs}

[10] Cf P. Duhem, *Le Système du Monde*, Vol.VIII, Paris 1956, p.128; see also R. Hooykaas, *Science in Manueline style*, note 456.

[11] Thus, Peter of Auvergne, a pupil of Thomas Aquinus, repeats Philo's opinion that in the experiment with the candle the air is changed into fire which either escapes through the glass or finally becomes

water, so that drops are deposited against the bottom of the inverted vessel. Peter reports that Averroës [a famous Arabian follower of Aristotle] had sometimes found them. Cf Duhem, *Le Système du Monde*, Vol.VIII, p.132.

12 Aristotle, *Physica*, Bk.IV, ch.6; 213 l.16-20 and Bk.IV, ch.7 214 l.5-8.

13 Buridan, o.c. Bk.IV, qu.7; fol.LXXIII r. col 1 - vs. col 2.

14 Macrobius, *In Somnium Scipionis Expositio*, Bk.II, ch.1, 9-13. The work was first printed in Venice in 1472.

15 Arnaldus Fabricius Aquitanus, *De Liberalium Artium Studiis Oratio Conimbricae Habita in Gymnasio Regio quâ Ludus Operiretur. IX Cal. Martis MDXLVII.* Conimbricae 1548.

16 Vincenzo Galilei, *Discorsi,* 1589, pp.103-104.

17 Thos. Bradwardine, *Tractatus de Proportionibus*, ch.III, pars 2 ; H.L. Crosby ed. Madison 1961, p.122.

18 P. Duhem, *Le Système du Monde,* Vol.VII, Paris 1956, p. 589.

19 Nicolaus von Kues, *Werke* I, Berlin 1567 (new edition of the ed. of Strassbourg 1488), p.286.

20 *Ibidem,* p.283

21 *Ibidem,* pp. 283, 286.

22 *Ibidem,* p.278.

23 *Ibidem,* p.277: '... nihil in hoc mundo precisionem attingere queat'.

24 'Destas operacoes fica claro que a variacaoque fazen as agulhas nao he por differenca de meridianos' (D. João de Castro, *Obras Completas*. A. Cortesão and L. de Albuquerque ed. Coimbra 1968, Vol.I, p.184).

25 *Ibidem.*

26 *Ibidem* (*Obras*, II, pp.81, 78, 72).

27 R. Hooykaas, *Science in Manueline Style*. Coimbra 1980, pp.111-112; 136-139. A work on the history of Scotland by G. Buchanan (1582) mentions the magnetic properties of the large piece of basaltic rock on which Dumbarton castle is built (—, *Rerum Scoticarum Historia*. Edinburgh 1582, Bk.XX, fol 241^{r}).

28 D. João de Castro, *Tratado da Sphaera per Perguntas e Repistas a Modo de Dialogo* (written probably before 1538). Reprint in: *Obras* I, pp.23-114.

29 Hooykaas, *Science in Manueline Style*, pp.191-196.

30 Garcia de Orta, *Coloquios dos Simplos e Drogas*. Goa 1563, coloq.XIV, pp.164^{vs}.

31 *Ibidem,* pp.163^{r} - 163^{vs}.

32 For quotations and more elaborate discussion of this controversy, see C. de Waard, *L'Expérience barométrique*. Thouars 1936, pp.27-28.

33 J.C. Scaliger, *Exotericarum Exercitationum Liber XV de Subtilitate, ad Hieronymum Cardanum* (1557). Francofurti 1612, exerc.XXVIII, pp.129-130.

34 G. Gilbertus, *De Magnete, Magneticisque Corporibus, et de Magno Magnete Tellure, Physiologia Nova Pluribus et Argumentis, et Experimentis Demonstrata*. Londini 1600.

35 *Ibidem,* Prefatio, fol IIr - IIvs. and IIIr.

36 Isaac Beeckman, *Journal tenu par Isaac Beeckman de 1604 à 1634*. C. de Waard ed. 4vols. La Haye 1939, 1942, 1945, 1953.

37 *Ibidem,* Vol.III, pp.331-332.

38 *Ibidem*, Vol.I, p.58.

[39] *Ibidem*, Vol.III, p.197.

[40] *Ibidem*, Vol.I, pp.261-264 When he became headmaster of the Latin school in Dordrecht he persuaded the magistrate to build a tower on the school, in which he installed apparatus for measuring wind currents.

[41] *Ibidem*, Vol.I, pp.21-22.

[42] *Ibidem*, Vol.I, p.60 (1618).

[43] J. Kepler, *Mysterium Cosmographicum*.

[44] G. Gilbert, *De Magnete*, preface.

[45] Galileo Galilei, *Dialogo sopre i Due Massimi Sistemi del Mondo, Tolemaico e Copernicano*. Reprint in: *Opere*. VII, pp.27-546; —, *Discorsi i Dimostrazioni alla Mecanica e i Movimenti Locali*. Leyden 1638. reprint in: *Opere* VIII, pp.43-318. We quote both works from Thos. Salusbury, *Mathematical Collections and Translations*, I and II, London 1664, henceforth here denoted as Salusb. I or II.

[46] Simon Stevin, *Weeghdaet* (an annex to *Beginselen der Weeghconst*, Leyden 1586). See: E.J. Dijksterhuis, *Simon Stevin*.

[47] Galileo, *Discorsi* (*Opere* VIII, p.109; Salusb.II, p.53).

[48] *Ibidem* (*Opere* VIII, pp.107ff; Salusb.II, pp.52ff).

[49] *Ibidem*, Third day; theor. II (Salusb.II, pp.144-145).

[50] *Ibidem* (Salusb.II, pp.147-148).

[51] Quoted by E.J. Dijksterhuis, *Val en Worp*. Groningen 1924, pp.375, 405.

[52] Letters from Baliani to Galileo 23 April 1632 (*Opere* XIV, p.346); Baliani to Galileo, 1 July 1639 (*Opere* XVIII, p.68); Galileo to Baliani, 1 Aug 1639 (*Opere* XVIII.) See Dijksterhuis, *Val en Worp*, pp.336-337 and 393; —, *De Mechanisering van het Wereldbeeld*. Amsterdam 1950, p.400.

[53] Galileo, *Dialogo*, II (Salusb.I, pp.125ff).

[54] *Ibidem* (Salusb.I, p.124).

[55] *Ibidem* (Salusb.I, p.367).

[56] *Ibidem* (Salusb.I, p.370).

[57] *Ibidem* (Salusb.I, p.422).

[58] E. Wohlwill, *Galileo und sein Kampf für die Copernicanische Lehre*. Bd I, Hamburg und Leipzig 1909, pp.587-594, 603-604; Bd II, Leipzig 1926, pp.119-129. In Vol.I, p.587: 'in Wahrheit auf der Grundlage eines Irrtums, wie ihn nur der Genius zu denken vermag, mit den Hilfsmitteln einer neuen Wissenschaft ein luftiges Truggebilde aufführt.'

[59] 'De Fluxu et Refluxu Maris', sent as a letter to Cardinal Ursini, 6 January 1616 (*Opere* V, pp.376ff.)

[60] Beeckman, *Journal*, Vol.III, p.206.

[61] Galileo, *Dialogo* IV (Salusbury I, pp.379-399, 404-424).

[62] Galileo, *Dialogo* VI (Salusbury I, p.393).

[63] Galileo, *Discorsi*, First Day (Salusbury II, pp.12-14).

[64] Galileo, *Discorsi* (Salusbury II, p.13).

[65] E. Gerland and F.Traumüller, *Geschichte der physicalischen Experimentierkunst*. Leipzig 1899.

[66] *Ibidem*, p.120.

[67] Galileo, *Dialogue* II (Salusbury I, p.148).

[68] Galileo, *Discorsi* (*Opere* VIII, p.268, Salusbury II, p.206).

[69] *Ibidem* (Salusbury II, p.213).

[70] N. Carpenter, *Geographie Delineated forth*, 2nd ed. Oxford 1635, Bk.II, ch.14, p.234.

[71] Thomas Norton, *Ordinal of Alchemy*, ch.I (ed. Thos. Ashmole, p.21).

[72] See, on Sala, Béguin and Billich: R. Hooykaas, *Het Begrip Element*, Utrecht 1935, pp.145-159.

[73] R. Hooke, *Micrographia, or some Physiological Descriptions of Minute Bodies Made by Magnifying Glasses. With Observations and Inquiries thereupon.* London: Martyn 1665, pp.333, 225.

[74] *Ibidem*, pp.225-226.

[75] *Ibidem*, p.333. 'Of Spring'.

[76] D. Mendeléev, *Grundlagen der Chemie*. St Petersburg 1891, p.7.

[77] *Ibidem*, p.17.

[78] *Ibidem*, p.7.

[79] Cf R. Hooykaas, 'De Wet van Massabehoud', in: *Chemisch Weekblad* 43 (1947), pp.244-248.

[80] Lavoisier, *Oeuvres*. Paris 1792, p.103.

[81] Lavoisier, *Oeuvres* I, p.663.

[82] 'Je me suis fait la loi rigoureuse de ne jamais suppléer au silence des faits' (Lavoisier, *Traité de Chimie*. 2nd ed. Paris 1793, Introduction, p.XIII).

[83] Lavoisier, *Oeuvres* I, p.662.

[84] Lavoisier, *Traité de Chimie*, p.45.

[85] Lavoisier, *Oeuvres* II, pp.627, 339, 591.

[86] *Ibidem*, p.235.

[87] Lavoisier, *Traité de Chimie*, p.115.

[88] Lavoisier, *Oeuvres* I, p.563.

[89] Lavoisier, *Traité de Chimie* III, pp.62, 250.

[90] Lavoisier, *Oeuvres* II, p.337.

[91] *Ibidem*, p.339.

[92] *Ibidem*, p.594.

[93] *Ibidem*, p.591.

[94] *Ibidem*, p.598.

[95] Lavoisier,*Traité de Chimie*, pp.139-140.

[96] *Ibidem*, p.15.

[97] A.L. Lavoisier and P.S. Laplace, 'Sur la Chaleur', in: *Mém. Acad. des Sci.* 1780, pp.355ff. Reprint in: Lavoisier, *Oeuvres* II, pp.283ff.

[98] 'Il est difficile quand on cherche les éléments d'une science nouvelle, de ne pas de commencer par des à peu près' (Lavoisier, *Traité de Chimie*, p.115).

[99] J.R. Mayer, *Bemerkungen über das mechanische Aequivalent der Wärme.* Heilbronn 1851. Reprint in: R. Mayer, *Die Mechanik der Wärme*. J.J. Weyrauch ed. Stuttgart 1893, p.266.

[100] Mayer, *ibidem*, pp.267 - 268.

[101] J.R. Mayer, *Die organische Bewegung in ihrem Zusammenhang mit dem Stoffwechsel.* Heilbronn 1845; reprint in: —, *Die Mechanik der Wärme.* J.J. Weyrauch 3. ed., Stuttgart 1893, pp.51-52: 'Wenn hier eine Verwandlung der Wärme in mechanischen Effect statuiert wird, so soll nur ein Tatsache ausgesprochen, die Verwandlung selbst aber keineswegs erklärt werden ... Die echte Wissenschaft begnügt sich mit positiven Erkenntnis und überlässt es willig dem Poeten und Naturphilosophen, die Auflösung ewiger Rätsel mit Hilfe der Phantasie zu versuchen.'

[102] Mayer, *ibidem*, p.74.

[103] This assumption is not correct, since — in the constant pressure case — there is also work done internally against the mutual attraction of the particles of air as they are moved to a greater distance from each other. Mayer, however, could refer to Gay-Lussac who had allegedly demonstrated that expansion of a gas without performance of external work does not change the temperature. R. Mayer, *Bemerkungen über das mechanische Aequivalent der Wärme.* Heilbronn 1851 (—, *Die Mechanik der Wärme*, p.269). Also: R. Mayer, 'Autobiographische Aufzeichungen' (—, *Kleinere Schriften und Briefe.* J.J. Weyrauch ed. Stuttgart 1893, p.379).

[104] Mayer, *Die Mechanik der Wärme*, Vorwort zur 2. Auflage, 1874. (Anmerkungen des Herausgebers). The accepted values of specific heat and expansion coefficient differed from those current in 1842.

[105] Mayer, 'Bemerkungen über die Kräfte der unbelebten Natur', in: *Ann. d. Chemie und Pharmacie*, 1842; quoted from reprint in: —, *Mechanik*, p.23: 'causa equat effectum', 'ex nihilo nihil fit'. Mayer, *Die organische Bewegung (—, Die Mechanik der Wärme.* Stuttgart 1893, pp. 46-48; 53-56).

[106] Mayer, *Die Mechanik der Wärme*, p.71.

[107] 'Was subjectiv richtig gedacht ist, ist auch objectiv wahr. Ohne diese von Gott zwischen subjectiven und objectiven Welt prästabilierte Harmonie wäre all unser Denken unfruchtbar' (Mayer, *Über notwendige Konsequenzen und Inkonsequenzen der Wärmemechanik.* In: —, *Mechanik der Wärme*, p.357). Cf —, *Kleinere Schriften und Briefe*, p.453.

[108] J.P. Joule, Lecture read at St Ann's Church reading-room and published in: *Manchester Courier*, 5 and 12 May 1847. Reprint in: —, *Scientific Papers.* London 1884-1887, Vol.I, pp.268-269.

[109] Joule, 'Calorific Effects of Magneto-electricity and the Mechanical Value of Heat', in: *Phil. Mag. Ser. 3*, Vol.XXIII, pp.263, 347 and 435. (*Scientific Papers* I, p.156).

[110] Joule, *Scientific Papers* I, pp.157-8 .

[111] Paper read 21 June 1849, publ. in: *Phil. Trans. Part I*; reprint in: —, *Scientific Papers* I, p.328.

[112] 'On Matter, Living Force and Heat' 1847; —, *Scientific Papers* I, pp.269, 271, 273.

[113] Th. Sprat, *The History of the Royal Society of London*, 4th ed. London 1734, p.434.

[114] Isaac Newton, *Principia Mathematica Philosophiae Naturalis*, rule III (ed. F. Cajori, *Sir Isaac Newton's Mathematical Principles of Natural Philosophy and his System of the World.* Berkeley 1947, p. 398)

CHAPTER VIII

[1] G. Berkeley, *Siris; Or a Chain of Philosophical Reflections and Inquiries Concerning the Virtues of Tar Water* (1744), §§249, 292, 293; —, *De Motu* (1721), §28; —, *Three Dialogues between Hylas and Philonous* (1713), dial.III.

[2] P. Mansion, 'Note sur le Caractère géométrique de l'ancienne Astronomie', in: *Abh. Gesch. Math.* IX (1899); P. Duhem, 'Essai sur la Notion de Théorie physique de Platon à Galilée', in: *Annales de Phil. Chrétienne*, Paris 1908.

[3] Ptolemy, *Almagest*, Bk.III, ch.4. In general, however, he is on the 'mathematical' side (Bk.XIII, ch.2).

[4] Nicolaus Copernicus, *De Revolutionibus Orbium Caelestium*, Norimbergae 1543, Bk.V, ch.32, fol.172r: 'Prius autem quam recedamus à Mercurio, placuit alium adhuc modum recensere priore non minus credibilem, per quem accessus et recessus ille fieri ac intelligi possit'; fol.173r: '... modum ... non minus rationabilem priori ...'

[5] *Ibidem,* Praefatio Authoris, fol.IVr: '... ut experirem, an posito terrae aliquo motu firmiores demonstrationes, quam illorum essent, inveniri in revolutione orbium coelestium possent.'

[6] Osiander to Rheticus, 20 April 1541. Cf Kepler, *Apologia Tychonis contra Nicolaum Ursum* (1601). *Opera Omnia* (ed. Frisch) Vol.I, Frankfurt-Erlangen 1858, p.246.

[7] Osiander to Copernicus, 20 April 1541. Cf L. Prowe, *Nicolaus Coppernicus*, Vol.I.2, Berlin 1883, p.522: 'De hypothesibus ego sic sensi semper, non esse articulos fidei, seu fundamenta calculi, ita ut, etiam si falsae sint, modo motuum phainomena exacte exhibeant, nihil referat ...'

[8] [A. Osiander], Ad Lectorem de Hypothesibus Huius Operis. In: Copernicus, *De Revolutionibus* fol.Ivs-IIr.

[9] *Ibidem.*

[10] Kepler, *Apologia Tychonis (Opera Omnia* I, pp.238-248).

[11] Kepler, *Mysterium Cosmographicum*, Tubingae 1596 Praefatio (*Werke* I, p.9).

[12] Plato, *Timaios,* 30a.

[13] Th. Paracelsus, *Liber Meteororum* (circa 1525), ch.2; Th. Paracelsus, *Sämtliche Werke* (ed. K. Sudhoff) Vol.XIII, München-Berlin 1931, p.135: 'so hat got drei für sich genomen und aus dreien alle ding gemacht ... nun ist das wort auch dreifach gewesen, dan die trinitet hats gesprochen ... und fürhin seind alle ding in drei gesezt, und nichts ist auf erden, es hat und ist in drei speciebus.'

[14] Kepler, *Mysterium Cosmographicum,* ch.2 *(Werke* I, p.27).

[15] *Ibidem,* 2nd ed. Francofurti 1621. Notae in ch.2 (*Werke* VIII, 1963, p.63).

[16] Kepler, *Astronomia Nova (*1619), Pt.III, ch.34 (*Werke* III, p.246).

[17] Kepler, *Apologia Tychonis (Opera* I, p.246).

[18] Kepler, *Epitome Astronomiae Copernicanae, libri I-III*, Lentiis ad Danubium 1618, Bk.I. 'Adeoque hic est ipsissimus liber Naturae, in quo Deus conditor suam essentiam, suamque voluntatem erga hominem ex parte, et alogoi quodam scriptionis genere propalavit atque depinxit ...' (*Werke* VII, p.25). Cf Newton, *Optice, sive de Reflexionibus, Refractionibus, Inflexionibus et Coloribus Lucis Libri Tres*. Londini 1706. (Latine reddidit Samuel Clarke), qu.23, pp.343, 345-346.

[19] Kepler, *ibidem,* Vol.II, Lentiis 1620, Bk.IV (*Werke*. VII, p.254).

[20] René Descartes, *Discours de la Méthode.* Leyde 1637 (*Oeuvres* VI, pp.63-64). Descartes will be quoted from *Oeuvres de Descartes*, Ch. Adam and P. Tannery ed., XII vols. Paris 1899 - 1910.

[21] Descartes, *Principia Philosophiae*, Amstelodami 1644, P.II, §36 (*Oeuvres* VIII, pp.61-62).

[22] See R. Hooykaas, '*Experientia ac Ratione', Huygens tussen Descartes en Newton.* Leiden 1979.

[23] Descartes, *Les Principes de la Philosophie*, Paris 1647; Pars II,§52 (*Oeuvres* IX B, p.193).

[24] Descartes, *Principia Philosophiae*, P.II, §36 (*Oeuvres* VIII, p.329).

[25] Descartes, *La Dioptrique* (1637) (*Oeuvres* VI, p.83).

[26] *Ibidem,* p.83.

[27] *Ibidem,* p.84.

[28] *Ibidem,* p.98.

[29] *Ibidem,* p.100.

[30] *Ibidem,* p.101.

[31] *Ibidem,* p.103.

[32] *Ibidem.*

[33] *Ibidem.*

[34] *Ibidem*, p.84. On the difference between 'motion' and 'action' see Descartes' letter to Morin, 13 July 1638 (*Oeuvres* II, p.215).

[35] 'Au reste, sçachant ainsi la cause des refractions qui se font dans l'eau et dans le verre ...' (Descartes, *La Dioptrique*; *Oeuvres* VI, p.104).

[36] Morin to Descartes, 22 February 1638 (*Oeuvres* I, pp.538-539); Descartes to Morin, 13 July 1638 (*Oeuvres* II, p.198).

[37] According to Chr. Huygens and Is. Vossius, Descartes must have known the sine law from a manuscript of Willebrord Snellius.

[38] 'Dixi nuper, cum una essemus, lumen in instanti ... à corpore luminoso ad oculum pervenire ... hoc mihi esse tam certum, ut si falsitatis ergui posset, nil me prorsus scire in Philosophia confiteri paratus sim ... Contra ego, si quae talis mora sensu perciperetur, totam meam Philosophiam funditus eversam fore inquiebam' (Descartes to Beeckman, Amsterdam 22 August 1634; *Oeuvres* I, pp.307-308). Beeckman had proposed an experiment to test whether light needs time for transmission. If that experiment had been executed indeed, Descartes would have got his way, because the time lag between the emission of light and its return after reflexion by a mirror at half a mile distant would have been imperceptible. Yet Beeckman was right in principle as was shown later by Olaus Römer (1676).

[39] Of course this should not be taken too seriously, for elsewhere he declared that if the senses taught something contrary to what he had demonstrated (by reasoning), then one should give more credit to his reasoning (see also page 235). Descartes had an unshakeable trust in his own reason ('God does not deceive us').

[40] Descartes, *Principia Philosophiae*, P.III, 'De mundo adspectabili' (On the visible world), §§19 (*Oeuvres* VIII, pp.26, 28). 'Terram in coelo suo quiescere, sed nihilominus ab eo deferri.'

[41] *Ibidem,* §43 (*Oeuvres* VIII, p.99).

[42] *Ibidem,* §44 *(Oeuvres* VIII, p.99).

[43] *Ibidem,* §45 (*Oeuvres* VIII, pp.99-100).

[44] *Ibidem,* §47 (*Oeuvres* VIII, p.101).

[45] *Ibidem*, Pt.IV, §1 (*Oeuvres* VIII, p.203).

[46] *Ibidem,* §§ 204-6 (*Oeuvres* VIII, pp.327-9).

[47] *Ibidem,* Pt.III, §73 (Oeuvres VIII, p.99).

[48] Mme C. Serrurier — who did not doubt the sincerity of Descartes' religious faith — nevertheless concluded that 'sa véritable religion a été le cartésianisme'. (C. Serrurier, 'Descartes, l'Homme et le Croyant', in: E.J. Dijksterhuis a.o. ed., *Descartes et le Cartésianisme Hollandais.* Paris-Amsterdam 1950, p.62.

[49] 'Larvatus prodeo' (Descartes, *Cogitationes Privatae*, *Oeuvres* X, p.213).

[50] 'Nos nunquam falli, cur soles clari et distincti percepti assentimur' (Descartes, *Principia Philosophiae*, Pt.I, §43; *Oeuvres* VIII, p.21).

[51] Cf R. Hooykaas, 'The First Kinetic Theory of Gases', in: *Arch. Intern. Hist. Sci.* 2 (1948), pp.180-184. Also in: *Actes Ve Congrès Intern. Hist. Sci., Lausanne, 30 Sept - 6 Oct 1947.* Paris 1948, pp.125-129; and in: R. Hooykaas, *Selected Studies*, pp.253-258.

[52] Isaac Newton, *Opticks: or, a Treatise on the Reflexions, Refractions, Inflexions and Colours of Light.* London 1704. Bk.II, Pt.III, prop.12, pp.78-86.

[53] Isaac Newton, *Principia Mathematica,* Bk.II, prop.23, theor.17, p.301. 'Particulae viribus quae sunt reciproce proportionales distantiis centrorum suorum se mutuo fugientes componunt Fluidum Elasticum, cujus densitas est compressioni proportionalis. Et vice versa, si Fluidi ex particulis se mutuo fugientibus compositi densitas sit ut compressio, vires centrifugae particularum sunt reciproce proportionales distantiis centrorum.'

[54] Isaac Newton, *Optice,* Quaestio 23, p.339.

[55] Newton, *Principia Mathematica*, p.303: 'Quod si particulae cujusque virtus in infinitum propagetur, opus erit vi majori ad aequalem condensationem majoris quantitatis Fluidi.'

[56] *Ibidem:* 'A vero Fluida Elastica ex particulis se mutuo fugantibus constent, Quaestio Physica est. Nos proprietatem Fluidorum ex ejusmodi particulis constantium Mathematice demonstravimus, ut Philosophis ansam praebeamus Quaestionem illam tractandi'.

[57] 'Die *volle* Wahrheit wird sich in dieser Weise nicht erreichen lassen ... Die an sich einfachste Vorstellung ... wird man ... sogar als relativ, und, man darf sagen, als menschlich wahr bezeichnen müssen' (A. Kékulé, *Die wissenschaftlichen Ziele und Leistungen der Chemie.* Bonn 1878, p.27).

[58] Thomas Young, article 'Chromatics', *Suppl. Encycl. Brit.* written 1817. In: G. Peacocke and J. Leitch ed., *Th. Young, Miscellaneous Works*, Vol.I, London 1855, p.335.

[59] Young to Arago, 12 January 1817 (*Miscellaneous Works* I, p.383).

[60] Young (*Miscellaneous Works* I, p.234).

[61] *Ibidem*, p.383.

[62] *Ibidem,* p.334.

[63] *Ibidem.*

[64] Fresnel recognized that 'M. Young est le premier qui ait énoncé positivement la possibilité d'une telle propriété [*viz.* transverse vibrations] dans un fluide élastique.' Young concluded from the optical properties of biaxial crystals discovered by D. Brewster, 'que les ondulations de l'éther pourraient bien ressembler à celles d'une corde tendue d'une longueur indéfinie, et se propager de la même manière' (A. Fresnel, 'Sur le Calcul des Teintes que la Polarisation Développe dans les Lames Cristallisées', in: *Ann. Chimie et Phys.* 17 (1821); *Oeuvres* I, pp.629 ff. We quote Fresnel from: Augustin Fresnel, *Oeuvres complètes d'Augustin Fresnel,* Emile Verdet a.o. ed., 3 vols, Paris 1866 - 1870.

[65] Fresnel (1822), *Oeuvres* II, Paris 1868, pp.4-5.

[66] 'M. Poisson nous a présenté le système de l'émission comme une espèce de Protée qui échappe aux objections en prenant toutes les formes, en adoptant toutes les hypothèses dont il a besoin. La multiplicité des hypothèses n'est pas une probabilité en faveur d'un système, et il peut d'ailleurs arriver, si on le multiplie trop, qu'elles deviennent difficiles à concilier entre elles, quand on les suit un peu avant dans leurs conséquences' (Fresnel, 'Sur les Accès de Facile Réflexion et de Facile Transmission des Molécules Lumineuses dans le Système d'Émission' (1821). From: *Oeuvres* II, p.155).

[67] 'Il est certaines lois si compliquées ou si singulières, que la seule observation aidée de l'analogie ne pourrait jamais les faire découvrir. Pour deviner ces énigmes, il faut être guidé par des idées théoriques appuyées sur une hypothèse vraie' (A. Fresnel in 1823 (*Oeuvres* II, p.484); cf *Oeuvres* II, p.161 (1821)).

[68] Fresnel in 1823 (*Oeuvres* II, p.485).

[69] The motto of his publication on the diffraction of light (1818) is: 'Natura simplex et fecunda' (Fresnel, *Oeuvres* I, p.247): 'Il est sans doute bien difficile de découvrir les bases de cette admirable économie, c'est-à-dire les causes les plus simples des phénomènes envisagés ...'. But this 'principe

général de la philosophic des sciences physiques' makes the human spirit adopt systems which 'appuyés sur le plus petit nombre d'hypothèses, sont les plus féconds en conséquences' (p.249). That is to say that methodological simplicity implies the greatest chance of hitting the ontological simplicity of nature: 'Mais, dans le choix d'un système, on ne doit avoir égard qu'à la simplicité des hypothèses' (p.248); 'la nature paraît s'être proposé de faire beaucoup avec peu' (p.249). 'Les physiciens, qui ont étudié avec attention les lois de la nature sentiront que cette simplicité et ces relations intimes entre les diverses parties du phénomène offrent les plus grandes probabilités en faveur de la théorie qui les établit' (Fresnel, 'Second Mémoire sur la double Réfraction' (1823). From: *Oeuvres* II, p.593).

[70] 'Wer etwas anderes als eine Welle mit dieser Eigenschaft sich denken kann, mag es seiner Ansicht anpassen' (J. Fraunhofer, *Gilb. Ann.* 74 (1823), p.369).

[71] R.J. Haüy, *Traité de Minéralogie* Vol.I, Paris 1801, pp.29ff; —, *Traité de Cristallographie* Vol.I, Paris 1823, pp.X, 50.

[72] R.J. Haüy, *Essai d'une Théorie sur la Structure des Crystaux*, Paris 1784, pp.140-141.

[73] Haüy, *Traité de Minéralogie* Vol.I, p.31.

[74] Haüy, *Traité de Minéralogie* Vol.I, p.7; —, *Traité élémentaire de Physique* Vol.I, Paris 1803. p.71.

[75] Haüy, *Traité élémentaire de Physique* Vol.I, p.72: '... s'ils ne nous donnent pas la figure des véritables molécules intégrantes des cristaux, méritent d'autant mieux de les remplacer dans nos conceptions.'

[76] 'Si ces formes ... ne sont pas celles des vraies molécules intégrantes employées par la nature, elles méritent du moins d'autant mieux de les remplacer dans nos conceptions, que c'est avec une aussi petite dépense de moyens que nous parvenons à établir une théorie qui embrasse tant de résultats divers' (Haüy, *Traité de Minéralogie* Vol.I, p.31). See also note 113 of chapter 6.

[77] Haüy, *Traité de Cristallographie* Vol.I, p.50.

[78] Haüy, *Traité élémentaire de Physique* Vol.I, p.70; —, *Traité de Cristallographie* Vol.I, p.IX: 'On reconnaît ici ce qui caractérise en général les lois émanées de la puissance et de la sagesse de Dieu qui l'a créée et la dirige. Economie et simplicité dans les moyens, richesse et fécondité inépuisables dans les résultats.'

[79] R.J. Haüy, 'Sur la Manière de ramener à la Théorie des Parallelepipèdes, celle de toutes les autres Formes primitives des Cristaux', in: *Mém. Acad. d. Sci. pour 1789*, Paris 1793, pp.519-533.

[80] Haüy, *Traité élémentaire de Physique* Vol.I, p.89; cf *Traité de Minéralogie* Vol.I, p.98.

[81] Haüy, *Traité de Minéralogie* Vol.I, p.98; —, *Traité de Cristallographie* Vol.I, p.52.

[82] H.J. Brooke, *A Familiar Introduction to Crystallography*. London 1823, p.52.

[83] G. Delafosse, *Recherches sur la Cristallisation, considérée sous les Rapports physiques et mathématiques*. Paris 1843 (Extrait des *Mém. Acad. Sci.* Paris 8 (1843), pp.641-692).

[84] A. Bravais, *Mémoire sur les Systèmes formés par des Points distribués régulièrement sur un Plan ou dans l'Espace* (présenté à l'Académie des Sciences le 11 déc.1848), pp.2-3: 'On peut, sans inconvénient et pour fixer les idées, attribuer à ces sommets des dimensions très-petites, en faire de véritables molécules, et attacher spécialement le nom de sommet aux centres de figure de ces molécules dont la forme polyhédrale restera d'ailleurs indéterminée.'

[85] Cf 'Report on the Development of the Geometrical Theories of Crystal Structure 1666 - 1901', in: *Reports Brit. Assoc. Adv. Science* 1901, pp.297-337.

[86] 'La notation des formules est quelque chose de purement conventionnel, comme le sont les équivalents chimiques eux-mêmes; mais telle qu'elle a été généralement adoptée, elle offre l'inconvénient d'habituer l'esprit à y voir des choses absolues, tandis qu'elle ne devrait exprimer que de simples rapports' (Ch. Gerhardt, *Précis de Chimie organique* Vol.I, Paris 1844. Avant-propos, p.IX).

[87] 'C'est un préjugé si généralement répandu qu'on peut, par les formules chimiques, exprimer la constitution moléculaire des corps, c'est-à-dire le véritable arrangement de leurs atomes, que j'aurai peut-être de la peine à persuader du contraire quelques-uns de mes lecteurs ...' (Ch. Gerhardt, *Traité de Chimie organique Vol.*IV, Paris 1856, §2450, 'Sens des formules', p.561)

[88] *Ibidem,* p.561.

[89] *Ibidem,* p.563

[90] 'Une notation est d'autant meilleure qu'elle rappelle à l'esprit plus d'analogies, qu'elle lui suggère plus de pensées fécondes' (*ibidem,* p.564).

[91] 'Les formules chimiques ... ne sont pas destinées à représenter l'arrangement des atomes; mais elles ont pour but de rendre évidentes, de la manière la plus simple et la plus exacte, les relations qui rattachent les corps entre eux sous le rapport des transformations' (*ibidem,* p.566).

[92] Gerhardt (and Williamson before him) had abandoned the simple and static idea that an initial structure immediately turns into a different final structure: 'Nous ne savons pas ce qui se passe en réalité dans l'intérieur de la molécule d'un corps lorsqu'il se transforme.' He recognized that many intermediate stages may occur of which the chemist has no idea at all and that, consequently, the behaviour of the molecule cannot be wholly predicted from the image we make of it. He was aware of the fact that chemistry deals with interactions. What is called a double decomposition is 'simply an image', an interpretation of similar relations. He denied that a 'rational formula' of a body, once given, 'is immovable — or in other words — that each body has only one rational formula' (*ibidem,* p.576).

[93] *Ibidem,* p.586; cf pp.561-566.

[94] '... mes types signifient tout autre chose que les types de M. Dumas, ceux-ci se rapportant à l'arrangement supposé des atomes dans les corps, arrangement qui, dans mon opinion, est inaccessible à l'expérience (*ibidem,* p.586).

[95] 'Dieselben [*viz.* die Gerhardtschen Typen] sind der Entwicklung der Wissenschaft ausserordentlich förderlich gewesen. Sie konnten aber nur als Theile eines Gerüstes betrachtet werden, das man abbrach, nachdem der Aufbau des Systemes der organischen Chemie weit genug gediehen war, um seiner entbehren zu können' (Lothar Meyer, *Die modernen Theorien der Chemie.* 5e Aufl. Breslau 1884, p.220).

[96] Gerhardt, *Traité de Chimie organique* Vol.IV, p.563.

[97] Meyer, *ibidem,* p.198. Once alcohol is conceived as

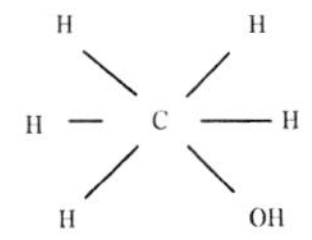

the use of various 'rational' formulae for this same substance ($C_2H_5 \cdot OH$ and $C_2H_4 \cdot H_2O$) has become obsolete.

[98] G.W. Wheland, *The Theory of Resonance and its Application to Organic Chemistry.* NewYork-London 1944. p.5.

[99] L. Pauling, *The Nature of the Chemical Bond and the Structure of Molecules and Crystals.* Ithaca-NewYork 1960, p.217. The objection that the two benzenes are equivalent is answered by Wheland, *The Theory of Resonance,* p.4.

[100] Wheland, *The Theory of Resonance,* p.3.

[101] *Ibidem,* ed.1955, p.4.

[102] *Ibidem,* p.5.

[103] W. Hückel, *Theoretische Grundlagen der organischen Chemie*, 2. ed. Vol.I, Leipzig 1934, p.439: 'Erschöpfung der formalen Ausdrucksmittel.'

[104] *Ibidem*, p.137.

[105] L. Pauling, *The Nature of the Chemical Bond*, 2nd ed. 1945 Quoted by I.M. Hunsberger, 'Theoretical Chemistry in Russia', in: *J. Chem. Educ.* 31 (1954), p.509.

[106] Cf Hunsberger, *ibidem*, note 13.

[107] Wheland, *The Theory of Resonance* ed. 1944, p.28.

[108] We borrow our data from the reports in *Questions scientifiques III, Chimie*, published by 'Les Éditions de la Nouvelle Critique', Paris 1953. The French editors were, as their introduction to these reports unambiguously declares, faithful Stalinists. Cf pp.7-9: 'Car s'il est évident que 'la lutte des opinions', selon l'expression de Staline, est indispensable entre savants de tous les pays ... qui oserait dire, après avoir lu le présent travail, qu'une telle liberté n'existe pas en Union Soviétique?' A commission of the Academy made a preliminary report on the theory of resonance, criticizing its Russian adherents. Kedrov had a large share in the discussion about that report. His 'intervention' and the more elaborate article in which he repeated it (B.M. Kedrov, 'Contre l'idéalisme et le mécanisme en chimie organique', in: *Questions scientifiques III*, pp.47ff) are the main sources for my quotations. See especially pp.32, 51, 52, 54, 55.

[109] See: Hunsberger, 'Theoretical Chemistry in Russia', pp.510-511. (This paper had been seen by Pauling). See also J.K. Syrkine, 'Critique de mes erreurs', in: *Questions scientifiques* III, pp.73ff.

[110] See L.R. Graham, 'A Soviet Marxist View of Structural Chemistry: The Theory of Resonance Controversy', in: *Isis* 55 (1964), p.24 n.20.

[111] Mendeléev, however, unnecessarily connected the notion of 'element' with that of 'atom'. Cf R. Hooykaas, 'De Wet van Elementenbehoud', in: *Chemisch Weekblad* 43 (1947), p.527 (Engl. transl. 'The Law of Conservation of Elements', in: R. Hooykaas, *Selected Studies* pp.121-143.)

[112] G. Urbain, *Les Notions fondamentales d'Elément chimique et d'Atome*. Paris 1925, p.9. The 'element' oxygen (O) is 'une idée pure, et défie toute description positive'.

[113] Kedrov, 'Contre l'Idéalisme', pp.70-71.

[114] *Questions scientifiques III*, pp.34-36.

[115] *Ibidem*, p.35.

[116] Kedrov, 'Contre l'Idéalisme', p.61. Kedrov was of the opinion that the defenders of resonance had made only a first step on the way towards recognizing their errors (p.57) and that the report was too much of a compromise, whereas organic chemistry should be 'liberated from this reactionary burden', by developing the theory of Butlerov and thus opening 'the way to its free development', out of the 'blind alley of the most inveterate mysticism and scholasticism' (pp.62-63).

[117] J.C. Maxwell, 'On Faraday's Lines of Force' (1855 - 1856). From: *Scientific Papers* I, London 1890, p.156.

[118] *Ibidem*, p.159.

[119] Pierre Duhem, *La Théorie physique, son Objet, sa Structure*. 2nd ed. Paris, p.30.

[120] *Ibidem*, pp.31-32.

[121] *Ibidem*, p.34.

[122] *Ibidem*, p.36.

[123] Henri Poincaré, *La Science et l'Hypothèse*. Paris 1902, ch X.

[124] *Ibidem*, pp.190-1.

[125] Sir James Jeans, *The Mysterious Universe*. Cambridge 1930, p.142: 'The laws of nature ... the laws of thought of a universal mind'. 'The uniformity of nature proclaims the self-consistency of this mind' (p.140).

CHAPTER IX

[1] Aristotle, *Metaphysica*, Bk.XII, ch.3, 2070a.

[2] Aristotle, *Metaph*ysica, Bk.VII, ch.7, 1032a-b.

[3] Aristotle, *Metaph*ysica, Bk.VII, ch.9, 1034a.

[4] Aristotle, *Metaph*ysica, Bk.VII, ch.7, 1032a.

[5] Aristotle, *Physica*, Bk.II, ch.8, 199a.

[6] Aristotle, *Physica*, Bk.II, ch.8, 199b.

[7] Cicero, *De Re Publica*, Bk.III, ch.22

[8] Cicero, *Orator* II, 7.

[9] P. de la Ramée, *Dialectique*, Paris 1555, p.4; quotation from edition M. Dassonville, Genève 1964, p.63. Cf his posthumous work: P. Ramus, *Commentationes de Religione Christiana Libri Quattuor*. Francofurti 1576, Bk.I, ch.1. See also: R. Hooykaas, *Humanisme, Science et Réforme – Pierre de la Ramée (1515 - 1572)*. Leiden: Brill 1958, p.25.

[10] P. Ramus, *Geometria* (1569), Bk.I. (Lazarus Schoner ed.) Francofurti 1627, p.1. Hooykaas, *Humanisme*, p.25.

[11] '... non pas l'art seullet mais beaucoup plus l'exercice d'icelluy et la practique faict l'artisan' (Ramus, *Dialectique*, p.136 (ed. Dassonville p.153)).

[12] 'Et vauldroit beaucoup mieux avoir l'usage sans art que l'art sans usage' (Ramus, *Dialectique*, Bk.II, p.139 (ed. Dassonville p.155)).

[13] Cf Ramus, *Scholae Mathematicae* 1569 (Lazarus Schoner ed. Francofurti 1599), Bk.IV, p.109.

[14] '... la souveraine lumière de raison' (Ramus, *Dialectique*, Bk.II (ed. Dassonville p.155)).

[15] Ramus, *Dialecticae Institutiones* (1543), fol.3vs.

[16] 'Naturalis autem dialectica, id est, ingenium, ratio, mens, imago parentis omnium rerum Dei, lux denique beatae illius, et aeternae lucis aemula, hominis propria est, cum eoque nascitur' (*ibidem*, fol.6r).

[17] Ramus, *Dialectique*, Bk.II, p.135 (ed. Dassonville p.153).

[18] *Ibidem*, p.139 (ed. Dassonville p.155).

[19] Ramus, *Dialecticae Institutiones*, fol.6r, 5vs.

[20] *Ibidem*, fol.6r.

[21] *Ibidem*, fol.6vs. The consultation of the sponaneous use of dialectics was recommended by some later Ramists. So the New England philosopher Alexander Richardson asserted that logical reasonings are correct when they prove themselves 'true by the practice of common people'. Cf P. Miller, *The New England Mind*. New-York 1939, p.144.

[22] *Ibidem*, fol.15r, 44r.

[23] Ramus, *Actio pro Regia Mathematicae Professionis Cathedra, Habita in Senatu 3 Id. Martis anno 1566*. In: *Collectaneae Praefationes, Epistolae, Orationes* 1577, p.522. Cf Hooykaas, *Humanisme*, p.94.

[24] This has been demonstrated by J.J. Verdonk, *Petrus Ramus en de Wiskunde*. Dissertation Amsterdam VU. Assen 1966, pp.117-118.

[25] Johannes Kepler, *Harmonice Mundi*. Linciae Austriae 1619. (*Gesammelte Werke* VI, p.82). Ramus and his follower Willebrord Snel (1591 - 1626) made, in Kepler's opinion, 'an architect into a wood merchant' (p.19). Kepler himself, on the other hand, did not want the tenth book of Euclid for 'making up the account of merchandise but for explaining the causes of things.' Cf Hooykaas, *Humanisme*, p.63.

[26] Ramus has been considered as anticipating Descartes as well as Francis Bacon, but both statements should be taken with much reservation. Descartes's deductive rationalism and Bacon's experimentalist empiricism are quite different from Ramus' utilitarianism. Bacon, who studied in Ramist Cambridge, had little sympathy for Ramism, though he shared its predilection for the applied sciences.

[27] Olivier de Serres, *Théâtre de l'Agriculture*. Préface, p.6. Quoted from the edition Lyon 1675.

[28] *Ibidem*, p.4.

[29] *Ibidem*.

[30] *Ibidem*, p.5.

[31] In particular the more refined viniculture he advised to be entrusted only to educated people and not to ignorant peasants whose taste is as rude as their understanding (*ibidem*, lib.III, ch.6., p.177).

[32] Ramus, *Oratio de sua Professione*. In: *Collectanea*, p.526. For Ramus' astronomical ideas see Hooykaas, *Humanisme*, ch.9 and Hooykaas, *G.J. Rheticus' Treatise*, ch.8.

[33] See: R. Hooykaas, *G.J. Rheticus' Treatise*, ch.8: 'Rheticus, Ramus and the Copernican Hypotheses'.

[34] Kepler, *Harmonice Mundi*, Bk.IV, ch.1, p 223.

[35] Miller, *The New England Mind*, pp.146-9.

[36] That the 16th century Puritans were 'somehow grotesque, elderly people, outside the main current of life' the late prof. C.S. Lewis called 'an absurd idea': 'In their own day they were considered, of course, the very latest thing...' (C.S. Lewis, *English Literature in the Sixteenth Century, excluding Drama*, Oxford 1954, p.43).

[37] Milton's re-writing of Ramus' *Dialectica* appeared in 1672 (Miller, *The New England Mind*, p.118).

[38] 'Fundatur igitur *Geometria* in praxi Mechanica — et nihil aliud est quam *Mechanicae universalis* pars illa quae artem mensurandi accurate proponit ac demonstrat' (Isaac Newton, *Principia Mathematica*. Praefatio ad lectorem).

[39] 'At eius picturam, non poësim videmus ... qui motus hominum, qui ferarum non ita epictus est, ut quae ipse non viderit nos ut videremus effecerit', Cicero, *Tusculan Disputations* Bk.V, 39, 114 (quoted from Hooykaas, 'Humanities', p.9 n.35).

[40] Horace, *De Arte Poetica*, line 361. Reference from Hooykaas, 'Humanities', p.9 n.36.

[41] Plato, *Republica* Bk.X, 596-598

[42] *Ibidem*, 596, 597b-e, 598b. We should realize that Plato regarded manual workers as inferior to philosophers (men of science), while the Renaissance artists yearned to be recognized as cultivators of a 'science' rather than of a mere (manual) art.

[43] L.B. Alberti, *De Re Aedificatoria*, Bk.VI, ch.2; ed. Parisiis 1512, fol.81.

[44] Ramus, *Dialecticae Institutiones* (1543), 7vs. Apelles, according to Plutarch the only painter whom Alexander the Great allowed to make his portrait, was considered the greatest painter of Antiquity. Ramus (and also Francisco de Holanda) speaks as if he is sure of the quality of Apelles' work, but it should be realized that he had to resort to Pliny's reports.

45 'imitari pingendo conemur' (*ibidem*).

46 *Ibidem*, 56vs.

47 Francisco de Holanda, *Da Pintura Antiqua* (J. de Vasconcellos ed. Porto 1918) Bk.II, dial.II, p.209.

48 *Ibidem*.

49 *Ibidem*, p.211

50 *Ibidem*, Bk.II, dial.II, p.208.

51 *Ibidem*, Bk.I, ch.14, p.98.

52 *Ibidem*, Bk.I, ch.15, p.99.

53 *Ibidem*, Bk.I, ch.2, p.66.

54 Francisco de Holanda, *Ao Rei Dom Sebastião, De quanto Serve a Sciencia do Desenho e Entendimento da Arte da Pintura na Republica Cristã, asi na Paz como na Guerra*. In: *Da Fabrica que Fallece á Cidade Lisbao*. Da Sciencia do Desenho, ed. J. de Vasconcellos, Porto 1879, p.6 (fol.16).

55 *Ibidem*, p.7 (fol.37v).

56 Aristotle, *Physica*, Bk.II, ch.8, 199a15ff.

57 For Sala see R. Hooykaas, *Het Begrip Element in zijn historisch-wijsgeerige Ontwikkeling*. Utrecht 1933, pp.148-153, 155-157.

58 Henri Langenstein, *Tractatus de Reductione Effectuum Specialium*. Quoted by P. Duhem, *Le Système du Monde*, Vol.VII, Paris, repr. 1954, pp.597-598.

59 Francis Bacon, *Novum Organum*, Bk.I, aph.88. (In: *Works*, Spedding, Ellis and Heath ed., London 1857 - 1874; Vol.I, p.195.

60 Francis Bacon, *New Atlantis* (*Works* III, pp.157-159).

61 Bacon, *Novum Organum* I, aph.3 (*Works* I, pp.157, 144).

62 *Ibidem*, aph.4 (*Works* I, p.157).

63 Bacon, *De Augmentis* II, ch.2 (*Works* I, p.196). Also: —, *Descriptio Globi Intellectualis*, ch.2 (*Works* III, p.730); —, *Novum Organum* I, aph.66, 75 (*Works* I, pp.177, 184).

64 Bacon, *Historia Naturalis et Experimentalis (Works* II, p.14).

65 'Et sane nullae sunt in Mechanica rationes, quae non etiam ad Physicam, cujus pars vel species est, pertineant: nec minus naturale est horologio, ex his vel illis rotis composito, ut horas indicet, quam arbori ex hoc vel illo semine ortae, ut tales fructus producat' (Réné Descartes, *Principia Philosophiae*, P.IV, sect.203; *Oeuvres* VIII, p.326).

66 J.F. Henckel, *Pyrytologie, ou Histoire naturelle de la Pyrite*. Transl. from German original. Paris 1760, p.298. In fact the direct interaction of iron and sulphur yields (black) ferrous sulphide [FeS] and not brass-coloured pyrite [FeS_2].

67 *Ibidem*, ch.13, p.297.

68 *Ibidem*, p.298. Elementary particles form a 'mixture'; mixta form a 'compositum'; composita form a 'decompositum'. Henckel refuses, however, to enter into the problem which are the ingredients that form iron (or sulphur) and which are those forming natural pyrite (p.294).

69 *Ibidem*, pp.295-297.

70 *Ibidem*, ch.14, p.331.

71 *Ibidem*, pp.330-1.

72 *Ibidem*, p.363.

[73] *Ibidem*, ch.13, p.293.

[74] G.G. Leibniz, *Protogaea, sive de Prima Facie Telluris Antiquissimae Historiae Vestigiis in ipsis Naturae Monumentis Dissertatio.* Göttingae 1749, par.9-10: 'Opera pretium autem facturum arbitror, qui naturae effecta ex subterraneis eruta diligentius conferat cum foetibus laboratoriorum (sic enim Chymicorum officinas vocamus) quando mira persaepe in ratis et factis similitudo apparet.' Also sect.9, p.18, sect.10, p.28.

[75] *Ibidem*, sect.10, p.22: 'cui montes sunt pro Alembicis, Vulcani pro furnis.'

[76] *Ibidem*, sect.9, p.18: 'neque enim aliud est natura, quam ars quaedam magna, nec semper toto genere a nativis factitia distinguuntur; nec refert eandemne rem Daedalus aliquis vulcanius in furno invenibus an lapicida ac terrae visceribus proferat in lucem.'

[77] Nicolas Leblanc, *De la Cristallotechnie, ou Essai sur les Phénomènes de la Cristallisation* ... Paris, an X - 1802. Leblanc was the inventor of the industrial process for making soda (1791 - 92, p.72).

[78] *Ibidem*, p.65; cf p.VIII.

[79] *Ibidem*, p.82.

[80] *Ibidem*, p.VI-VII.

[81] *Ibidem*, p.X.

[82] *Ibidem*, p.73.

[83] Gay-Lussac, 'Réflexions sur les Volcans', in: *Ann. Chim. Phys.* 22 (1823), pp.415-429.

[84] E. Mitscherlich, 'Über künstliche Krystalle von Eisenoxyd', in: *Ann. der Physik und Chemie* 1829, pp.630-632.

[85] Letter quoted by K.C. von Leonhard, *Hüttenerzeugnisse und andere auf künstlichem Wege gebildete Mineralien als Stütz-Puncte geologischer Hypothesen.* Stuttgart 1858, p.63.

[86] C.W.C. Fuchs, 'Die künstlich dargestellten Mineralien ...', in: *Natuurkundige Verhandelingen der Hollandsche Maatschappij der Wetenschappen.* 3e reeks dl.I, Haarlem 1872, p.3.

[87] Cf H. de Sénarmont, 'Expériences sur la Formation des Minéraux par Voie humide dans les Gîtes métallifères concrétionnés', in: *Comptes Rendus Ac. Sci* 32 (1851), p.409.

[88] Durocher, *Comptes Rendus Ac. Sci* 32 (1851), p.8: 'C'est en combinant les résultats obtenus en laboratoire avec l'étude géologique des caractères propres aux divers gîtes, que l'on peut apprécier la manière dont ces phénomènes se sont passés dans l'intérieur de la terre.'

[89] J. Hall, *Transact. RSE* 3 (1790), pp.9-ii; 'Experiments on Whimstone and Lava', in: *Transact. RSE* 5 (1798), p.43, 59.

[90] The phenomenon of devitrification of glass had been observed before, e.g. by Réaumur, but it had not been recognized as 'cristallisation' (Cf Dartigues, 'Mémoire sur la Dévitrification du Verre', in: *J. d. Physique* 59 (1804), p.6-8).

[91] J. Hutton, *Theory of the Earth with Proofs and Illustrations*, Vol.I, Edinburgh 1795, p.25. Quoted by Hall, 'Experiments on Whimstone and Lava', *Trans. RSE* 5 (1798), p.45.

[92] Hall, 'Experiments on Whimstone and Lava', p.45.

[93] J.F. d'Aubuisson de Voisins, *Traité de Géognosie,* Vol.I, Strasbourg-Paris 1814, pp.XXX-XXXI.

[94] Hall, 'Experiments on Whimstone and Lava', pp.48, 56, 59, 43, 45.

[95] *Ibidem*, p.68.

[96] Hall, *Transact. RSE* 6 (1805), or 5 (1802), p.74.

[97] *Ibidem*, p.76.

[98] J. Hall, 'Account of a Series of Experiments, Shewing the Effects of Compression in Modifying the Action of Heat', in: *Transact. RSE* 6 (1812).

[99] *Ibidem*, pp.152, 173.

[100] Gregory Watt, 'Observations on Basalt, and on the Transition from the Vitreous to the Stony Texture, which Occurs in the Gradual Refrigeration of Melted Basalt; with some Geological Remarks', in: *Phil. Trans.* London (1804), pt.II, pp.279 ff.

[101] Dartigues, 'Mémoire sur la Dévitrification du Verre. Et les Phénomènes qui arrivent pendant sa Cristallisation', in: *J. d. Physique* 59 (1804), p.13.

[102] Fleuriau de Bellevue, 'Mémoire sur l'Action du Feu dans les Volcans, sur divers Rapports entre leurs Produits, ceux de nos Fourneaux, les Météorites, et les Roches primitives', in: *J. d. Phys.* 60 (1805), an XIII, pp.409-470.

[103] *Ibidem*, p.411.

[104] *Ibidem*, p.418.

[105] *Ibidem*, p.453.

[106] *Ibidem*, p.459.

[107] *Ibidem*, p.412.

[108] J.A.L. Hausmann, 10 February 1816 in the 'Versammlung der K. Wissenschaftlichen Sozietät zu Göttingen' (quoted by Von Leonhard in: *Hüttenerzeugnisse*).

[109] Von Leonhard, *Hüttenerzeugnisse.*

[110] Fleuriau de Bellevue, 'Mémoire sur l'Action', p.411.

[111] Th. Scheerer, 'Discussion sur la Nature Plutonique du Granite et des Silicates cristallisés qui s'y rallient', in: *Bull. Soc. Géol. de France* 4, pp.468-496. Cf K.A.von Zittel, *Geschichte der Geologie und der Paleontologie.* München 1899, p.749. See also: Th. Scheerer, *Der Paramorphismus*, Braunschweig 1854.

[112] Daubrée, *Études et Experiences synthétiques sur le Métamorphisme et sur la Formation des Roches cristallisées.* Paris 1859.

[113] *Ibidem*, p.IX.

[114] *Ibidem*, p.113-115.

[115] *Ibidem*, p.147. It should be noticed that it had long been recognized that the same minerals and rocks could have dissimilar origins. As H.H. Read (*The Granite Controversy.* London 1957) put it: 'there are granites and granites.'

[116] F. Fouqué and Michel Lévy, *Synthèse des Minéraux et des Roches*, Paris 1882, p.6: '... l'union de la cristallographie, de la chimie et de la géologie ... cette triple alliance.'

[117] *Ibidem*, p.63.

[118] N. Desmarest, 'Sur l'Origine et la Nature du Basalte à grandes Colonnes polygones, déterminées par l'Histoire naturelle de cette Pierre, observée en Auvergne', in: *Mém. Acad. Sci. Paris 1771,* 87 (1774), pp.705-775. It is interesting that this staunch defender of the igneous (volcanic) origin of basalt maintained, against Hutton, the neptunistic conception of the origin of granite. See: *Encyclopédie Méthodique*, Vol.I, Paris, an III (1794), pp.749, 752, 756.

[119] A.F. Fourcroy, *Système des Connaissances chimiques, et leurs Applications aux Phénomènes de la Nature et de l'Art.* Vol.VII, Paris, an IX, sect.7, pp.5-7, 54-55.

[120] Thenard, *Traité de Chimie élémentaire théorique et pratique*, Vol.III, 3.ed. Paris 1821, pp.3-4.

[121] F. von Kobell, *Vergleichende Betrachtungen über die Mannigfaltigkeit in der organischen und anorganischen Natur.* München 1836, p.12. Lamarck went much further: according to his 'pyrotic theory' all compounds tend to disintegrate into their components; the existence of composite bodies is due to their organic origin. The 'pouvoir de la vie' is a force acting against the general 'tendance de la nature' to decomposition (J.B. Lamarck, *Réfutation de la Théorie pneumatique ou de la*

nouvelle Doctrine des Chimistes modernes. Paris an IV, p.12). Also: J.B. Lamarck, *Recherches sur les Causes des principaux Faits physiques.* Paris an II (1795), Vol.II, p.273, 289, 27. Also: J.B. Lamarck, *Hydrogéologie*, Paris an X, p.100: '.... les Principes de tout composé quelconque ont une tendance à se dégager'. 'L'action organique des corps vivans forme sans cesse des combinaisons qui n'eussent jamais existé sans cette cause' (pp.105, 117).

122 F. Wöhler, 'Über die künstliche Bildung von Harnstoff', in: *Pogg. Ann. Phys.* 12 (1828), p.25. Cf Wöhler to Berzelius, 22 February 1828 (quoted by C. Graebe, *Geschichte der organischen Chemie* I, Berlin 1920, p.55).

123 Ch. Gerhardt, *Précis de Chimie organique*, Vol.I, Paris 1844, pp.1-3.

124 Ch. Gerhardt, *Traité de Chimie organique*, Vol.I, Paris 1853, p.1.

125 *Ibidem*, p.1.

126 *Ibidem*, p.4.

127 *Ibidem*, p.3.

128 R. Hooykaas, 'Die Chemie in der ersten Hälfte des 19. Jahrhunderts', in: *Technikgeschichte* 33, nr.1 (1966), pp.1-24.

129 J.F. Daniell, *An Introduction to the Study of Chemical Philosophy.* 2. ed., London 1843, p.3.

130 In: Marcellin Berthelot, *Science et Philosophie.* Paris 1886, p.66.

131 *Ibidem*, p.67.

132 *Ibidem*, p.64.

133 Galileo Galilei, *Discorsi* (1638). Quoted from —, *Discourses and Demonstrations Touching Two New Sciences* In: Thomas Salusbury, *Mathematical Collections and Translations*, Vol.II, London 1665, p.3.

134 A.L. Lavoisier, *Traité de Chimie*, 2. ed., Vol.I, Paris 1793, p.69.

135 Ptolemy, *The Almagest*, Bk.XIII, ch.2. Quoted after the translation in ed. *Encyclopaedia Britannica*, Chicago 1952, p.429

136 Willem Jansz. Blaeu, *Tweevoudigh Onderwijs van de Hemelsche en Aerdsche Globen; Het een na de Meyning van Ptolemeus met een vasten Aerdkloot; Het ander na de natuerlijcke Stelling van N. Copernicus met een loopenden Aerdkloot.* Amsterdam: Joan. Blaeu 1666. The Latin translation appeared as: *Philolai, sive Dissertationis de Vero Systemate Mundi* (4 vols.). Amsterdam: Guil. & Iohannem Blaeu 1639. See also chapter VI, 'And the Sun stood still'.

137 Gulielmus Gilbertus, *De Magnete*, Londini 1600, Bk.I, ch.3, p.12: '... forma sphaerica perfectissime et cum terra globosa maxime consentit.'

138 *Ibidem*, p.13.

139 *Ibidem*, lib.VI, ch.1, ch.4.

140 *Ibidem*, lib.VI, ch.4, p.223: 'Omitto quod Petrus Peregrinus constanter affirmat, terrellam ... moveri circulariter integra volutatione 24 horis! Quod tamen nobis adhuc videre non contigit.'

141 *Ibidem*, lib.II, ch.2, p.60: 'The electric motion is the motion of conservation of matter; the magnet motion is that of arrangement and order. The matter of the terrestrial globe is brought together and held together by itself electrically. The earth's globe is directed and revolved magnetical.'

142 *Ibidem*, lib.IV, ch.2, p.155.

143 *Ibidem*, Bk.II, ch.35, p.103.

144 *Ibidem*, lib.VI, ch.4, pp.223-224.

145 Beeckman had seen this through Andreas Colvius (Kolff), the Reformed minister in the Netherlands' embassy in Venice. The work was not printed before 1780, but the theory was

mentioned in Galileo's *Dialogues on the Two World Systems* (1632). Cf ed. Salusbury, Vol.I, Dialogue IV, p.380.

[146] C. de Waard ed., *Journal tenu par Isaac Beeckman*, Vol.III, Den Haag 1945, p.206 (12 April 1631). Beeckman proposed to make groves on the surface, representing the Atlantic Ocean, in order to check whether the revolutions cause ebb and flow *twice* every 24 hours. It should be noticed that Galileo rejected the explanation by influence of the moon and the sun, and that he even criticized Kepler for adhering to such a non-mechanical explanation: 'I more wonder at Kepler than any of the rest, who being of a free and piercing wit, and having the motion ascribed to the Earth, before him, hath for all that given his ear and consent to the Moon's predominancy over the Water, and to occult properties and such like trifles' (Dialogue IV, ed. Salusbury, Vol.I, p.422). Descartes, too, wanted a mechanistic explanation which implied low tide when, in fact, there is high tide. Both cases show that mechanicists who scorned 'occult qualities' were led to absurdities no less than those they rejected.

[147] Aristotle, *Meteorologica*, Bk.II, ch.8, 367a10.

[148] Albertus Magnus, *Meteororum,* Bk.III, tr.2, ch.17: 'Dico autem qualitatem moventem caliditatem solam, cujus exemplum in artificialibus sit ... generatur vapor in vase, quem fortificatum retro erumpit per alterum foramen obstructum: et si irrumpit superius, longe projicit aquam sparsam in ignem, et impetu vaporis projicit ... carbones et cineres calidos longe ab igne super circumstantia loca ...'

[149] Nicolas Lémery, *Mém. Acad. Royale d. Sciences* (1700), pp.131 ff.

[150] 'Extrait de quelques Lettres du Docteur Paccard, Sur les Causes ... de la Direction oblique perpendiculaire, horizontale des Couches ornées et apparentes, etc., et sur la Manière d'imiter artificiellement les Mines', in: *Observations sur la Physique de Rozier et Mongez* 18 (1781), pp.184-192.

[151] *Ibidem,* p.186.

[152] *Ibidem,* pp.187-189.

[153] *Ibidem,* p.192.

[154] Cf Bailey Willis, 'The Mechanics of the Appalachian Structure', in: *U.S. Geol. Survey*, 3d Annual Report, Washington 1893, pp.210-283.

[155] A. Favre, *Comptes Rendus Ac. Sci.* 86 (1878), pp.1092-1094.

[156] A. Daubrée, *Études synthétiques de Géologie expérimentale.* Paris 1879, ch.IV, p.288.

[157] *Ibidem*, p.294.

[158] L. de Launay, La Science géologique, ses Méthodes, ses Résultats, ses Problèmes, son Histoire. 2. ed., Paris 1913, pp.27-28.

[159] James David Forbes (20 April 1809 - 11 December 1868) was professor of natural philosophy in Edinburgh 1833 - 1859; Principal of United College St Andrews 1859 - 1868. After 1840 his main interest shifted from physics to geology. He discovered the polarization and double refraction of radiant heat, and investigated the movements and structure of glaciers (Alps, Norway) and their causes.

[160] J.D. Forbes, *Travels through the Alps of Savoy*. Edinburgh 1893, p.365.

[161] J.D. Forbes, 'Experiments on the Flow of Plastic Bodies and Observations on the Phenomena of Lava Streams', in: *Philosophical Transactions* 1846; quoted from: J.D. Forbes, *Occasional Papers*, Edinburgh 1859, XI, p.77.

[162] *Ibidem*, p.78.

[163] *Ibidem*, p.82.

[164] Étienne Geoffroy St Hilaire, in: *Mém. Ac. d. Sci. Paris* 12 (1832), p.80.

[165] *Ibidem*, p.82.

[166] Cf R. Hooykaas, *Natural Law and Divine Miracle. A Historical-Critical Study of the Principle of Uniformity in Geology, Biology and Theology*. Leiden [1]1959, [2]1963, pp.117-118. Also R. Hooykaas, *Continuïté et Discontinuïté en Géologie et Biologie*, Paris 1970, pp.202-203.

[167] Étienne Geoffroy St Hilaire, *Mém. Musée Hist. Naturelle* 17 (1828), p.213.

[168] Charles Darwin, *On the Origin of Species by Means of Natural Selection*. London 1859, ch.3, p.61.

[169] J.D. Forbes, 'Theoretical Investigations, Intended to Illustrate the Phenomena of Polarisation', in: *Suppl. Encyclop. Britt.* (1823), p.415.

[170] 'To form what Gauss called a 'construirbare Vorstellung' of the invisible process of electrical action is the great desideratum in this part of science' (*Nature* 11 (1874) (*The Scientific Papers of James Clerk Maxwell*. W.D. Niven ed. London 1890, Vol.II, p.419)).

[171] Maxwell, 'On Faraday's Lines of Force', in: *Transact. Cambr. Phil. Soc.*10 (1855-56) pt I (*Scientific Papers* I, p.155).

[172] *Ibidem*, p.160.

[173] Maxwell, 'On Physical Lines of Force', in: *Phil. Mag.* 21 (1861) and 23 (1862) (The theory of molecular vortices applied to electric currents); *Scientific Papers* I, pp.468 ff.

[174] Maxwell himself called his conception 'somewhat awkward' (*Scientific Papers* II, p.486).

[175] *Ibidem*.

[176] William Thomson, 'Steps towards a Kinetic Theory of Matter', in: *Brit. Assoc. Report* 1884. Quoted from: W. Thomson, *Popular Lectures and Addresses*, Vol.I, London 1889, pp. 235-236.

[177] *Ibidem*, p.240.

[178] W. Thomson, *Mathematical and Physical Papers*, Vol.III, London 1890, pp.505-507.

[179] Cf S.P. Thompson, *The Life of William Thomson*, Vol.II, London 1910, p.830. With Thomson the confidence in mechanical models competed with the acknowledgement that as yet no adequate models and theories had been formed. He was concerned that 'the scales will fall from our eyes; that we shall look on things in a different way — when that which is now a difficulty will be the only common sense and intelligible way of looking at the subject' (Thomson, *Math. and Phys. Papers*, Vol.III, p.511; cf p.465).

[180] Thomson, *ibidem*, p.484.

[181] O.J. Lodge, *Modern Views on Electricity*, London 1892, ch.X (Mechanical models of a magnetic field), p.202.

[182] *Ibidem*, p.206.

[183] *Ibidem*, p.59.

[184] *Ibidem*, p.66.

[185] *Ibidem*, p.67.

[186] Ludwig Boltzmann, *Vorlesungen über Maxwells Theorie der Elektrizität und des Lichtes*. Tl.II, Leipzig 1893, p.13.

[187] *Ibidem*, p.14.

[188] *Ibidem*, p.35.

[189] *Ibidem*, p.35.

[190] *Ibidem*, p.44.

[191] *Ibidem*, p.45.

[192] *Ibidem*, pp.46-48.

[193] *Ibidem*, p.49.

[194] Maxwell, *Brit. Assoc. Reports*, Liverpool 1870; *Scientific Papers* II, p.219.

[195] Lodge, *Modern Views on Electricity*, p.67.

[196] Maxwell, 'On Faraday's Lines of Force' (*Scientific Papers* I, p.156).

[197] Maxwell, 'Address Math. Phys. Section Brit. Assoc 1870'; *Scientific Papers* II, p.220.

[198] James Jeans, *The Mysterious Universe*, Cambridge 1930, p.141.

[199] *Ibidem*, p.142.

[200] *Ibidem*, p.146.

[201] *Ibidem*, p.148.

[202] *Ibidem*, p.149.

[203] *Ibidem*, p.127.

CHAPTER X

[1] Blaise Pascal, *Pensées*, fr.162.

[2] Voltaire, *Elémens de la Philosophie de Neuton*, nouv. éd. Londres 1741, pp.8-9: 'Confidens du Très-Haut, Substances éternelles,/Qui brûlez de ses feux, qui couvrez de vos aîles/Le Trône où votre Maître est assis parmi vous,/Parlez, du grand Neuton n'étiez-vous point jaloux?'

[3] Herman Boerhaave, *De Comparando Certo in Physicis*, Lugduni Batavorum 1715: '... *Newtonum* nostri miraculum seculi' (p.13); '... a Principi omnium Philosophorum Isaaco Newton' (p.8).

[4] D. Mornet, *Les Sciences de la Nature en France au XVIIIe Siècle*, Paris 1911, p.85.

[5] Henry Brougham in: *Edinburgh Review* 24 Oct 1803, vol III; 6th ed.1810, p.3.

[6] R.J. Haüy, *Traité élémentaire de Physique*, 3. ed., Vol.II, Paris 1821. Newton's emission theory is to be preferred (§999, p.149); Biot's 'polarisation mobile' is accepted (§1426, p.406).

[7] John Dalton, *A New System of Chemical Philosophy*, Vol.I, Pt.I, Manchester 1808; Vol I, Pt.II, Manchester 1810; Vol II, Pt.I, Manchester 1827.

[8] Isaac Newton, *Philosophiae Naturalis Principia Mathematica*, Londini 1687, Bk.II, prop.23, theorema 17, p.301. See chapter VIII, notes 53 - 56.

[9] See for Euler and Daniel Bernoulli below, note 37.

[10] William Higgins, *A Comparative View of the Phlogistic and Antiphlogistic Theories*, London 1789.

[11] John Dalton, 'A New Theory of the Constitution of Mixed Aeriform Fluids and Particularly of the Atmosphere', in: *Nicholson's Journal* 5 (1801), p.241.

[12] 'Consequently when a vessel contains a mixture of two such elastic fluids, each acts independently upon the vessel, with its proper elasticity, just as if the other were absent, whilst no mutual action between the fluids themselves is observed' (Dalton, *A New System* I, p.154).

[13] Dalton, *A New System* I, p.168. Of course the term 'demonstrated' is not correct! In a lecture at the Royal Institution on 27th Jan 1810 Dalton stressed again the Newtonian basis of his theory. See the reprint in H.E. Roscoe and A. Harden, *A New View of the Origin of Dalton's Atomic Theory*, London 1896, pp.13-18.

[14] Dalton, *A New System* I, p.168.

[15] *Ibidem*, p.214

[16] Dalton's Notebook I, d.d. 21 March 1803 (Roscoe and Harden, *A New View*, p.34).

[17] Dalton's Notebook I (Roscoe and Harden, *A New View*, p.38).

[18] Dalton did not use such formulae; he represented elementary atoms by pictorial symbols as in figure 50 and not by letters as was done afterwards.

[19] Dalton's Notebook, 6 September 1803 (Roscoe and Harden, *A New View*, p.28, pl.3).

[20] Dalton, *A New System* I, p.188. Cf Dalton's Notebook, 14 September 1804 (Roscoe and Harden, *A New View*, p.65).

[21] '... gases do not unite in equal or exact measures in any one instance; when they appear to do so, it is owing to the inaccuracy of our experiments. In no case, perhaps, is there a nearer approach to mathematical exactness, than in that of 1 measure of oxygen to 2 of hydrogen; but here, the most exact experiments I have ever made, gave 1.97 hydrogen to 1 oxygen' (Dalton, *A New System* I.2, p.559); Dalton's Notebook I, 6 September 1803 (Roscoe and Harden, *A New View*, p.28).

[22] Dalton's Notebook II, 1809. See W.W.H. Gee, H.F. Coward and A. Harden on Dalton's lectures and lecture illustrations, in: *Manchester Memoirs* 59 (1915), p.46.

[23] Dalton, *A New System* I.2, pp.558, 433.

[24] Dalton, *A New System* I.2, p.550.

[25] Like Dalton, Avogadro meant by the volume of a particle that of the particle proper plus its heat mantle. 'L'hypothèse ... qui paroît même la seule admissible, est de supposer que le nombre des molécules intégrantes dans des gaz quelconques, est toujours le même à volume égal, ou est toujours proportionnel aux volumes' (A. Avogadro, 'Essai de déterminer les Masses relatives des Molécules élémentaires des Corps, et les Proportions selon lesquelles elles entrent dans ces Combinaisons', in: *Journal de Physique, de Chimie et d'Histoire Naturelle* 73 (1811), p.58).

[26] '... les molécules constituantes d'un gaz simple quelconque ... ne sont pas formés d'une seule molécule élémentaire, mais résultent d'un certain nombre de ces molécules réunies en une seule par attraction' (Avogadro, 'Essai', p.60).

[27] *Ibidem*, p.61. The possiblility that splitting the molecule of a simple body might yield 4, or 6, etc. atoms is left open.

[28] Dumas, in: *Ann. Ch. Phys.* [2] 33 (1826), pp.337ff; 49 (1832), p.210; 50, p.170.

[29] Avogadro now, by analogy, supposed that *all* simple bodies consist of biatomic molecules (S_2, C_2, P_2, etc): 'L'analogie tirée des autres combinaisons dont nous avons déjà parlé, où il y a en général redoublement de volume, ou partage de la molécule en deux, nous porte à supposer qu'il en est de même de celle dont il s'agit, c'est à dire, que le volume du gaz de soufre est la moitié de celui de l'acide sulfureux' (Avogadro, 'Essai', p.66).

1 volume sulphur vapour + 2 volumes oxygen gas → 2 volumes sulphurous acid gas;
[i.e. $S_2 + 2\ O_2 \rightarrow 2\ SO_2$]

[30] M.A. Gaudin, *Ann. Ch. Phys.* [2] 52 (1833), p.113.

[31] Ch. Gerhardt, *Ann. Ch. Phys.* [3] 7 (1843), p.129; 8, p.238.

[32] In 1846 A. Laurent (*Ann. Ch. Phys.* [3] 18, 2, 66) based the molecular weights and the molecular formulae of all substances (elementary as well as compound) on the assumption that the molecular weight is twice the vapour density, but (like Avogadro himself) he pushed analogy too far, and regarded as biatomic not only the molecules of hydrogen, nitrogen, oxygen and chlorine, but also those of potassium (K_2), zinc, mercury, etc, so that the formulae of their oxides were sometimes correct (H_2O, K_2O) but in other cases wrong (Zn_2O instead of ZnO; Hg_2O for red mercury oxide instead of HgO). In 1853 - 1856 Ch. Gerhardt followed Laurent.

[33] Bineau (1839) found for the vapour of ammonium chloride only half the expected density, which seemed to contradict Avogadro's law. He rightly explained this by dissociation, which led to duplication of the volume ($NH_4Cl \rightarrow NH_3 + HCl$).

[34] A. Kékulé, *Lehrbuch der organischen Chemie oder der Chemie der Kohlenstoffverbindungen* I, Erlangen: Enke 1861, p.58. J.J. Berzelius (*Jahresberichte* 13 (1834), p.185) made a distinction between the 'empirical' formula (i.e. the direct result of chemical analysis into the simple bodies) and the 'rational' formula, which said something about the grouping of the parts of a (compound) molecule. For example, the empirical formula of the alcohol molecule is C_2H_6O, whereas the rational formula depended on the author's theoretical views: either $C_2H_4O + H_2O$, or: $C_2H_6 + O$. He recognised that the choice is difficult and that, besides the uncertainty about the molecular weight, this was another reason for the confusion in organic chemistry. (See also Gerhardt's conception on pp.254-257 of chapter VII.)

[35] A. Kékulé, *Die wissenschaftlichen Ziele und Leistungen der Chemie.* Bonn 1878, p.7.

[36] Stanislao Cannizzaro, *Sunto di un Corso di Filosofia Chimica* ('Epitome of a Course of Chemical Philosophy'). German ed. by Lothar Meyer, *Abriss eines Lehrganges der theoretischen Chemie, von S. Cannizzaro,* Ostwalds Klassiker der exacten Wissenschaften, nr.30.

[37] Such mathematical-physical kinetic theories had been put forward by Leonhard Euler, 'Tentamen Explicationis Phaenomenorum Aeris', in: *Comm. Acad. Sc. Imp. Petropol. 1727,* 2 (1729) pp.347ff (cf R. Hooykaas, 'The First Kinetic Theory of Gases', in: *Arch. Intern. Hist. Sci.* 2 (1948) pp.180-4). Shortly afterwards Daniel Bernoulli (*Hydrodynamica*, Argentorati 1738, sect.X, pp.200-3), supposed the 'air' to consist of perfectly elastic particles in translatory motion. Their impact on the wall of the vessel caused the pressure. Like Euler he could derive Boyle's law on the basis of his hypothesis. The kinetic theory of gases was revived by J. Herapath (*Annals of Phil.* (1821), pp.273ff.; 340ff.; 401ff.) and by J.P. Joule (1848) (*Manchester Memoirs* 9 (1851), pp.107ff).

[38] A. Krönig (*Pogg. Ann.* 99 (1856), pp.315ff.) from his hypotheses derived not only Boyle's law but also the thesis that equal volumes of different gases contain the same number of molecules. In the article 'Grundzüge einer Theorie der Gase' he expresses it as follows: 'von verschiedenen Gasen sind bei gleichem Druck und gleicher Temperatur in gleichem Raum gleich viele Atome enthalten' (*ibidem,* p.318).

[39] Rudolph Clausius, 'Über die Art der Bewegung, welche wir Wärme nennen', in: *Pogg. Ann.* 100 (1857), pp.497-507. Clausius' theory was much more sophisticated than Krönig's. He attributed not only a translatory motion to the particles but also a rotational motion and, within poly-atomic molecules, intramolecular vibrations.

[40] *Ibidem*, p.369. Clausius admitted the possibility that molecules of some simple bodies in the gaseous state (e.g. sulphur, phosphorus) contain more than 2 atoms; he hoped that chemistry would shed light on the abnormal vapour pressures and the numbers of atoms per molecule (p.370).

[41] Clausius, in: *Pogg. Ann.* 103 (1858), p.645.

[42] J.J. Waterston, 'On the Physics of Media that are Composed of Free and Perfectly Elastic Molecules in a State of Motion' [Received by the Royal Society 11 Dec. 1845 and published in the] *Phil. Transact.* 183 (1892), A, pp.1ff, by Lord Rayleigh, secr. R.S. The quotation is from: J.S. Haldane ed., *The Collected Scientific Papers of John James Waterston.* Edinburgh 1928, pp.207-317, *q.v.* p.230.

[43] Examples of partition of molecules into 2 atoms were hydrogen, nitrogen, oxygen, chlorine, bromine, iodine; into 4 parts, arsenic and phosphorus; into 6 parts, sulphur (Waterston, 'On the Physics' (*Scientific Papers,* pp.230-1)).

[44] J.J. Waterston, 'On the Theory of Sound', in: *Phil. Mag. Suppl.* 16 (1858) (reprint in *Collected Scientific Papers,* p.348. The references are to Newton's *Principia Mathematica*, Bk.II, prop.23 and to the beginning of Bk.III. See also chapter VII). The paper having been presented to the Royal Society, a referee was of the opinion that 'the paper is nothing but nonsense, unfit even for reading before the Society' (*Collected Scientific Papers,* p.209). J.S. Haldane, the editor of Waterston's *Collected Scientific Papers*, considered it a misfortune that the paper was not printed when it was written, for it shadows forth many of the ideas of modern chemistry and it might have hastened their reception by chemists (p.211).

[45] Clausius, in: *Pogg. Ann.* 103 (1858), p.645.

[46] Lothar Meyer in his German edition of Cannizzaro's *Sunto*.

[47] Acids (principes salifiants [SO_3, N_2O_5, etc.]) consist of oxygen plus a non-metallic element; their opposites are 'bases' (bases salifiables [e.g. CaO, CuO]). The combination of acids and bases (in modern terms one would say, acid anhydrides and basic oxides) yields salts. The names of the salts express the genus (the acid) and the species (the metal) in the Linnaean fashion: sulfate de cuivre [$CuSO_4$], sulfate de plomb [$PbSO_4$], nitrate de plomb [$Pb(NO_3)_2$]. See A.L. Lavoisier, *Traité élémentaire de Chimie*, 2. ed., Vol.I, Paris 1793, Ch.XVI, p.163.

[48] Dumas found that the transformation of acetic acid into chloro-acetic acid is not an essential change (*Comptes Rendus Ac. Sci.* 7 (1838), p.444). He further expounds this (*Comptes Rendus Ac. Sci* 8 (1839), p.629): '... le chlore en prenant la place de l'hydrogène, n'a rien changé aux propriétés du composé, qu'il fut acide, corps neutre ou base ...' It is interesting that with Dumas the *morphological* approach plays an important role: in inorganic chemistry we are guided by 'isomorphism' (which does not have to agree with electrodualism), and in organic chemistry the substitution theory plays the same role. (For a brief survey of the development of chemical theory in the first half of the 19th century, see R. Hooykaas, 'Die Chemie in der ersten Hälfte des 19. Jahrhunderts', in: *Technikgeschichte* 33 nr.1 (1966), pp.1-24; reprinted in: W. Treue and K. Mauel ed., *Naturwissenschaft, Technik und Wirtschaft im 19. Jahrhundert*, Vol.II, Göttingen 1976, pp.587-613; and in: R. Hooykaas, *Selected Studies in the History of Science*, Coimbra 1983, pp.215-51).

[49] R.G. Collingwood, *The Idea of History* [circa 1940], Oxford 1961, p.80.

[50] Robert Mayer, 'Über Auslösung', in: *Staatsanzeiger für Württemberg*, 1876. Repr. in: J.J. Weyrauch ed., *Die Mechanik der Wärme*, in: *Gesammelte Schriften von Robert Mayer*, 3. ed., Stuttgart 1893, pp.440ff.

[51] Blaise Pascal pointed out that a little 'grain of sand' in the 'ureter' caused Cromwell's death (and the restoration of the British monarchy) (*Pensées*, fr.176).

[52] Collingwood, *The Idea of History*. Shortly after the first publication of Collingwood's book, the 'general historian' Herbert Butterfield wrote (1949) that the so-called scientific revolution 'outshines everything since the rise of Christianity and reduces the Renaissance and Reformation to the rank of mere episode ...', looming large 'as the real origin both of the modern world and of the modern mentality ...' (Butterfield, *The Origins of Modern Science*, London 1950, p.VIII). A similar claim (without the reservation of the rise of Christianity) has been made for the rise of 19th century evolutionism (oddly enough by a Jesuit priest circa 1940).

[53] Collingwood, *The Idea of History*.

[54] Pascal, Preface to *Traité du Vide* (1647); quoted from: L. Brunschvicg ed., *Oeuvres de Blaise Pascal*, Vol.II, Paris 1914, 2.ed. 1923, pp.127-145.

[55] Pascal, *ibidem* (*Oeuvres* II, p.145).

[56] Forbes's response was given in two lectures at Edinburgh University (1848): J.D. Forbes, *The Danger of Superficial Knowledge*, London 1849.

[57] Robert Boyle (1626 - 1691) declared that he collected experiments 'for more philosophical heads to explicate' (Boyle, 'Proëmial Essay' to his *Physiological Essays* (1661)).

CHAPTER XI

Quotations are from *Oeuvres de Blaise Pascal*, ed. L. Brunschvicg etc. Paris 1914; 2. ed. 1923. The *Pensées* are quoted from 'Brunschvicg minor'; in general the number of the fragment is put between brackets after the quotation, unless there is reference to a note. [*Editors' note*: For passages from the *Pensées* Krailsheimer's translation in the *Penguin Classics* has been an ongoing source of inspiration and, here and there, emulation].

1 'L'homme n'est qu'un roseau, le plus faible de la nature, mais c'est un roseau pensant. Il ne faut pas que l'univers entier s'arme pour l'écraser: une vapeur, une goutte d'eau, suffit pour le tuer. Mais, quand l'univers l'écraserait, l'homme serait encore plus noble que ce qui le tue, puisqu'il sait qu'il meurt, et l'avantage que l'univers a sur lui, l'univers n'en sait rien. Toute notre dignité consiste donc en la pensée' (fr.347); 'L'homme est visiblement fait pour penser ... Or l'ordre de la pensée est de commencer par soi, et par son auteur et sa fin' (fr.146).

2 '... par l'espace l'univers me comprend et m'engloutit comme un point, par la pensée je le comprends' (fr.348).

3 Charles Darwin, *The Descent of Man,* 2. ed., London 1874, ch.21, p.619.

4 'S'il se vante, je l'abaisse; s'il s'abaisse, je le vante, et le contredis toujours, jusqu'à ce qu'il comprenne qu'il est un monstre incompréhensible' (fr.420).

5 He was very close to the Jansenists, who defended the Augustinian doctrine of grace against the Jesuits. The latter made much of his final estrangement from the Jansenists, which came about, however, because they were not consistent enough in their Jansenism. Pascal never recanted because there was nothing to recant: he held that the 'Jansenistic' heresy was a construction of the enemies of Port-Royal. Cf R. Hooykaas, 'Pascal: his Science and his Religion' (original in Dutch 1939), in: *Tractrix* 1 (1989), pp.115-139. See also the most illuminating work: E. Cailliet, *Pascal, Genius in the Light of Scripture*, Philadelphia s.d., pp.354-358.

6 Pascal, *Traités de l'Equilibre des Liqueurs et de la Pesanteur de la Masse de l'Air* (1654); *Oeuvres* III, p.199.

7 Pascal, *Traités de l'Equilibre des Liqueurs*, Conclusion des deux Traités (*Oeuvres* III, p.265).

8 Réponse de Blaise Pascal au Père Noël S.J., 19 October 1647 (*Oeuvres* II, pp.96-8). See on the void: Pascal, *Expériences nouvelles touchant le Vide* 1647 (*Oeuvres* II, p.75).

9 Réponse au Père Noël; (*Oeuvres* II, p.100).

10 Pascal, *Traités de l'Equilibre des Liqueurs* (*Oeuvres* III, p.255).

11 Lettre à M.le Pailleur, au sujet du Père Noël, Jésuite, February 1648 (*Oeuvres* II, p.183).

12 Fragment de Préface sur le *Traité du Vide* (October - November 1647) (*Oeuvres* II, pp.132-134).

13 *Lettres Provinciales* XVIII, au Révérend Père Annat, Jésuite, 24 mars 1657 (*Oeuvres* VII, p.53).

14 *Ibidem,* p.54.

15 *Ibidem,* p.55. When, under pressure from his enemies, his *Lettres Provinciales* had been put on the index of prohibited books, Pascal wrote: 'I am alone amongst thirty thousand? Not at all ... I have the truth, and we will see who will win'. No other court of appeal was left to him than that of the truth: "Ad tuum, Domine Jesus, tribunal appello" (fr.920).

16 'Inquisition et la Société, les deux fléaux de la vérité' (fr.920).

17 'Ecrire contre ceux qui approfondissent trop les sciences. Descartes' (fr.76). Pascal used the term 'roman de la nature' for the Cartesian system. Cf Brunschvicg, in Pascal, *Pensées* I, Paris 1924, p.97 n.7.

18 'Manque d'avoir contemplé ces infinis, les hommes se sont portés témérairement à la recherche de la nature, comme s'ils avaient quelque proportion avec elle' (fr.72).

19 Pascal, *Traités de l'Equilibre des Liqueurs* (*Oeuvres* III, p.263).

20 'Ce n'est pas par notre capacité à concevoir ces choses que nous devons juger de leur vérité' (*De l'Esprit géométrique* (1658 - 59); *Oeuvres* IX, pp.258-259).

21 Pascal, *De l'Esprit géométrique*; *Oeuvres* IX, p.259.

[22] It is interesting that G.W. Leibniz, who was much more rationalistic, and besides a great protagonist of natural theology, did not make this reservation in the case of mathematics, but held that man's depraved will would trifle even with the principles of mathematics if it saw any profit in it. 'Si la géométrie s'opposait autant à nos passions et à nos intérêts présents que la morale, nous ne la contesterions et ne la violerions guère moins, malgré toutes les démonstrations d'Euclide et d'Archimède, qu'on traiterait de rêverie, et croirait pleines de paralogismes' (Leibniz, *Nouveaux Essais sur l'Entendement humain*, bk.I, ch.2, par.12). This is said by 'Théophile' who represents Leibniz's own standpoint.

[23] 'There is a certain model of agreement and beauty, which consists in a certain relation between our nature — weak or strong, such as it is — and the thing that pleases us' (fr.32).

[24] 'We are born with a character of love in our hearts ..., which leads us to love what seems beautiful to us without anybody's ever having said what it is ... Man finds in himself the model of that beauty which he seeks outside himself' (*Discours sur les Passions de l'Amour*, circa 1653). According to Brunschvicg this work has been written by Pascal, whereas J. Chevalier (*Oeuvres de Pascal*, éd. La Pléiade, Paris 1950, p.313) considers that only certain maxims in it should be attributed to him.

[25] Pascal, *De l'Esprit géométrique* (*Oeuvres* IX, p.270). Also: fr.72.

[26] 'Le silence éternel de cet espace infini m'effraie' (fr.206).

[27] 'L'auteur de ces merveilles les comprend. Tout autre ne le peut faire ... Quand on est instruit, on comprend que la nature ayant gravé son image et celle de son Auteur dans toutes choses, elles tiennent presque toutes de sa double infinité' (fr.72).

[28] Hugo Grotius, *De Veritate Religionis Christiana*, bk.I, ch.7; Joannes Clericus ed., London 1794, p.10; Cf a literal translation: *On the Truth of the Christian Religion, Specially Intended for the Use of Self-taught Students of Divinity*, by Thomas Sedger, M.A. curate of Fundenhall, Norfolk; 3. ed., London 1865, pp.7-8. Hugo de Groot (1583 - 1645) wrote this work in the Netherlands' language circa 1620 in rhyme; it was intended to be used by sailors for their own edification and for enabling them to propagate Christian religion when meeting pagans, Mahometans and Jews on their voyages. He translated it into Latin (ed. Paris 1639). J. Leclerk's editions were provided with notes to make it more accessible to the (then) modern readers (Amsterdam 1709).

[29] 'Haec habet vir summus ex Peripateticorum Philosophia, quae aquam mero vacui per antlias adscendere statuebat quod pondere aëris fieri jam omnibus constat. Sed in gravitate, quemadmodum a Recentioribus explicatur, non minus ordo Universi, et Conditoria sapientia elucent' (Note of Leclerk to Grotius, *De Veritate Religionis Christiana*, bk.I, ch.7).

[30] 'Ce qui y paraît ne marque ni une exclusion totale, ni une présence manifeste de divinité, mais la présence d'un Dieu qui se cache' (fr.556)

[31] '... c'est même le nom qu'il se donne dans les Ecritures, *Deus absconditus* ...' (fr.194). Also fr.242.

[32] Newton, *Optice, sive de Reflexionibus, Refractionibus, Inflexionibus et Coloribus Lucis Libri Tres*. Latine reddidit Samuel Clarke. Londini 1706, qu.20 fin., pp.314-315; qu.23 fin., p.348.

[33] Newton, *Philosophiae Naturalis Principia Mathematica*, 3. ed. Londini 1727; Scholium Generale, p.528.

[34] 'Il est donce juste qu'elle se soumette, quand elle juge qu'elle se doit soumettre' (fr.270); 'Il n'y a rien de si conforme à la raison que ce désaveu de la raison' (fr.272); 'Tout ce qui est incompréhensible ne laisse pas d'être' (fr.430).

[35] 'Les hommes prennent souvent leur imagination pour leur coeur' (fr.275).

[36] Fr.581, 282, 284. *De l'Esprit géométrique* (*Oeuvres* IX, p.273).

[37] 'Toute religion qui ne dit pas que Dieu est caché, n'est pas véritable' (fr.585); 'Il faut, pour faire qu'une religion soit vraie, qu'elle ait connu notre nature. Elle doit avoir connu la grandeur et la petitesse, et la raison de l'une et de l'autre. Qui l'a connue, que la chrétienne?' (fr.433). Cf fr.442.

[38] 'Nous avons une impuissance de prouver, invincible à tout le dogmatisme. Nous avons une idée de la vérité, invincible à tout le pyrrhonisme' (fr.395); 'Deux choses instruisent l'homme de toute sa nature: l'instinct et l'expérience' (fr.396).

[39] '... je ne'entreprendrai pas ici de prouver par des raisons naturelles, ou l'existence de Dieu, ou la Trinité, ou de l'immortalité de l'âme, ... non seulement parce que je ne me sentirais pas assez fort pour trouver dans la nature de quoi convaincre des athées endurcis, mais encore parce que cette connaissance, sans Jésus-Christ, est inutile et stérile' (fr.556). Cf fr.504.

[40] 'Dieu seul peut les mettre [*viz.* les vérités divines] dans l'âme, et par la manière qu'il lui plaît. Je sais qu'il a voulu qu'elles entrent du coeur dans l'esprit, et non pas de l'esprit dans le coeur, pour humilier cette superbe puissance du raisonnement, qui prétend devoir être juge des choses que la volonté choisit ...' (*De l'Art de persuader*; *Oeuvres* IX, p.273). See also fr.434 on pyrrhonism and dogmatism and the inability of reason to make a choice.

[41] 'Le coeur a ses raisons que la raison ne connaît point' (fr.277).

[42] 'C'est le coeur qui sent Dieu, et non la raison' (fr.278).

[43] 'La foi est un don de Dieu; ne croyez pas que nous disions que c'est un don de raisonnement' (fr.279).

[44] 'Le pyrrhonisme est le vrai' (fr.432).

[45] 'Incroyable que dieu s'unisse à nous ...' (fr.430).

[46] 'Car je voudrais savoir d'où cet animal, qui se reconnaît si faible, a le droit de mesurer la miséricorde de Dieu, et d'y mettre des bornes que sa fantaisie lui suggère' (fr.430).

[47] 'Tout ce qui est incompréhensible ne laisse pas d'être' (fr.430).

[48] *Lettre à Mlle de Roannez* (1656) (*Oeuvres de Pascal*, ed. Pléiade, p.287)

[49] John Donne, *The Works of John Donne*, H. Alford ed., Vol.V, London 1839, p.431.

[50] Pascal sees Jesus' suffering prolonged throughout the ages because many of his disciples are asleep: 'Jésus sera en agonie jusqu'à la fin du monde: il ne faut pas dormir pendant ce temps-là' (fr.553).

[51] 'L'Evangile ne parle de la virginité de la vierge que jusques à la naissance de Jésus-Christ. Tout par rapport à Jésus-Christ' (fr.742).

[52] On the semi-deism of adherents of the mechanistic philosophy, see: R. Hooykaas, *Religion and the Rise of Modern Science*, Edinburgh-London 1972, pp.16, 19. Cf also: R. Hooykaas, *Natural Law and Divine Miracle. A Historical-Critical Study of The Principle of Uniformity in Geology, Biology and Theology*. Leiden: Brill 1959, [2]1963, pp.192-206.

[53] 'On se fait une idole de la vérité même, car la vérité hors de la charité n'est pas Dieu, et est son image et une idole' (fr.582).

[54] R. Hooykaas, 'Pascal, zijn Wetenschap en zijn Religie', in: *Orgaan* 1939, p.32. English translation as: 'Pascal, his Science and his Religion', in: *Free Univ. Qu.* 2 (1952), p.124, and in: *Tractrix* 1 (1989), pp.115-139.

[55] Pascal, Préface sur le *Traité du Vide* 1647 (*Oeuvres* II, p.135). References to the dispute on the void with Descartes and the Jesuit P. Noël as well as to the laxity in ethics and the anti-Augustinian doctrines that were at stake in the disputes between Pascal (on behalf of the Jansenists) and the Jesuits.

[56] *Ibidem (Oeuvres* II, p.136).

[57] Pascal, Préface (*Oeuvres* II, pp.136-137).

[58] *Ibidem (Oeuvres* II, p.142).

[59] *Ibidem (Oeuvres* II, pp.143-144).

[60] *Ibidem* (*Oeuvres* II, p.145, p.36).

[61] 'Nous n'avons ni vrai ni bien qu'en partie, et mêlé de mal et de faux' (fr.385).

[62] 'Il est bon d'être lassé et fatigué par l'inutile recherche du vrai bien, afin de tendre les bras au Libérateur' (fr.422). Cf fr.737.

[63] Possibly the arrangement of the *Pensées* would have corresponded to a large extent with that of the *Editio Princeps* of Brunschvicg a.o.

[64] '... gloire et rebut de l'univers' (fr.434).

[65] '... vous êtes embarqué ...' (fr.233). This long fragment shows clearly that Pascal is in dialogue with an imaginary partner.

[66] *Oeuvres* II, pp.298-314. Cf R. Taton, 'Sur l'Invention de la Machine arithmétique', in: R. Taton ed., *L'Oeuvre scientifique de Pascal*, Paris 1960, p.214.

[67] *Oeuvres* II, pp.401-404.

[68] Anagram of 'Louis de Montalte', the pseudonym under which he had written the *Lettres Provinciales*.

[69] Pascal to Fermat, 10 August 1660 (*Oeuvres* X, pp.4-5).

[70] Pascal to Huygens, 6 January 1659 (*Oeuvres Complètes de Christiaan Huygens* II, p.310).

[71] In fr.793 the infinite distance from the material to the intellectual is compared with the even larger distance between 'esprit' and 'charité' and wisdom. Archimedes was in the kingdom of learning what Jesus was in that of sainthood. Both of them were princes, but without outward, worldly glory. Archimedes did not fight battles that catch the eye, but made great inventions and shone in 'esprit'. Jesus made no inventions and had no worldly power, but was a saint in the eyes of God and glorious to the eyes of the 'heart' which see wisdom.

[72] 'Je vois mon abîme d'orgueil, de curiosité, de concupiscence' (fr.553). This fragment is from a dialogue with Jesus.

[73] '... nous nous cachons et nous déguisons à nous-mêmes' (fr.377).

[74] R. Taton, art.'Pascal' in: *Dictionary of Scientific Biography,* Vol.X, New-York 1974, p.337.

[75] 'Il y avait un homme qui, à douze ans, avec des *barres* et des *ronds*, avait créé les mathématiques; ... qui, dans les courts intervalles de ses maux, résolut par distraction un des plus hauts problèmes de la géométrie, et jeta sur le papier des pensées qui tiennent autant du Dieu que l'homme. Cet effrayant génie se nommait *Blaise Pascal*.' These words of the early 19th century romanticist Châteaubriand probably gave more satisfaction to the laudator than they would have given to the lauded.

[76] 'L'éloquence continue ennuie' (fr.355).

[77] 'Il faut se renfermer le plus qu'il est possible dans le style naturel' (fr.16). His 'pensées' concerning this 'natural style' and its simplicity show that there was much 'art' in it — as in a 'natural' English garden.

[78] 'Quand on voit le style naturel, on est tout étonné et ravi, car on s'attendait de voir un auteur, et on trouve un homme' (fr.29).

[79] 'On ne s'imagine Platon et Aristote qu'avec de grandes robes de pédants. C'étaient des gens honnêtes et, comme les autres, riant avec leurs amis ...' (fr.331).

LIST OF ILLUSTRATIONS

[*Editors' note:* We have done our utmost to identify and trace photographers of the illustrations collected over the years by the author for use in this book. This has not in every case proved possible; for those cases we invite institutions and individuals concerned to contact us through the publisher.]

Frontispiece. Preparation of Experiments, from Thomas Norton, *Ordinall of Alkimy* (British Library, Norton additional ms 10302, fol. 37v).

Figure 1 Grotesque (Peter Floetner 1546)........ 10

Figure 2 Tetraktys........ 28

Figure 3 Woodcut of Pythagoras' 'experiments' with strings tended by weights, flutes with various lengths, bells filled with different amounts of water, etc. (Franchino Gaffurio, *Theoria Musica* 1492, p.127)........ 30

Figure 4 Kepler's five regular geometrical bodies, representing the dimensions and the distances of the planetary orbs (Kepler, *Mysterium Cosmographicum* 1597 (Werke I))........ 38

Figure 5 The music of the planetary spheres according to Kepler (*Harmonice Mundi* 1619). From: –, *Gesammelte Werke VI*........ 40

Figure 6 Calc spar prisms and scalenohedrons; stages of formation of primitive rhombohedron. From: Haüy, *Traité de Cristallographie I* (1822), pl.3: fig. 33-37........ 47

Figure 7 Law of decrescence: development of a dodecahedron with rhombic faces, resulting from a primitive cubic form (Haüy, *Traité de Cristallographie I* (1822), pl.5: fig.70-71)........ 48

Figure 8 Haüy's contact goniometer (—, *Traité de Cristallographie* 1822, pl.2: fig.23)........ 48

Figure 9 Gell-Mann triangle (From: P.T. Matthews, 'Order out of sub-nuclear Chaos' in: *New Scientist*, 379 (20-2-1964), fig.3)........ 76

Figure 10 Philosopher's Stone. From: Jean-Conrad Barchusen, *Elementa Chymiae, quibus Subjuncta est Confectura Lapidis Philosophici Imaginibus Repraesentata*. Leyden 1718, plate 19, fig.78. Courtesy Bibliotheca Philosophica Hermetica, Amsterdam........ 82

Figure 11 Illustration from the 'Goldmaking Art of Cleopatra'. From: M. Berthelot, *Les Origines de l'Alchimie*, Paris, 1885........ 85

Figure 12 'In the sweat of thy face shalt thou eat thy bread'........ 86

Figure 13 Allegorical representation of the two principles. (From: *Rosarium Philosophorum* (1470) Francofurti: Jacobi 1550)........ 88

Figure 14 Allegorical representation of the generation of the Philosopher's stone........ 89

Figure 15 'Two fishes swimming in our sea'. From: Lambsprinck (pseudonym of Lampert Spring), *Traité de la Pierre Philosophale*. (Translated from German)........ 97

Figure 16 Table of elements according to Lavoisier. From: Lavoisier, *Traité Élementaire de Chimie I*, Paris 1793, p.192. 112

Figure 17 The Aristotelian system of the universe according to Petrus Apianus' *Cosmographia* (1539)..... 120

Figure 18 The Aristotelian system of the universe according to Rembertus Dodonaeus' *Cosmographia* (1548) 121

Figure 19 Diagram of the universe with hell in the centre (14th century). 123

Figure 20 Graphs representing the relation between qualities and distances, according to the geometrical method of Oresme. 129

Figure 21 The Ptolemaic system of the Universe (from: Joannes Luyts, *Astronomica Institutio*. Trajecti ad Rhenum: Halma 1692, p. 205) 150

Figure 22 Copernicus' system (Copernicus, *De Revolutionibus Orbium Caelestium*) 151

Figure 23 The 'Pythagorean' [= Copernican] system of the universe according to Thomas Digges's *A Perfit Description of the Caelestiall Orbes*, London 1576. 156

Figure 24 Vitruvius' Humani corpori mensura (proportion of a man as he ought to be) from: Vitruvius, *De Architectura*. Ed. Como 1521.. 169

Figure 25 Tycho Brahe's Diagram of the universe (From the 2nd issue of his *Progymnasmata* (1610)..... 176

Figure 26 Hero's mechanism for temple doors. From: F. Dannemann, *Die Naturwissenschaften in ihrer Entwicklung und in ihrem Zusammenhang*. Leipzig ^{2}1920, p.195). 185

Figure 27 Wreath of spheres ('Clootcrans') of Simon Stevin. From: Simon Stevin, *The Principal Works* I (Amsterdam 1955), p.176. 191

Figure 28 Title page of Simon Stevin's *Wisconstige Gedachtenissen* ('Mathematical Memoirs'). Leiden: Bouwensz. 1608 (^{1}1605). 193

Figure 29 Device to demonstrate 'impetus' (Scaliger, *Exotericarum Exercitationum* 1557)...201

Figure 30 The fabrication of magnetized iron. From: William Gilbert, *De Magnete* (1600)....203

Figure 31 Illustration of cylindrical vessel used to measure the weight of vacuum. From: Galileo, *Discorsi*, First Day (OpereVIII, p.62). 212

Figure 32 Béguin's interpretation of the reaction of 'antimonium' [Sb_2S_3] with mercury sublimate [$HgCl_2$]. 215

Figs. 33-36 Descartes' refraction law compared with a mechanical model I-IV (Descartes, *La Dioptrique* (from: *Oeuvres* VI, p.98-103). 236-237

Figure 37 Descartes's 'whirlpool of celestial matters'. (From: Descartes, *Principia Philosophiae* 1644, p.106). 241

Figure 38 Haüy's 'molécules soustractives'. 251

Figure 39 Diagrams of benzene rings 258

Figure 40 Tellurium Willem Jansz. Blaeu. (Courtesy Netherlands Maritime Museum, Amsterdam) 301

Figure 41 Paper stereochemical models of molecules with carbon atoms made by J.H. Van 't Hoff (1875). (Museum Boerhaave, Leiden) 302

Figure 42 Gilbert's *terrella* (Gilbert, *De Magnete*, lib.IV, p.57)..........303

Figure 43 Gilbert's analogy between (static) electricity and the attraction between a wet and a dry rod (Gilbert, *De Magnete*, lib.II, p.57)..........304

Figure 44 Daubrée's models of mechanical geological phenomena (A. Daubrée, *Études Synthétiques de Géologie Expérimentale*. Paris 1879, ch.IV, fig.78-80 (p.294)..........307

Figure 45 J.D. Forbes, 'Experiments on the Flow of Plastic Bodies and Observations on the Phenomena of Lava Streams', in: *Philosophical Transactions* 1846; Figures from the reprint in: J.D. Forbes, *Occasional Papers*, Edinburgh 1859, Plate I, figs.1, 2..........308

Figure 46 Maxwell's model of electromagnetic force. James Clerk Maxwell, *On Physical Lines of Force* 1861-2. From: *The Scientific Papers of James Clerk Maxwell*. Ed. W.D. Niven. Cambridge 1890. Vol.I, Plate VIII..........311

Figure 47 Thomson's gyrostatic system imitating the properties of a spring. From: William Thomson, *Popular Lectures and Adresses* vol.I, London 1889, p.239 (fig. 45)..........312

Figure 48 Cogwheels representing electricity. From O.J. Lodge, *Modern Views of Electricity* (1892), p. 203, fig. 34..........313

Figure 49 Boltzmann's machine to demonstrate self-induction of electrical currents (From: Ludwig Boltzmann, *Vorlesungen über Maxwells Theorie Elektrizität und des Lichtens,* fig. 15, Taf. II)..........315

Figure 50 The atmosphere consisting of the four separate gases of water vapour, oxygen, nitrogen and carbon dioxide (law of partial pressures of mixed gases). From: Dalton, *Manchester Memoirs* 52, p.602..........324

Figure 51 Particles of gas atoms with their caloric atmospheres. Lecture sheet used by Dalton circa 1807. (From H.F. Coward and A. Harden, *Manchester Memoirs* 59, nr.12 (1915) pl. 5, p.49)..........325

Figure 52 Mechanistic explanation of the mutual repulsion of similar particles and of the mutual indifference of particles of different species. (Dalton, *New System* I, 2 (1813), pl.7, p.548)..........328

Figure 53 Schematic model of reaction of $Cl_2 + H_2 \rightarrow 2HCl$; $O_2 + 2H_2 \rightarrow 2H_2O$; $N_2 + 3H_2 \rightarrow 2NH_3$ (from: M.A. Gaudin, *Ann.Ch. Phys.* [2] 52 (1833), p.113)..........330

Figure 54 Rational formulae for acetic acid ($C_2H_4O_2$) current about 1860. From: A. Kékulé, *Lehrbuch der Organischen Chemie oder der Chemie der Kohlenstoffverbindungen* I, Erlangen: Enke 1861, p.58..........332

Figure 55 Dualistic nomenclature of salts according to Lavoisier (from: A.L. Lavoisier, *Traité élémentaire de Chimie*, 2. éd. T.I, Paris 1793 (1789), p.238.)..........336

Figure 56 Title page of J.D. Forbes's refutation of Macaulay..........340

Figure 57 Portrait of Pascal as a young man. Artist unknown..........345

THE TEXT AND EDITORIAL ACTIONS WITH REGARD TO IT

Professor R. Hooykaas delivered his Gifford lectures 'Fact, Faith and Fiction in the Development of Science' in the University of St Andrews in 1976, partly from a typed text and partly extempore. The lectures were:

1. Introduction: Fact, Faith and Fiction
2. The harmony of nature
3. The Philosopher's Stone
4. The undying fire
5. A tunnel through the earth
6. 'And the sun stood still'
7. Old and New in Renaissance Science (i): Fact against fiction
8. Old and New in Renaissance Science (ii): Discovery or rediscovery?
9. Thinking with the hands
10. Utility versus Reality (i): in theories
11. Utility versus Reality (ii): in classification systems
12. Works of Nature, Works of Art
13. Cleopatra's Nose
14. Reconstructing the Past
15. 'Civil and Natural History'
16. The 'thinking reed'.

From about 1985, having been urged by friends in St Andrews and elsewhere, to publish, RH embarked on a process of revising and expanding his material — a process which continued, though interrupted by other researches and activities, to within a few weeks of his death in January 1994.

Of lectures 7, 8, 11, 14, 15, we have found no trace apart from some photographs. Some of the matters they touched on can be surmised from R. Hooykaas's other writings (see the Selected Bibliography on page xv). For lectures 7 and 8 see, for example, his account of the Portuguese voyages of discovery in *Science in Manueline Style* and his chapter on 'The Reception of Copernicanism in England and the Netherlands', in *Selected Studies*; for lectures 14 and 15 see his chapters on geology and on evolution in the latter volume as well as *Natural Law and Divine Miracle*, and 'Pitfalls in the Historiography of Geological Science'.

The present volume contains the rewritten texts, for eleven of the above lectures. They were edited and finalized in the manner described below.

The Editorial Process

RH's English was rather good, but he recognized that it was not perfect. Down the years he often sent manuscripts in English (for example those of *Religion and the Rise of Modern Science* 1972, and *G.J. Rheticus' Treatise on Holy Scripture and the Motion of the Earth* 1984) to his friend Professor Donald M. MacKay at Keele University in England for checking over.

In this way his Gifford lectures 3 and 13 (chapters 3 and 10 of this volume) were checked and smoothed by DMMcK and finalized by RH before the Autumn of 1986. After Donald MacKay's death in 1987, his widow Dr Valerie MacKay took on the task of typing further Gifford chapters and suggesting modifications to both the English and the presentation.

By late 1993 nine of the sixteen lectures had thus attained a near final form. These originated as lectures 1, 2, 3, 4, 5, 6, 10, 13 and 16 (lecture 1 having been split up to become the present Preface, Introduction and chapter 1.)

Since 1994 it has been possible to add two further chapters to those listed above (chapters 7 and 9).

During his last months RH had worked heroically on a rewriting of chapter 7 and the second section of chapter 9, including a full reference list and a specification of the figures to be used. The section on Joule had been typed by VMcK and finalized by RH in 1993. But for the remainder we have worked (in 1998) from RH's apparently completed manuscript.

In particular chapter 9 gave us problems. Firstly because for the pages which now form section 2, we cannot be sure that RH had finished working on his revision (part of it existing in manuscript only) and secondly because we have two versions of the changes RH wished to make. Having deleted everything that RH had deleted from *both* versions, we have aimed at what would give a harmonious result, making use of RH's 'Humanities, Mechanics and Painting' (1991) to fill remaining gaps. The first and third sections of chapter 9 gave only minor problems. Though several copies existed they could rather easily be combined.

For the book as a whole, the editorial modifications and additions which have not been seen for approval by RH are as follows:

1) some or all of the section headings in chapters 5, 7, 9, and 11;

2) the figure captions for all but chapters 3 and 11;

3) the translations of Pascal have in some cases been modified (see comment in notes heading for chapter 11);

4) in the whole text, but particularly in chapters 4, 7, parts of 8, and 9, which had not previously been seen by an English speaker, numerous alterations to word order and vocabulary have been made to smooth the English or improve the clarity of one

or two of the scientific explanations — while sticking as close as possible to RH's text.

5) in 1992 RH withdrew the section on the law of Titius-Bode in chapter 2 to rewrite it, but never did so. However, we found RH's account of the law so apposite to the themes of the chapter that we have included it.

6) for the epilogue to the final chapter see the editors' note at the head of that section.

ACKNOWLEDGEMENTS

To secure the publication of a posthumous, almost-but-not-quite-completed book surely involves a special kind of labour. Some loose ends are bound to remain when time and resources are available in necessarily less than unlimited quantities; yet the pleasure of making significant ideas finally see the light of day has easily prevailed.

Many institutions and individuals have helped to make our burden lighter, and we are happy to record them here.

The Netherlands Organisation for Scientific Research NWO made possible JCB's contribution by means of a generous, 3-months' grant. The Agricultural University Wageningen permitted him to organise his work accordingly. The North-Holland 'Rijksarchief' allowed him to check the contents of its R. Hooykaas collection of documents. The University of Chicago Press gave permission to reprint, as part of our Editors' Foreword, portions of HFC's éloge of Hooykaas in *Isis* 89 (nr.1), March 1998, pp.181-184.

We further extend our thanks to the following individuals for a variety of somewhat invisibly good offices: Wim Gerritsen, Annie Kuipers, Stephanie Harmon, Wijnand Mijnhardt, and Nancy J. Nersessian. Further, José Peulers typed a large part of chapter 9. Hans Peters found a well-fitting 'representation' of the Philosopher's Stone. Fokko-Jan Dijksterhuis found an able computer company — Mapej, Shropshire — ready to convert Amstrad files into Word files. Gabie Ebbing deserves very special thanks for mastering a difficult handwriting and typing and interpreting parts of chapter 9. Finally, we thank the Hooykaas family for their trust and their support and their help whenever needed.

INDEX

A

Accademia del Cimento 213
accommodation principle *see* theology, exegesis, accommodation principle
acoustics 33; 71 *see also* harmony, musical
Adelard of Bath125
aether *see* ether
agriculture 272
Albert the Great 118; 184; 304–5
Albert of Saxony 134; 135
Alberti, Leon Battista 164; 276; 279
alchemy *see* chemistry, alchemy
Alexander the Great 277
Alexandrian engineers 184; 194; 279
Alhazen 197
Ames, William 274
Ampère, André Marie 333
analogical reasoning *see* nature, analogy of
Ångström 69
Apelles 277–78
Apianus, Petrus 119–20; 155
Aquinas, Thomas *see* Thomas Aquinas
Archimedes 37; 138; 172; 184; 197; 208–9; 300; 338; 368
Archytas 279
Aristarchus 39; 172
Aristotle & Aristotelianism 8; 11; 29; 39; 138; 139; 141; 275; 338
  commented on
    by Francis Bacon 284
    by Boyle 99; 217
    by Kepler 233
    by medieval commentators 117; 119; 304
    by Pascal 370
  compared
    with ancient rivals 79–80
    with Copernicanism 158–62
  critic
    of atomism 216
    of Pythagoreanism 28; 167; 172
    of Zeno 125
  influence exerted by 79; 142
    Christian Aristotelianism 20
  marked by
    observation and common sense 156; 192; 194; 212
    organistic world view 280
  scientific doctrine: overview 79–83
    motion 137
    system of the heavens 230
  scientific doctrine: specific
    compound and mixture 81; 132; 281
    continuity of bodies 194
    distinction sublunar and supralunar 159
    Earth, in daily rotation or stationary 156; 158
    elements 120; 133; 217; 220

falling bodies 192; 206
force and resistance 127
function of gills 298
heating up of projectiles 124
homocentric spheres 162
mathematics in nature 242
Mechanica 20; 280
natural place 124
position of sun 167; 170
projectile motion 131; 192
rotatory motion 132
views
on form 277
on gods and nature 18
on healing 161
on manual work 184
on mathematical arts 275
on middle 168
on nature and art 267–68
on substantial form 266
Aristoxenus 31–32
Arndt, F. 258
art (in the sense of artificial) *see* nature, and art
art and letters 276–79; 351; 356
Aston 66
astronomy
cosmology 36–46; 119
models 300–304
as evidence of order 18; 21
mathematical hypotheses in 230–35
planetary distances 37; 43–46
planetary and electronic orbits compared 74
planetary motion 39; 141; 230–35
irregularity of 33–34; 37
planetary systems
geocentric *see* Ptolemy, astronomy; *also under* Aristotle, scientific doctrine
heliocentric *see* Copernicus & Copernicanism; *also under* Galilei, Galileo *and under* Kepler
Athens and Jerusalem 228; 269
atomic number 66–68; 74
atomic weight 58–68; 327
atomism 11 *see also* corpuscular theories
ancient 17–18; 20; 22; 80; 129–30
modern 20; 178; 216; 310; 346
attraction
of likes 102
and tunnel through the Earth 141–45
Aubuisson de Voisins, J.F. d' 291
Augustine (bishop) 18; 119; 150; 152; 358
automaton 279
Averroës 124; 131; 162; 195
Avogadro, Amedeo 319–20; 329–37

B

Bacon, Francis 141, 214; 217; 284–85; 296; 308; 355
Bacon, Roger 91; 186; 196
Balmer, Johann Jakob 27; 68–74
Barlaeus, Caspar 300
Barlow 53
barnacle goose 7
barometer 346
Barth, Karl 2
Bartholomeus Anglicus 87
Beaumont, Elie de 294; 306
Beeckman, Isaac 142–43; 178–79; 183; 191; 204–5; 210; 304; 309; 337

Béguin, Nicolas 215
Bel, Joseph Achille le 337
Bellay, Joaquin du 271
Benedetti, Gianbattista 139–40; 192; 207
Benzenberg 180
Bergman, Torben 53
Berkeley, George 310
Bernard de Trévisan 85, 88, 90
Bernoulli, Daniel 219; 242; 322
Bernoulli, Jacob 219
Berthelot, Marcellin 85; 297; 331
Berzelius, Jöns Jakob 63–4; 109–10; 331; 335; 337
Bessel, Friedrich 115; 180
Billich, Anton 99; 216
Biot, Joseph Baptiste 247; 321
Black, Joseph 105; 110; 221; 292; 320
Blaeu, Willem Jansz. 300
blind alleys *see* science, blind alleys in
blind windows *see* Pascal, general views, blind windows
Bode, J.E. 27; 43–45
Bodin, Jean 186; 214; 281
Boerhaave, Herman 222; 226; 320
Boethius 91; 198
Bohr, Niels 14; 27; 68; 72–74; 77; 179; 261; 286
Boisbaudran, de 61; 64
Boltzmann, Ludwig 314–15
Bonnet, Charles 43
Born, Max 74
Bostock 189
Boyle, Robert 79; 246
  air pump and void 186; 217
  corpuscular theorist 100
  critic of
    Aristotle 99; 217
    Descartes 226
    Paracelsus 98–99
  experimental chemist 216
  law 76; 217; 242–44; 321-22; 332
  mode of explaining 240
  nature
    and art 282
    purpose in 20
  religion
    'awe and veneration' 352; 371
    God's ultimate incomprehensibility 226
    natural theology 17; 21; 367
  scepticism 218; 240; 347
  'underbuilder' 340
Bradley, James 115; 180
Bradwardine, Thomas 127–28; 135–36; 196
Bragg, L. 253
Bragg, W.H. 253
Brahe, Tycho *see* Tycho Brahe
Brauner, B, 62; 64
Bravais, A. 53; 252
Breithaupt, A. 294
Brewster, David 5–6; 248
Broglie, Louis de 76
Brooke, H.J. 50; 252
Brougham, Lord Henry 1; 21; 320; 351
Brouncker, Lord 218
Buchanan, George 157
Büchner, Ludwig 22
Buonarotti *see* Michelangelo
Bürgi, Jost 300
Buridan, Jean 117; 124; 130–36; 139; 156; 158; 162; 194–95; 202; 285
Butlerov 258–62

C

Calcidius 170
Calculators 126; 128; 137; 145; 198
calendar reform 148
Calvin, Jean 151–52; 157
Cannizzaro, Stanislao 332–33
Carangeot 46; 53; 187
Cardano, Girolamo 138
Carpenter, Nathaniel 214
Cassirer, Ernst 57
Castro, João de 11; 123; 183; 198–201; 205
Cavendish, Henry 103; 110; 217
Ceva, Thomas 152
Chambers, Robert 8
Chancourtois, A.E. Béguyer de 58–61
chemistry 80–98; 189
 affinity 105
 alchemy 35–36; **83–98**; 100; 102; 184–85; 214; 267; 281–84
 as art in mechanicist world view 281–84
 atoms and molecules 322–37
 caloric 99; 105; 109–13; 222–23; 326
 conservation of mass 221–24
 element/compound 57; 81; 87–90; 94–98; 111; 214; 242; 281–82 *see also* Aristotle, scientific doctrine, elements *and* idem, compound and mixture
 experimental 214–17
 formulae 254–62
 impurities 216
 instruments 217
 isotopes 67
 Karlsruhe congress 333
 medieval geometrization of 129
 nomenclature 331–34
 organic compounds, synthesis of 295–97
 origin of rocks 293
 oxidation 102–03; 105; 109–10; 114 *see also* phlogiston
 pneumatic 105–8; 110; 217; 222
 qualitative theories **99–115**
 reduction 102; 114
 resonance theory 257–62
 revolution 110
 scientific status of 216
 series of compounds 53–57
 series of elements 57–68; 95; 337
 spirits 216
 types 254–57
Chesne, Joseph de 216
Chwolson 72–73
Cicero 17–21; 155; 168–71; 184; 268–71; 275; 277
Clausius, Rudolph 242; 333; 337
Cleopatra (queen of Egypt) 15; 85; 183; 319; 335–40
clock, as a metaphor 21
Colding 338
Collingwood, R.G. 338
Columella 272
combustibility, principle of *see* phlogiston
compass needle *see* magnetism
conservation laws 219–28
 energy 225–28; 338
 mass 220–24
 momentum 235
conventionalism *see* hypotheses, realist and/or conventional
Cooke, J.P. 56
Copernicus & Copernicanism 11; 39–40; 79; **147–81**; 213; 234; 347
 absurdity of 156–58

going against testimony of senses 212
ancient sources for 171–73
extent of Copernicus' Aristotelianism 158–60
controversy over 175–77; 273 *see also under* Galilei, Galileo *and* Kepler
acceptance prior to direct proof 115
confirmation by direct proof 180
greatness in contemporary eyes 174; 180
triumph 179
explanatory range 232
impiety, expected charge of 149–55
life and character 148
publication delay 148; 172
punning physician 161
as student 118
motions of the Earth 149
annual revolution 164–66
daily rotation 156–62; 208–10; 232
third motion 166; 174
novelty 170–75
place in history 180
realism 164
universe
man's place in 153–55
models of 300
planetary order in 165
size of 155; 177
structure of 162–66
sun central in 167–70
'symmetria' of 163–65; 168; 277
technical issues and devices 274
books II-VI of *De Revolutionibus* 173; 180
epicycles 231
equant, removal of 163; 173–5; 177
falling bodies 141
precession 166; 174
Coresio, Giorgio 206
corpuscular theories 9; 99–103; 173; 214; 217; 240; 282 *see also* atomism *and* mechanistic world view
Cotes, Roger 235
Coulomb, Charles Augustin 73; 190
Couper 258–61
Cromwell, Oliver 338
crystallography 46–53; 187; 189; 249–53; 257; 289; 372
Cusanus, Nicolaus 183; 195–98

D

Dalton, John 12
atom and element 322
atomic theory 322–29
model of 300
compounds 132
experiments by Gay-Lussac 189; 327–29
inspired by Newton 320–29; 334–37
misunderstanding Newton 323
multiple proportions, law of 68; 114; 189; 324–29
Daniell, J.F. 252; 296
Dartigues 292
Darwin, Charles 22; 62; 261; 309; 335; 337; 344
Daubrée, G.A. 294–95; 306
Davy, Humphry 104
Delafosse, G. 252; 257
Democritus 17

Descartes & Cartesianism 8; 142; 178; 235–41; 242; 320
mode of science
corpuscularianism 100
deductive method 349
experiment 192
'roman de la nature' 349
'scientific story' 360
religion/faith 239–41
divine deception 226
semi-deism 360
views
on Beeckman 204
on conservation of matter 224
on Copernicanism 348
on living beings 83
on nature and art 285
on quantity of matter 221
on refraction of light 236–41
on void 346
Desmarest 295
Digby, Everard 274
Digges, Thomas 155–56; 372
Dirac, Paul 76; 261
Döbereiner, J.W. 58
Dodoens, Rembert 119; 155
Donne, John 154; 358
Duhem, Pierre 263–64
Dumas, J.B. 55–56; 255; 330; 333; 335
Dürer, Albrecht 164; 276–77; 280

E

Earth, tunnel through the *see* tunnel through the Earth
Ecphantus 171
Einstein, Albert 77; 249
electricity 235; 286; 302; 309 *see also next entry*
electromagnetism 310–17
energy, conservation of 225–28, 338
Engels, Friedrich 261
Ennius 18
Epicurus 11; 17
Erasistratos 221
Erasmus, Desiderius 124; 138
ether 178; 246–49; 264; 309–17; 346–47
Euclid 138; 271–72
Eudoxus 34; 230
Euler, Leonhard 242; 322
evolution *see* Darwin
and religion 22
experiment **183–228**
advocated by Francis Bacon 284
carried out (or not) by
Alexandrian engineers 184; 194
Beeckman 204
Béguin 215
Billich 216
Boyle 218
Buridan 194
Castro 199
de Chesne 216
Copernicus 160
Cusanus 197
Dalton 189
Darwin 309
Daubrée 294
Descartes 239
Fizeau & Foucault 249
Foucault 180
Galilei, Galileo 205–13
Galilei, Vincenzo 195
Gay-Lussac 327
Gay-Lussac & Thenard 289

Geoffroy St Hilaire 309
Gilbert 203
Hall 291–95
Haüy 189
Hooke 218
Joule 227
Kirwan 291
Langenstein 130
Lavoisier 103; 110; 222–24
Leblanc 289
Lémery 305
Newton 145
Norman 202
Orta 200
Pascal 346
Petrus Peregrinus 196
Ptolemy 188
Richter 54
Sala 214
Scaliger 202
Stahl 101–02; 111
Stevin 191
Stevin & de Groot 206
Watt (Gregory) 292
causes of error 199
classification of 186–87
defined 183
domains
ancient science 280
alchemy 86; 214
mechanical geology 305–8
medieval kinematics 135
mineralogy 288
music 31; 33
and magic 183
and organistic world view 301
outcomes checked 199
pseudo 194; 196; 200–202; 210
subjects
animal monsters 309
artificial selection 309
atmospheric air 106
conservation of energy 225–28
conservation of mass 220–24
crystal theories 253
pendulum 180
phlogiston 101–2
rock formation 291–95
spectral lines 69
speed of light 249
void 346
testing Aristotle 201
thought 31; 118; 124; 134; 190–95; 204; 207; 211
model *see* models in science
tunnel through the Earth 123–45

F

Fabricius, Arnaldus 195
facts *see also next entry*
and authority 348
decisive in Copernican controversy 177
nature of 5–9
properties of metals 92–94
sacrosanct 371
spurious 7–9; 104; 174; 181 *see also* experiment, pseudo
submission to 361
facts, faith and fiction, interplay between **5–14**; 29; 31; 76; 187–88
in Balmer's hypothesis on spectral lines 68–69

in Bohr's early model of the electron 73
in Copernicanism 147; 175; 180–1
in Dalton's atomic theory 322
in Fresnel's optical wave theory 257
in Gell-Mann's omega-minus particle 29
in Gerhardt's organic chemistry 257
in Haüy's crystal theory 257
in Joule's gas theory 228
in Kepler's laws 42–43
in Lavoisier's 'murium oxide' 104
in Paracelsus' three principles 97
in Pascal's theory of the void 346–47
in Plato's conception of circular planetary motion 34
in Pythagorean musical harmony 31
faith
religious *see* religion
within science 9–12 *see also previous entry*
Faraday, Michael 59; 314
Faust 90; 98
Favre, Alphonse 306
Fedorov, E. von 53; 253
Fermat, Pierre de 368
'fictions' in science 12–13; 229; 317 *see also* theories *and* science, imagination in
Fizeau, Armand Hippolyte Louis 239; 249
Fleuriau de Bellevue 292–94
Forbes, James David 5–6; 307–10; 339–40
Foucault, Léon 180; 239; 249
Fouqué, F. 295
Fourcroy, Antoine François de 295
Fraunhofer, Joseph 249
Freiberg, Dietrich von 197
Fresnel, Augustin 12; 178; 229; 247–49; 256; 264; 310; 321; 338
Froidmont, Libert 152
Fuchs, C.W.C. 290

G

Gaffurio, Franchino 30; 195
Galen 21; 282
Galilei, Galileo 12; 40; 79; 218; 295
1616 decree 175
defence of Copernicanism 153–54; 158; 177; 206; 212; 304
experiment 187; 192; 205–13; 217
influencing Mersenne 143
prose style 372
reason and experience, on balance 208
telescopic observations 176; 205
theories and discoveries
falling bodies 144; 204; 206; 337
infinity of the universe 155
motion of the Earth 208–10
musical harmony 33
natural motion 160
projectile motion 213
tides 210; 304
tunnel through the Earth 140
vibrational motion 144
void 211
trial 239; 348
views
on Copernicanism in Holland 152
on 'ideal reality' of physics 279
on method 206; 212–13
on nature and art in scale models 298
Galilei, Vincenzo 33; 195

gas **321–37** *see also* chemistry, pneumatic *and* Boyle, law
kinetic theory 242; 322; 333; 337
law 217–19
Gassendi, Pierre 142; 152; 205; 346; 348
Gaudin, M.A. 330; 333–34
Gay-Lussac, Joseph Louis 189–90; 289; 327; 329; 335
Geber 84; 87; 91–3
Gellibrand, Henry 177; 179
Gell-Mann, Murray 45; 75
Geoffroy St Hilaire, Etienne 308–09
geography, discoveries 148–49; 175
geology 200; 290–95; 304–8
continental drift 179; 364
uniformity of nature 12
Gerhardt, Charles 56; 229; 254–61; 296; 330; 333; 335
Gerland, E. 211
Gheraert, Brother 122
Giese, Tiedemann 148–49
Gifford, Lord 1–3
Gilbert, William 117; 141–42; 187; 201–05; 209; 302–04
Giraldus Cambrensis 7
Goethe, Johann Wolfgang von 90; 314
Gogh, Vincent van 335
Goodere, Henry 154
Gray, Asa 22
Groot, Hugo de 353
Groot, Johan de 206
Grosseteste, Robert 197
Guericke, Otto von 186; 217

H

Hagenbach, E. 68
Hales, Stephen 110; 217
Hall, James 291–94; 305–06
hands, thinking with the *see* experiment
Harkins, W.D. 114
harmony 233; 248 *see also* nature, harmony in
Cicero's view on 18
idea of 12
musical 29–33; 36–43; 51–52; 59; 68–74; 195; 198; 274
Harriot, Thomas 205
Hartmann, Georg 202
Harvey, William 170
Hausmann, J.B.L. 293
Haüy, René Just 12; 46–51; 53; 189; 249–53; 257; 289; 295; 321; 372
'heart' *see* Pascal, general views, heart
heat, mechanical equivalent of 225–28
heaven *see* theology, heaven, location of
Hegel, G.W.F. 75
Heisenberg, Werner 2; 76
hell *see* theology, hell, location of
Helm, G. 113
Helmholtz, Hermann 338
Helmont, Johan Baptista van 221
Henckel, J.F. 287–88
Henry, W. 189
Herakleides of Pontus 171
Hermes Trismegistus 84; 96; 167; 172–73
Hero 185; 194; 280
Herodotus 275
Herschel, William 44
Hertz, Heinrich 249
Hicetas 171
Hiero (king of Syracuse) 198
Higgins, William 322
Hipparchus 34; 172; 230–31

history
  causality in 338
  grand patterns in 341
  law-like or contingent 319; **337–41**
history of science 14; 16; 52; 338–41
  and history in general 335
  inner tension of 372
  our ancestors and ourselves 341; 372
  progress in 336
  as re-enactment of the past 6; 92; 147; 181; 341
  task of 371
  by tracing back 137
Hoff, J.H. van't 300, 337
Holanda, Francisco 278
Hollandish chemists 109
Holmes, Arthur 180; 259
Homer 275
Hooke, Robert 186; 217; 218; 337
Hooykaas, R. ix–xiii; 1–3; 75; 150; 161; 340–41; 361
Horace 276
Hückel, Walter 259
Hugh of St Victor 119
humanism, opposed to scholasticism 138; 269
Humboldt, Alexander von 200
Hutton, James 17; 291–94; 306
Huxley, Thomas Henry 371
Huygens, Christiaan
  atomism 178
  crystal
    measurements 49
    structure 252
  method 219
  in Newton's *Principia* 179
  place in historical development 180
  views
    on Copernicanism 179
    on Descartes 9; 235
    on Pascal 368
hydrogen, structure of 68–74
hypotheses, realist and/or conventional **229–65**

I

Ibn-Ostrul-Oshul 185
idealization
  astronomy
    Copernicus 173–74
    Plato 33
  crystallography
    Haüy 53; 372
    Romé de Lisle 76
  ether
    Boltzmann 315
    Fresnel 178
    Thomson 312
  experiment generally 187; 190
  gas law
    Boyle 218
  and mathematical theory 264 *see also* mathematics in nature
  mechanics 118
    Galileo 140; 209;213
    medieval (Swineshead, Oresme, Bradwardine) 126–27; 134–36
  musical harmony
    Plato 33
  resonance theory
    Pauling 258; 260
imagination *see* science, imagination in
impetus 14; 130–35; 138–40; 202–03; 209

Index Librorum Prohibitorum 23
inertia 132; 204; 209; 224
infinity 155; 351–2; 361
Ingold, C.K. 258; 260–61
Inquisition 348–49

J

Jabir ibn-Hayyan 27; 35–36; 87
Jansenius (bishop) 348
Jeans, James 76; 264; 317; 355
Jesuits 175; 201; 348; 367; 369
Jesus Christ 2; 25; 153; 356–61; 368
Joule, James Prescott 227–28; 337
Jungius, Joachim 79; 99; 282

K

Kedrov, B.M. 261
Kékulé, Friedrich August 257–61; 331–32
Kepler 35; 115; 248
  empiricism 39
  in history 180
    among the greatest scientists 371
  intellectual resources
    Christianity 39
      trinity 22; 233
    Copernicanism 177
    Platonism 39; 42; 70
    Pythagoreanism 57
  laws 39–41; 73–4; 233; 243
    fulfilling Ramus' task 273
    role of, in Copernicanism 177
  'psalm' in *Harmonice Mundi* 41; 372
  universe, structure of 36-43
    finite 155
    geometric 37; 53; 355
    harmonic 36–43; 52
    magnetism in 205
    model of 300
    purpose of 155
    undiscovered planets assumed 44
  views
    on discovery 16
    on hypotheses 232–35
    on natural theology 17; 22; 41–43
    on Ramus 271
    on scientist's grasp of nature 317
al-Khwazimi 125
Kingsley, Charles 22
Kirwan, Richard 291
Krönig, August Karl 242; 333; 337
Ktesibios 194
Kues *see* Cusanus

L

La Mettrie, Julien Offray de 83
Lactantius 149
Lamarck, Jean Baptiste de Monet 261; 296
Langenstein, Heinrich von 129; 197; 283–84
Lansbergen, Philips van 152; 178
Laplace, Pierre-Simon 109; 246
Larmor, Joseph 317
Latini, Brunetto 43; 123
latitudes of forms 128
Laue, Max von 253
Launay, L. de 306
Laurent 333; 335
Lavoisier, Antoine Laurent 12; 103–15
  analogy between elements and organic radicals 56

compared with Stahl 106–7; 111–13
conservation of matter 220–24
creating his own myth 110
critic of phlogiston 248
definition of element 57; 111; 323
experiment 222–24
instruments 217
method and application compared 104
nomenclature 335
phlogiston and oxygen 103
prose style 372
revolution in chemistry 331
using analogical reasoning 299
Lawrence of Lindores 117–18; 129; 131
Leblanc, Nicolas 289
Leclerc, Jean 354
Leeuwen, Albert van 157; 174
Leeuwenhoek, Antoni van 209
Leibniz, Gottfried Wilhelm 265; 288; 337
Lémery, Nicolas 305
Lenin, V.I. 261
Leonardo da Vinci 164; 205; 280
Leonhard, K.C. von 293
Lewis, G.N. 114
Libavius, Andreas 217
Liebig, Justus von 296
light *see* optics
Lodge, Oliver 313; 316–17
longitude determination 199
Lucianus 221
Lucretius 17
Luke (evangelist) 359
Lyell, Charles 22; 335
Lysenko, Trofim D. 261
Lysis 172

M

Macaulay, Thomas Babington 6; 339–40
Macrobius 168–69; 195; 198
macrocosm/microcosm 163; 167
Maestlin, Michael 41
magic 83–84; 98; 130; 183–84; 191; 196; 283–86; 297 *see also* supernatural
magnetism 11; 141; 196–97; 199–200; 202–4; 209; 234; 244; 301–4; 347 *see also* electromagnetism
Malebranche, Nicole 20; 105
Malus, Etienne 246
manual work 83; 85 *see also* experiment
Maricourt, Pierre de *see* Petrus Peregrinus
Mark Antony 319; 337
Markovnikov 262
Marlowe, Christopher 98
Marsilius van Inghen 194
Martianus Capella 172
mathematics 269; 275; 337; 368
fed by practice 271
in nature 27; 32; 76; 230–35 *see also* idealization
astronomical hypotheses 230–35
Balmer 70–71
Calculators 137–38
chemistry 216
Galileo 140; 206; 209; 213
Haüy 249–53
Jabir ibn-Hayyan 35
Kepler 38–39; 42
link between model and theory 315
medieval view 135
modern view 263
in need of picture 316
Newton 242; 245

Oresme's latitudes 128
Pascal 355
regular solids 37
Swineshead's rod 127
wave theory of light 246
reason in 350
Matthews, P.T. 75
Maurolyco, Federico 139
Maxwell, James Clerk 73; 179; 249; 263–64; 310–17; 371
Mayer, Julius Robert 113; 225–27; 338
measurement 187; 197; 204; 211; 218; 227
absolute values 198; 208
ancient 221
Dalton 329
medieval 135–37
Mersenne 144
problematic results 48
pseudo-exactness 224
mechanics (ancient) 184
mechanistic world view 20; 83; 87; 126; 135; 142–45; 187; 190; 205; 249; 281–83; 286; 301 *see also* organistic world view
Melanchthon, Philippus 148
Mendeléev, Dimitri 45; 59–67; 95; 220; 262; 337
Mersenne, Marin 142–44; 205; 208
Meyer, Lothar 59–65; 256–57; 333; 337
Michelangelo Buonarotti 277–278; 280
Michel-Lévy 295
microscope 186; 209
Miller 253
Milton, John 274
mineralogy 286–90; 293
mining 184
miracle *see* supernatural
Miriam (Moses' sister) 84
missing links *see* vacant places
Mitscherlich, F. 290
models in science 297–317
cosmographical 300–304
elastic fluids 322
ether 309–17
general problems 297–99
geological 304–8
obsolescence of 317
Moitrel d'Elément 217
Monge, Gaspard 108; 305
Montaigne, Michel de 23
Moses 84
Mulerius, Nicholas 175; 177; 300

N

natural selection *see* Darwin
nature
analogy of 7; 10; 245
Bohr 72
its breakdown in saccharine 94
Gilbert 304
Haüy 47; 49
Lavoisier 104
Maxwell 310
in models generally 298–99
modern atomic theory 74
Newton 72
Pascal 346; 372
series of elements 59
Weiss 52
and art 81; 89; 184; **265–317**
animal species 308–9
alchemy/chemistry 281–84
daily life 266

experimental rock formation 291–95
mineral synthesis 286
models *see* models in science
possibilities 130; 267; 286
synthesis of organic compounds 295–97
views on
antiquity 266–68; 280
Aristotle 267–68
Descartes 285
mechanical philosophy 283–86
Renaissance 268–80
book of 23–25; 42; 206; 234; 272; 367
circular course 184
constancy 220; 224; 228
contingency 131, 355
design 19
God in *see* theology, natural
harmony in **27–77**
interfered with by experiment 187
mathematization of *see* mathematics, in nature
not identical with science of nature 6; 264; 354; 371
order 11–12; 18–21
and power 284–85; 296
scientist in dialogue with 9; 45; 76; 181
simplicity 11–12; 67; 75; 115; 157; 166; 189; 229; 247–51
theology *see* theology, natural
Naturphilosophie 51; 71; 225
new star 148; 175
Newlands, J.A.R. 27; 58–62
Newton & Newtonianism 5; 8; 12; 71; 115; 127; 178; 320
in history 180
among the greatest scientists 371
status 320; 335
science and method 219
analogy of nature 12; 72
critic of Descartes 9; 235; 349
experimental method 219; 228
'hypotheses non fingo' 243; 321
geometry and mechanics 275
Ramism in *Principia* 274–75
task of science 355
theses and conjectures distinguished 320
theology
God's ultimate incomprehensibility 226
natural theology 17; 21; 355
theories and discoveries
attraction inside the Earth 144
calculus 337
elastic fluids 242–45; 321–23
gravitation 160; 337
incorporation of Copernicanism 179
light 102; 248
spectrum and musical scale 68
Nilson 61; 64
nominalism *see* scholasticism, via moderna
Norman, Robert 202
Norton, Thomas 27; 89–93; 214
numerology 35–36; 51
Nunes, Pedro 199

O

Ockham, William of 12; 130; 192
octaves, law of 58 *see also* harmony, musical

Octavian (emperor Augustus) 319; 337
omega-minus particle 29; 75
optics 68–74; 188; 197; 236–41; 243
  wave theory 68; 246–49; 256; 263; 310; 321; 338 *see also* electromagnetism
Oresme, Nicole 128–36; 140; 156–58; 168; 285
organistic world view 81–84; 87; 125–26; 129; 141; 158; 187; 234; 279–83; 301
  see also mechanistic world view
  contrasted with mechanistic world view 83
  transition towards mechanistic world view 87; 135; 281-83
Orta, Garcia de 183; 200–01
oscillation *see* vibrational motion
Osiander, Andreas 148; 231–34; 244
Ouroboros (snake) 84; 184

P

Paccard 305
Pacioli, Luca 164
Paley, William 22
Palissy, Bernard 273
Papin, Denis 217
Pappus 184; 280
Paracelsus 22; 96–100; 196; 233
parallax *see* Copernicanism, universe, size
Partington. J.R. 57; 82
Pascal 23; **343–70**
  effect upon non-Christian readers 344
  general views
    authority 348; 362
    beauty 351; 356
    blind windows 29; 349; 366
    Cleopatra's nose 319; 338–39
    heart 13; 350–51; 354; 356; 370
    infinity 351–52; 361
    'learned ignorance' 371
    man 343–44; 352; 356; 367
    realism 344; 347; 350
    reason 344; 349; 355
    scepticism and dogmatism 364
    submission to facts 361
    system, none 344; 349; 351; 366
  historian of science 338; 362–64
    'robes de pédant' 15; 370
  life and character
    1654, before and after 361; 367
    character 369
    practical affairs 368
    style 369–70; 372
  science and method *see also above*: general views
    analogical reasoning 346; 372
    compared with Boyle 347
    conditional faith in science 219
    Copernicanism 179; 347; 364
    critic of Descartes 9; 235; 349; 360
    experimentally inclined 219
    facts of nature 348
    limits of science 355
    loss of immediate reality in science 310
    openness of physics 349
    sciences of two kinds 361
  theology *see also above*: general views
    agnosticism 356
    apologetics 353; 366–67
    biblical exegesis 357
    Christ hidden 358
    dual test for religion 355

foundations of Christianity 356–60
God hidden 354
God revealed in nature 353
God's ultimate incomprehensibility 226
Jesus and Archimedes compared 368
natural theology 351–55
style of gospels 358
theories and discoveries
mathematics 368
void 346–47; 350; 353; 360; 363
Patrizzi, Francesco 201
Paul (apostle) 23
Paul III (pope) 148; 162
Paul of Middelburg (bishop) 148
Pauli, Wolfgang 36; 68
Pauling, Linus 258; 260–62
pendulum *see* vibrational motion
Périer, Marguérite 368
periodic system *see* chemistry, series of elements
Petreius 148
Petrus Apianus *see* Apianus, Petrus
Petrus Peregrinus 183; 186; 196; 301–03
Philo 194; 280
Philolaos 171–73
Philoponos, Johannes 131; 206
philosopher's stone *see* chemistry, alchemy
phlogiston 7; 99–115; 181; 221–22; 248
see also chemistry, oxidation
Piazzi 44
Piccolomini, Alessandro 125; 139
Pico della Mirandola, Francesco 138
Piltdown case 8
Planck, Max 72–74; 77; 179
Plato & Platonism 3; 20; 28; 79–80; 170; 338
alleged visit to Philolaos 173
cave myth 317
Christian Platonism 20
laughing with friends 370
organistic world view 280
views
on art 276
on astronomy 33–35; 39; 230
on atomism 18
on geometrical form 53
on manual work 184
on music 31–33
on similar things 160
on substance 268
Plutarch 34; 124; 171; 184
Poincaré, Henri 264
Poisson, Siméon Denis 246
Pope, Alexander 320
Power, Henry 218
Prechtl 252
Priestley, Joseph 106; 110; 217
Proust, Louis Joseph 114
Ptolemy 11; 34; 39; 173; 174; 177; 179–80; 347
astronomy
cosmographical model 300
daily rotation of the Earth 159; 161
eccentrics and epicycles 163; 230
equant 163; 171; 173; 234
explanatory range of 232
multitude of circles 174
physical hypotheses 233
simplicity 12
geography 175
optics 197–98
experiments 188; 198

Pythagoras & Pythagoreanism 28; 30–31; 33; 35–36; 75; 79–80; 173; 195; 198; 230
central fire 167
counter-earth 28; 29
experimental discovery of musical harmony 31
middle 168
musical theory 29
position of sun 170
source for Copernicanism 171
tetraktys 28

Q

quantum 72; 179
Quercetanus *see* Chesne, Joseph de

R

Ramus & Ramism 15; 196; 265; 269—79
Raven C.E. (canon) 7
Ray, John 367
realism 164; 344; 347 *see also* hypotheses, realist and/or conventional
religion 70; 192; 351–61; 365–67 *see also* theology
Rembrandt van Rhijn 335
Renaissance, new and old in 170
Rheticus, Georg Joachim 148–53; 231; 273
Richter, J.B. 27; 53–57
Roberts, J.D. 259
Robinson, R. 258
Romé de Lisle, J.B.L. 46; 53; 76; 187; 253
Ronsard, Pierre 271
Roscoe, H.E. 62
Rozier 305
Rutherford, Ernest 66; 72; 286
Rydberg 67; 72–73

S

Sá de Miranda 185
Sabundus (Raymond de Sébonde) 17; 23–24; 155; 354
Sacrobosco (John of Holywood) 154
Sala, Angelo 99; 214–17; 221; 281–82
saving the phenomena 34; 114; 131; 171; 230
Scaliger, Julius Caesar 138; 202–03
Scheerer, Th. 294
Schelling, Friedrich Wilhelm 70
Schiel, J. 55
scholasticism 119; 271; 323
via antiqua 117–18; 124
via moderna 117; 130; 283
Schönberg, Nicolas (cardinal) 149; 171
Schönflies, A. 53; 253
Schrödinger, Erwin 36; 76; 261
Schroeder van der Kolk 113
science
authorities in 8; 320; 335; 348; 362
blind alleys in 36; 44; 52; 94
facts, faith and fiction in *see* facts, faith and fiction, interplay between
and faith *see* faith
imagination in 104; 178; 202; 210; 229; 240; 310; 350; 355; 361; 371 *see also below*: intuition in

'heart' in 370
as human construct 371
and human values 15–16; 24; 371–72
idealization in *see* idealization
instruments in 228
intuition in 219
loss of concreteness 310
models in *see* models in science
modern and premodern compared 36; *see also below*: recurrent patterns in
alchemy 98
conservation of mass 224
dynamics 143
gravitation 160
impetus 132
machine mechanisms 286
mathematics in nature 32; 127–29; 136–37; 145
medieval kinematics 118; 127–29; 132; 136–37
models 317
thought experiments 118
multiple discovery 337
natural and human 299
new conceptions in 178–79
openness of 349
as power over nature 36
and reality of nature 310; 317
recurrent patterns in 3; 13; 29; 36; 80; 98; 114; 219; 224; 245
and truth 14; 92; 94; 179–81; 218; 226; 229–64; 317; 335; 361; 364
unending excitement of 76
scientism 371
scientist
in dialogue with nature 9; 45; 76; 181
prose style of 372
upon shoulders of giants 338; 372
undecided about status of hypotheses 245
Scorel, Jan van 280
Sébonde *see* Sabundus
Seeber, L. 252; 257
Seneca 184; 268; 269
Sennert, Daniel 99; 214; 217
Serres, Olivier de 265; 272
Shakespeare, William 372
Snellius, Willebrord 237; 271; 300
Sohncke, L. 53; 253
Sommerfeld, Arnold 69; 73
Sorby, H.C. 295
Soviet science 261–62
spectral lines 68–74
spontaneous generation 7; 84
Sprat, Thomas 228
Stahl, Georg Ernest 100–03; 105–13; 287
Stalin, J.V. 261
Stevin, Simon 191; 193; 206
Stoicism 2; 19–20; 24; 79; 269; 277; 280
Strabo 339
Suisseth *see* Swineshead
sulphur-mercury theory *see* chemistry, alchemy
supernatural 119; 121–22; 130 *see also* magic
sweetness, defying analogy of nature 94
Swineshead, Richard 117; 126; 138
symmetry 163–65; 168; 276–77
synthesis *see* nature, and art

T

Tartaglia, Niccolò 139–40
Tauladanus, Robertus 93

Taylor, A.E. 3
Telesio, Bernardino 201
Tempier, Etienne (bishop) 131; 136
Temple, William 274
Termier, P. 314
Thenard 289; 295
theology
  apologetics 353; 366–67
  Copernicanism 149–55
    Galileo's trial 239; 348
  exegesis 357
    accommodation principle 150–53
  God as Prime Mover 120
  heaven, location of 119
  hell, location of 118–24, 155
  natural **1–3**; 12; **17–25**; 34; 41–43; 226; 367
    and revelation 3; 22; 23–25; 71; 351–61
  trinity 22; 37; 96; 233
  voluntarist 19; 24; 39; 130–31; 285
theories
  nature of *see* hypotheses, realist and/or conventional
  their structure a lasting feature 114
  whether or not scientific 7; 14; 92; 114–15; 180–81
thinking with the hands *see* experiment
Thomas Aquinas 117–19; 122; 130–131
Thomson, William 312–13; 317
Titius (Johann Daniel Tietz) 43–5
Townley, Richard 217–18; 243
transmutation 80–98
Traumüller, F. 211
Trevisanus *see* Bernard de Trévisan
triads, law of 58
truth *see* science, and truth
tunnel through the Earth **118–45**; 190
Turner, William 7
Tycho Brahe 39–40; 71; 147; 152; 175–81; 232; 273; 347–48

U

University of Paris 117; 124
University of St Andrews 1–3; 5; 8; 117
Urey 66
Ursus (Reymers Bär) 232

V

vacant places
  in periodic table 45; 60; 65
  in range of planets 44–45
  in series of chemical substances 54–57
vacuum *see* void
Varro 18
Vavilov 261
vibrational motion 133–35; 139–45; 178; 190; 210; 246–49; 263; 310
Vincent of Beauvais 125
Virgil 272; 278
Vitruvius 163; 168–69; 185; 276
Vives, Luís 138; 196
Voet, Gisbert 152
Vogelsang, W. 295
void
  Alexandrian engineers (siphon) 184
  Aristotle 80
  Boyle (air pump) 186; 217
  Buridan 133; 194
  Descartes 221; 346–47
  Galileo 207; 211
  Gassendi 346–47

de Groot (Grotius) 353
Guericke (air pump) 186; 217
Lavoisier (phlogiston) 106; 108
Marsilius van Inghen 194
Pascal 346–47; 350; 353; 360; 363
Patrizzi 201
Telesio 201
Tempier 136
volcano 288–90; 293–94; 305
Voltaire 320

W

Waals, Johannes Diederik van der 219
Waard, Cornelis de 204
Wallace, Alfred 337
Walter of Burley 117; 125
Waterston, J.J. 333; 337
Watt, Gregory 292
Wegener, Alfred 179
weight *see* phlogiston *or* conservation laws, mass
Weiss, C.S. 51–53; 253
Wheland, G.W. 258; 260
Wilkins, John 152–53; 367
Winkler 61; 64
Witelo 197
witness accounts 7; 199
Wöhler, Friedrich 290; 295; 331
Wohlwill, Emil 210
Wollaston, W.H. 49–50; 189; 252
Würtz, Adolphe 110

Y

Young, Thomas 229; 246–47; 321; 338

Z

Zachary (pope) 348
Zarlino, Gioseffo 33
Zeno 125
Zeuxis 276–77
Zirkel, F. 295
Zittel, K.A. von 113
zoology 195; 259; 308–9
as evidence of order 21

Boston Studies in the Philosophy of Science

37. H. von Helmholtz: *Epistemological Writings.* The Paul Hertz / Moritz Schlick Centenary Edition of 1921. Translated from German by M.F. Lowe. Edited with an Introduction and Bibliography by R.S. Cohen and Y. Elkana. [Synthese Library 79] 1977
ISBN 90-277-0290-X; Pb 90-277-0582-8
38. R.M. Martin: *Pragmatics, Truth and Language.* 1979
ISBN 90-277-0992-0; Pb 90-277-0993-9
39. R.S. Cohen, P.K. Feyerabend and M.W. Wartofsky (eds.): *Essays in Memory of Imre Lakatos.* [Synthese Library 99] 1976 ISBN 90-277-0654-9; Pb 90-277-0655-7
40. Not published.
41. Not published.
42. H.R. Maturana and F.J. Varela: *Autopoiesis and Cognition.* The Realization of the Living. With a Preface to "Autopoiesis' by S. Beer. 1980 ISBN 90-277-1015-5; Pb 90-277-1016-3
43. A. Kasher (ed.): *Language in Focus: Foundations, Methods and Systems.* Essays in Memory of Yehoshua Bar-Hillel. [Synthese Library 89] 1976
ISBN 90-277-0644-1; Pb 90-277-0645-X
44. T.D. Thao: *Investigations into the Origin of Language and Consciousness.* 1984
ISBN 90-277-0827-4
45. F.G.-I. Nagasaka (ed.): *Japanese Studies in the Philosophy of Science.* 1997
ISBN 0-7923-4781-1
46. P.L. Kapitza: *Experiment, Theory, Practice.* Articles and Addresses. Edited by R.S. Cohen. 1980 ISBN 90-277-1061-9; Pb 90-277-1062-7
47. M.L. Dalla Chiara (ed.): *Italian Studies in the Philosophy of Science.* 1981
ISBN 90-277-0735-9; Pb 90-277-1073-2
48. M.W. Wartofsky: *Models.* Representation and the Scientific Understanding. [Synthese Library 129] 1979 ISBN 90-277-0736-7; Pb 90-277-0947-5
49. T.D. Thao: *Phenomenology and Dialectical Materialism.* Edited by R.S. Cohen. 1986
ISBN 90-277-0737-5
50. Y. Fried and J. Agassi: *Paranoia.* A Study in Diagnosis. [Synthese Library 102] 1976
ISBN 90-277-0704-9; Pb 90-277-0705-7
51. K.H. Wolff: *Surrender and Cath.* Experience and Inquiry Today. [Synthese Library 105] 1976
ISBN 90-277-0758-8; Pb 90-277-0765-0
52. K. Kosík: *Dialectics of the Concrete.* A Study on Problems of Man and World. 1976
ISBN 90-277-0761-8; Pb 90-277-0764-2
53. N. Goodman: *The Structure of Appearance.* [Synthese Library 107] 1977
ISBN 90-277-0773-1; Pb 90-277-0774-X
54. H.A. Simon: *Models of Discovery* and Other Topics in the Methods of Science. [Synthese Library 114] 1977 ISBN 90-277-0812-6; Pb 90-277-0858-4
55. M. Lazerowitz: *The Language of Philosophy.* Freud and Wittgenstein. [Synthese Library 117] 1977 ISBN 90-277-0826-6; Pb 90-277-0862-2
56. T. Nickles (ed.): *Scientific Discovery, Logic, and Rationality.* 1980
ISBN 90-277-1069-4; Pb 90-277-1070-8
57. J. Margolis: *Persons and Mind.* The Prospects of Nonreductive Materialism. [Synthese Library 121] 1978 ISBN 90-277-0854-1; Pb 90-277-0863-0
58. G. Radnitzky and G. Andersson (eds.): *Progress and Rationality in Science.* [Synthese Library 125] 1978 ISBN 90-277-0921-1; Pb 90-277-0922-X
59. G. Radnitzky and G. Andersson (eds.): *The Structure and Development of Science.* [Synthese Library 136] 1979 ISBN 90-277-0994-7; Pb 90-277-0995-5

Boston Studies in the Philosophy of Science

60. T. Nickles (ed.): *Scientific Discovery.* Case Studies. 1980
ISBN 90-277-1092-9; Pb 90-277-1093-7
61. M.A. Finocchiaro: *Galileo and the Art of Reasoning.* Rhetorical Foundation of Logic and Scientific Method. 1980 ISBN 90-277-1094-5; Pb 90-277-1095-3
62. W.A. Wallace: *Prelude to Galileo.* Essays on Medieval and 16th-Century Sources of Galileo's Thought. 1981 ISBN 90-277-1215-8; Pb 90-277-1216-6
63. F. Rapp: *Analytical Philosophy of Technology.* Translated from German. 1981
ISBN 90-277-1221-2; Pb 90-277-1222-0
64. R.S. Cohen and M.W. Wartofsky (eds.): *Hegel and the Sciences.* 1984 ISBN 90-277-0726-X
65. J. Agassi: *Science and Society.* Studies in the Sociology of Science. 1981
ISBN 90-277-1244-1; Pb 90-277-1245-X
66. L. Tondl: *Problems of Semantics.* A Contribution to the Analysis of the Language of Science. Translated from Czech. 1981 ISBN 90-277-0148-2; Pb 90-277-0316-7
67. J. Agassi and R.S. Cohen (eds.): *Scientific Philosophy Today.* Essays in Honor of Mario Bunge. 1982 ISBN 90-277-1262-X; Pb 90-277-1263-8
68. W. Krajewski (ed.): *Polish Essays in the Philosophy of the Natural Sciences.* Translated from Polish and edited by R.S. Cohen and C.R. Fawcett. 1982
ISBN 90-277-1286-7; Pb 90-277-1287-5
69. J.H. Fetzer: *Scientific Knowledge.* Causation, Explanation and Corroboration. 1981
ISBN 90-277-1335-9; Pb 90-277-1336-7
70. S. Grossberg: *Studies of Mind and Brain.* Neural Principles of Learning, Perception, Development, Cognition, and Motor Control. 1982 ISBN 90-277-1359-6; Pb 90-277-1360-X
71. R.S. Cohen and M.W. Wartofsky (eds.): *Epistemology, Methodology, and the Social Sciences.* 1983. ISBN 90-277-1454-1
72. K. Berka: *Measurement.* Its Concepts, Theories and Problems. Translated from Czech. 1983
ISBN 90-277-1416-9
73. G.L. Pandit: *The Structure and Growth of Scientific Knowledge.* A Study in the Methodology of Epistemic Appraisal. 1983 ISBN 90-277-1434-7
74. A.A. Zinov'ev: *Logical Physics.* Translated from Russian. Edited by R.S. Cohen. 1983
[*see also* Volume 9] ISBN 90-277-0734-0
75. G-G. Granger: *Formal Thought and the Sciences of Man.* Translated from French. With and Introduction by A. Rosenberg. 1983 ISBN 90-277-1524-6
76. R.S. Cohen and L. Laudan (eds.): *Physics, Philosophy and Psychoanalysis.* Essays in Honor of Adolf Gr~nbaum. 1983 ISBN 90-277-1533-5
77. G. Bhme, W. van den Daele, R. Hohlfeld, W. Krohn and W. SchÎfer: *Finalization in Science.* The Social Orientation of Scientific Progress. Translated from German. Edited by W. SchÎfer. 1983 ISBN 90-277-1549-1
78. D. Shapere: *Reason and the Search for Knowledge.* Investigations in the Philosophy of Science. 1984 ISBN 90-277-1551-3; Pb 90-277-1641-2
79. G. Andersson (ed.): *Rationality in Science and Politics.* Translated from German. 1984
ISBN 90-277-1575-0; Pb 90-277-1953-5
80. P.T. Durbin and F. Rapp (eds.): *Philosophy and Technology.* [*Also* Philosophy and Technology Series, Vol. 1] 1983 ISBN 90-277-1576-9
81. M. Marković: *Dialectical Theory of Meaning.* Translated from Serbo-Croat. 1984
ISBN 90-277-1596-3
82. R.S. Cohen and M.W. Wartofsky (eds.): *Physical Sciences and History of Physics.* 1984.
ISBN 90-277-1615-3

Boston Studies in the Philosophy of Science

83. É. Meyerson: *The Relativistic Deduction.* Epistemological Implications of the Theory of Relativity. Translated from French. With a Review by Albert Einstein and an Introduction by Milić Čapek. 1985 ISBN 90-277-1699-4
84. R.S. Cohen and M.W. Wartofsky (eds.): *Methodology, Metaphysics and the History of Science.* In Memory of Benjamin Nelson. 1984 ISBN 90-277-1711-7
85. G. Tams: *The Logic of Categories.* Translated from Hungarian. Edited by R.S. Cohen. 1986 ISBN 90-277-1742-7
86. S.L. de C. Fernandes: *Foundations of Objective Knowledge.* The Relations of Popper's Theory of Knowledge to That of Kant. 1985 ISBN 90-277-1809-1
87. R.S. Cohen and T. Schnelle (eds.): *Cognition and Fact.* Materials on Ludwik Fleck. 1986 ISBN 90-277-1902-0
88. G. Freudenthal: *Atom and Individual in the Age of Newton.* On the Genesis of the Mechanistic World View. Translated from German. 1986 ISBN 90-277-1905-5
89. A. Donagan, A.N. Perovich Jr and M.V. Wedin (eds.): *Human Nature and Natural Knowledge.* Essays presented to Marjorie Grene on the Occasion of Her 75th Birthday. 1986 ISBN 90-277-1974-8
90. C. Mitcham and A. Hunning (eds.): *Philosophy and Technology II.* Information Technology and Computers in Theory and Practice. [*Also* Philosophy and Technology Series, Vol. 2] 1986 ISBN 90-277-1975-6
91. M. Grene and D. Nails (eds.): *Spinoza and the Sciences.* 1986 ISBN 90-277-1976-4
92. S.P. Turner: *The Search for a Methodology of Social Science.* Durkheim, Weber, and the 19th-Century Problem of Cause, Probability, and Action. 1986. ISBN 90-277-2067-3
93. I.C. Jarvie: *Thinking about Society.* Theory and Practice. 1986 ISBN 90-277-2068-1
94. E. Ullmann-Margalit (ed.): *The Kaleidoscope of Science.* The Israel Colloquium: Studies in History, Philosophy, and Sociology of Science, Vol. 1. 1986 ISBN 90-277-2158-0; Pb 90-277-2159-9
95. E. Ullmann-Margalit (ed.): *The Prism of Science.* The Israel Colloquium: Studies in History, Philosophy, and Sociology of Science, Vol. 2. 1986 ISBN 90-277-2160-2; Pb 90-277-2161-0
96. G. Mrkus: *Language and Production.* A Critique of the Paradigms. Translated from French. 1986 ISBN 90-277-2169-6
97. F. Amrine, F.J. Zucker and H. Wheeler (eds.): *Goethe and the Sciences: A Reappraisal.* 1987 ISBN 90-277-2265-X; Pb 90-277-2400-8
98. J.C. Pitt and M. Pera (eds.): *Rational Changes in Science.* Essays on Scientific Reasoning. Translated from Italian. 1987 ISBN 90-277-2417-2
99. O. Costa de Beauregard: *Time, the Physical Magnitude.* 1987 ISBN 90-277-2444-X
100. A. Shimony and D. Nails (eds.): *Naturalistic Epistemology.* A Symposium of Two Decades. 1987 ISBN 90-277-2337-0
101. N. Rotenstreich: *Time and Meaning in History.* 1987 ISBN 90-277-2467-9
102. D.B. Zilberman: *The Birth of Meaning in Hindu Thought.* Edited by R.S. Cohen. 1988 ISBN 90-277-2497-0
103. T.F. Glick (ed.): *The Comparative Reception of Relativity.* 1987 ISBN 90-277-2498-9
104. Z. Harris, M. Gottfried, T. Ryckman, P. Mattick Jr, A. Daladier, T.N. Harris and S. Harris: *The Form of Information in Science.* Analysis of an Immunology Sublanguage. With a Preface by Hilary Putnam. 1989 ISBN 90-277-2516-0
105. F. Burwick (ed.): *Approaches to Organic Form.* Permutations in Science and Culture. 1987 ISBN 90-277-2541-1

Boston Studies in the Philosophy of Science

106. M. Almsi: *The Philosophy of Appearances.* Translated from Hungarian. 1989 ISBN 90-277-2150-5
107. S. Hook, W.L. O'Neill and R. O'Toole (eds.): *Philosophy, History and Social Action.* Essays in Honor of Lewis Feuer. With an Autobiographical Essay by L. Feuer. 1988 ISBN 90-277-2644-2
108. I. Hronszky, M. FehÇr and B. Dajka: *Scientific Knowledge Socialized.* Selected Proceedings of the 5th Joint International Conference on the History and Philosophy of Science organized by the IUHPS (VeszprÇm, Hungary, 1984). 1988 ISBN 90-277-2284-6
109. P. Tillers and E.D. Green (eds.): *Probability and Inference in the Law of Evidence.* The Uses and Limits of Bayesianism. 1988 ISBN 90-277-2689-2
110. E. Ullmann-Margalit (ed.): *Science in Reflection.* The Israel Colloquium: Studies in History, Philosophy, and Sociology of Science, Vol. 3. 1988 ISBN 90-277-2712-0; Pb 90-277-2713-9
111. K. Gavroglu, Y. Goudaroulis and P. Nicolacopoulos (eds.): *Imre Lakatos and Theories of Scientific Change.* 1989 ISBN 90-277-2766-X
112. B. Glassner and J.D. Moreno (eds.): *The Qualitative-Quantitative Distinction in the Social Sciences.* 1989 ISBN 90-277-2829-1
113. K. Arens: *Structures of Knowing.* Psychologies of the 19th Century. 1989 ISBN 0-7923-0009-2
114. A. Janik: *Style, Politics and the Future of Philosophy.* 1989 ISBN 0-7923-0056-4
115. F. Amrine (ed.): *Literature and Science as Modes of Expression.* With an Introduction by S. Weininger. 1989 ISBN 0-7923-0133-1
116. J.R. Brown and J. Mittelstrass (eds.): *An Intimate Relation.* Studies in the History and Philosophy of Science. Presented to Robert E. Butts on His 60th Birthday. 1989 ISBN 0-7923-0169-2
117. F. D'Agostino and I.C. Jarvie (eds.): *Freedom and Rationality.* Essays in Honor of John Watkins. 1989 ISBN 0-7923-0264-8
118. D. Zolo: *Reflexive Epistemology.* The Philosophical Legacy of Otto Neurath. 1989 ISBN 0-7923-0320-2
119. M. Kearn, B.S. Philips and R.S. Cohen (eds.): *Georg Simmel and Contemporary Sociology.* 1989 ISBN 0-7923-0407-1
120. T.H. Levere and W.R. Shea (eds.): *Nature, Experiment and the Science.* Essays on Galileo and the Nature of Science. In Honour of Stillman Drake. 1989 ISBN 0-7923-0420-9
121. P. Nicolacopoulos (ed.): *Greek Studies in the Philosophy and History of Science.* 1990 ISBN 0-7923-0717-8
122. R. Cooke and D. Costantini (eds.): *Statistics in Science.* The Foundations of Statistical Methods in Biology, Physics and Economics. 1990 ISBN 0-7923-0797-6
123. P. Duhem: *The Origins of Statics.* Translated from French by G.F. Leneaux, V.N. Vagliente and G.H. Wagner. With an Introduction by S.L. Jaki. 1991 ISBN 0-7923-0898-0
124. H. Kamerlingh Onnes: *Through Measurement to Knowledge.* The Selected Papers, 1853-1926. Edited and with an Introduction by K. Gavroglu and Y. Goudaroulis. 1991 ISBN 0-7923-0825-5
125. M. Čapek: *The New Aspects of Time: Its Continuity and Novelties.* Selected Papers in the Philosophy of Science. 1991 ISBN 0-7923-0911-1
126. S. Unguru (ed.): *Physics, Cosmology and Astronomy, 1300–1700.* Tension and Accommodation. 1991 ISBN 0-7923-1022-5

Boston Studies in the Philosophy of Science

127. Z. Bechler: *Newton's Physics on the Conceptual Structure of the Scientific Revolution.* 1991 ISBN 0-7923-1054-3
128. É. Meyerson: *Explanation in the Sciences.* Translated from French by M-A. Siple and D.A. Siple. 1991 ISBN 0-7923-1129-9
129. A.I. Tauber (ed.): *Organism and the Origins of Self.* 1991 ISBN 0-7923-1185-X
130. F.J. Varela and J-P. Dupuy (eds.): *Understanding Origins.* Contemporary Views on the Origin of Life, Mind and Society. 1992 ISBN 0-7923-1251-1
131. G.L. Pandit: *Methodological Variance.* Essays in Epistemological Ontology and the Methodology of Science. 1991 ISBN 0-7923-1263-5
132. G. MunÇvar (ed.): *Beyond Reason.* Essays on the Philosophy of Paul Feyerabend. 1991 ISBN 0-7923-1272-4
133. T.E. Uebel (ed.): *Rediscovering the Forgotten Vienna Circle.* Austrian Studies on Otto Neurath and the Vienna Circle. Partly translated from German. 1991 ISBN 0-7923-1276-7
134. W.R. Woodward and R.S. Cohen (eds.): *World Views and Scientific Discipline Formation.* Science Studies in the [former] German Democratic Republic. Partly translated from German by W.R. Woodward. 1991 ISBN 0-7923-1286-4
135. P. Zambelli: *The Speculum Astronomiae and Its Enigma.* Astrology, Theology and Science in Albertus Magnus and His Contemporaries. 1992 ISBN 0-7923-1380-1
136. P. Petitjean, C. Jami and A.M. Moulin (eds.): *Science and Empires.* Historical Studies about Scientific Development and European Expansion. ISBN 0-7923-1518-9
137. W.A. Wallace: *Galileo's Logic of Discovery and Proof.* The Background, Content, and Use of His Appropriated Treatises on Aristotle's *Posterior Analytics.* 1992 ISBN 0-7923-1577-4
138. W.A. Wallace: *Galileo's Logical Treatises.* A Translation, with Notes and Commentary, of His Appropriated Latin Questions on Aristotle's *Posterior Analytics.* 1992 ISBN 0-7923-1578-2
Set (137 + 138) ISBN 0-7923-1579-0
139. M.J. Nye, J.L. Richards and R.H. Stuewer (eds.): *The Invention of Physical Science.* Intersections of Mathematics, Theology and Natural Philosophy since the Seventeenth Century. Essays in Honor of Erwin N. Hiebert. 1992 ISBN 0-7923-1753-X
140. G. Corsi, M.L. dalla Chiara and G.C. Ghirardi (eds.): *Bridging the Gap: Philosophy, Mathematics and Physics.* Lectures on the Foundations of Science. 1992 ISBN 0-7923-1761-0
141. C.-H. Lin and D. Fu (eds.): *Philosophy and Conceptual History of Science in Taiwan.* 1992 ISBN 0-7923-1766-1
142. S. Sarkar (ed.): *The Founders of Evolutionary Genetics.* A Centenary Reappraisal. 1992 ISBN 0-7923-1777-7
143. J. Blackmore (ed.): *Ernst Mach – A Deeper Look.* Documents and New Perspectives. 1992 ISBN 0-7923-1853-6
144. P. Kroes and M. Bakker (eds.): *Technological Development and Science in the Industrial Age.* New Perspectives on the Science–Technology Relationship. 1992 ISBN 0-7923-1898-6
145. S. Amsterdamski: *Between History and Method.* Disputes about the Rationality of Science. 1992 ISBN 0-7923-1941-9
146. E. Ullmann-Margalit (ed.): *The Scientific Enterprise.* The Bar-Hillel Colloquium: Studies in History, Philosophy, and Sociology of Science, Volume 4. 1992 ISBN 0-7923-1992-3
147. L. Embree (ed.): *Metaarchaeology.* Reflections by Archaeologists and Philosophers. 1992 ISBN 0-7923-2023-9
148. S. French and H. Kamminga (eds.): *Correspondence, Invariance and Heuristics.* Essays in Honour of Heinz Post. 1993 ISBN 0-7923-2085-9
149. M. Bunzl: *The Context of Explanation.* 1993 ISBN 0-7923-2153-7

Boston Studies in the Philosophy of Science

150. I.B. Cohen (ed.): *The Natural Sciences and the Social Sciences.* Some Critical and Historical Perspectives. 1994 ISBN 0-7923-2223-1
151. K. Gavroglu, Y. Christianidis and E. Nicolaidis (eds.): *Trends in the Historiography of Science.* 1994 ISBN 0-7923-2255-X
152. S. Poggi and M. Bossi (eds.): *Romanticism in Science.* Science in Europe, 1790–1840. 1994 ISBN 0-7923-2336-X
153. J. Faye and H.J. Folse (eds.): *Niels Bohr and Contemporary Philosophy.* 1994 ISBN 0-7923-2378-5
154. C.C. Gould and R.S. Cohen (eds.): *Artifacts, Representations, and Social Practice.* Essays for Marx W. Wartofsky. 1994 ISBN 0-7923-2481-1
155. R.E. Butts: *Historical Pragmatics.* Philosophical Essays. 1993 ISBN 0-7923-2498-6
156. R. Rashed: *The Development of Arabic Mathematics: Between Arithmetic and Algebra.* Translated from French by A.F.W. Armstrong. 1994 ISBN 0-7923-2565-6
157. I. Szumilewicz-Lachman (ed.): *Zygmunt Zawirski: His Life and Work.* With Selected Writings on Time, Logic and the Methodology of Science. Translations by Feliks Lachman. Ed. by R.S. Cohen, with the assistance of B. Bergo. 1994 ISBN 0-7923-2566-4
158. S.N. Haq: *Names, Natures and Things.* The Alchemist J̄abir ibn Ḥayyān and His *Kitāb al-Aḥjār* (Book of Stones). 1994 ISBN 0-7923-2587-7
159. P. Plaass: *Kant's Theory of Natural Science.* Translation, Analytic Introduction and Commentary by Alfred E. and Maria G. Miller. 1994 ISBN 0-7923-2750-0
160. J. Misiek (ed.): *The Problem of Rationality in Science and its Philosophy.* On Popper vs. Polanyi. The Polish Conferences 1988–89. 1995 ISBN 0-7923-2925-2
161. I.C. Jarvie and N. Laor (eds.): *Critical Rationalism, Metaphysics and Science.* Essays for Joseph Agassi, Volume I. 1995 ISBN 0-7923-2960-0
162. I.C. Jarvie and N. Laor (eds.): *Critical Rationalism, the Social Sciences and the Humanities.* Essays for Joseph Agassi, Volume II. 1995 ISBN 0-7923-2961-9
Set (161–162) ISBN 0-7923-2962-7
163. K. Gavroglu, J. Stachel and M.W. Wartofsky (eds.): *Physics, Philosophy, and the Scientific Community.* Essays in the Philosophy and History of the Natural Sciences and Mathematics. In Honor of Robert S. Cohen. 1995 ISBN 0-7923-2988-0
164. K. Gavroglu, J. Stachel and M.W. Wartofsky (eds.): *Science, Politics and Social Practice.* Essays on Marxism and Science, Philosophy of Culture and the Social Sciences. In Honor of Robert S. Cohen. 1995 ISBN 0-7923-2989-9
165. K. Gavroglu, J. Stachel and M.W. Wartofsky (eds.): *Science, Mind and Art.* Essays on Science and the Humanistic Understanding in Art, Epistemology, Religion and Ethics. Essays in Honor of Robert S. Cohen. 1995 ISBN 0-7923-2990-2
Set (163–165) ISBN 0-7923-2991-0
166. K.H. Wolff: *Transformation in the Writing.* A Case of Surrender-and-Catch. 1995 ISBN 0-7923-3178-8
167. A.J. Kox and D.M. Siegel (eds.): *No Truth Except in the Details.* Essays in Honor of Martin J. Klein. 1995 ISBN 0-7923-3195-8
168. J. Blackmore: *Ludwig Boltzmann, His Later Life and Philosophy, 1900–1906.* Book One: A Documentary History. 1995 ISBN 0-7923-3231-8
169. R.S. Cohen, R. Hilpinen and R. Qiu (eds.): *Realism and Anti-Realism in the Philosophy of Science.* Beijing International Conference, 1992. 1996 ISBN 0-7923-3233-4
170. I. Kuçuradi and R.S. Cohen (eds.): *The Concept of Knowledge.* The Ankara Seminar. 1995 ISBN 0-7923-3241-5

Boston Studies in the Philosophy of Science

171. M.A. Grodin (ed.): *Meta Medical Ethics*: The Philosophical Foundations of Bioethics. 1995 ISBN 0-7923-3344-6
172. S. Ramirez and R.S. Cohen (eds.): *Mexican Studies in the History and Philosophy of Science.* 1995 ISBN 0-7923-3462-0
173. C. Dilworth: *The Metaphysics of Science*. An Account of Modern Science in Terms of Principles, Laws and Theories. 1995 ISBN 0-7923-3693-3
174. J. Blackmore: *Ludwig Boltzmann, His Later Life and Philosophy, 1900–1906* Book Two: The Philosopher. 1995 ISBN 0-7923-3464-7
175. P. Damerow: *Abstraction and Representation.* Essays on the Cultural Evolution of Thinking. 1996 ISBN 0-7923-3816-2
176. M.S. Macrakis: *Scarcity's Ways: The Origins of Capital.* A Critical Essay on Thermodynamics, Statistical Mechanics and Economics. 1997 ISBN 0-7923-4760-9
177. M. Marion and R.S. Cohen (eds.): *QuÇbec Studies in the Philosophy of Science.* Part I: Logic, Mathematics, Physics and History of Science. Essays in Honor of Hugues Leblanc. 1995 ISBN 0-7923-3559-7
178. M. Marion and R.S. Cohen (eds.): *QuÇbec Studies in the Philosophy of Science.* Part II: Biology, Psychology, Cognitive Science and Economics. Essays in Honor of Hugues Leblanc. 1996 ISBN 0-7923-3560-0
Set (177–178) ISBN 0-7923-3561-9
179. Fan Dainian and R.S. Cohen (eds.): *Chinese Studies in the History and Philosophy of Science and Technology.* 1996 ISBN 0-7923-3463-9
180. P. Forman and J.M. Snchez-Ron (eds.): *National Military Establishments and the Advancement of Science and Technology.* Studies in 20th Century History. 1996 ISBN 0-7923-3541-4
181. E.J. Post: *Quantum Reprogramming*. Ensembles and Single Systems: A Two-Tier Approach to Quantum Mechanics. 1995 ISBN 0-7923-3565-1
182. A.I. Tauber (ed.): *The Elusive Synthesis: Aesthetics and Science.* 1996 ISBN 0-7923-3904-5
183. S. Sarkar (ed.): *The Philosophy and History of Molecular Biology: New Perspectives.* 1996 ISBN 0-7923-3947-9
184. J.T. Cushing, A. Fine and S. Goldstein (eds.): *Bohmian Mechanics and Quantum Theory: An Appraisal.* 1996 ISBN 0-7923-4028-0
185. K. Michalski: *Logic and Time.* An Essay on Husserl's Theory of Meaning. 1996 ISBN 0-7923-4082-5
186. G. MunÇvar (ed.): *Spanish Studies in the Philosophy of Science.* 1996 ISBN 0-7923-4147-3
187. G. Schubring (ed.): *Hermann G~nther Graßmann (1809–1877): Visionary Mathematician, Scientist and Neohumanist Scholar.* Papers from a Sesquicentennial Conference. 1996 ISBN 0-7923-4261-5
188. M. Bitbol: *Schrdinger's Philosophy of Quantum Mechanics.* 1996 ISBN 0-7923-4266-6
189. J. Faye, U. Scheffler and M. Urchs (eds.): *Perspectives on Time.* 1997 ISBN 0-7923-4330-1
190. K. Lehrer and J.C. Marek (eds.): *Austrian Philosophy Past and Present.* Essays in Honor of Rudolf Haller. 1996 ISBN 0-7923-4347-6
191. J.L. Lagrange: *Analytical Mechanics.* Translated and edited by Auguste Boissonade and Victor N. Vagliente. Translated from the *MÇcanique Analytique, novelle Çdition* of 1811. 1997 ISBN 0-7923-4349-2
192. D. Ginev and R.S. Cohen (eds.): *Issues and Images in the Philosophy of Science.* Scientific and Philosophical Essays in Honour of Azarya Polikarov. 1997 ISBN 0-7923-4444-8

Boston Studies in the Philosophy of Science

193. R.S. Cohen, M. Horne and J. Stachel (eds.): *Experimental Metaphysics*. Quantum Mechanical Studies for Abner Shimony, Volume One. 1997 ISBN 0-7923-4452-9
194. R.S. Cohen, M. Horne and J. Stachel (eds.): *Potentiality, Entanglement and Passion-at-a-Distance*. Quantum Mechanical Studies for Abner Shimony, Volume Two. 1997 ISBN 0-7923-4453-7; Set 0-7923-4454-5
195. R.S. Cohen and A.I. Tauber (eds.): *Philosophies of Nature: The Human Dimension*. 1997 ISBN 0-7923-4579-7
196. M. Otte and M. Panza (eds.): *Analysis and Synthesis in Mathematics*. History and Philosophy. 1997 ISBN 0-7923-4570-3
197. A. Denkel: *The Natural Background of Meaning*. 1999 ISBN 0-7923-5331-5
198. D. Baird, R.I.G. Hughes and A. Nordmann (eds.): *Heinrich Hertz: Classical Physicist, Modern Philosopher*. 1999 ISBN 0-7923-4653-X
199. A. Franklin: *Can That be Right?* Essays on Experiment, Evidence, and Science. 1999 ISBN 0-7923-5464-8
200. Reserved
201. Reserved
202. Reserved
203. B. Babich and R.S. Cohen (eds.): *Nietzsche, Theories of Knowledge, and Critical Theory*. Nietzsche and the Sciences I. 1999 ISBN 0-7923-5742-6
204. B. Babich and R.S. Cohen (eds.): *Nietzsche, Epistemology, and Philosophy of Science*. Nietzsche and the Science II. 1999 ISBN 0-7923-5743-4

Also of interest:
R.S. Cohen and M.W. Wartofsky (eds.): *A Portrait of Twenty-Five Years Boston Colloquia for the Philosophy of Science, 1960-1985*. 1985 ISBN Pb 90-277-1971-3

Previous volumes are still available.

KLUWER ACADEMIC PUBLISHERS – DORDRECHT / BOSTON / LONDON